PRISM AND LENS MAKING

Frank Twyman FRS (1876–1959). Portrait courtesy of The Royal Society.

PRISM AND LENS MAKING

A textbook for optical glassworkers

F TWYMAN FRS

Adam Hilger, Bristol and New York

© IOP Publishing Ltd 1988

All rights reserved. No part of this publication may be reproduced, stored in a retrieval system or transmitted in any form or by any means, electronic, mechanical, photocopying, recording or otherwise, without the prior permission of the publisher.

British Library Cataloguing in Publication Data
Twyman, F.
 Prism and lens making.——2nd ed.——(The Adam Hilger series on optics and optoelectronics).
 1. Lenses   2. Prisms
 I. Title   II. Series
 681'.42   TS513

ISBN 0–85274–150–2

Library of Congress Cataloging-in-Publication Data

Twyman, F. (Frank), 1876–1959.
 Prism and lens making.

  Reprint. Originally published: 2nd ed. London: Hilger & Watts, 1952.
  Bibliography: P.
  Includes index.
   1. Prisms——Design and construction.   2. Lenses——Design and construction.   I. Title.
QC385.2.D47T88   1988   681'.42   87–23762
ISBN 0–85274–150–2 (pbk.)

First edition 1942, reprinted 1943, 1944
Second edition 1952, reprinted 1957, 1988 (with minor corrections), 1989
Published under the Adam Hilger imprint by IOP Publishing Ltd
Techno House, Redcliffe Way, Bristol BS1 6NX, England
335 East 45th Street, New York, NY 10017-3483, USA

First printed and bound in Great Britain by
Butler & Tanner Ltd, Frome and London
Reprinted in Great Britain by
J W Arrowsmith Ltd, Bristol

> But this rough magic
> I here abjure ; and when I have require'd
> Some heavenly music, which even now I do
> To work mine end upon their senses that
> This airy charm is for, I'll break my staff,
> Bury it certain fathoms in the earth,
> And deeper than did ever plummet sound,
> I'll drown my book.
>
> WILLIAM SHAKESPEARE : *The Tempest*

PREFACE TO FIRST EDITION

This book describes methods which are in use in the optical workshops of Adam Hilger Limited for making high quality prisms and lenses.

Optical firms are reticent about their methods and it may well be that we are considerably behind some of our fellow-manufacturers in just those directions where we suppose ourselves to be leading. Certain of the methods here described are certainly antiquated, even if others have only recently been developed by us, but of one thing the reader can be assured—namely, that these methods and machines will enable even an unskilled worker, after a short period of training under competent supervision, to produce work of first quality as far as definition and accuracy of angle are concerned.

I hope that those manufacturers who may be using other, and perhaps better, methods may be willing to give me the benefit of a full knowledge of them. A reliable and complete text might in that way become available which would raise opticians to a level of proficiency not attained in our generation.

The bibliographical references have been incorporated in the Name Index where the system used in the Royal Society's Publications has been followed (*i.e.* references in the text are by the author's name and date of publication—thus : Jones & Homer (1941)). There are, as exceptions to this, some brief Bibliographical Notes at the close of Chapter I, and a special Bibliography concerning the Hilger Interferometers, placed for convenience at the close of Chapter IX.*

The section on the deposition of metals was kindly written for me, by Dr. K. M. Greenland of the British Scientific Instrument Research Association. During the preparation of the book I have had much help from co-workers of mine at Hilgers, in particular Mr. S. J. Underhill, Head of the Optical Department ; Mr. J. W. Perry, Chief Computer and Head of Applied Optics Department ; Mr. A. Green, Mr. A. Pope and Mr. A. S. Henderson, who are in charge of various optical shops ; but it would not be correct to assume that any errors or grave omissions are other than my own. My thanks are also due to Mr. T. L. Tippell, whose careful proof-reading has resulted in the correction of many errors and crudities of diction. My acknowledgments are also due to the publishers of the *Encyclopaedia Britannica* for permission to use extensively my article in the 14th Edition on " The Manufacture and Testing of Lenses ", and to my co-Directors on the Board of Adam Hilger Ltd. for permission to publish much information hitherto private to the Company.

F. Twyman

* Corresponding to Chapter 12 in the second edition.

PREFACE TO SECOND EDITION

THE reception given to the first edition made me feel that it would be worth while to extend its scope to cover types of work other than those with which I was personally familiar. The present volume therefore includes information on the manufacture of spectacle lenses, on fine dividing, and on the working of large object glasses and mirrors. It also deals with developments to which the firms, now amalgamated to form Hilger & Watts Ltd, have devoted much attention in the last few years, *e.g.* making artificial crystals and the production of non-spherical surfaces. I am also able to include a description, issued from the Kodak Research Laboratories, of the recent important developments in optical glasses.

The help of the collaborators who have assisted me in thus extending the scope of the book is acknowledged in the text ; but besides these, and those mentioned in the preface to the first edition, I must add my recognition of the help given me by Mr H. W. Yates, Head of the Optical Workshop at the Hilger Division of Hilger & Watts Ltd, in bringing the practical instructions into accord with the latest procedure. Mr Yates, as Head of the Hilger Optical Workshops, has taken the place held for so many years by Mr A. Green, now retired (though often consulted), with whom I worked for over fifty years. I must also thank Mr A. H. Winterflood, of the laboratory concerned with dioptric materials, for experiments on annealing carried out to check and extend the work done by me many years ago.

I shall be grateful to any reader who will bring to my notice errors or omisssions, and I would add that no reviews are so helpful as those which are critical.

October 1951   F. TWYMAN

PREFACE TO REPRINT OF SECOND EDITION

Frank Twyman (1876–1959) had no formal academic qualification and he acquired his skills in all aspects of optics by learning on the job, which was not very unusual in the early years of this century. Twyman joined Otto Hilger, a maker of simple spectroscopes at £10 apiece, in 1898, and when Otto died in 1902 Twyman took over. In 1904 the firm of Adam Hilger Ltd was formally born and for some 50 years the name Hilger was known for the finest quality optical and mechanical work. Twyman was Managing Director for 44 years and during the whole of that time he was intimately concerned with the scientific and technical aspects of the company; he designed many of the instruments himself and was constantly concerned with improving the techniques of optical grinding and polishing. I sometimes saw him in one of the optical shops working at a fixed post polisher, trying out a new idea or testing a new polishing medium. He was always prepared to discuss a new instrument or a new application with a customer and many of the developments in industrial and in biological spectroscopy came through Hilger instruments.

Twyman's name is, of course, best known for the lens-testing interferometer which he invented in collaboration with Alfred Green, the then foreman of the optical shops, but he also carried out fundamental studies of the annealing process for glass and invented new spectrophotometers and spectrographs. In 1943 he published *Prism and Lens Making*, a slim book which packed in a great amount of technical information never before available in print. He produced a considerably enlarged edition in 1952 and the present reprint is almost unchanged from that one, apart from the correction of minor misprints. Some of what is in it has undoubtedly been superseded by more recent developments but there is much that will be of lasting value; there are almost certainly things in the book which have been forgotten but which badly need rediscovery and use. I am sure this classic of applied optics will give pleasure as well as help to all concerned with making good optical systems.

June 1987   WALTER WELFORD

CONTENTS

Chapter		Page
1	HISTORICAL	5

Ancient optical work—The invention of spectacles—Telescope and microscope lenses—Newton and the reflecting telescope—Early grinding and polishing machines—Bibliographical notes

2	SINGLE SURFACE WORKING	21

Making a single prism—Making a single lens

3	THE NATURE OF GRINDING AND POLISHING	49

4	TOOLS AND MATERIALS IN GENERAL USE	66

Optical tools—Polishers—Mallets—Abrasives—Manufacture of carborundum—Cleaning liquids—Cements, plasters, and varnishes—Experiments on the shrinkage of cements

5	DIOPTRIC SUBSTANCES	118

Introduction—Chromatic aberration of lenses and its correction—Durability—Mechanical properties—Hardness—Coefficient of expansion and variation of refractive index with temperature—The nature of glass—Development of optical glasses since 1880—New optical glasses—Problems arising in the manufacture of optical glass—Artificial crystalline materials—Composition and hardness of various polishing and abrasive powders—Decolorizing coloured crystals—Binary mixed crystals as optical materials—Cutting, grinding and polishing crystals, plastics and metals—Further notes on the treatment of rocksalt and other difficult crystals—Notes on the idiosyncrasies of crystals from the point of view of the working optician—Sundry notes on artificial crystals—Optical plastics—Combined Optical Industries: Report on progress—Moulded lenses of glass

6	PRODUCTION OF LENSES IN QUANTITY	191

Introduction—Moulding, trepanning and roughing by hand: machine roughing by abrasive wheels and diamond laps—Process layouts for lenses—Sequence of operations—Procedure after polishing—Balsaming—Cylindrical lenses

7	THE MANUFACTURE OF SPECTACLE LENSES	251

Spherical lenses—Toric lenses—Manufacture of fused bifocals—Solid bifocals—Planning—Prescription work

8	MICROSCOPE LENSES	278

General considerations—Construction of a hemispherical or hyperhemispherical front lens—The metal work—Testing the completed lens

CONTENTS

9 THE PRODUCTION OF PRISMS IN QUANTITY . . . 293
Machining—Polishing—Polishing blocks of prisms—Plane parallel glasses

10 NON-SPHERICAL SURFACES 323
Introduction—The position in 1918—Some developments from 1918 to 1948—Machine used by Zeiss to grind aspherical lens surfaces—Dr Burch's angle-control aspherizing machine

11 TESTING OPTICAL WORK 364
Introduction—Tests concerning definition—Testing lenses—Testing surfaces—Testing polished surfaces—Tolerances in the specification of optical components—Testing angles—Application of the Angle Dekkor—An absolute standard of planeness—light sources used in optical manufacture and testing—Notes on parallelism of glass plates

12 THE HILGER INTERFEROMETERS FOR TESTING PRISMS AND LENSES, AND OTHER INTERFEROMETERS COGNATE THEREWITH 422
Introduction—The Hilger prism and lens interferometer—The Hilger microscope interferometer—Ultra-violet microscope—Other applications of the Hilger interferometers—Bibliography—Other interferometers for testing optical work derived from Hilger's—Phase contrast method

13 SURFACE TREATMENTS 456
The preparation of reflecting surfaces and anti-reflection films—"Physical" processes—Deposition of thin films on optical surfaces by evaporation in vacuo—Plants for producing surface coatings by the evaporation process—Glass graticules in optical instruments

14 TESTING OPTICAL GLASS : ANNEALING AND NORMALIZING 499
Faults in glass—Transmission and refractive indices of glasses—The Hilger-Chance Precision Refractometer for transparent solids—First scientific control of annealing and normalizing glass—Twyman's method of determining annealing temperatures—Determination of the best cooling schedule for annealing

15 LARGE OBJECT GLASSES AND MIRRORS 538
The development of astronomical telescopes—Sir William Herschel's telescopes—The work of Draper—The work of Sir Howard Grubb—The work of Ritchey—Testing—Notes on the 200-inch mirror at Mount Palomar—The work of Lyot, Couder, Texereau

16 REFERENCE BOOKS ON OPTICS AND OPTICAL GLASSWORKING 588
Theoretical—Practical—Tables of lens curves

CONTENTS

APPENDIXES 591

  (i) Derivation of the formula for the Twyman annealing schedules

  (ii) Making polarizing prisms

  (iii) The present state and future trends of the manufacture of optical glass

  (iv) Glossary of terms used in the optical industry

BIBLIOGRAPHY 618

INDEX 623

CHAPTER 1
HISTORICAL
Ancient optical work
1 As with most of the useful arts, the development of lens making has been in an order the reverse of what the scientific man feels to be logical. Instead of first studying the principles of the process, then putting the process into operation, and then finding a use for the product—a course followed by the science-born electrical industry—man first discovered some optical uses of accidentally produced lens-shaped bodies, and only then set himself deliberately to make lenses, leaving till quite recent times the study of the process of lens polishing.

2 It is a pleasing fancy that the possibility of using a lens as a burning glass may be related to the supposed ability of the priestly classes during the Nilotic and Mesopotamian civilizations to " bring down fire from heaven " during religious ceremonies. The word *focus* (Latin) meant originally a hearth or burning place, and its etymology goes back to roots which suggest that originally it had associations with temple altars or places where sacrifices were burned (see the larger Oxford Dictionary). It may be noted that the modern French word " foyer " is used for both " hearth " and " focus."

Gunther (1923) points out that in England, the practice of kindling the new fire on Easter Eve by a burning glass was not uncommon in the Middle Ages ; an entry to that effect occurs in the Inventory of the Vestry, Westminster Abbey, in 1388.

The references of Pliny and other ancient writers show quite clearly that burning glasses were known to them in the shape of glass spheres filled with water ; and passages from Greek and Roman writers have been cited as showing that they knew of the magnifying properties of lenses, or at least of such glass spheres filled with water. The very thorough account of the subject by Wilde (1838-43)[1]* denies to the ancients all knowledge of spectacle lenses whether for short or long sight, or indeed of any kind of lenses, if we except the spheres of glass filled with water referred to above ; and maintains that the lens-shaped glasses or crystals which have been found from time to time among the relics of departed civilizations were made by polishers of jewels for purposes of ornament. Mach (1926), on the other hand, seems to tend, on the whole, to the opinion that a few archaeological

* The references by number are to the Bibliographical Notes at the end of this Chapter ; those without numbers are in the General Bibliography.

objects which have been found were made, and intended to be used, as lenses.

3 Beck (1928) adduces further evidence to show that lens-shaped objects were used as magnifying and burning glasses from very early times. He points out that the usual varieties of glass have been continuously made from the time of the Eighth Egyptian Dynasty, whilst a piece of glass in the Ashmolean Museum is claimed to be First Dynasty, if not pre-dynastic. A large piece of blue glass from Abu Shahrein in Mesopotamia dates from about 3000 B.C. The author continues—

> But whatever may have been the original date of the invention of glass we know that by the fourteenth century B.C. there was a well-established centre of glass manufacture in Egypt, and a totally different one in the Aegean, where the technique was in use which did not penetrate into Egypt until a very much later date. Also, although the amount of transparent glass made in Egypt at that time was only a trifling proportion of the total output, much of the Aegean glass was transparent and a considerable amount colourless.
>
> The date of the first manufacture of colourless glass need not however, limit us in finding a possible date for early magnifiers, as crystal was always to hand and the earliest magnifiers known are in that material.
>
> The first reference to a lens that I know of in literature is in The Comedy of the Clouds by Aristophanes, which was performed in 434 B.C. In the second act comes the following passage—
>
> STREPSIADES: You have seen at the druggists that fine transparent stone with which fires are kindled?
> SOCRATES: You mean glass?
> STREPSIADES: Just so.
> SOCRATES: Well what will you do with that?
> STREPSIADES: When a summons is sent me, I will take this stone and placing myself in the sun I will, though at a distance, melt all the writing of the summons.
>
> *Note.*—The point of the remark of Strepsiades in the above passage is that the writing was a summons for debt. Such a writing would be traced on wax, and the suggestion is that, if he melted it with his burning glass, the record would be lost and he would thus be freed from his debts.
>
> ... Lactantius in A.D. 303 says that a glass globe filled with water and held in the sun could light a fire even in the coldest weather.
>
> Now lenses sufficiently good to make burning glasses, would make magnifying glasses . . .

There are in the Egyptian department of the British Museum two magnifying glasses which would make excellent burning glasses, except for the tarnish. They are about $2\frac{1}{2}$ in. diameter and about $3\frac{1}{2}$ in. focus and would magnify three diameters. They have been ground and are not merely cast. The flat surface has been ground against another flat surface with a rotary motion as at the present. For example ; one of these glasses (22522 Egyptian Department) was found at Tanis, and definitely dated A.D. 150. (See Fig. 1.)*

. . . but the most conclusive proof as to the early magnifying glasses is the discovery this year (1927) by Mr. E. J. Forsdyke, in Crete, of two crystal magnifying lenses that date back at least as early as 1200 B.C., and probably 1600 B.C., as most of the small objects from the tombs where they were found are of that date.

The author adds finally—

The excellence and minuteness of some of the work in the objects recently discovered by Mr Woolley, and supposed to date before 3000 B.C., make it seem very probable that magnifiers were used, and the fact that crystal was then used makes one hope that further discoveries will enable us definitely to place the manufacture of magnifying lenses, or crystals earlier even than these very important ones just found in Crete.

About half-a-dozen other examples of objects which, whether intended to be used as magnifying glasses or not, were—and indeed, in some cases, are—*capable* of being so used are given in Chapter 3 of Greeff's book (1921). One of them is illustrated in Fig. 1. This is in the Volkermuseum in Berlin among the well-known objects excavated at Troy by Schliemann ; it is supposed to date from the second half of the third century B.C. Nevertheless, the author quotes with approval the opinion of Furtwängler (*translation*) : " That the ancients used magnifying glasses in their work has been asserted and disputed. It cannot be proved, however, much as one may consider it probable."

It will be seen that the methods of polishing jewels for ornaments were available when men first felt the urge to polish lenses for use. Even then the first use was probably a ritual in connection with religious rather than secular purposes and possibly kept in close secrecy ; a secrecy which still tends to linger in the " mystery " of the trade.

---

* I should have liked to examine this glass myself to see whether my judgment confirmed that of the author, but unfortunately this particular specimen had not yet been returned to the British Museum after its removal during the war to a place of safety. I have been able, however, to insert as part of Fig. 1 a fresh photograph from the British Museum which illustrates the nature of the object more clearly than the illustration in the reference cited. (F.T.)

8 PRISM AND LENS MAKING

Fig. 1—Ancient optical work
Above: Lens-shaped pieces of rock crystal from ancient Troy.
Left: An early Egyptian lens from Tanis (A.D. 150).

4 The name of Alhazen is often mentioned in connection with the early history of Optics.

Alhazen (Abu Ali Al-Hasan Ibu Alhasan) was born at Basra and died in Cairo in A.D. 1038. He solved the problem of finding the point on a convex mirror at which a ray coming from one given point shall be reflected to another. His treatise on optics was translated into Latin by Witelo (1270) and published by F. Risner in 1572 with the title *Optical thesaurus Alhazeni libri VII cum ejusdem libro de crepusculis et nubium ascensionibus* (*Enc. Brit.*, 11th Ed., 1910–11, " Alhazen "). The Latin translation will be found in the same volume of the *Bulletino di Biographia*, etc., as the dissertation of Th. Henri Martin referred to in ref. 1. The only mention Alhazen made of lenses, however, appears to be his statement that if an object is placed at the base of the larger segment of a glass sphere, it will appear magnified.*

The invention of spectacles

5 We must come to the end of the thirteenth century for the first authentic mention of the use of spectacles, which appears to be that of Meissner (1260–80) when he expressly states that old people derive advantage from spectacles (Bock, 1903).[2] In the archives of the old Abbey of Saint-Bavon-le-Grand, the statement is found that Nicolas Bullet, a priest, in 1282 used spectacles in signing an agreement (Pansier, 1901).[3] The first picture in which spectacles are known to have appeared is by Tomaso de Modena, in the Church of San Nicola in Treviso, and is of date 1362 (Oppenheimer, 1908). This is illustrated in the Frontispiece.

Martin[1] says—

The invention of spectacles for the short and the long sighted is mentioned as a *quite recent* discovery in a manuscript of 1299 in Florence. Bernard Gordon, Professor at Montpelier, in his work " Lilium medicinae " begun in 1305 alludes to spectacles as a means of remedying visual defects. Giordano da Rivalto, in a sermon given on February 23rd, 1305, remarks that this invention *is not yet twenty years old*. The man who copied the sermon says he himself saw and conversed with the inventor. Thus it was about 1285—no earlier—that spectacles were invented. We know besides that the inventor was the Florentine, Salvino d'Armato degli Armati, who died in 1317. He hid his secret to keep it as a monopoly. But Brother Alessandro Spina of Pisa, who died in 1313 having seen spectacles made by Salvino d'Armati and having succeeded in making similar ones, hastened to make public the secret. (*Translation.*)

In a note at the bottom of page 237, Martin states—

Primitive spectacles consisted of two pieces of leather which were fastened on to a cap, worn low over the forehead. (*Translation.*)

Greeff (1921) has examined a great mass of data about the invention of spectacles. He points out that, although we can scarcely neglect the evidence of pictures, it would be naive to conclude that any spectacles shown were of the epoch the picture represents. Even in depicting very ancient scenes, painters from the time of the van Eycks used to introduce a pair of spectacles to add verisimilitude when they wished to

---

* Much interesting information on the origins of optical science will be found in a paper by J. P. C. Southall (1922, *J.O.S.A.*, **6**, 293).

represent a person sunk in study or meditation. The author instances many cases of such anachronisms, for example, Moses is furnished with spectacles in a miniature painted in Heidelberg about 1456. The painter aimed not at historical accuracy in detail, but at representing things as their contemporaries saw and felt them ; spectacles in their pictures were, therefore, of their own period.

In another chapter Greeff examines the suggestions put forward from time to time that spectacles originated in India or in China, but he finds them groundless. He concludes, after citing many authorities, that there is absolutely no evidence that the Chinese had spectacles before they originated in Europe and shows good ground for supposing that they came in through Malacca in the early 16th century. Dr. Greeff attributes, however, to Prof. Hirth of Columbia University, New York, the statement that the Chinese had mirrors both concave and convex, of bronze, in the first century B.C.

It may be accepted, from this and like evidence, that the use of spectacles dates from about A.D. 1280. The picture reproduced in the Frontispiece was not painted from life, since the Cardinal died in 1262 and the picture was painted in 1352.

Brockwell (1948) says—

As for the making of the lenses, Roger Bacon, in his *Opus Majus* of about 1266, and in his *Perspectivae Pars Tertia*, showed that " by placing a segment of a sphere on a book with its plane side down, one can make small letters appear large."* He communicated his knowledge of optics to his friend Heinrich Goethals, who, travelling in Italy in 1285, handed on his information to Alessandro della Spina, a monk in Pisa, who " could make anything he liked " [*operava di sua mano ogni cosa che volesse*], and who died in 1313.

6 The present writer has found no account of how lenses were made before William Bourne's (*c*. 1585)[4] account ; very imperfect, but sufficient to show that processes were then in use very like those still extant. He says (of spectacle lenses)—

These sortes of glasses ys grounde upon a toole of Iron made of purpose, somewhat hollowe, or concave inwardes. And may be made of any kynde of glasse, but the clearer the better. And so the Glasse, after that yt ys full rounde, ys made fast with syman uppon a small block, and so ground by hand untill yt ys bothe smoothe and allso thynne, by the edges, or sydes, but thickest in the middle.

Nothing else is said by him of the materials, tools, or method of working.

* Exactly what Alhazen had pointed out more than 200 years earlier.

Baptista Porta

7 Very different is the account given by Baptista Porta of Naples (1591) in his famous book on Natural Magic—a technical encyclopaedia embracing subjects as diverse as optics, magnetism, cosmetics, cooking, alchemy, pharmacy, and practical jokes. Among much that is trivial, debased and revolting is also to be found much, like his description of optical polishing, which shows a keen quest after knowledge and accurate knowledge of a singularly wide range of subjects. The following extract is from Book 17, Chapter XXI in the English translation published in 1658, but the matter is identical with the Latin edition of 1591. The translation has, however, rendered the original " pilae vitreae "—the phrase employed (as by Pliny) to describe a *hollow* glass ball—as " Glass-balls," and the reader must bear this in mind if he wishes to follow the description correctly.

In Germany there are made Glass-balls, whose diameter is a foot long, or there abouts. The Ball is marked with the Emrilstone round and is so cut into many small circles, and they are brought to Venice. Here with a handle of wood are they glewed on, by Colophonia melted. And if you will make Convex Spectacles, you must have a hollow iron dish, that is a portion of a great sphaere, as you will have your spectacles more or less Convex ; and the dish must be perfectly polished. But if we seek for concave spectacles, let there be an Iron ball, like to those we shoot with Gun-powder from the Great Brass Cannon ; the superficies whereof is two, or three foot about. Upon the Dish or Ball, there is strewed whitesand, that comes from Vincentia, commonly called Saldame, and with water it is forcibly rubbed between our hands, and that so long until the superficies of that circle shall receive the form of the Dish, namely a Convex superficies or else a Concave superficies upon the superficies of the Ball, that it may fit the superficies of it exactly. When that is done heat the handle at a soft fire, and take off the spectacle from it, and join the other side of it to the same handle with Colophonia, and work as you did before, that on both sides it may receive a Concave or Convex superficies, then rubbing it over again with the Powder of Tripolis that it may be exactly polished ; when it is perfectly polished, you shall make it perspicuous thus. They fasten a woollen-cloth upon wood ; and upon this they sprinkle water of Depart, and powder of Tripolis ; and by rubbing it diligently, you shall see it take a perfect glass. Thus are your great Lenticulars and spectacles made at Venice.

Telescope and microscope lenses

Cherubin d'Orléans (1671)

8 In 1671 appeared the well-known book by a Père Cherubin d'Orléans which not only deals with optics, with telescopes (including binoculars) and microscopes, their theory and construction and use, but with lens making and the various machines invented by himself to lessen the labour and increase the speed and accuracy of polishing lenses for telescopes. This latter part of the book consists of eighty-two pages and many illustrations (approx. 35,000 words). These descriptions are so good, and show such thoughtful personal knowledge of the subject, that they would be suitable to place in the hands of an optical apprentice today.

The materials which he used were so nearly like those which are still widely used and are described so well that a free translation of what he says will not be out of place, if only that one may realise how little that is fundamentally new has been discovered in the interim if we except the processes which have been introduced in the twentieth century.

His materials were—for the tools, iron and brass; for the mallets (molettes) for holding the lenses, lead, tin or (which he prefers) copper. He gives full particulars for making the patterns for casting, making the moulds, and turning and grinding the tools. He also describes lathes which he invented for turning the tools.

Of the cement for attaching the lens to the mallet, he says that some make it of best black pitch (which must not be burnt) and sifted ashes of vine cuttings; but that he prefers to add (to the pitch) a fourth part of good grape jelly and, in place of the ashes, finely ground ochre or whiting.

As a good material for grinding he recommends broken grindstones. These are graded by putting the powder into a large vessel full of water, agitating it well, letting it settle a little for the coarse particles to settle out, and then pouring off quickly into another vessel most of the liquid, which will carry with it the finer material. This one allows to settle entirely, gently pouring off the remaining liquid. The sediment (containing the useful grains) is treated in this way several times, and in this way one separates out the grains of several degrees of " strength " which are kept separately for use according to the nature of the work. For polishing material he used Tripoli (preferably that of Germany) or " potée d'estain " (putty powder).

Of the Tripoli he says that the lightest is the best. If of good quality it can be used in the lump, as Nature produces it, otherwise ground with brandy or (failing that) white wine, and kept in a closed jar of water to soften for four or five months. It can be sun-dried, and used in lumps, or used wet straight from the jar.

The putty powder he prefers is that made by calcining tin (he gives detailed instructions for its preparation). The glass was provided by broken Venice mirrors, and was thus in a form polished on both sides and suitable for examination.

He claims to have made an improvement in the mode of polishing in that he stretched across the concave tool, which he had used for grinding, a soft thin piece of leather of uniform thickness, fixing it in position with a ring which just fitted the circumference of the tool. On this he rubbed his lens, pressing it down so that the leather was forced to fit the surface of the grinding tool. Another way he used was to coat the surface of the grinding tool with paper, the latter being pasted in position with many precautions to avoid wrinkling or other inequalities of the surface. This paper he moistened with Tripoli powder and so used for polishing. Of the machines he describes as invented by himself several may well have been the progenitors of some in use today. One of them may be mentioned in which the optical tool, mounted on a vertical spindle, is rotated by means of a rope and suspended weight so that the operator's hands are free. The tool is turned accurately to the desired radius by means of a radius arm pivoted at one end, and bearing a turning tool at the other, by means of which the tool is turned to the desired concavity. The same machine is used in grinding and polishing the glasses, the latter being held in the hand as in free-hand working. The present writer has found no earlier description of machines actually produced for turning the tools or for grinding and polishing lenses.

Hooke

9 About this time Hooke was working on the microscope (Hooke 1667). He describes a way of making microscope objective lenses; he drew a piece of broken Venice glass in a lamp into a thin thread, then held the end of this thread in a flame till a globule of glass was formed. He then polished a flat surface on the thread side of the globule, first on a whetstone and then on a smooth metal plate with Tripoli; but these lenses being too small he used good plano-convex object glasses, and there is no indication that he made these himself.

Leeuwenhoek

10 The great Dutch microscopist, Leeuwenhoek (1719), made his own lenses, but left no account of his methods. He says in a letter to Leibnitz dated the 28th of September, 1715—

As to your idea of encouraging young men to polish glass—as it were to start a school of glass polishing—I do not myself see that would be of much use. Quite a number, who had time on their hands at Leyden, became keen on polishing glasses, owing to my discoveries; indeed there were three masters of that art in that town, who instructed students who were interested in such things. But what was the result of their labour? Nothing at all, so far as I have learnt.

Now to every study the proposed object is this : to acquire wealth by knowledge, or celebrity by reputation for learning. But that is not to be gained either by polishing glasses or by discovering abstruse things. And then I am convinced hardly one in a thousand is properly fitted to take up this study for much time is consumed in it and money is wasted, and if one is to make any progress in it one's mind must be for ever on the stretch, thinking and speculating. The majority of men are not sufficiently inflamed with the love of knowledge for that. Indeed, many whom it by no means becomes, do not hesitate to ask, What does it matter whether we know these things or not?

Newton and the reflecting telescope

11 Newton was the first to make a successful reflecting telescope. It is true that James Gregory had proposed his Gregorian telescope in 1663 and the next year came to London to commission the manufacture of such a telescope from Reive, a famous London optician. It was to be 6 ft focus but the figure proved so bad that the attempt was abandoned. Gregory was of the opinion that the failure was due to Reive trying to polish the mirrors with cloth ; Gregory must therefore have had in mind the possibility of a more perfect medium for a polisher having been known before Newton's work. The passage from Newton (1721) is reputed to be the first *publication* of the use of pitch for a polisher.

Newton completed his first reflecting telescope in 1668. It had an aperture of 1 in. only and focal length of 6 in.

The reasons why Newton adopted the reflecting form of telescope will be mentioned in Chapter 10.

In the list of contents of the *Philosophical Transactions* for March 25, 1672, one reads " An accompt of a new kind of Telescope, invented by Mr. Isaac Newton." In the " Accompt " it is stated that for metal he tried various mixtures of copper, tin and arsenic. With one of them, consisting of copper 6 oz, tin 2 oz, arsenic 1 oz, a friend of his said he had " polish't better than he did the other." He used putty powder for polishing his metal mirror.

It will be seen that the metal he used was very like speculum metal, defined, in *Encyclopaedia Britannica* IXth Ed., Article " Bronze," as consisting of 2 of Copper to 1 of Tin. This is the mixture which I myself have cast and used for making many hundreds of mirrors.

HISTORICAL

Newton (1721) makes some important remarks on polishing, which though referring to mirrors are also applicable to lenses, and appears to have been the first to use pitch for polishing, an innovation of the very greatest importance Newton says in the reference cited—

The Polish I used was in this manner. I had two round Copper Plates each five inches in diameter, the one convex, the other concave, ground very true to one another. On the convex I ground the Object-Metal or Concave which was to be polished, till it had taken the figure of the Convex and was ready for the Polish. Then I pitched over the convex very thinly, by dropping melted Pitch upon it and warming it to keep the Pitch soft, whilst I ground it with the concave copper, wetted to make it spread evenly all over the convex. Thus by working it well I made it as thin as a Groat,* and after the convex was cold I ground it again to give it as true a figure as I could. Then I took Putty which I had made very fine by washing it from all its grosser particles, and lay a little of this upon the pitch, I ground it upon the Pitch with the concave Copper till it had done making a noise ; and then upon the pitch I ground the Object-Metal with a brisk motion, for about two or three minutes of time, leaning hard upon it. Then I put fresh putty upon the Pitch and ground it again till it had done making a noise, and afterwards ground the Object-Metal upon it as before. And this work I repeated till the Metal was polished, grinding it the last time with all my strength for a good while together, and frequently breathing upon the pitch to keep it moist without laying on any more fresh Putty. The object-metal was two inches broad and about one third of an Inch thick, to keep it from bending. I had two of these Metals, and when I had polished them both I tried which was best, and ground the other again to see if I could make it better than that which I kept. And thus by many trials I learn'd the way of polishing, till I made those two reflecting Perspectives I spake of above. For this Art of Polishing will be better learned by repeated Practice than by my description. Before I ground the Object-Metal on the Pitch, I always ground the Putty on it

---

* I am indebted to Mr. E. S. G. Robinson, Deputy Keeper of Coins and Medals of the British Museum, London, for the following information.

The groat in Newton's time was approximately half a millimetre thick. It was, however, not in wide currency, but only a Maundy piece, and the thickness of the earlier *currency* groats, which in his day would be considerably worn, would hardly have been more than one-third of a millimetre.

If, then, Newton intended his phrase to be taken literally and was not merely using the familiar phrase " thin as a groat "—most unlikely in view of his usual precision of statement—he was using a thickness of pitch not more than one-third of what is customary today. There is much to be said for such practice, since it would make possible the use of a much softer pitch without distortion—thus obtaining less liability to scratch or sleek.

with the concave Copper till it had done making a noise, because if the Particles of the Putty were not by this means made to stick fast in the Pitch, they would by rolling up and down grate and fret the Object-Metal and fill it full of little holes.

But because metal is more difficult to polish than Glass and is afterwards very apt to be spoiled by tarnishing and reflects not so much Light as Glass quick-silvered over does : I propound to use instead of the Metal, a Glass ground concave on the foreside, and as much convex on the back-side, and quicksilvered over on the convex side. The Glass must be everywhere of the same thickness exactly. Otherwise it will make objects look coloured and indistinct. By such a Glass I tried about five or six Years ago to make a reflecting telescope of four Feet in length to magnify about 150 times, and I satisfied myself that there wants nothing but a good Artist to bring the design to perfection. For the glass being wrought by one of our London Artists after such a manner as they grind Glasses for Telescopes, tho' it seemed as well wrought as the object-glasses use to be, yet when it was quick-silvered, the Reflexion discovered innumerable Inequalities all over the Glass. And by reason of these Inequalities, Objects appeared indistinct in the Instrument. For the errors of reflected Rays caused by an Inequality of the Glass are about five times greater than the Errors of refracted rays caused by the like Inequalities. Yet by this Experiment I satisfied myself that the Reflexion on the concave side of the Glass, which I feared would disturb the Vision, did no sensible prejudice to it, and by consequence that nothing is wanting to perfect these Telescopes but good Workmen who can grind and polish Glasses truly spherical. An Object-Glass of a fourteen Foot Telescope made by an Artificer at London, I once mended considerably by grinding it on Pitch with Putty, and leaning very easily on it, in the grinding, lest the Putty should scratch it. Whether this way may not do well enough for polishing these reflecting Glasses, I have not yet tried. But he that shall try either this or any other way of polishing which he may think better, may do well to make his Glasses ready for polishing by grinding them without that violence wherewith our London workmen press their Glasses in grinding. For by such violent pressure, Glasses are apt to bend a little in the grinding, and such bending will certainly spoil their figure.

Early grinding and polishing machines

Herschel

12 Some optical grinding and polishing machinery appears to have been made and used early in the seventeenth century. Dr C. A.

PLATE 1 [*Facing page* 16]

Sir Isaac Newton (1642-1727)

His many contributions to practical and theoretical optics rank in his work second only to his great *Principia*

PLATE 2 [*Facing page* 17

Sir William Herschel (1738–1822)

Supreme among observational astronomers his great discoveries were made entirely with telescopes optically and mechanically of his own contrivance, which were superior to any previously made

Joseph von Fraunhofer (1787–1826)

His careful study of refractive indices and dispersive powers enabled him to make the finest achromatic object-glasses of his time

Crommelin (1929) describes and illustrates machines made or proposed by Descartes, Huygens, Hooke, Helvelius and others.

Herschel (*Collected Papers*, 1912) in 1774 used a pitch polisher for polishing the speculum mirrors, some of them very large, for his telescope. He mentions the polishing operation as having been carried out by ten men on one occasion. He gave an account of the polishing of a large speculum by a machine which he made to avoid the necessity of employing so many men; but published no details of his working methods, as he evidently looked upon these as trade secrets. He did, however, contribute a vague paper (quite short) to the Royal Society in 1789, in which he states that he has at last completed a polishing machine that really works, whereas he had tried and failed 6 years before. I think we may take it that from 1789 onwards he used machines for polishing and figuring *all* his mirrors. The only details that have ever been published are incorporated in an article ("Telescope") written by J.F.W.H. for the 8th edition of the *Encyclopaedia Britannica*. Some extracts from this article are given by Dreger in p. XLIX of Vol. I of the *Collected Scientific Papers*. Here there is a diagram showing the principle employed in the movement and rotation of mirror and polisher at each stroke.*

Fraunhofer

13 Fraunhofer, who made telescope lenses of great excellence, is said to have been first to use proof spheres for testing the accuracy of surfaces, but Dévé (1936) attributes their earliest use to the French firm of Laurent (now Messrs Jobin et Yvon) about fifty years ago.

Lord Rosse

14 Finally, in what we may still call the historical period, Lord Rosse (Parsons, 1926) (then bearing the title of Lord Oxmantown—he succeeded to the title of Earl of Rosse in 1841) described before the Royal Society a machine for polishing large specula. (See Fig. 2)—

A is a shaft connected with a steam-engine; B an eccentric, adjustable by a screw-bolt to give any length of stroke from 0 to 18 inches; C a joint; D a guide; E, F a cistern for water, in which the speculum revolves; G another eccentric, adjustable like the first to any length of stroke from 0 to 18 inches. The bar D, G passes through a slit, and therefore the pin at G necessarily turns on its axis in the same time as the eccentric. H, I is the speculum in its box immersed in water to within one inch of its surface, and K, L the polisher which is of cast iron, and weighs about two

* W. H. Steavenson, 1924–25 (*Trans. Opt. Soc.*, **26** (4), 210) refers to four unpublished volumes by Herschel on telescopes.

and a half hundred weight. M is a round disc of wood connected with the polisher by strings hooked to it in six places, each two-thirds of the radius from the centre. At M there is a swivel and hook, to which a rope is attached, connecting the whole with the lever N so that the polisher presses upon the speculum with a force equal to the difference between its own weight and that of the counterpoise O. For a speculum three feet in diameter I make the counterpoise ten pounds lighter than the polisher. The bar D, G fits the polisher nicely, but without tightness, so that the polisher turns

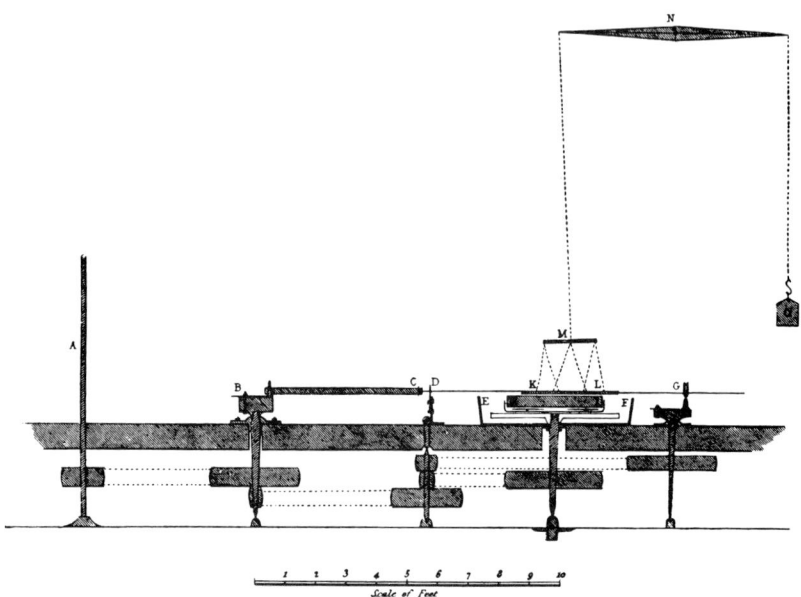

Fig. 2—Lord Rosse's polishing machine

freely round, usually about once for every fifteen or twenty revolutions of the speculum, and it is prevented by four guards from accidentally touching the speculum, and from pressing upon the polisher by the two guides through which its extremities pass. I have tried a variety of contrivances for connecting the machinery with the polisher, but the one I have described is by far the best. The wheel B makes, when polishing a three-feet speculum, sixteen revolutions in a minute; to polish a smaller speculum the velocity is increased by changing the pulley on the shaft A.

In the course of the same papers he describes the method of preparation of his rouge (Parsons 1926, p. 98) by calcination, at a dull red heat, of peroxide of iron produced as a precipitate with ammonia water from a dilute solution of iron sulphate. Lord Rosse (father of Sir Charles Parsons of steam turbine fame) used " ammonia soap " with the water used in polishing. This probably retarded tarnishing— it is known that swabbing speculum mirrors with dilute ammonia will remove tarnish.

15 These steps in the development of existing methods of lens making must suffice, but it should be added that surviving accounts very possibly leave unnamed the workers to whom the methods (probably originally derived from the ancient art of polishing stones) are due.

Bibliographical notes

1. Th. Henri Martin, " Sur des instruments d'Optique faussement attribués aux anciens par quelques savants modernes," *Bulletino di Bibliografia e di Storia delle Scienze matematiche e fisike* (Roma). Vol. IV., pp. 165–238, 1871. The evidence discussed by Martin is in considerable part the same as that to be found in E. Wilde, *Geschichte der Optik* (Berlin, 1st vol. 1838, and 2nd vol. 1843) : but Martin's treatment is by far the fuller. M. Martin, in this entertaining paper, reviews comprehensively the evidence that the ancients were acquainted with certain optical instruments, and opines that there is no reliable evidence that they knew anything of telescopes, microscopes, reading glasses, or lenses for short or long sight. The many lenticular crystals and glasses which have been found he supposes to be ornamental only ; and the utmost he concedes (p. 213) is that lens-shaped transparent stones were used as burning glasses, and were, for instance at the time of Aristophanes, sold by druggists as burning glasses to light the fire by rays from the sun (these references from Aristophanes and elsewhere are very clear and free from ambiguity), and that lenses (globes of glass filled with water) could be used to enlarge objects.

2. E. Bock, *Die Brille und ihre Geschichte* (Vienna, 1903). Repeats part of the matter dealt with so fully by Martin, though without reference thereto. " Der grosse Schotte Winfried, den die Kirche den hl. Bonifazius nennt, (680–755) kannte die Wirkung der Vergrosserungsgläser." " Erst Roger Bacon . . . erwähnt 1276 die Brille." " Zu derselben Zeit aber waren die Brillen in Deutschland schon bekannt, denn in der Sammlung der Minnesänger sagt der Meissner (1260–1280) ausdrücklich, dass alte Leute sich der Brille bedient hätten." He gives no details of manufacture.

3. P. Pansier, *Histoire des Lunettes* (Paris, 1901), p. 21. Contains little that is not discussed by Martin, although he does not acknowledge it. On p. 5, " En tout cas, les Romains ignoraient l'usage des verres convexes pour pallie aux inconvenients de la presbytie, puisque,

au temoignage de Manni, Cicéron, Cornélius Nepos, Suétone, attestent que, lorsque, en vieillissant, la vue s'affaiblit, on n'a pas d'autre ressource que de se faire faire la lecture par un esclave " ; and later, " Les verres concaves paraissent avoir été complètement inconnus des anciens," and " Aucune de ces citations ne nous permet de conclure à l'usage des verres chez les anciens pour pallier aux anomalies de la vision."

4. William Bourne, *A Treatise on the properties and qualities of glasses for optical purposes* (MS. Lansd. Mus. Brit. 121). The treatise is undated, but in all probability was written in 1585 since the author refers to a book written by him seven years previously, which is thought to be his *Treasure for Travellers*, published in London 1578. A brief mention only of grinding in a hollow iron tool.

CHAPTER 2

SINGLE SURFACE WORKING

16 This chapter deals with methods which have been in use for several centuries as applied to the making by hand of a single prism or lens. For the benefit of laboratory artificers and amateur lens polishers, descriptions have been included of machines which, although not of the highest efficiency, involve only a minimum of equipment while still serving to reduce the tedium formerly associated with hand glass-working.

The processes and materials are subject to numerous modifications in the manufacture of prisms and lenses in quantity; nevertheless, the fundamental principles remain the same, and are directly descended from the earliest known methods of glass polishing. The manufacture of lenses and prisms in quantity is described in Chapters 6 and 7.

Just a word first of all about the accuracy of surface which has to be achieved. A piece of window glass about 1 foot square will usually have surfaces flat to within about one hundredth part of an inch—to be precise, if you put it on a flat plate, there will be places where there is a separation between the two surfaces of perhaps one hundredth of an inch. A piece of good plate glass of the same size would have errors up to something like one thousandth of an inch. In a second-rate pair of binoculars, the prisms have a flatness to within one ten-thousandth of an inch; this also is about the accuracy to which spectacle lenses are made by a good firm. In binoculars of the best quality, the inaccuracies will not amount to more than one hundred-thousandth of an inch and, in work of the utmost accuracy required for certain scientific purposes, the surfaces must not depart from flatness by more than one millionth of an inch: sometimes, indeed, even higher accuracy must be attained (see §§227·1, 336 and 337).

17 The manufacture of single optical components, whether prisms, lenses or parallel plates, consists of four main operations—

(a) Sawing the blank from the raw (slab) glass,
(b) Rough grinding to size and angle (or radius),
(c) Fine grinding while maintaining angles (or radii),
(d) Polishing and finishing the part.

Making a single prism

18 We will suppose that it is desired to make a glass prism for use in a spectroscope, and that a slab of suitable glass has been obtained from the glass-maker.

Sawing the blank

Three methods of sawing glass are available, and in order of increasing efficiency they are—

1. Remove the teeth from a hacksaw blade on a tool grinder, insert the blade in the usual frame, and feed with a mud of carborundum and water while cutting.
2. Use carborundum cutting discs, as used by engineers for cutting off carbide tips. It is important that the grit and bond are suitable; the makers will supply the correct disc if told the purpose for which it is required.
3. Use diamond charged saws. These are obtainable commercially in several forms, or may be improvised with the help of a small power-driven lathe, using for the blade soft iron from 0·03 to 0·04 in. thick.

Regarding the respective merits of the three methods, it can be said that the hacksaw method, although slow and tedious, is for these very reasons fairly safe in unskilled hands. Both the carborundum saw and the diamond blade need to be rotated at high speed to be effective, the actual speeds of the former being prescribed by the makers. If a small lathe is available, it can, by the addition of a suitable rest, be used for the turning, notching and charging of the blade, and also for the actual sawing.

Instead of by sawing, slabs can be divided by scoring and breaking, and a special breaking press makes this method available for very thick pieces of raw stock. The glass is first scored by a hard steel roller and placed, with the scored surface uppermost, on the table of the press. It is so positioned that the score lies parallel to, and vertically above, a knife-edge a little above the plane of the table. Pressure is then brought to bear on the upper surface of the glass by two more knife-edges, parallel to the first and roughly equidistant on either side of the score (the distance between them subtending an angle of about 60 degrees at the lower knife-edge). These two are free to assume positions in the plane of the upper surface of the slab to compensate for any lack of parallelism between the two glass surfaces: they are also lined with padding to prevent extreme pressure points and resultant local fractures. Their pressure breaks the glass cleanly along the line of the score and, in general, the fractured surfaces produced by this means are flat to within about 8 per cent of the thickness of the glass. The equipment can be used for breaking slabs of glass from $\frac{1}{4}$ in. up to 8 in. thickness.

Charging the blade with diamond

19 In early times, the blade was charged with diamond in a crude but effective manner. A few largish diamonds of the type known as "Industrial Boart" were crushed, say between a couple of flat irons, and then mixed with the fingers into a paste formed by moistening the ash of brown paper with paraffin oil. The diamond pieces were not more than about seven thousandths of an inch in diameter.

This black paste was wiped on the edge of the blade with the fingers, while the blade was being used to cut a piece of hard glass or quartz. This procedure drove the diamonds into the edge of the saw, which was then ready for use.

A way of making the diamond powder, more refined than the rough procedure described above, is to break the diamonds in a small cylinder of hardened steel fitted with a hard plunger. Having produced a powder in this manner it should be washed in a way similar to that described for the washing of emeries in §72. Diamond powder is needed for grinding such hard substances as diamond itself, corundum, ruby, jasper and other hard crystals (carborundum, being harder than emery, is more suitable than the latter for working quartz and crystals of a like hardness).

About 1910, or thereabouts, the following technique was introduced. The blade, a flat disc of soft iron or copper, from 6 to 12 inches in diameter and from 1 to $2\frac{1}{2}$ mm thick, was turned true, marked by a coarse knurling tool and notched with a small chisel of hard steel, the notches being about 2 mm apart and a millimetre deep. It will be noted that in adopting this procedure we used a thicker blade, since the thinner blades are liable not to cut straight. But as the thickness increases so does the wastage of glass and also the cost of the diamond.

Diamond dust (preferably graded to 80 grit) was then mixed with vaseline and worked with the finger into the notches which were then simultaneously closed with a hard steel roller, so as to grip the diamond.

If much cutting is to be done, particularly of quartz, there are several ways in which this simple charging procedure can be modified to extend the cutting life of the blade. Among them is giving a saw-like "set" to alternate pairs of notches to increase cutting clearance and, at the same time, giving the notches a slight angle of attack so that the diamond is forced further into them rather than pulled out. The notches are then closed by an additional pair of rollers at 45 degrees to the peripheral roller, so that lateral escape of the diamond is prevented.

In use the blade should be rotated at a speed of cut of 1000 ft per minute, although some advocate 3000 ft per minute ; a well-prepared blade of this kind should cut about 1000 sq. in. of glass.

The durability of diamond may be illustrated by the fact that a small diamond was brazed into a cavity at the tip of a steel rod, and was used for years in Hilger's workshop to turn a large grindstone true. Eventually the brazing failed and the diamond was lost, but no trace of wear had ever appeared on it. This makes it probable that diamond grains on laps, etc., do not *wear* out but *fall* out after losing their hold.

The " notching " technique was brought to a higher standard by the Felker Manufacturing Co. in the U.S.A., and by Impregnated Diamond Products Ltd in the United Kingdom. Using a copper alloy blade, and mechanical devices for the notching, charging and rolling operations, blades are produced which will cut five thousand

Fig. 3—Glass slab marked for cutting six constant-deviation prisms

square inches of glass each ; but for best results and economy of time it is desirable to employ a specially designed machine, the bearings and general construction of which permit the blade (if 14 in. diameter) to run at about 1000 r.p.m. Notched blades have, however, now been discarded in favour of the impregnated diamond blades of the sintered type described in §127.

Cutting the prisms to shape

20 We will assume that the required shapes have been marked out on the slab with a grease pencil ; the prism is now sawn out leaving one-eighth of an inch to be removed during the ensuing processes. Fig. 3 shows a slab marked for cutting six prisms of the type known as " constant deviation." Three of its surfaces are to be polished, the fourth left grey. If the maker's melting number is removed during this stage, it should immediately be engraved with a marking diamond on the piece or pieces that are left. The melting number is the only accurate indication of the optical properties of the glass, and glass of unknown properties is of very little use.

Fig. 4—Protractor for setting to angles of prisms

Rough grinding to angle

21 The rough grinding to angle can be carried out with the aid of a flat disc of iron screwed to a firm support, on which is spread the usual mud of carborundum and water, the grade of carborundum being 80 followed by 160. If the disc, or tool, is screwed to a power-driven vertical shaft, and surrounded by a metal tank to catch the abrasive and water thrown off, it then becomes a conventional hand roughing machine (see Fig. 61) and the rate of removal of material is very much greater. If it is not possible to power-drive the spindle, then a post should be erected around which the worker can walk, taking a small pace to one side after every half-dozen strokes ; the wear on the tool

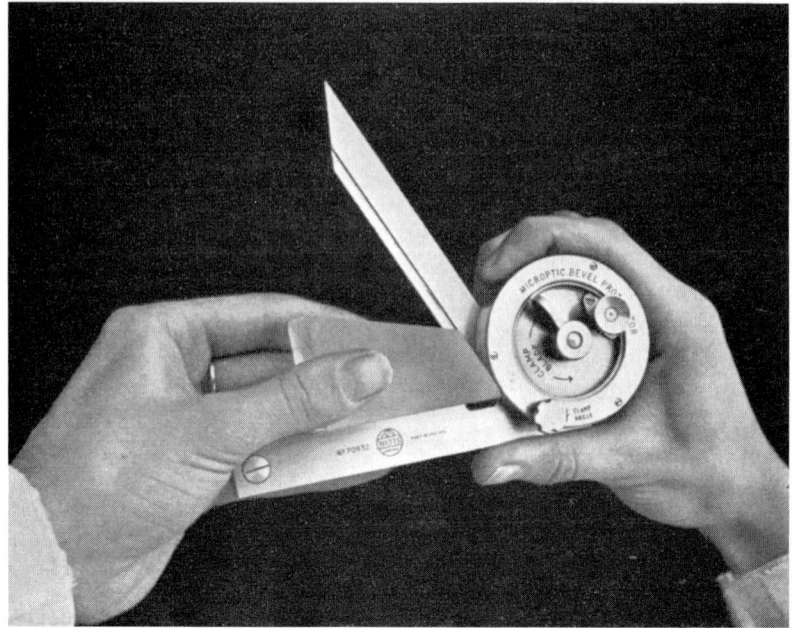

Fig. 5—Watts "Microptic" bevel protractor

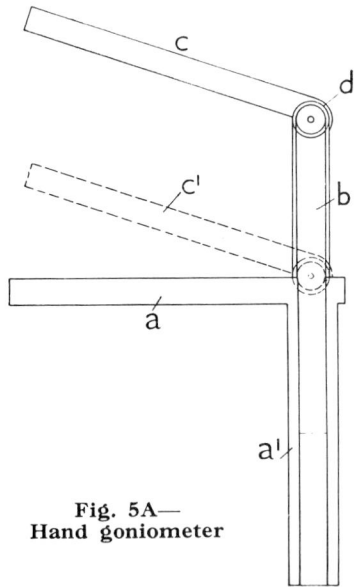

Fig. 5A—
Hand goniometer

surface is then distributed symmetrically round the centre, and is easier to correct.

Any one rectangular prism face is now ground square to the chosen base, using an engineer's square to test for squareness ; with care an accuracy of 2 minutes of arc can be achieved. This surface, and the chosen base, are then marked with a grease pencil to avoid re-grinding in error. A good protractor, such as those shown in Figs. 4 and 5, is then set to the required angle of the prism and a second rectangular face ground to this angle and perpendicular to the base. Mark this face also with a grease pencil and proceed to grind the third face, keeping it also perpendicular to the base and to angle.

As the exact amount of material removed during the fine grinding process which follows depends on the skill of the worker, it is customary to leave 1/100th of an inch full on each ground surface. With modern methods of sawing and grinding (see Chapters 6 and 7) one can, however, treat the sawn surface in every respect as a trued one, and an allowance of about 5/1000ths of an inch is ample.

Fig. 5A shows a hand goniometer which is useful for setting to angles in the measurement of prisms of acute angle when they are truncated at the pointed end. Obviously, it may be quite impossible to measure these with the form of goniometer in Fig. 4.

Trueing, fine trueing, smoothing, and testing the flatness of trueing tools
22 The rough ground faces are now " trued " and " smoothed " with fine and finer abrasives, a suitable series having the following grain sizes—

Trueing and fine trueing   Aloxite 320 D followed by Aloxite 400
Smoothing   ...   ...   Aloxite 500 and 600

The first essential for these processes is a set of accurately flat tools.

The rate of removal of glass by pitch polishers is so slow that much time can be saved by smoothing the surface flat to nearly the same precision as is ultimately required in the polished state. For this reason the final smoothing of small single flat surfaces is done on the true tools. But the amount of smoothing on these should be limited to as few " emeries " as possible as each " emery " destroys their state of flatness a little, entailing the reflattening of the true tools at more or less frequent intervals, depending on the skill of the worker to grind down all parts of the tool at the same rate. It is therefore advantageous to keep the trueing tools in nearly as good condition as the true tools. A quick way of gauging the flatness of tools is—

Select from plate glass (a $\frac{1}{4}$-inch thickness is suitable for a tool of 9-inch diameter) a piece 1 inch longer than the diameter and $\frac{1}{2}$ inch wider than the radius of the tool, i.e. a piece 10 inches long and 5 inches wide for a tool of 9-inch diameter. The only area of the plate which need be really flat is a strip about $\frac{3}{4}$ inch wide along one side of its length ; the remaining area can be several wavelengths out of flat.

Having cleaned and dusted this side of the testplate and the surface of the tool, place the two together so that the flat strip on the testplate is over the centre of the tool and insert the end of a visiting card or bus ticket under the opposite edge of the testplate, thus forming a small wedge angle of air between tool and plate, tapering to zero along the flat edge of the testplate in contact with the centre of the tool.

Now take a swab of cotton wool wound on to the end of a dogwood peg or a small camel-hair brush, load it with methylated spirit and stroke it quickly along the contact edge of the testplate and tool. Capillarity will cause the liquid to run between testplate and tool. Repeat till the wedge of liquid from the zero edge is the same width as the flat strip on the testplate, *i.e.* $\frac{3}{4}$ inch. Allow to stand for 5 minutes.

Owing to surface tension the liquid will in this interval have adjusted itself to a uniform thickness along its thicker edge and the departure from flatness of the tool can be judged in the same way as though this thicker edge were a Newton's fringe between two glass surfaces inclined at this same angle, *i.e.* if the said liquid edge is straight the tool is flat; if its middle dips towards the zero end of the wedge, the tool is concave; if away from the edge the tool is convex. A hump at each end and one in the centre is a sign of a zonal depression, and so forth. Astigmatism is disclosed by slowly rotating the proof plane on the tool allowing rest intervals between rotation and reading times.

For trueing tools of 9 in. diameter a London 'bus ticket is sufficiently thin and is convenient in that its thickness is about 0·008 inch, so that when inserted $\frac{1}{2}$ inch between glass and tool it gives a wedge angle of 100 wavelengths per inch, which is easy to remember. There is the further advantage that the methylated spirit evaporates more quickly than with a thinner separating piece. The assessment of sign and degree of curvature can be made in a few seconds by lowering the eye to nearly tool level and looking along the liquid wavefront; viewed thus, end-on, a departure from straightness of 0·02 inch (equal to an out-of-flatness of 2 wavelengths) can be detected. As a check on this the beginner can place a light-weight straight edge on the testplate and measure the sag, but he must be careful not to press on the testplate as this will cause the liquid to flow and it may take a few minutes for it to settle down to a state of equilibrium again.

To revert, now, to the trueing and smoothing.

During this fine grinding process, the angles of the prisms are maintained to the required precision, and care must be taken to rub the prism in circular strokes around the centre and over the whole surface of the tool, especially the region near the extreme edge, so as to keep the tool a good shape. It must be remembered that more than one half of the total area of the tool is included in a marginal area only one sixth of the diameter of the tool in width. The face which is not to be polished is then trued. The final smoothing, in 500 and 600, of the faces which are to be polished must be performed on the flattest tool available, which should preferably be slightly convex rather than concave.

Throughout this book a "grinding," or as the workshops would say, an "emery," is taken to mean that the workpiece is ground with the specified abrasive grade until the emery has distinctly lost its "bite" or cut. Using a tool of 9 in. diameter this may take from 5 to 15 minutes, depending on how much emery is applied in the first place.

If much trouble is encountered with scratches when smoothing with Aloxite 600, polishing can proceed straight from a worked-down smooth with 500, though polishing will naturally take a little longer. As an aid to avoiding scratches, the emery can be "bruised" before use; this consists merely of smearing the tool with the usual emery paste, and then rubbing over it a piece of flat glass, using a gentle pressure and working from the centre to the edge of the tool in a slow spiral. Any very large particles or aggregates of emery will, by this means, be pushed to the edge of the tool, whence they can be wiped with one sweep of a wet sponge. With each grade of abrasive the aim is to remove the pits left by the previous grade, and the series detailed above is so chosen that three grindings in each grade should generate a fresh surface.

Unless the prism faces are very flat in the grey, polishing them flat can be a very lengthy process. It is possible to get Newton's fringes by using a testplate directly on a smoothed surface, by observing at an oblique angle, but there are several objections to this procedure. First, there is a risk of scratching both the testplate and the smoothed surface. Secondly, the weight of the testplate may deform the surface being tested.

A useful method in which these disadvantages are minimized is to run a suitable ink—Stephen's blue condensed ruling ink is the one used by us—between the surface to be tested and a testplate. In the regions of contact there is a white patch. This is surrounded by a pink area and this change of colour enables differences of flatness of as little as 2λ to be detected. Such departures from flatness are easily removed in polishing.

This ink possesses the advantage of being dichromatic. Increasing thickness is evidenced by the following sequence of colours: white, pink, blue and black with intermediate hues. To improve the sensitiveness a denser and more mobile liquid might be sought.

It is probable that this test could be used with advantage for metal surfaces which are bright enough to give a 20 per cent reflection even when they are so scored as to render the Newton's ring test out of the question—for example, ground stock, scraped surfaces, lathe slides, etc.

Fuller information about abrasives is given in Chapter 4

When the faces have been smoothed, all the edges and corners are chamfered with 500 abrasive. This chamfering of edges and corners has a double utility in that it increases the resistance of the edge or corner to accidental damage, and also prevents the minute splinters in a ground corner from breaking out during the polishing process and causing scratches.

The prism is now ready for polishing, and after a thorough clean-up of the working space, the preparation of the polisher can be commenced.

Preparing the polisher

23 Opticians usually keep plane tools in sets of four, a set comprising three accurately flat, and a fourth—which need not be free from pores or other surface defects—for use as a backing tool for a pitch polisher.

Pure Swedish wood pitch is heated in a thick saucepan over a low heat until it is of such a hardness that at room temperature it can be readily but not deeply indented with the thumbnail.

Laboratory artificers will no doubt be inclined to devise more scientific means of determining the viscosity of pitch (see §§ 60 and 61) ; a surprising degree of consistency can nevertheless be achieved with nothing more than the thumbnail test and a little experience.

The Rev. C. L. Tweedale (1943), an accomplished mirror maker, recommended the following test for polishing pitch—

> Having with a gentle heat melted the pitch, then with an old spoon drop a little on to a small tin lid, and place in a basin of water tested by the thermometer to a temperature of 60 degrees Fahrenheit. Now obtain a NEW shilling and rest its edge on the smooth level surface of the pitch in the tin lid for 15 minutes. If it shows 3 " nicks " of the milled edge clearly, it is of the right hardness. If it shows only 2 nicks it is too hard and must be slightly re-heated and about 15 DROPS (no more) of pure turpentine must be well stirred into the 2 lb. of melted pitch. Test again, and if still too hard add a few more DROPS of turpentine and so on until right. If more than 3 nicks show, it is too soft and must be gently simmered for a few minutes to drive off the excess of turpentine. WATCH IT CAREFULLY. It is very liable to fire. All this will take a little time, but it *has to be done*.

While the pitch is being hardened, a small sample is taken every half-hour or so, cooled rapidly in cold water to room temperature, and tested for viscosity. When this is satisfactory, a wooden handle, Fig. 7a, is screwed into the back of the polisher holder, which is then put on the gas ring and heated until it is just too hot to be comfortable to the hand. The hot tool is lifted from the gas ring and the molten pitch, which should be of the consistency of a stiff treacle (molasses), is then poured over the hot tool so as to form an even layer $\frac{1}{4}$-inch thick (see Fig. 6). The polisher is then pressed face down on to one of the flat tools, which is itself screwed to the bench or post by means of a nose, Fig. 7b.

If the polisher is inclined to stick to the cold tool, it can easily be dislodged by a smart tap with a hammer on the boss of the tool. Irregularities in the polisher surface can be removed by warming the polisher face downwards about a foot above a gas-ring ; this will also help to remove air bubbles near the surface. After each warming it should be re-pressed on the flat tool which, if necessary,

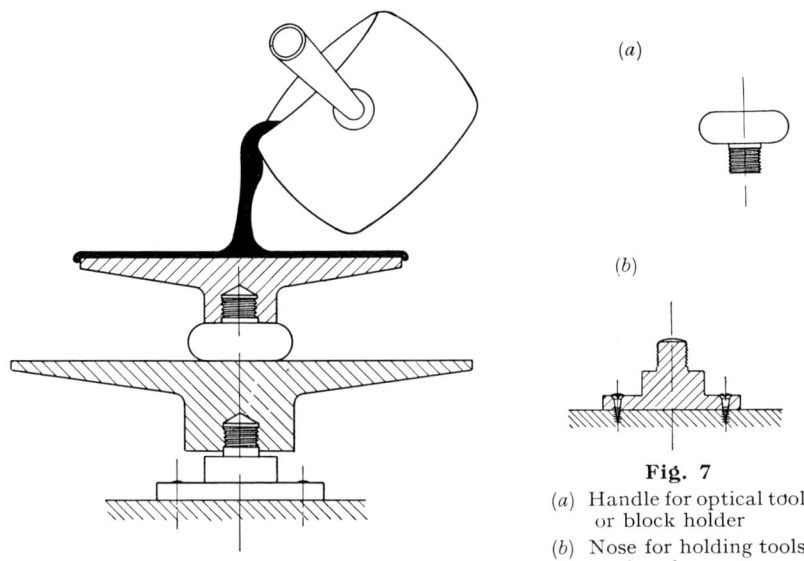

Fig. 6—Making a flat polisher
Pouring out the pitch

Fig. 7
(a) Handle for optical tool or block holder
(b) Nose for holding tools on bench

can be cooled with water as repeated pressings raise its temperature near the point at which the pitch will readily adhere.

It is not out of place to remark here that although, with much experience, it may be possible to obtain very flat surfaces using tools that are themselves far from true, it is not to be recommended. Time spent in the flattening of working tools will be saved in the polishing stage many times over.

When the pressing is completed, and the polisher is smooth and flat all over, it is placed in a tank of cold water, where it should remain for a quarter of an hour ; pitch is a poor conductor of heat.

Reticulations are now cut in the polisher surface, leaving facets of three-eighths to three quarters of an inch square. For making these reticulations one uses a strip of brass about one-half inch wide by 1/32nd inch thick. This is pressed firmly against the surface of the polisher with the fingers of the left hand and acts as a ruler to guide a graver, consisting of a three-square file sharpened to a triangular point, with which a series of grooves is made to form the desired checkered surface. If many polishers are to be made, then a grooving tool (Fig. 8) can be used just before the final cooling.

The polisher, while still hot is pressed twice on the grooving tool, being turned through a right angle between the pressings.

After a final brief pressing on the flat tool the polisher is cooled in cold water.

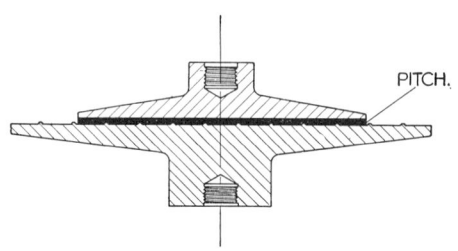

TOOL FOR RETICULATING POLISHER.
Fig. 8—Making a polisher
Grooving the surface

Stamping and rubbing up

24 Whilst the polisher is cooling off, a piece of Brussels netting (cotton net having about 10 meshes to the linear inch) or mosquito netting is soaked in water. The flat tool, on which the polisher was pressed, is heated over a low gas to about 80–90°C; the polisher is then screwed to the post and painted with a fair quantity of wet rouge, the wooden handle being transferred to the hot flattening tool.

The wet Brussels netting, about an inch larger all round than the polisher is now stretched evenly over the polisher surface, and the hot tool pressed firmly on top. If the tool temperature is about correct, the net will be forced into the pitch by a pressure of 20 or 30 pounds applied for about half a minute.

If the tool is not hot enough, it must be further heated and painted with more wet oxide; if too hot it will stick to the polisher and will need cleaning, first with petrol (gasoline) and then with methylated spirit. Until some experience has been gained it will repay the novice to start with the flattening tool too cold rather than too hot.

When the netting is pressed right into the pitch, so that tiny flat facets are visible between the meshes, the tool is slid off and the netting raised simultaneously, when it should be found that the netting comes away from the polisher without any pitch adhering to it.

PLATE 3   [*Facing page* 32]

Albert A. Michelson (1852–1931)

His interferometer—famous in connection with the Michelson-Morley experiment —is the basis of many modern optical testing instruments

[From *La Science, des Origines à nos Jours* by permission of Urbain et Boll (Paris, Larousse)]

PLATE 4 [Facing page 33

Jean Bernard Léon Foucault (1819–1868)

He gave to practical opticians the first precise method for testing optical surfaces—the knife-edge test

[From *La Science, des Origines à nos Jours* by permission of Urbain et Boll (Paris, Larousse)]

Bernhard Schmidt (1879-1935)

He invented and made a form of camera which is revolutionizing astronomy and finding many industrial applications

SINGLE SURFACE WORKING   33

The tool is now cooled until only "hand warm," painted again with rouge, and rubbed firmly over the surface of the polisher until the action is quite smooth and noiseless.

The finished polisher is shown in Fig. 9.

Polishing the prism

25 Polishing of the prism can now be commenced.

Fig. 10 shows the way in which the prism is held for flattening the surface. The best practice is generally to hold the prism quite

Fig. 9—The finished polisher

near the base, but it must be remembered that the process of flattening a surface is extremely complicated; although a better adjustment of pressure may be obtained by holding the prism very near the base, this is offset by the heat of the hand which, in that location, has a marked effect on the distortion of the surface by temperature. One has to consider the polishing of materials so different in heat conductivity as glass and quartz (crystalline quartz has approximately fifteen times the conductivity of glass). Another variable factor is the temperature of a man's fingers; they differ very much in this respect from time to time and from individual to individual. The man of whose hands a photograph is shown in Fig. 10 is one of Hilger's best single-surface workers, and he selected his hold deliberately because the prism he is polishing is of quartz ; such a man will instinctively be on his guard against this effect of temperature if the prism feels cold to his touch. It would be unwise to prescribe to a first-class worker that he should alter his procedure.

A long oval polishing stroke will be found the easiest to start with, and a pressure of about 1–2 lb/sq. in. should be applied. When

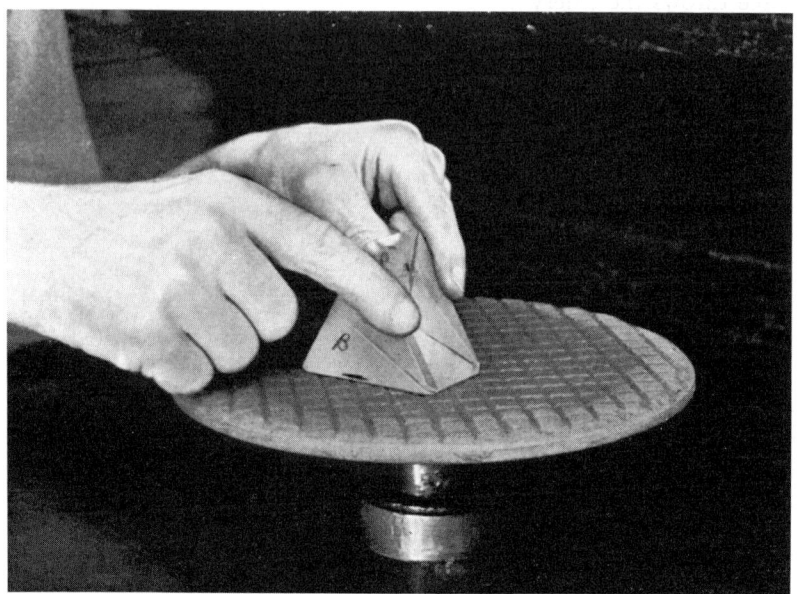

Fig. 10—Polishing the prism

hand polishing, it is desirable to have the polisher rotating slowly on a vertical shaft, say at 1 r.p.m., as this helps to keep the polisher a symmetrical shape and avoids astigmatism. If the rotating post is not available, good work can be done by erecting a post around which the worker can walk, taking a step to the side after every half-dozen strokes.

If while polishing it is decided to re-smooth the surface, a little time should elapse first to allow the prism to cool down.

If the necessary conditions have been observed, that is if—
  (a) the tools are flat,
  (b) the prism is sufficiently ground with each grade of abrasive, from the coarsest to the finest,
  (c) the polisher is well formed on the flattening tool,

then the prism surface should commence polishing evenly all over, and all trace of grey should be removed inside the hour.

It is the general practice to apply wet rouge to the surface of the polisher with a soft bristle brush from time to time. In shops where lenses are being polished in large quantities, various means of automatically applying the emery or rouge are in use. For example, in one shop (United Kingdom Optical Company) the tool and work are enclosed in a bowl which is attached to the spindle. The centrifugal force throws the emery or rouge and water mixture to the side of the bowl, from whence a collector directs it in a small stream on to the lens (see Chapter 7). Up to 50 spindles are thus operated by one man whose only responsibility is to change the lenses after the normal time has elapsed. Hilger's most recent practice, however, is that, once polishing is well under way, fresh oxide is not added to the polisher, neither is the polisher itself brushed. Sufficient oxide is retained by the " stamping marks " left by the Brussels net, and it is not at all clear that the loose oxide is in fact mainly responsible for the polishing action. (See Chapter 3 for a consideration of the nature of grinding and polishing.)

Clean water should be added to the tool with a brush and the tool rubbed on the polisher for one or two minutes at the end of each " wet ". A " wet " is the time elapsing between the rubbing of the polisher and the point where the rouge begins to dry and considerable resistance to polishing becomes evident ; it varies from two or three minutes to a quarter of an hour, depending on the amount of water added to the polisher when rubbing up.

Particular care must be taken during the end of a wet. Owing to the adhesion of the glass to the polisher it is necessary to use a stronger force to move the glass or polisher as the case may be. Consequently the pitch is liable to become a little soft and go out of shape. This effect can be minimized by a frequent change in the direction of movement of the polisher. If, however, the polisher does become non-uniform in surface there is no remedy but to clean the forming tool, warm the polisher and form it up again.

For the final polishing strokes after being assured that the forming tool is the correct shape (see below) and the surface of the prism free from grey, though probably not perfectly free from faint sleeks, warm the tool very slightly and wet the surface with the brush without rouge. Do not brush the polisher, but place the moistened tool on it ; take short strokes for a minute or two, and then draw the tool off. Without delay proceed with the final polishing of the prism, working slowly over the polisher which ought to feel nice and smooth until it is dry.

Maintaining tool flatness

26 Optical workers who do a lot of hand polishing keep their tools flat in the following way.

If two discs are ground together, the upper one will tend to become concave, and the lower one convex. This principle is well known to amateur telescope makers, who continue the process until the desired radius of curvature is produced, and then use the other disc as a backing tool for the polisher.

If the tool which is used as the former is rubbed on top of the polisher it also will become more concave, until the stage is reached where flat surfaces can no longer be obtained, and the tool must be corrected by grinding with emery.

If, however, the polisher is rubbed on top of the tool the previous trend will be reversed; in fact it will even, in time, make the tool definitely convex. Thus by discriminating reversal of tool and polisher during forming, both can be kept flat for very long periods.

The simplest way to test a polished surface for flatness is by means of Newton's fringes, using a proof plate which is known to be flat; the fringes, by revealing the thickness of the air gap between the surface and the proof plane, show what is virtually a contour map of the surface. The thicknesses at points on two adjacent fringes differ by one-half the wavelength of light used, that is by about one hundred-thousandth of an inch. More precisely for the green radiation from a mercury lamp the difference is 10·5 millionths of an inch per fringe.

The laying of a test plate on an iron tool to test the latter, even when it has acquired a degree of polish, is to be deprecated unless more than one test plate is available; test plates used regularly on tools become deeply and generally scratched to an extent that may endanger the glass surface on which they are subsequently used.

One strong objection to the use of proof planes is the difficulty of avoiding scratching the work. It is true that scratching can be avoided by meticulous cleaning, but this very often takes quite a time and is particularly difficult when the work is taken off the machine for examination. It is good practice, therefore, to arrange an interferoscope, broadly on the design of that described in §225, the only difference being that the interference is observed between the surface to be tested and a proof plane with which the surface does not come into contact. This has a further advantage in that, since a gap of as much as $\frac{1}{4}$-in. can be used between the two surfaces, there is not that tendency to promote the harmful effects of placing together a proof plane and a piece of work which are not of identical temperature.

Primary test plates can be made without reference to a master flat by use of the principle generally attributed to Sir Joseph Whitworth—

In 1840 he attended the meeting of the British Association in Glasgow, and read a paper on the preparation and value of true planes, describing the method which he had successfully used for making them when at Maudslays, and which depended on the principle that if any two of three surfaces exactly fit each other, then all three must be true planes. The accuracy of workmanship thus indicated was far ahead of what was contemplated at the time as possible in mechanical engineering. (*Encyclopaedia Britannica*, 11th Edn, 28, 616.)

To make proof plates three discs of plate glass, say four inches in diameter and three-quarters of an inch thick, are ground together with the last three abrasives in the second series (§74) until all are of a fine smooth finish and free from scratches. Each in turn is partly polished on the polisher, and when bright enough to see through they are laid together a pair at a time in monochromatic light.

The resulting appearance of each combination (*ab*, *bc*, and *ca*) should be recorded, calling a convex resultant positive and vice versa. Taking the simplest possible case, if all three combinations show two rings concave, then obviously all three plates must be one ring concave each, and if any one plate is worked to show with either of the others only one fringe concave then that plate will be nearly flat. A similar improvement is now carried out on the other two and so on until each combination shows uniform coloration over the whole surface— or as near to that perfection as circumstances require.

The other possible types of appearance which may be exhibited by the three combinations are not so simple; there are no less than ten varieties in all. These are described in §186 in connection with keeping tools flat. The skilled optician gets to know instinctively the steps to take to flatten three test plates or tools, but even he would do well to master the procedure set forth in §186. Careful study for an hour or two will save many months of ineffective work in the course of a life-time.

In order to obtain good fringes and few, optical parts laid together for this test should be as clean and free from dust as possible. When it is realised that a piece of dust only a ten-thousandth of an inch in diameter will introduce 10 fringes of green light, the importance of scrupulous cleanliness will be appreciated.

Finishing the prism

27 Reverting to the prism which we left in the final stages of polishing the surfaces should now be tested with one of the test-plates, the making of which is described in the last few paragraphs.

When laying down the test-plate on a piece of finished or nearly-finished optic, the utmost care should be taken to avoid scratching the glass. Test-plates once in position should not be moved about the surface by sliding; every piece of dust may scratch the test-plate or prism.

When the prism surface shows one circular fringe or less, it will be good enough for all but the most exacting work.

All that remains to do now is to grind the grey bases and the unpolished back of the prism with a fine abrasive, and to chamfer the edges and corners. Needless to say, great care will be taken whilst so doing to avoid damaging the polished surfaces; with dense flint glasses this is all too easily done.

Before passing on to consider the making of a single lens, it is well to realise that to polish single surfaces of the highest quality is generally considered the most severe test of the skill of the optician. This art will not be acquired except by the expenditure of many hours of patient trial, nor is it the most vigorous worker who necessarily achieves success most quickly. In the end, it is a question rather of patient thought than of manual dexterity. There is one advantage in hand polishing which is lost when a machine is used, namely that the optician can feel exactly whether the polisher is " taking "—on the outside or in the middle—and at once alter his procedure accordingly. If it is the middle of the polisher which is operative, the piece being polished rotates freely as it is moved ; if the reverse, it binds and offers more resistance to movement.

Making a single lens

28 The process used for the making of a single lens is the same as that outlined for making a prism, with the appropriate modifications to take account of the fact that lens surfaces are usually very far from flat. The first stage is the same as before ; a blank is sawn from the slab of glass about one-quarter of an inch over size, and usually in the form of an octagon, or square with the corners removed. This is ground with coarse carborundum to a roughly circular shape, the thickness—in the case of a bi-convex lens—being left about one-tenth of an inch more than the finished size. If one or both surfaces are concave they must, of course, be roughed before the centre thickness can be ascertained.

Where the glass is in plate form, or has been reduced to plate form by slitting or grinding, an alternative to the method just described is " shanking." This is a very old-fashioned method, but it compares well in speed with modern methods where the number of discs to be made is small. In the use of the shanks it is customary to mark out in pencil the diameter of disc required, and then to break away around the margin, keeping just outside the circle (Fig. 11). Where

SINGLE SURFACE WORKING

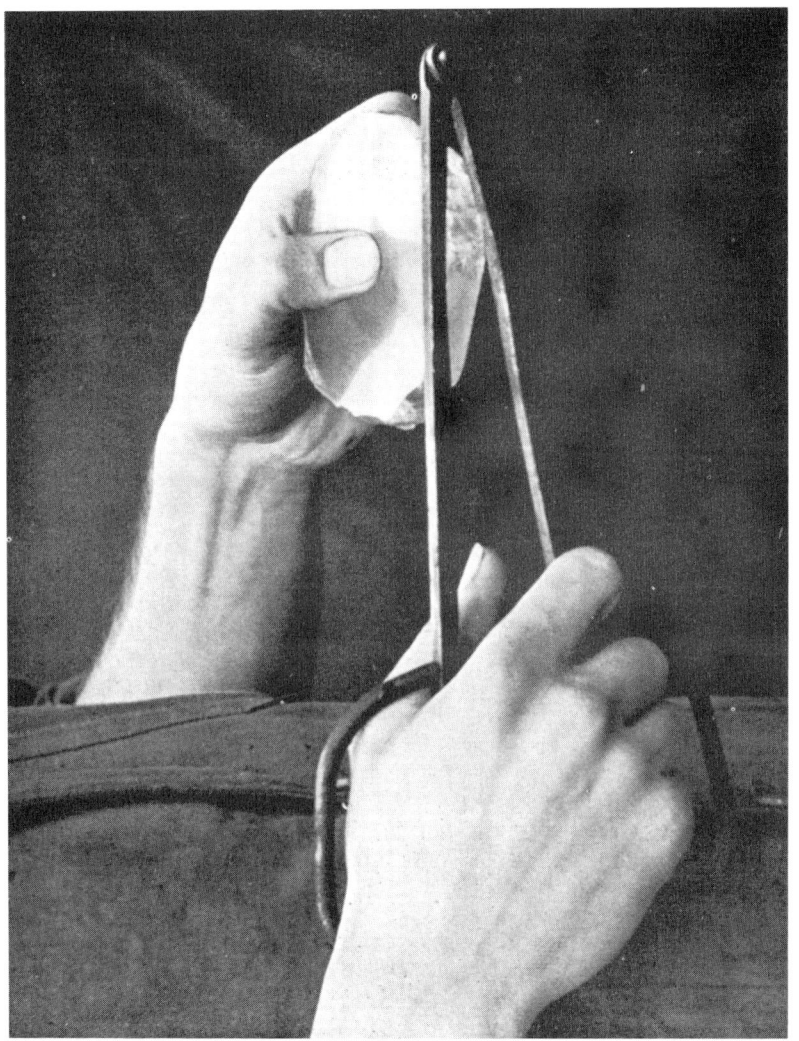

Fig. 11 – Shanking a disc

quantity production is in question the quickest method is trepanning (see §125).

Fig. 12 shows a convenient form of calipers for measuring the centre thickness of lenses during the various stages of roughing, trueing and smoothing.

Let us assume that we are to make a bi-convex lens, one surface being considerably deeper than the other. It will be necessary to provide a pair of " true tools," accurately turned and lapped to the required radius, for each surface. In addition, it is highly desirable to provide a roughing tool and a polisher holder for each surface, and

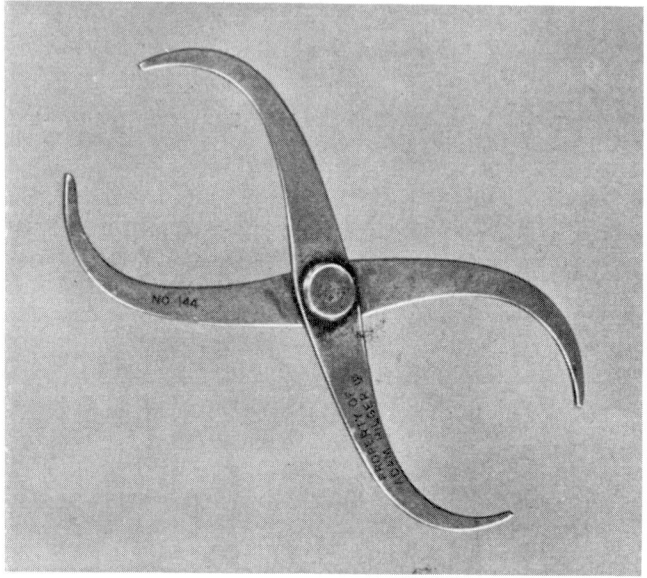

Fig. 12—Calipers for measuring the centre-thickness of a lens

thus avoid using the true tool as support for the polisher and for roughing, which latter will tend to alter the shape of the tools. The preparation of curved optical tools is described in §§46 to 48.

If the depth of curve is not great, that is if the radius of curvature is five times the tool diameter or greater, the roughing tool and polisher holder can be of the same radii as the true tools, though not, of course, lapped together. If the radius of curvature is less than this the curves should be modified as shown in Fig. 68.

The lens blank is ground in the roughing tool with coarse carborundum until the flat surface has nearly disappeared, leaving a small circular patch or " witness " in the centre about $\frac{1}{4}$ in. diameter.

The other surface is ground in a similar manner on the other roughing tool, leaving a similar witness.

Now both surfaces are ground in turn with fine carborundum (160) until each witness just disappears ; it can then be assumed that the pits of the coarse carborundum have been removed.

After thoroughly cleaning out the roughing tool and working space, grinding is carried on with 320 Aloxite, or other abrasive of like grain still in the roughing tool, until the blank is about 0·02 in. over the finished centre thickness.

Two or three grindings are then carried out in the true tool, using the same abrasive as the last and taking care to work as evenly as possible over the whole tool surface.

The thickness of the lens should be measured at each stage either with simple calipers (Fig. 12) or with a micrometer. If micrometers are used in the optical shop constantly, they should be tested and adjusted at least once a month ; the presence of grinding material in the optical shop causes a good deal of wear.

The fine smoothing of the lens is performed in the same way, the tool being carefully cleaned each time the grade of emery is changed.* One should avoid using more abrasive than necessary, particularly with the very fine grades, for if too much is used the work " rides " on the surface, the pressure on each grain being insufficient to crush the glass.

When the fine smoothing process is complete, the lens should be of a fine, even grey, free from scratches, and within the allowed tolerance for centre thickness ; the polishing removes very little material.

The importance of scrupulous cleanliness cannot be over-stated : when passing from one grade of abrasive to a finer one, the tool hands, sponge and bench must be thoroughly cleansed of all traces of the coarser grit, otherwise scratches are very likely to appear, for which the abrasive supplier may be wrongly blamed. It is indeed desirable to use a separate sponge for each grade of abrasive.

Before proceeding to the preparation of the polisher, the lens should be chamfered on both sides ; if the blank still exhibits the irregular edges of the original lump, the chamfering should be carried on until the intersection of the surface and the chamfer is a full circle

Preparing curved polishers

29 The method of pressing the polisher is very similar to that detailed for the prism (see §23), except that of course the polisher holder is now curved to the radius of the surface. Into the concave

---

* See §74, Chapter 4, for the abrasives to use for trueing, smoothing and fine smoothing.

tool, which is warmed for this purpose, is poured melted pitch. While this is still warm the convex tool (used cold, and if necessary dipped into water once or twice to cool it during the process) is used to press the pitch over the surface of the concave tool till it forms a thin coating adhering to the latter, a little more than $\frac{1}{4}$ up to $\frac{3}{8}$ inch thick. The convex tool is moved about in the socket thus formed and wetted occasionally with cold water to prevent its sticking, while the concave tool is allowed to cool down.

The reticulation is then cut in the surface of the pitch so that its surface is broken into squares of about $\frac{1}{2}$ inch side.

The grooves of the reticulation can be made in one of two ways, according to individual preference. If square reticulations are preferred, as on the flat polisher, then a more flexible straight edge than the normal steel rule is used to guide the cutting tool. This is pressed down on the curved surface of the polisher with the thumb and fingers of the left hand while the grooves are cut with a sharpened triangular file as described in §23. Many workers prefer circular rings because they are much easier to renew when necessary. Ringed polishers are, however, not suitable for large lenses, as they are liable to stick; further they are liable to cause zoning, which although not usually severe enough to matter on a transmitting surface is not permissible on a mirror. To generate circular grooves, the polisher is rotated on a lathe or polishing machine at a speed of say 200 r.p.m. and the grooves cut as before. A little practice will ensure that the grooves are clean-edged and deep enough to remain through the stamping process.

Polishing the lens

30 Polishing is carried out as for a prism (§25). Some difficulty may be experienced in holding the lens whilst polishing, particularly if the curves are deep, and it is advisable in such case to stick the lens to a suitably shaped support ("mallet") by means of pitch.

For the best work it is necessary to re-form the polisher at frequent intervals by rubbing it on the true tool. If the weather be cold, or the pitch unduly hard, the tool may be slightly warmed to assist the polisher to take up the desired shape more quickly.

When the polishing of both surfaces has been completed, we have a bi-convex lens which needs only to be edged.

Centring and edging the lens

31 The method of edging to be described here is ancient and crude, but is nevertheless in fairly wide use even today, particularly in connection with optical elements of large size and high quality.

A piece of brass tubing is chucked in a power lathe, and turned true, parallel, and to fit exactly into the same cell as the lens. The end of this tube is trued by chamfering the inside and outside at 30° until the chamfers meet at an angle of 60°, and a very small flat, say 0·003 in. wide, made on the intersection of the chamfers. The tube is warmed with a Bunsen burner until pitch will adhere to it, and the lens, also warmed, is pressed on to the warm pitch. If the lens is correctly set on the chuck the brass rim of the latter will be uniformly visible through the lens. The lathe is now rotated and the reflections in the lens observed; if it happens to be rotating about its

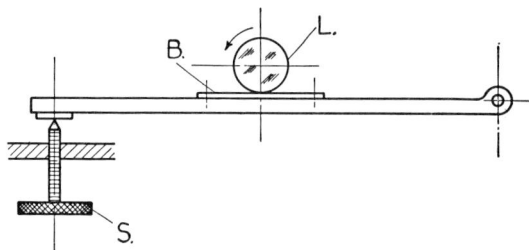

Fig. 13—Edging the lens

optical axis both the reflections will be stationary; if this is not the case the lens must be adjusted for position by pushing it sideways (warming the pitch again if necessary) until this condition is satisfied.

Cold water is then dripped from a sponge on the chuck, causing the pitch to become relatively hard. A hinged brass plate is then brought against the edge of the lens (Fig. 13) and 320 grade abrasive* fed in until the lens is edged truly circular and almost down to the diameter of the chuck. Then, with a section of brass tube whose internal diameter is the same as the outside diameter of the chuck, the edging should be completed by hand, using trueing emery and moving the tube to and fro axially to avoid grooving.

Before removing the lens from the tube, the exposed edge can be chamfered, by means of a deep concave optical tool, using the same abrasive. The lens is then warmed off the tube and allowed to cool. The pitch is cleaned off and the other edge can then be chamfered. The pitch is removed by soaking the lens in turpentine.

The main disadvantage of this method is that when edging a convex lens, the edge thickness increases as the edging proceeds. Thus, by the time the full width of the final lens edge is reached, the brass grinding plate, which itself has been ground away in the

---

* Madame la Comtesse de Maigret informs me that she uses boron carbide instead of emery or carborundum for edging ruby, as the lap lasts a good deal longer.

earlier stages, will produce a rounded edge. This shortcoming may be remedied either by frequently moving the piece of brass to a fresh place, or arranging to traverse it mechanically.

Cementing lenses is described in §144 *et seq*.

32 This simple method of hand polishing of prisms and lenses, although long discarded for quantity production, is still in use when work of high class has to be produced in small quantities, and is likely to become more widely used as the larger research organizations determine that provision should be made in their laboratories for making small optical elements which may at any time be required for some urgent project. If specially made outside, months of valuable time may be lost.

A problem which may occasionally occur is the regrinding and polishing of a damaged surface of a lens for which the optician has no tool. If the matter is urgent the surface can be reground in the following way—

(a) Warm the lens and press it into a surface of molten Woods metal, which melts at about 100°C. The metal, which can be in any kind of container, *e.g.* an egg cup, will soon set and this can be used as a grinding tool.

(b) When the lens has been trued, smoothed and fine smoothed, the same tool is used as polisher holder, the polisher being formed on the lens itself. Care must be taken in pouring the pitch into the Woods metal tool that the pitch is not hot enough to melt the Woods metal.

Naturally, this is not a method for a novice, and of course should be used only in case no proper tools are available.

Edging machines for production shops are described in §143, but the typical reader of this chapter probably works in a shop where the amount of work does not justify the purchase of such a machine.

33 The following unconventional notes on polishing deserve respectful attention, coming as they do from a man of great originality and independance of thought and a keen observer—A. M. Graeme-Thom (*Laboratory Journal of Australia*, 1938, **1**, 128).

> There seems to be little doubt but that all high-grade optical surfaces in glass, and most optical surfaces in metal, should be finished off with a spell of polishing or figuring using one of the black polishing media. . . . If an optical surface is polished with a white medium such as tin or zinc oxides its surface fissures become charged with that medium, and the surface when dry reflects or transmits diffused white light in direct proportion to the ratio of

fissures to optically perfect surface. The same occurs in the longer wavelengths when rouge is the polishing medium ; but when a black medium has been used, such light as is not reflected or transmitted is absorbed, and herein lies the advantage of . . . finishing off the polishing and figuring processes with black media.

This is quite noticeable even by rough and ready eye tests in the case of glass surfaces, but is particularly the case in reflecting surfaces of metals and alloys, which, viewed by oblique light, invariably show the colour of the polishing medium used. . . .

There is a very useful black polishing medium—a proprietary product called " Glassite."* The writer has found, however, that the ordinary quality of Windsor and Newton's Indian Ink, is a most excellent and readily obtainable polishing and figuring medium for both metal and glass surfaces. It is of particular utility for cleaning optical surfaces in glass and metal (not ruled gratings of course), which through wear and tear have become scratched, or possibly pitted by chemical and atmospheric action. . . .

If such optical surfaces be thoroughly cleaned and then lightly coated with Indian ink, gently applied with a clean soft camel-hair brush, dried in a dust-free atmosphere, and the dry ink gently wiped off with a pad of the softest washed chamois leather, the improvement in optical performance resulting from the surface being relatively free from diffusion, is most marked.

It cannot be too strongly impressed, however, that even the carbon particles held in colloidal suspension in Indian ink (the mildest polishing medium known to the writer) have an abrasive action under pressure, and therefore care and intelligence is necessary in charging defective optical surfaces with black media.

For the same reason, the writer had found in polishing off the " bloom " from chemically silvered astronomical glass speculae, that a most marked improvement in the optical performance becomes evident if " Glassite " or Indian ink is used instead of the rouge recommended in the standard text-books for burnishing the silver film—the star images being beautifully bright against a velvety black background.

In finishing optical surfaces on speculum metal and stainless steel, the writer's predilection is for Indian ink as having a finer, if slower, polishing action than " Glassite " (magnetic oxide of iron, black in colour). Tentative tests indicate that Manganese dioxide (MnO_2) shows promise as a polishing medium for both glass and metal, and that the commercial lubricants having a colloidal sus-

---

* Magnetic oxide of iron, black in colour, very popular in Australia in 1943 as a polishing powder.

pension of graphite in mineral oil as base, are well worth investigation as polishing media for the softer metals. The action of the latter seems to be the formation of a " Beilby layer " on the metal surface.

In general, metal surfaces are very much more difficult to bring to optical perfection than glass, and when Liebig discovered an easy chemical method of depositing silver on glass surfaces some eighty years ago, glass came into use for most purposes where metal previously had been used. Consequently, the intervening three generations of opticians had little practice in working metal optical surfaces, and the art thereof was largely forgotten.

Of late years it has again come into prominence owing to the increasing use in science and industry of instruments of control where metal optical surfaces are a *sine qua non*. While much had been written in the interim regarding the technique of glass working for optical purposes, practically nothing new had been written regarding the optical working of metals, and to gain an understanding of the technique of optically working metals one has to refer to the old text-book " Compleat System of Opticks," by Robert Smith, which was published in 1738, and which by the way, was translated into French in the year 1767 under the title " Cours complet d'optique, traduit de l'anglais, de Robert Smith." . . .

In optically working metal surfaces, the actual use of polishing media on pitch laps should be preceded by a pre-polishing course of fine grinding with thoroughly washed glass powder, prepared in exactly the same way as described by G. W. Ritchey (Smithsonian Contributions to Knowledge, Vol. 34, 1904) for grading fine emery.

The advantage of the use of glass is that it does not " charge " the metal surface, and cause it to darken, to the same extent as is the case with both carborundum and emery. The final spell of this prepolishing fine grinding should be with a " 60 minute " glass. This is particularly efficacious in the case of stainless steel, and is best used on a lead lap, into which the glass particles embed themselves deeply, leaving only a small projecting " cutting edge." Under no circumstances either in polishing or pre-polish fine grinding, should a lap be used which is nearly as hard as, or harder than, the metal being optically worked; the lap must always be much softer.

Metals generally, but more especially Stainless Steel and Speculum Metal, should be polished with zinc oxide in the early stages and with a mixture of zinc oxide and Indian ink in the intermediate stages, completing the polishing and figuring with Indian ink only.

The carbon particles in colloidal suspension in the Indian ink will be found to have sufficient hardness to polish slowly, but to an intense brilliancy, which shows only the metallic lustre when critically examined by oblique light.

SINGLE SURFACE WORKING

Under certain conditions of use, it is possible at times to protect an optical surface in metal by coating it with cellulose lacquer, as is regularly done with silver-on-glass astronomical speculae. (*Amateur Telescope Making* and *Amateur Telescope Making, Advanced*, both published by Munn & Co., New York, for the *Scientific American*.) Under other conditions such surfaces may be protected by the method of Professor Charles Féry (*Discussion on the Making of Reflecting Surfaces*, Physical Soc. of London and Optical Soc. 26/11/1920, Fleetway Press, page 24) which makes use of a varnish of Judée bitumen in French turpentine. This varnish does not absorb heat radiations, and has the additional advantage of hardening under the influence of light.

(*Note*.—The opinion in England is that the virtue of Glassite lies in its disguising imperfect polishing, and that rouge is preferable as it is easy to prepare in a standard form.)

The following Obituary Notice, by Mr. J. W. Perry, from the *Annual Report of the Royal Astronomical Society*, 1945, **105,** 90, is included by permission of the Society and of the writer.

Alfred Marshall Graeme-Thom (b. 1886, d. 1943), a twin, came of a Scottish Highland crofter family with a long-standing military tradition. At the early age of fourteen years he volunteered for military service, enlisted in the Gordon Highlanders, took part in the Boer War and in the same year received the D.C.M., being probably the youngest holder of that decoration. Continuing on active service he distinguished himself in the field and in the air, playing a prominent part in the early days of the R.F.C., persisting in hazardous exploits in spite of serious wounds (and well-intentioned dissuasion), attaining at length the rank of Major and receiving, over the period of his service, numerous further military honours and decorations, including those of some of the highest European orders, namely, the Order of the Legion of Honour, the Order of St. Anne of Russia, and the Order of Leopold of Belgium. His military career was terminated in 1918 by the effects of a severe head wound which, on his return to civil life, caused him eventually to seek refuge in a solitary existence on his own estate, an extensive virgin terrain in a remote part of the East Gippsland " bush," Australia. Here he first devoted himself to the cultivation and curing of tobacco and to research in this subject, in which he was known to have a specialized knowledge. Here also he could indulge his love of nature and profit by his skill as an accomplished archer, using only bows of his own fashioning, while developing an interest in astronomy, especially as an observer of meteors.

For reasons of health Thom was eventually compelled to abandon the production of tobacco and thereupon decided to give full rein to his passion for astronomical observation. He then resolved to construct and equip a second and larger observatory. Having already considerable mechanical skill, he proceeded to perfect his knowledge of optics and cultivated this subject assiduously from both its theoretical and practical aspects. He experimented with success in the production of specula for mirrors and gratings and had already achieved a considerable measure of optical and instrument-making proficiency when, impelled by his craftsman's interest and thirst for optical knowledge, he was able, by virtue of sustained self-denial—in which he was aided by his wife, a lady of Gippsland descent—to arrange a world tour for the purpose of visiting the most important centres of optical instruction and production. This brought him in 1938 to Europe, where he worked for a time at various optical firms, including those of Messrs. Chance Bros. of Birmingham and Messrs. Hilger of London, after having travelled widely in Britain, France, Germany, Italy and Russia. The first contemplated completion of his plan by a visit to the great optical centres of the U.S.A. had to be abandoned owing to the exigencies of the present war. Thom and his wife lost their lives by enemy action on making the return journey to Australia, where Thom was to have taken up an appointment in which his newly developed optical ability would have been of value to the Empire's cause.

The discriminative knowledge gained from his intensive study and application to the craft and science of optics procured him many minor triumphs in this field, made him a keen judge of values therein and enabled him to contribute materially to the war effort in this country. Thom's capabilities in optics were not yet plumbed and his work in astronomy was not yet accomplished when in 1943, in his fifty-seventh year, he met his tragic death at sea, but his example as an amateur, in the true and literal sense of the word, is in the best traditions of astronomy and optics.

He was elected a Fellow of the Royal Astronomical Society on 1942 July 10.

Balsaming or Cementing

The process of balsaming or otherwise cementing lenses will be described in Chapter 6.

CHAPTER 3

THE NATURE OF GRINDING AND POLISHING

34 The object of polishing is to produce regular transparent surfaces on a piece of glass or other clear substance. The surfaces are usually required to be flat or spherical, although occasionally departures from these shapes are needed in order to obtain some optical advantage not otherwise attainable. The process is divided into two : grinding and polishing. They are commonly held to be quite different in character, although this opinion is not universal.

Polished surface

A polished surface is characterized by the absence of cavities and projections having dimensions in excess of the wavelength of the reflected light, the present practical limits for the most accurately prepared glass surfaces being that the deviations from a theoretical plane can be reduced to the order of 20 angstrom units : approximately the same accuracy can be obtained in polishing metal surfaces. In the case of polished glass surfaces, this order of accuracy corresponds to projections of dimensions not exceeding the thickness of one or two molecular layers.

It is pointed out by Grebenshchikov (1931 and 1935) that, apart from the property of reflecting light, polished surfaces exhibit other characteristics which differ from those associated with the bulk material. Thus, polished surfaces usually have an elevated mechanical strength and an increased mechanical resistance, and polished materials different heat and electrical surface conductivity, electro-magnetic characteristics, and crystalline structure, from those of the bright faces of a single crystal.

Lord Rayleigh (1903) states that the particles of emery in grinding glasses appear to act by pitting the glasses, *i.e.* by breaking out small fragments. He points out that surfaces may be ground so fine that a candle is seen reflected at an angle of incidence not exceeding 60° and, indeed, that at grazing incidence even coarsely ground surfaces behave as if polished. The wave theory, he says, shows that a regularly corrugated surface behaves as if absolutely plane, provided that the distance apart of the corrugations is less than a wavelength of light. He says—

> In view of these phenomena we recognise that it is something of an accident that polishing processes, as distinct from grinding, are needed at all ; and we may be tempted to infer that there is **no**

essential difference between the operations. This appears to have been the opinion of Herschel whom we may regard as one of the first authorities on such a subject. But, although perhaps no sure conclusion can be demonstrated, the balance of evidence appears to point in the opposite direction. It is true that the same powders may be employed in both cases. In one experiment a glass surface was polished with the same emery as had been used effectively a little earlier in the grinding. The difference is in the character of the backing. In grinding the emery is backed by a hard surface, *e.g.* of glass, while during the polishing the powder (mostly rouge in these experiments) is imbedded in a comparatively yielding substance, such as pitch. Under these conditions, which preclude more than a moderate pressure, it seems probable that no pits are formed by the breaking out of fragments, but that the material is worn away (at first, of course, on the eminences) almost molecularly.

35 The opinion of Herschel referred to by Lord Rayleigh is from *Enc. Met.*, Art. Light, p. 447, 1849. Herschel says—

. . . it may reasonably be asked, how any regular reflection can take place on a surface polished by art, when we recollect that the process of polishing is, in fact, nothing more than grinding down large asperities into smaller ones by the use of hard gritty powders which, whatever degree of mechanical comminution we may give them, are yet vast masses, in comparison with the ultimate molecules of matter, and their action can only be considered as an irregular tearing up by the roots of every projection that may occur in the surface. So that, in fact, a surface artificially polished must bear somewhat of the same kind of relation to the surface of a liquid, or a crystal, that a ploughed field does to the most delicately polished mirror, the work of human hands.

36 Lord Rayleigh continues—

The progress of the operation is easily watched with a microscope, provided, say, with a ¼-inch object glass. The first few minutes suffice to effect a very visible change. Under the microscope [Fig. 14] it is seen that little facets, parallel to the general plane of the surface, have been formed on all the more prominent eminences. The facets, although at this stage but a very small fraction of the whole area, are adequate to give a sensible specular reflection, even at perpendicular incidence.

And further on—

Perhaps the most important fact taught by the miscroscope is that the polish of individual parts of the surface does not improve in the process. As soon as they can be observed at all, the facets appear absolutely structureless. . . .

. . . Of course, the mere fact that no structure can be perceived does not of itself prove that pittings may not be taking place of a character too fine to be shown by a particular microscope or by any possible microscope. But so much discontinuity, as compared with the grinding action, has to be admitted in any case, that one is inevitably led to the conclusion that in all probability the operation is a molecular one, and that no coherent fragments containing a large number of molecules are broken out. If this be so, there would be much less difference than Herschel thought between the surfaces of a polished solid and of a liquid.

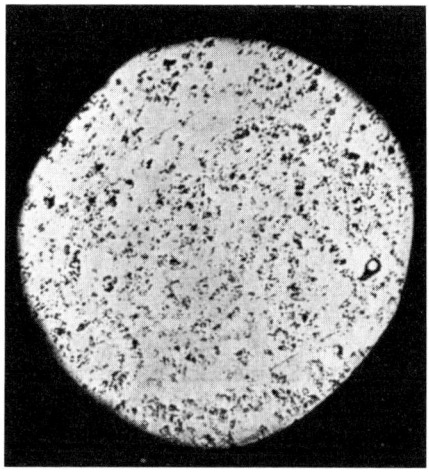

Fig. 14—Surface after a few minutes' polishing
(*From Rayleigh's " Scientific Papers." Reproduced by permission of the Cambridge University Press*)

These passages still refer to polishing by the same emery as was used in the grinding.

37 The nature of the ground glass surfaces was studied very thoroughly by Preston (1922, 1926). It is not possible after reading his account or the earlier one of French (1916) to think of a ground glass surface as merely a number of intersecting cuts or grooves, a view which was formerly held by some. It is well known that a smooth unbroken line can with care be drawn on glass with a diamond, and I have indeed seen under the microscope what appeared (though broken when observed) to have been originally continuous threads of glass shaved away from the polished surface by a lightly loaded diamond in the ruling of a fine line ; but Preston, as a result of a number of careful observations, confirms what Lord Rayleigh says, namely that the process carried out preparatory to polishing the surface produces a great number of conchoidal fractures from which pieces of glass have been broken. Conditions of stress under which such fractures could originate were studied by Dalladay and myself (Dalladay and Twyman, 1921). Preston also observed another peculiarity of ground surfaces, namely that below the obviously broken surface referred to above, there is a region of small cracks which must be removed by polishing if the surface is to be perfectly clear.

A phenomenon observed by myself (Twyman, 1905) (*Proc. Optical Convention*, 1905, 52) is apropos at this point. In describing the making of plane parallel plates for Michelson Echelons I say—

(B) Supposing the plate to have been corrected to the required accuracy, will the cutting of it up cause distortion ? The answer to this is ; under certain circumstances, yes ; but with due care, no. For instance, the plates must not, of course, be cut up with a diamond. Neither, in using the slitting wheel, must too great speed or force be used. As a matter of fact, we always use the old-fashioned hand saw fed with emery, that being the safest ; and with this we find that neither on the proof plane nor by the interference test described above can any distortion be, as a rule, detected.

(C) A third point worth mentioning is this. The edges of the plates are ground to a very fine matt surface. Now for some reason the region near a matt surface is in a state of strain. It seems probable that this is due to the grinding material which, in crushing pieces out of the glass surface, subjects the part near the surface to a permanent strain from which it does not completely recover. This is easily detected by the use of polarised light, in the case of small pieces, say 1 mm thick. With the case of fairly massive pieces of glass, however, which do not permit the strain to be transferred inward by the bending of the whole piece, the strain becomes confined to an extremely thin skin of glass, and does not seem to have the slightest deleterious effect on the action of the echelon.

A strain such as mentioned above is entirely removed by polishing, which shows that it originates extremely near the surface.

One may reasonably suppose, in fact, that as the thickness removed by polishing in a particular case was within about 1/5000 of an inch, this is the thickness in which the strain originates, and an approximate calculation seems to indicate that the strain is *of the order* of the crushing strain of glass. If the glass is thin, say 1 or 2 mm this pressure over the skin bends the glass as a whole, and consequently is transferred inwards to a considerable depth ; but if the glass is thick it does not bend by an appreciable amount and the strain is extremely local.

This last named observation of Preston may be supplemented by those of Dr J. A. Anderson of the Mt Wilson Observatory. Some of the Mt Wilson's old lenses showed surface cracks which were extremely fine. Dr Anderson measured the depths of these hair cracks optically and found them to be about 30-wavelengths deep. Such hair-cracks have also been observed on old telescope lenses in different parts of the world, and it is thought that the periodic (diurnal, especially) heating and cooling over a span of years might have caused them. This effect may possibly be ascribed to fatigue-failure of the surface layer if the layer was rendered weaker in the process of polishing.

Dr Anderson reproduced such surface cracks in the laboratory by heating ordinary lenses to 100-120° C and subsequently chilling them on a cool metal plate. The depths of such cracks were also found to be of the same order as the natural cracks.

These observations suggest that the skin of a rouge-polished lens has physical properties different from the bulk of the lens body. This may be possible if the polishing takes place through thermal (softening) action at the superficial layer. (*J.O.S.A.*, 1949, **39**, 92).

Attempts made by Rayleigh (1908) to discover whether the surface of polished glass is different in physical properties from the mass, resulted in the conclusion that while grease and moisture on the surface (though extremely difficult to avoid) did not have much effect on the optical properties of the surface, yet even a recently polished surface is in a highly complicated condition.

38 Beilby (1921) brought fresh light on the problem of polish by his observations on metals, glass and Iceland Spar. According to Beilby, in polishing, molecules are set in gliding motion by the polisher so that they form an extremely thin film of fluid subject to surface tension, and this, he thinks, accounts for the smooth surface which is left by polishing.

It is worth quoting some passages selected from pp. 107–110 of his book *Aggregation and flow of Solids* (Beilby, 1921, Macmillan, London)—

> When glass is marked by the passage over it of a hard point, it may be either scratched, furrowed or cleft. A scratch is caused by the splintering of the glass along the track over which the hard

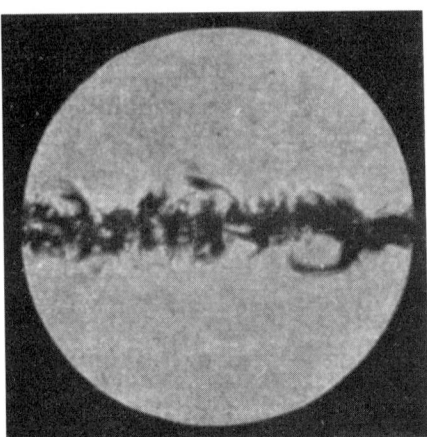

Fig. 15—A splintering scratch

(*Figs. 15–22 are from Beilby's "Aggregation and Flow of Solids." Reproduced by permission of Messrs Macmillan & Co. Ltd*)

body has moved. A furrow is ploughed when the tool is so formed and guided that its point or points lay hold of a layer only a few molecules in depth.

A certain amount of the glass is shaved off, but the perfectly smooth coating of the groove which is left shows that the surface layer has passed through the mobile or liquid condition. A cleft results when a fine wedge-like point is drawn along the surface. The entry and passage of the thicker part of the wedge may result in furrowing or splintering at the outer surface, but these are not essential features of the true glass-cutter's cleaving scratch which, to be effective, must have forced the glass apart till a cleft is started.

The photograph [Fig. 15] shows a splintering scratch at a low magnification. The conchoidal pits, which are irregularly broken out along the track, are very characteristic, and will be referred to again in connection with another photograph. Even in this rough scratch it is not difficult to detect traces of ploughing and flow.

Fig. 16 shows a series of grooves ploughed by drawing the edge of an uncut diamond along the surface under a slight pressure. The magnification in this case is high, about 700 diameters, and the resolution is also good; if reduced to the magnification of the preceding figure, it would only show as a narrow faint trace. . . .

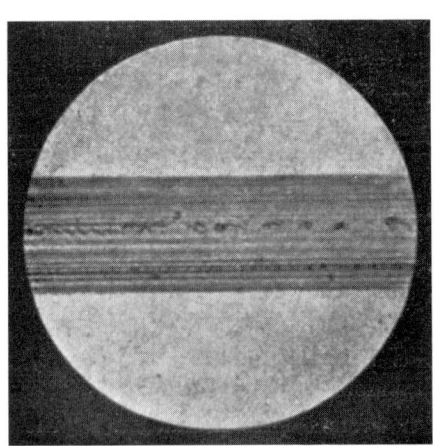

Fig. 16—Grooves ploughed by an uncut diamond

GRINDING AND POLISHING

The next two photographs, Figs. 17 and 18, show cleaving or cutting scratches. In Fig. 17 a certain amount of splintering has occurred as the diamond was pressed through the surface layer, but along the track there has been no free splintering of the edges. The broad dark band which follows the cut is the shadow caused by the total reflection from the cleft which has been opened in the substance of the glass below the surface . . .

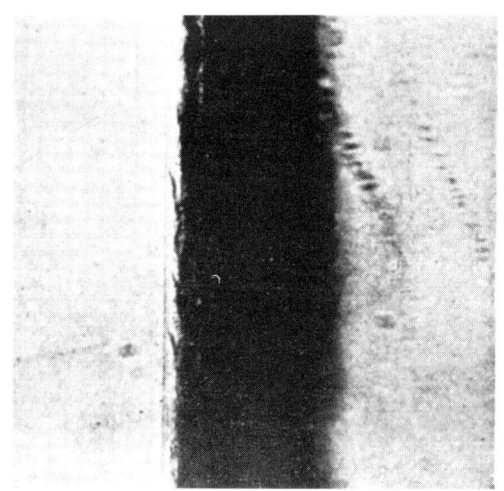

Fig. 17—A cleaving scratch
There is a certain amount of splintering

Grinding is mainly due to splintering

The large part played by splintering in the grinding of glass at once marks it off from the other examples of polishing which have already been illustrated. . . .

Fig. 18—A cleaving scratch

While splintering plays a very large part in the grinding of glass, ploughing plays only a small one, at any rate, till the later stages are reached. It has been shown that from the nature of the case only a very shallow furrow can be ploughed in so hard and brittle a material as glass, and there is great danger that splintering may occur even in the final stages of grinding.

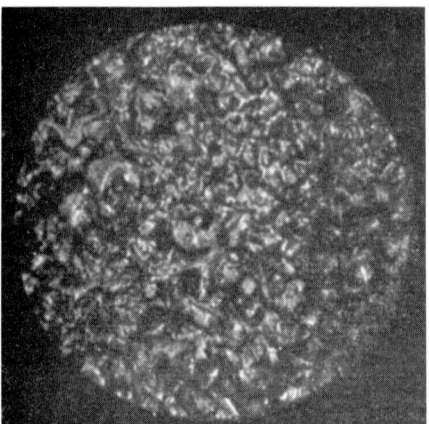

Fig. 19—Good commercial ground glass
Highly magnified

Steps in grinding and polishing

Fig. 19 is a photograph at a high magnification of a piece of good commercial ground glass. This is merely shown for comparison with the following specimens, which were all hand-ground with emery powders of known degrees of fineness. Fig. 20 shows the first effect, when moderately fine emery is applied between the two glass surfaces which are being ground together. The surface is indented in many places by the points of the emery grains, and in some cases it has been splintered and pitted. The greater part of the original skin is, however, still intact. In the next photograph, Fig. 21, the whole of the surface skin has disappeared, and the surface is pitted all over, some of the pits being of very large size.

The next photograph, Fig. 22, shows the remarkably quick effect of the rough polishing. The rocky-looking texture of the emery surface has disappeared, and has given place to larger patches of a structureless viscous-looking material. The larger pits are now brought out with a striking distinctness. These effects are soon carried to a further stage, and the number and size of the pits are considerably reduced. The rounded edges of the remaining pits show unmistakably that the flowed surface layer is covering over and masking the irregularities below, just

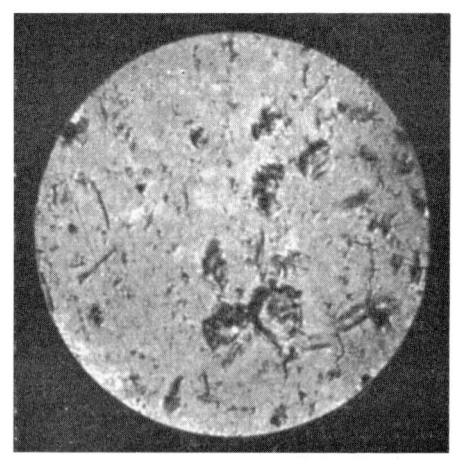

Fig. 20—Result of grinding with moderately fine emery

as it did in the case of antimony. The final stage of polishing is reached when all the pits have been covered over. At this stage there is no further structure to photograph . . .

In spite of this overwhelming evidence to the contrary Elihu Thomson (1922) still maintained that—

Some have most erroneously tried to explain the result of the process, by assuming that the glass has, during the polishing, actually flowed; or that there was some peculiar plastic condition brought about which allowed the glass surface being polished to take on the characteristics of a liquid surface.

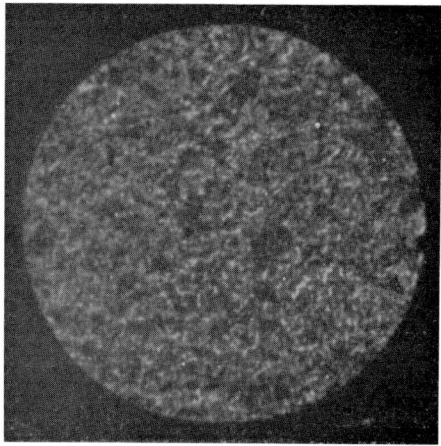

Fig. 21—**The surface of Fig. 19 after continued grinding**
The surface skin has completely disappeared

Confirmation of Beilby's observations were obtained by a modified form of X-ray analysis, devised by P. B. Hirsch and J. N. Kellar (1948),* of the Crystallographic Laboratory, the University of Cambridge, which can be used under certain conditions to measure the thickness of thin surface layers which differ in physical properties from the main crystal.

It has been applied as a test case to measure the thickness of a ground and polished layer on the surface of calcite, of the order of one wave-length of red light in depth, and indirect confirmation obtained of the approximate correctness of the measurement.

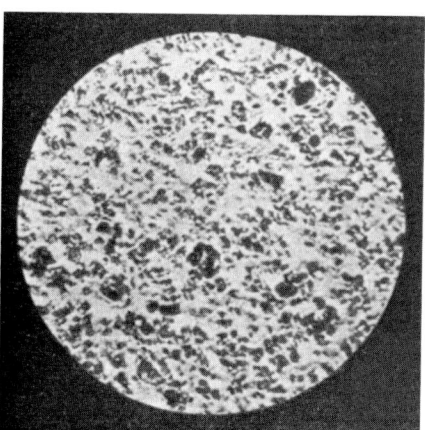

Fig. 22—**Effect of rough polishing**

* Hirsch and Kellar, 1948, *Nature*, **162**. 609.

39 French (1916–1917) made a number of experiments which led him to the conclusion—to a great extent in harmony with the views of Beilby—that the surface of glass is converted in polishing into a form having properties materially different from the remainder. It can, for example, receive smooth-sided scratches (" sleeks ") whereas scratches which are deep (called by the optician " cuts ") invariably consist of a series of conchoidal fractures. French actually went so far as to state that he believed the glass to become melted, a view which I, for one, regarded at the time as fantastic. In one illuminating passage (1917, p. 23) he draws a sharp distinction between two stages of wet and of dry polishing, " The function of the first stage is to remove material ; the function of the second is to fill up sleeks."

Preston (1926) on the other hand concluded, after a careful examination of the views of previous workers, that the process of polishing is principally one of microscopic abrasion, although " flow or fusion of some sort on a molecular scale may in fact be operative simultaneously with the more important phenomena of mechanical abrasion."

40 The view that the temperature is sufficiently high for surface melting in polishing was adopted by Macaulay (1926, 1927, 1931).* Macaulay was able to detect the products of thermal decomposition of the powder used for polishing glass plates. This evidence was questioned because many reactions may occur at a freshly exposed surface due to causes other than temperature. It seemed, however, that Macaulay was right in the view that he held in the light of experiments of Bowden and Ridler (1936). Working at the Laboratory of Physical Chemistry, Cambridge, these authors deduced from their experiments the temperature of the surface layers of bodies during their sliding on one another, by using the rubbing contact of the two substances—actually two different metals—as a thermocouple and determining the electromotive force generated on sliding.

The behaviour of readily fusible metals confirmed that the temperature measured was a real one, for with metals of low melting point, such as gallium, Wood's metal, or lead, the measured temperature rose to a constant value which could not be exceeded and which corresponded numerically to the melting temperature of each metal. With less fusible metals the local surface temperature may exceed 1000°C.

The matter was carried further by Bowden and Hughes (1937).

* Macaulay, J. M., 1926, *Nature*, **118**, 339. 1929–32, *J. R. Tech. Col. Glasgow*, **2**, 378.

TABLE I

Polishing material	Melting point of polishing material °C	Substance to be polished	Melting point of substance °C	Vickers hardness of substance	Results
Camphor	178	Wood's alloy	69	25	polish
Camphor	178	Tin	232	4	no polish
Oxamide powder or camphor	417	Tin	232	4	polish
		Speculum metal	745	?	no polish
		Copper	1083	?	no polish
Lead oxide powder	888	Speculum metal	745	505	polish
		Nickel	1452	164	no polish
		Molybdenum	2470	234	no polish
Chromic oxide	1990 }	produced polish on all the metals tried			
Ferric oxide	1560 }				
Oxamide	417	Lead glass	469		polish
		Soda glass	600		slight polish
		Pyrex glass	815		barely perceptible polish
		Quartz glass	1710		no polish

In this investigation the influence of the relative melting point of the polisher and the solid was determined. If high local temperatures really occur at the points of rubbing contact, as previous experiments (Bowden and Ridler, 1936) had proved, one would expect the relative melting point to be an important factor in the process, and the experiments showed that this is the case. Various materials were used as polishers, differing greatly in melting point. In some cases the polishing materials were massive substances such, for example, as camphor. In other cases they were in the more usual form of powders used on appropriate bases, camphor or otherwise. Table 1, compiled from the paper cited, shows some of the results. It will be observed that it is not the softness of the material that is the important factor in deciding whether it can be polished by a given polishing material, but whether its melting point is lower than that of the latter. For example, Wood's alloy is polished easily by camphor, although it is harder than tin which is not, the reason being that the Wood's alloy melts at a lower temperature, and tin at a higher temperature, than the camphor.

It will be seen that in the polishing of glass, the division is not so sharp; for example, all the glasses of course have a much higher melting point than the oxamide, but the nectart has a softening point almost the same; on the other hand, soda glass polishes slightly and the pyrex also takes a barely perceptible polish. It must be remembered, however, that glasses have a very long range over which they become of increasing softness with rising temperature, whereas the metals which have been considered have very definite melting points. When high melting oxides, such as chromium ferric-oxide, and zinc oxide are used they readily cause polish on quartz and all the glasses.

The paper also contains some interesting information on the loss of weight which accompanies polishing, showing that in the case of metals there is a definite removal of material. It is not the softest material which is necessarily removed most quickly, but the one with the lowest melting point—for example, Wood's alloy polished on thick filter paper loses more weight than lead, although it is five times as hard on the Vicker's scale, the reason being that the lead has the higher melting point. Gallium, with a Vicker's hardness only slightly greater than lead, loses nearly 90 times as much weight.

More positive evidence of the nature of the surface layer in certain cases has been attained by electron diffraction and I am indebted to Professor G. I. Finch of the Department of Chemical Technology, Laboratory of Applied Physical Chemistry, Imperial College of Science and Technology, for the following interesting particulars of observations made in that laboratory.

Electron diffraction, at grazing incidence, with 50 to 60 kV electrons, gives the following results—

Surface	*E. D. Pattern*	*Conclusion*
Fresh glass fracture conchoidal	Two diffuse haloes and much diffuse background scattering	Amorphous
Polished glass (microscope object glass)		
Polished plateglass (Pilkington's make)	,,	,,
Polished silica glass		
Polished fused quartz	,,	,,
Polished fused quartz after heating during 2 hours at 400°C	Faint spots and fainter arcs	Unidentified crystal structure
Lead glass polished	Diffuse haloes and much background scattering	Amorphous
Lead glass, after prolonged weathering	Ring and arc pattern. (See, also, Kamogawa, phys. Rev., **58,** 660, 1940)	Lead sulphide and lead sulphate crystals

Professor Finch also remarks that, in electron-microscopic examination of the results of polishing glass scratched by diamond, the " flowing over " of the scratches is " remarkably evident ". For this study he uses moist rouge with a soft polishing pad. Finch, Quarrell and Roebuck (1934) also, in studying the photoelectric properties and structure of certain surfaces, obtained results that the authors considered " confer an objective reality upon the Beilby layer which raises its existence from the realm of hypothesis to that of established fact ".

Hirsch and Kellar (1948) have measured the thickness of the surface layer resulting from the polishing of crystals, which in the case of ground and polished calcite (Iceland spar) was found to be 7500A (7.5×10^{-4} mm). The method depends on measuring the variation of intensity of X-ray refraction from the polished surface of the crystals at various angles of incidence, and is being applied by the authors to the study of the ground and natural surfaces of single crystals and polycrystalline aggregates. It could be applied, they say, to measuring the thickness of any surface layer for which the value of the refraction coefficient differs from that of the substrate material, for example a layer of one metal electrodeposited on a different metal.

Further information on the subject will be found in Thompson (1930) and Browning (1944).

41 I may add one or two personal observations of my own. Some glass is certainly removed in polishing. No one who has done any figuring by local retouching will doubt this. In correcting large plates for Michelson echelons I never found that retouching, which reduced the thickness of the area polished, caused any rise in the adjacent surface. Thus it is not usually just a matter of sweeping removed glass along the polished surface; the glass is taken up by the polisher or comes away with the rouge.

I used, in retouching, to keep account of the amount of rubbing and the quantity of glass removed. Counting the number of circular sweeps with a $1\frac{1}{4}$-inch diameter cloth polisher with the rouge fairly moist, and taking strokes of about 1 inch in diameter, I found that 100 strokes per inch of the area being polished removed about one Newton's ring (using Michelson's test, §226)—that is about $\frac{1}{150000}$ inch. It follows that a single sweep with such a polisher would remove one hundredth of this, that is $\frac{1}{15} \times 10^{-6} = 7 \times 10^{-8}$ in. Now if we consider a molecule of silica to have the dimensions of one lattice spacing of a silica crystal, the size of such a molecule will be about 10^{-7} in. Since one can apparently continue this process of reduction to any extent merely by continuing the rubbing (I have myself carried the process to a depth of about a dozen Newton rings) one can only come to the conclusion either that one is removing the glass in portions of less than molecular dimensions, which is scarcely consistent with the picture of a flowing liquid, or that one is effecting a closer packing of the molecules.

On the other hand, although in ordinary figuring the glass is removed, either being carried away with the polisher or becoming mixed with the rouge, yet in certain exceptional circumstances one can get a transfer of the glass of a comparatively massive character, to which no other word seems applicable except " flow." In some small prisms, of which at one time we polished a considerable number, there was an obtuse angle, and if one surface was being polished singly by the optician, occasionally when the rouge was allowed to dry up pretty thoroughly the prism was found suddenly to develop, on the surface not being polished, a small bulge. Speaking from memory, I should say that its height was something like $\frac{1}{4}$ mm. (French, in the paper cited, observed what may have been the same phenomenon, but he found the " lump " to consist of a mixture of rouge and glass.)

I am indebted to Dr W. E. Williams for the information that Dr Tillyer of The American Optical Company, has demonstrated that optical polishing can be done either with a removal of glass and loss of weight or with progressive gain in weight dependent merely on the conditions of polishing. In the latter case, spot tests after etching the top layer with hydrofluoric acid, show the presence of Fe_2O_3 which must have been in the glass in a colourless transparent form. The probability is that the many different explanations of polishing arise, since it can be brought into being in different ways.

Finally, on this question of the removal of glass by polishing, Ray (1949)* notes that the chemical examination of spent rouge shows the presence of silica (which must come out of the glass), but not enough to account for the total polish. He examined the polished facets of a half-polished glass surface on an electron microscope at magnifications of 8,000 to 20,000 but was unable to detect any structure such as would arise from the rouge particles acting as tiny cutters. He also examined fresh and spent rouge samples under the electron microscope and found that the spent rouge seemed to be better dispersed and had more rounded corners than the fresh rouge. The particle sizes were mostly around 0.25μ.

42 A paper by Lord Rayleigh† describes experiments in which he proved that the reflecting power of a polished silica surface varies materially according to the treatment.

These variations in reflecting power were measured by immersing the polished pieces in a liquid of the same refractive index as the interior of the material, and it was found that surfaces, polished by methods which do not quickly remove the material, may reflect in the

* Ray, Kamalesh, 1949, *J.O.S.A.*, **39**, 92.
† Rayleigh, 1937, *Proc. Roy. Soc.*, A, **160**, 507.

liquid as much as 0·28 per cent of the incident light. A kind of burnishing seems to take place in these instances which modifies the surface and may bring its refractive index up from 1·461 (the ordinary value of fused silica) to as much as 1·6—quite as high as light flint glass, and much higher than any known variety of silica. On the other hand, surfaces polished by a process which removed material rapidly, or surfaces washed in hydrofluoric acid, do not reflect appreciably in the liquid. These effects are found in a less degree in ordinary glass, and, in a very much less degree, in crystalline quartz. In normal cases the reflected light changes in tint from red to blue as the refractive index of the immersion fluid is increased through the critical value for minimum reflection. The modified silica surface is anomalous in this respect, reflection being red on either side of the minimum value. An explanation is suggested for this in the paper. The thickness of the modified layer was measured as $0·06\,\lambda$ where λ is the wavelength of green light in air.

These experiments explain the variable reflecting powers earlier found from the interfaces of fused silica or glass surfaces in optical contact. (Rayleigh, 1936).*

Contacted surfaces of crystal quartz give a reflection which is practically independent of the way in which the surfaces have been polished. The mean distance between the two crystals, when they are put in optical contact, is found to be about seven times the spacing of the layers of silicon atoms within the crystals.

In these earlier experiments (1936) Lord Rayleigh found that, using a power-driven pitch polisher very wet and pressing lightly, the reflecting power was increased ; allowing it to become nearly dry, so that it dragged heavily and tended to squeak, the reflecting power was diminished. It must be remembered that the higher reflecting power was, in the circumstances of this experiment, an indication that the refractive index of the surface layer was different from that of the interior. Now the dragging and squeaking polisher approximates to the condition under which commercial polishing is finished ; the heavy dragging of the polisher indicates, as is shown by the experiment, rapid removal of material and, therefore, effective working from the optician's point of view.

43 The practical optician, in polishing glass with rouge on pitch, finds that if the pitch is too hard he gets scratches. Further, he finds that when he requires to polish materials softer than glass, such as Iceland spar, he needs not only a soft pitch polisher, but a " soft " polishing material, such as putty powder.

* Rayleigh, 1936, *Proc. Roy. Soc.*, A, **156**, 326.

Two questions may therefore be involved from his point of view; the hardness of the polishing medium (pitch, or what-not) and the hardness of the polishing powder—or what is regarded as the hardness of the polishing powder.

Exactly on what ground he feels (as I suppose we all instinctively do) that putty powder is softer than rouge, I am unable to say, but the grounds for his doing so are undoubtedly shaky. For example, the following figures (*Rutley's Mineralogy*, H. H. Read, 24th Edition, Murby) for the hardness of minerals are very significant—

Haemetite (Ferric Oxide Fe_2O_3) Hardness 5·5 to 6·5 on Mohs' scale
Cassiterite (Stannic Oxide SnO_2) Hardness 6 to 7 on Mohs' scale

Although inferences based on natural minerals are probably not valid it seems at least likely that putty powder is somewhat *harder* than rouge. Certainly, therefore, hardness is not the sole factor. The paper by Bowden and Hughes (1937, cited above) makes it clear beyond all doubt, that it is essential for the polishing material to be of a higher melting point than that of the substance to be polished. There are thus two factors in question for satisfactory polishing, the relative "hardness" of the polisher and polished, and also the relative melting points of the polishing powder and the substance polished. Probably, in the avoidance of scratches, what is important is not the hardness of the individual crystals of the polishing material, but the force required to break down the aggregates which, in the case of certain samples of rouge, while liable in the initial stages of polishing to cause scratches, undoubtedly cause them to be faster polishing materials than the finer rouges.

Grebenschikov (1931, 1935), reviewing the position, points out that three basic theories of glass polishing have been proffered. The first of these draws an analogy between polishing and grinding, according to which polishing is produced by breaking down the normal crystalline structure of the material by a mosaic of scratches so fine that the destruction of continuity produced by them becomes invisible. The second theory assumes plastic deformation, flow and re-crystallisation of the glass, and views the polishing process essentially as a redistribution, as opposed to withdrawal of the surface layers. Thirdly, it has been suggested that the upper layers of the polished article are fused under heat generated during the polishing process, with subsequent solidification to form an amorphous glassy substance on the surface.

Objections can be drawn to each of the above explanations, but it is probable that all three effects take place to a greater or lesser extent, according to the degree of polish attained and the type of machine used to effect the polish. However, of late, much attention has been given to chemical aspects of the polishing process, in an endeavour to explain some of the known effects of varying the physical and chemical properties of the material to be polished, the properties of the polishing medium, the chemical composition of liquids introduced in polishing processes, and the characteristics of the polishing base.

This leads, says the author, to a theory which suggests that the liquid medium (usually water) first reacts with the glass to form a thin protecting film, which tends to absorb particles of the grinding material. Thus if an appreciable amount of grinding material is absorbed simultaneously both by the glass surface and the surface of the polishing base, this will act as a form of binder between the polishing base and the glass—and the motion of the polishing base will tend to tear the film from the projections left after grinding.

Similarly, in considering the mechanism of glass-grinding, it has been suggested that the coolant enters into the grinding process proper by virtue of chemical reaction with the glass at the base of the cavities formed by the scratching action of the abrasive; the supposition being that the presence of the reaction-product sets up stresses which tend to assist in the scouring off of the glass contained between adjacent surface cavities and scratches previously made by the abrasive.

This summary of Grebenschikov's view was kindly prepared for me by Mr T. H. Redding of the British Scientific Instrument Research Association.

N.B.—The above note is based on views proffered by I. V. Grebenschikov in the following papers—
(1) " Surface Properties of Glass "—*Keramika i Steklo* 7, No. 11–12, 36–41 (1931)
(2) " Part played by Chemistry in the Polishing Process," *Sotsialistischeskaya Reconstructsiya i Nauka*, No. 2, 1935, pages 22–23.

Summary

44 It appears clear, from the facts cited, that glass is to a certain extent removed in polishing; that the surface sometimes undergoes a physical change and sometimes does not; that, in certain circumstances, the surface is actually molten; and that the relative melting points of the polishing powder and material being polished is a factor of prime importance. Whether appreciation of these facts will help the optician to polish faster or better may be a matter of doubt, but it will serve to occupy his mind while he carries out this useful, but, it must be admitted, often tedious process.

CHAPTER 4

TOOLS AND MATERIALS IN GENERAL USE

Optical tools

45 The machines in use for polishing optical work are very simple and pretend to no accuracy of construction, though the accuracy of surface to be produced is extremely high. A piece of plate glass, say 6 in. square, may depart from flatness by one-hundredth of an inch; from a number of specimens a piece with errors no more than one-thousandth of an inch may be selected. The surfaces of good spectacle lenses depart from true spheres by amounts of the order of one ten-thousandth of an inch, while the optical work in good binoculars rarely has errors of as much as one hundred-thousandth; and, finally, the best optical work departs from the ideal aimed at by less than one millionth of an inch. To obtain such precision by nice mechanical guidance of the tool would be a hopeless task, even were the mechanism perfectly rigid, the polishers free from wear or flow, and the films of polishing substances of invariable thickness. None of these conditions is complied with, and the optician relies for the production of accurate polished surfaces on principles other than those used by the engineering machinist.

He is satisfied, then, with a machine which will move his polisher to and fro over the surface of his lens or block of lenses from say 6 to 250 times a minute according to the size of the work. Simultaneously the tool, or lens, whichever is undermost, rotates on a vertical spindle, the uppermost element being allowed to rotate freely as it will.

True tools

46 For the trueing, smoothing, and polisher-forming tools themselves the material most in favour is cast-iron. We find the most satisfactory type to be an iron in whose preparation the molten metal has been treated with calcium silicide, which acts as a graphitizer and also gives a very fine structure; this form of iron also casts very free from porosity and other defects. Such a cast-iron is often loosely described as Meehanite, a term which is in general use in the U.S.A., Germany and France. The name is applicable, however, to a wide range of cast-irons produced in factories licensed by the International Meehanite Metal Co. Ltd. Thus in ordering castings for optical tools the type of cast-iron required should be stated.

It is perhaps worth while to stress-relieve the tool, after it has been turned, by heating slowly to 550 to 600°C, over a period of about one hour for each square inch of section, holding at that temperature for about 30 minutes, and then cooling slowly over a period of 2½ hours to 100°C, when it may be removed from the furnace.

Aluminium tools

The thirteen-inch iron tools normally used for polisher holders are heavy, and make the work of handling the polishers mildly strenuous. Aluminium tools have been found entirely satisfactory; polishers built up on these tools appear to behave no differently from those built up on iron ones. Aluminium, however, is not suitable for trueing, smoothing, or polisher-forming tools.

I may here mention, that when I first went to Hilger's, in 1898, and indeed for many years afterwards, there were in use a number of tools 6 in. diameter and upwards, made of gun-metal. Gun-metal has the advantage that pitch does not stick to it as readily as to iron.

Curved tools for lenses are turned or milled to the radius required, and must then be lapped together with trueing emery until they exactly fit. Since the turning process is quicker than lapping, there is an obvious economy of time in producing as fine and accurately turned a surface as possible.

Preparation of gauges for the optical tools

47 Gauges for the smaller optical tools may be prepared by turning a disc of brass about two millimetres thick to the requisite diameter

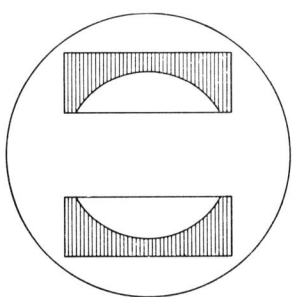

Fig. 23—Turning gauges for small convex optical tools

as measured by micrometer, the edge being reduced to about one half millimetre. This is used as a gauge for the turning of the corresponding gauge for convex tools, which may be turned from two strips of brass, in the manner indicated in Fig. 23. In this simple way sufficient accuracy can be obtained for most purposes.

For gauges of greater radius than can be dealt with on any lathe that may be available, one has to adopt other procedures.

For example, radius turning can be carried out on any lathe, fitted with an attachment which guides the tool point in a circular path ; as, for instance, by attaching it to a radius arm centred on the axis of the lathe. The disadvantage of the method is that the length of radius that can be turned is limited by the length of the lathe.

There is a simple mechanism, applicable to an ordinary lathe, which imposes no such limitation and can be used for turning either concave or convex surfaces of long radius.

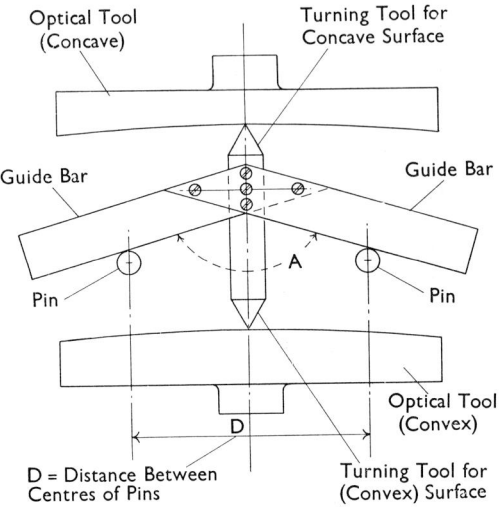

Fig. 24—Turning optical tools of large diameter

If two straight edges making an angle A with each other slide over two points whose distance apart is D the point of intersection will move in a circle to which the two straight edges will be tangential. The radius of the circle is given by the formula $R = D/2 \sin A$. In the arrangement shown in Fig. 24 the pins of course do not correspond ideally with the points referred to above, but with large radii the departure from the above formula is insignificant. The distance from the point of intersection of the guiding edges of the bar to the point of the cutting tool must—for turning concave tools—be subtracted from the desired radius to obtain the R from which D and A are to be evaluated ; vice versa for convex tools.

A good way of making gauges for large tools of medium curvature is to make them from plate glass. The usual form of bar compass, consisting of a rectangular bar of wood about 1 metre long carrying two riders, one of which bears a hard steel point and the other a good cutting diamond, is used to cut longitudinally in two a 6×2 in. strip of $3/16$ in. plate glass of the required radius of curvature. Thus two glass strips are provided, one having a convex the other a concave edge. The resulting two halves are then laid on a flat surface and the curved edges ground together (with frequent changes of side and end) until all the broken surfaces are ground away. The radius is then written on each piece with a marking diamond.

48 The best method of preparing optical tools is, however, by the now well-known process of radius generation used on modern optical grinders. If a milling machine is fitted with a tilting head, which is designed to swivel through 90° and carries a single-point tool rotating at an appropriate distance from the axis, then any radius from plane to hemispherical can be handled on the same machine.

A special machine is made by Adcock and Shipley, working on the same principle, but with an impregnated diamond ring. On this the lens surfaces themselves can be machined; it is illustrated in Figs. 63 to 66.

Two conditions must be satisfied in order to produce really spherical surfaces by this method—

(a) The cutting tool must pass exactly through the centre of the tool, *i.e.*, the milling marks must intersect in a point.

(b) The plane of rotation of the vertical head must pass through the axis of rotation of the tool. Correct adjustment can again be observed by reference to the milling marks, each of which should extend right across the centre of the tool face from the point of entry to the point of exit.

When the above conditions have been attained, the radius produced is given by

$$\frac{r}{R} = \sin A$$

where r is the radius of cutting tool

R is the radius generated

A is the angle of the vertical head

Using accurate radius gauges in the early stages, and some form of spherometer for finishing, it is possible to mill the tools to within one part in a thousand of their radius, which is near enough for most purposes. Careful setting of the milling or grinding tool will ensure that after an hour or two of lapping on a standard polishing machine, the tools will be ready to go into service.

In cases where such accuracy is not possible, and the machining of the tools leaves them too far from the required radius to be used, some time may be saved by rubbing them together with trueing emery, and then grinding down the places of contact with a bonded carborundum abrasive *wheel*. It is difficult to wean opticians from the use of abrasive *sticks*, which may do more harm than good. If a cupped abrasive wheel, of diameter approximating to the width of the high places, is pressed against the tool it will fit the latter no matter what their radii may be, but will grind away the elevations rapidly.

The final lapping must be carried out with trueing emery, and must be carried on until, when the emery is wiped away and the tools rubbed together, they touch over nearly the whole surface.

The tools must then be measured with a spherometer in order to ascertain whether the radius is actually near enough to that intended.

The spherometer

49 The instrument most generally used to measure radii of curvature is the spherometer; for alternative methods when the use of a spherometer is not convenient, or, as in some instances, is not possible, reference may be made to an informative article by Guild (1923).

The conventional tripod form of spherometer is not now used for serious work. This form consisted of a triangular metal frame, with three pointed legs fixed at the corners of an equilateral triangle. Equidistant from these legs was a pointed micrometer screw with divided head and scale. When the central screw was raised high enough the three legs stood firmly on the optical tool or other surface, the radius of curvature of which was to be measured. When the central point was screwed down to touch the surface, the instrument swung freely about the point of contact. The instrument was set for zero by placing it on a flat tool, or preferably, a proof plane.

The radius of curvature is given by the formula

$$r = \frac{s^2}{2d} + \frac{d}{2}$$

where s is the distance from the point of the micrometer screw to the points of each of the three legs when adjustment is made on a flat surface, while d is the distance which the screw must be raised or lowered just to touch the surface being measured.

The three-legged spherometer possesses two main sources of error. These are the difficulty of determining the effective radius of the circle passing through the points of contact of the three legs, and the exact location of the point of contact between the central screw and the test surface.

50 Some forty years ago, I reduced the former of these two uncertainties by replacing the tripod with a ring form of instrument (Fig. 25).

The ring was so ground as to have two truly circular, and concentric, sharp edges, one of which makes contact with concave, and the other with convex surfaces. It is a severe test for a spherometer if, when two spherical surfaces of fairly short radius and, say, about three times the diameter of the spherometer circle, carefully ground to be in contact, yield the same measurement of radius. The inside and outside diameters of my spherometer can be measured very accurately in spite of the simplicity of the instrument. With this form of instrument a far closer agreement is obtained, between the radii of paired surfaces, than is possible with the three-legged type.

Fig. 25—Ring-form spherometer

An attendant advantage of the ring form instrument is that the wear on the circle is very much less than the corresponding wear on the three points of the older form. It follows that the calibration is maintained for a longer period.

The original ring form spherometer has been in use in our optical testing room for over forty years, and is still in good condition.

In the ring form spherometer when it is used for testing proof glasses which are of exactly the same radius, as shown by Newton's rings, one can either measure the concave and the convex surfaces against a plane glass plate separately or take the reading with the spherometer, first on one and then on the other, and use in the formula the total distance so measured.

These readings never correspond exactly—at any rate I have never been able to make them correspond, even when I carefully ground the ring of the spherometer myself to ensure that the edges were sharp. The custom in the latter type of measurement is to take as the radius of the ring the mean between the outer and the inner edge.

A paper by Doering (Doering J., 1949, *Optik*, 5, p. 167) gives calculations showing that it is not correct to take the arithmetic mean of the linear values of the two edges but to take the square root of the mean of the squares of the two radii. Even this is only an approximation and a more exact formula is given. This latter, however, the author says is rarely necessary and can be neglected when the angle subtended at the centre of curvature by the ring of the spherometer is equal to a right-angle.

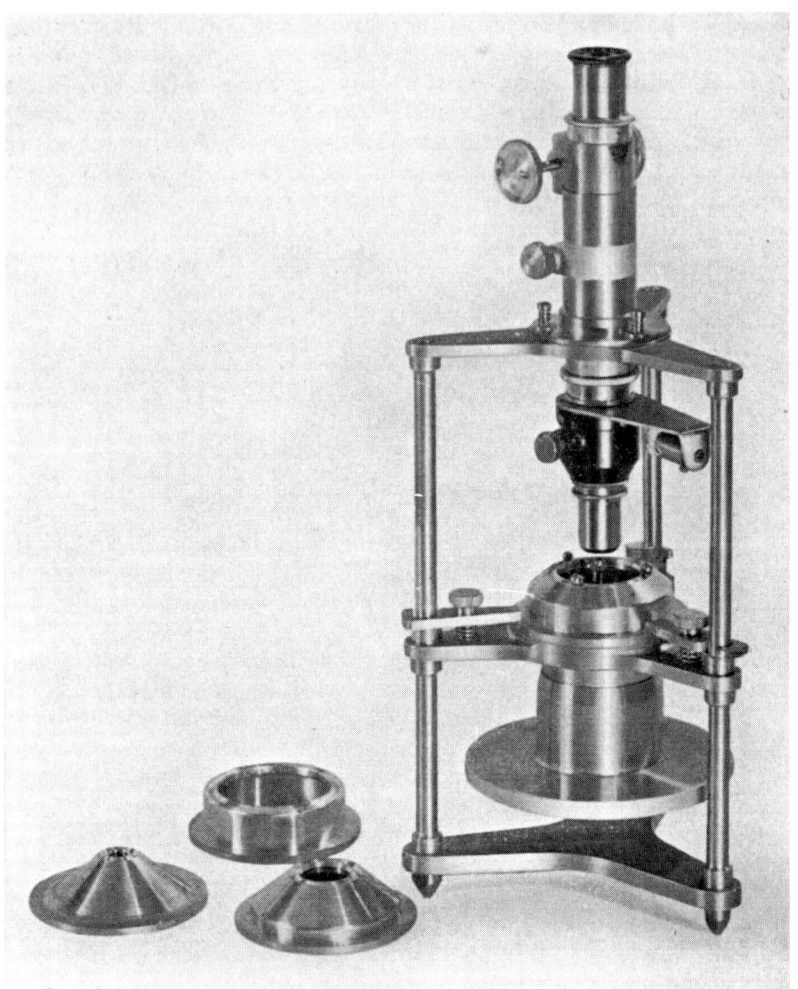

Fig. 26—The Guild spherometer

The second of the major sources of error, that due to the determination of the point of contact by the screw, is substantially reduced in the instrument due to Guild (1918).*

* Guild, J., 1918, *Trans. Opt. Soc.*, v. XIX, p. 103. Guild, J., 1923 *Dict. of Applied Physics*, v. IV, p. 786.

51 The Guild Spherometer is a robust and accurate spherometer but is suitable for use only with transparent test objects such as lenses and proof-spheres.

This spherometer is, in principle, similar to the older form of spherometer having three legs, or a circular ring, with a centrally mounted adjustable micrometer screw, but has, in addition, facilities for determining optically the exact point of contact of the screw with the surface under examination.

Fig. 26 gives a general view of the spherometer. The test surface is rested upon a support consisting of three small steel spheres of the same radius, or alternatively, a steel ring may be supplied for this purpose. A small quartz (or glass) sphere is attached concentrically to the end of a micrometer screw of high accuracy, and is adjusted by rotation of a divided drum to make contact with the test surface. By means of a microscope provided with a vertical illuminator, the system of Newton's interference rings formed between the test surface and the quartz sphere may be observed closely through the transparent object, in order to determine the exact moment of " contact," by which is meant in practice the position where some definite and easily recognizable configuration of the interference pattern occurs. The pattern generally chosen for the setting criterion is that in which the central black spot trisects the first ring. It is a help, in this connection, slightly to modify the light by inserting a coloured filter at the lamp. In this way the contrast of the rings is enhanced and the settings thereby made more critical. Readings may in this way be made under conditions capable of control and repetition to an accuracy measured in fractions of a wavelength of light, thus ensuring that the fullest advantage is taken of the high degree of mechanical perfection of the instrument.

The ball or ring supports are removable and accurately replaceable. For use in measuring curvatures of surfaces of varying diameters, four sizes of such ring or ball supports are supplied with each instrument, of the following diameters : 12·5 mm, 25·0 mm, 37·5 mm, and 50 mm for metric micrometers, and corresponding " inch " diameters for English micrometers.

In making a measurement with the Guild spherometer the procedure followed is, in general, the same as that with the three-leg spherometers of the older type but, in order to realize fully the accuracy of the instrument, a certain sequence of operations is recommended. The recommendation is that every measurement on a spherical surface is both preceded and succeeded by measurements on the plane reference surface. It is further advisable also, to use in every case the largest ring to suit the test object.

If then, the difference between the reading obtained for the optical flat and the reading for the surface under test be h, the radius of the surface is given by—

$$r = \frac{A}{h} + \frac{h}{2} \pm a$$

where the sign $+$ for the last term applies to concave surfaces and the sign $-$ to convex. The constant a is the radius of the steel balls, and is omitted from the formula when the ring supports are used. The constant A is given by $A = \frac{1}{2}R^2$, R being here the radius of the circle of contact with the steel ring, or the radius of the circle passing through the centres of the three steel balls. For this latter case, R is given by

$$R = \frac{\alpha \beta \gamma}{4\sqrt{s(s-\alpha)(s-\beta)(s-\gamma)}}$$

α, β and γ being the distances between the centres of successive balls, measured by means of a hand micrometer, and $s = \frac{1}{2}(\alpha + \beta + \gamma)$.

The formula given earlier for the simple ring form spherometer is seen to be a particular case of the more general one given here.

Recently, a modernized version of the Abbe spherometer has become available. This is the Watts Precision Spherometer. The test object, which may be either transparent or opaque, is located in the usual manner on three accurate steel balls which latter are fitted into a recess in an accurately turned ring. A plunger through the centre of the instrument has attached to it a very accurately divided glass scale, the position of which is read by a reading microscope to an accuracy of 0·00001 in. by estimation.

The plunger makes contact with the surface under test with constant pressure, this being achieved by a system of counterweights. Heavy objects likewise may be counterbalanced by a spring support from above so that the resultant load on the instrument is one pound.

The final accuracy of the results obtained for measurements of radii of curvature by spherometers, depends on three factors. These are, for the general case, errors in the constant A, errors in the reading of the " sag " h, and errors in the radii of the steel balls in the supports. If the percentage errors in these three factors are E_A, E_h and E_a respectively, then the percentage error in the radius of the test surface, E_r, is given by

$$E_r = E_A + \left(\frac{h-r}{r}\right) E_h + \left(\frac{a}{r}\right) E_a$$

If one takes an extreme case of a very deep surface, of short radius, then h may approximate to $r/2$, and a to $r/3$. For this case then

$$E_r = E_A - (\tfrac{1}{2}) E_h + (\tfrac{1}{3}) E_a$$

Suppose that a final accuracy of one tenth per cent is required in r, then an estimate of the tolerances in A, h and a may be found; for instance, one set of tolerances, for equal contributions of error, is 0·03 per cent, 0·06 per cent, and 0·09 per cent for E_A, E_h and E_a respectively.

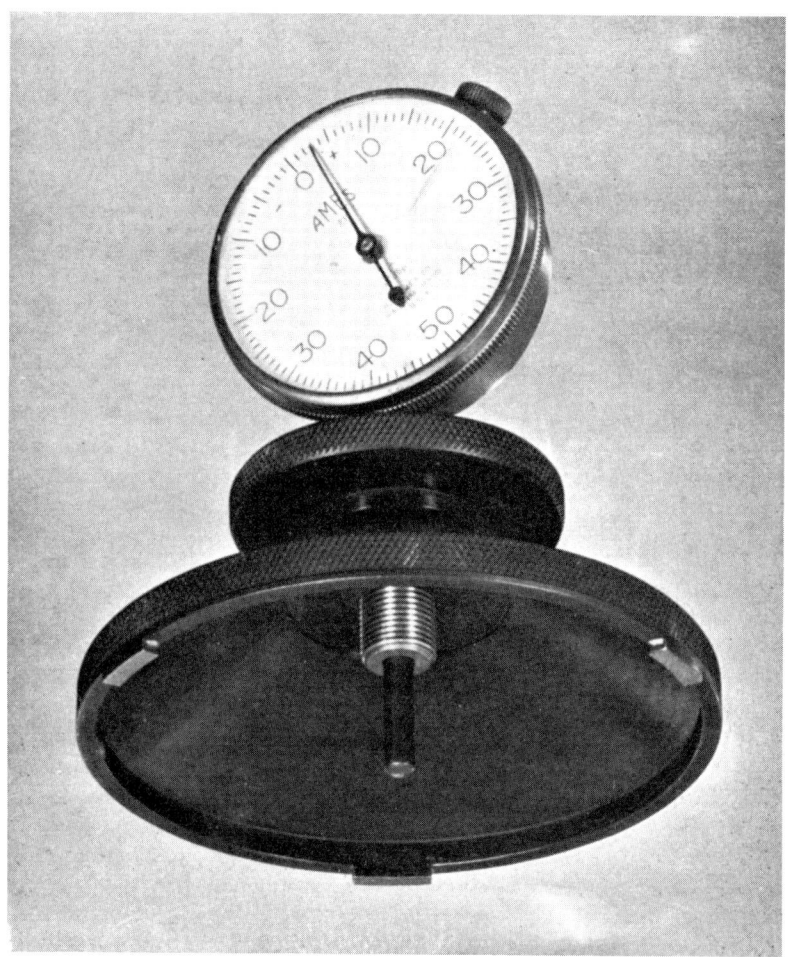

Fig. 27—The spheroscope

52 During the recent war, a comparator spherometer, called a spheroscope, was developed, the particular use of which was in the rapid checking of the radii of curvature of new tools (in the course of the milling operations) and of lens blanks when setting up the Bryant & Symons roughing machine.

The instrument, shown in Fig. 27, is essentially a ring form spherometer, on which the micrometer screw is replaced by a plunger attached to a dial gauge, reading to one ten-thousandth of an inch.

In operation, the spheroscope is applied to a proof sphere, the radius of which is near to that under consideration, the reading on the dial gauge being noted. A simple calculation, or a reading from a table, gives a correction to the dial gauge reading, the modified reading corresponding to the required radius. The curvature of the test surface is then modified until the correct setting is obtained.

A convenient expression for use with the spheroscope, of sufficient accuracy for the purpose, is derived from the usual error expression for spherometers. If r_1 and r_2 are the radius of curvature of the nearest test plate, and the desired radius of curvature respectively, and if A is the spheroscope constant, $s^2/2$, then the correction to the spheroscope reading is given by,

$$\text{Correction} = \frac{A}{r_1^2}(r_1 - r_2)$$

This expression is of use also when correcting proof spheres for radius. Should a correction be necessary, one proof sphere is altered away from its mate by the number of Newton's rings calculated from the above formula, it being remembered that a change of one ring is equivalent to a correction (a change in " sag "), of 0·00001 inch.

If made of large size (9 in. from ring to centre say) such a tool is extremely useful for testing the flatness of plane tools. In such large size, and for such purpose it is better, however, to make it in the three-legged form, with a ball end to each leg.

A paper by Jensen (1950) compares the relative accuracies with which radii of curvature can be determined by various methods, and summarizes the various ways of measuring radius of curvature of optical tools and surfaces. It mentions several forms of spherometer which are not mentioned above, including a precision ring spherometer, made by the Askania-Werke, of massive design which seems to be an excellent and reliable tool. The paper is well worth referring to by those wishing to overhaul their means of spherometry.

53 We will assume, then, that by measuring the tools with the spherometer, grinding them locally with a lead lap or abrasive wheel (see §48), and rubbing them together, they have at length been made of the correct radius of curvature.

The two tools so prepared are the " true " or finishing tools and in polishing blocks of lenses two others are required, a block holder and a polisher holder. These must be of radii respectively less and greater than the radius of the tool when a convex surface is to be polished, and vice versa when the surface to be polished is concave (see Fig. 68). This variation of radii only becomes important with the shorter radii, say with a 9 in. diameter tool for radii of curvature less than 12 in.

Thus referring to Fig. 68, if R is the radius of curvature of the lens surface polished to the thickness of the lens, t_0 the thickness of the lens, t_1 that of the polisher and t_2 that of the mallet, then the radius of polisher holder should equal $R + t_1$ and the radius of blockholder should be equal to $R - (t_0 + t_2)$. This ensures that the polisher and the mallets are of uniform thickness.

Chamfering tools

54 Following is a table detailing the radius of chamfering tool which will generate a bevel at approximately 45° on a blank of the stated diameter—

| Dia. of blank | ... D | 0·3 | 0·6 | 0·9 | 1·2 | 1·6 | 2·0 | 2·6 | 3·0 | 4·0 |
| Radius required, | ... R | 0·21 | 0·42 | 0·63 | 0·85 | 1·13 | 1·41 | 1·85 | 2·12 | 2·8 |

The general formula for this case is : $R = \dfrac{2D}{2}$

55 Although at one time tools were kept in pairs for the convenience of being able to correct the radius of the used half by lapping together with the unused tool, it is known to be possible to maintain the radius of spherical tools within the closest limits by a method analogous to to that detailed under hand polishing with flat tools.

In any case, one pair at least of each radius should be in the form of a closely lapped pair, if only to facilitate the making of curved test plates or proof spheres.

By the simple process of rubbing them together when they are clean and dry one can easily detect differences of 2/10,000ths of an inch. Where the tools touch they become shiny. On a pair of tools of large diameter, say nine inches, this represents a very high degree of accuracy.

Flat tools

56 Roughing tools and machines are described in Chapter 6. A set of flat true tools consists of four ; three of them are finishing tools, and are kept as flat as the worker knows how.

In general, the ideal shape of a final smoothing tool is very slightly convex, say about four fringes in nine inches, and the tool with which the polisher is rubbed up should then be about three fringes convex in nine inches.

To be fully equipped for any demand, a skilled worker should have a set each of 6 in., 9 in., and 13 in. tools available.

57 The polisher holders, convex, plano or concave, are similar to the true tools, but may contain casting defects of a nature and extent that would make them useless for trueing or smoothing.

Polishers

58 The function of the polisher is to provide an accurately shaped flat or spherical medium for applying the polishing material.

Polishers are usually of pitch, though felt or cloth is still used for some of the commoner work. Various wax mixtures are also used; they polish very much more slowly but are described in §65 *et seq.*

For hand polishing the writer's experience is that there is nothing better than wood pitch (Swedish), hot filtered through 100 mesh chiffon and boiled until at room temperature it can be readily but not deeply indented with the thumbnail.

(A mechanical device for measuring the viscosity of pitch is described in §60.)

When the polishing of quartz is contemplated, it may be found that the addition of half an ounce of beeswax and rosin to each pound of the hardened wood pitch will give cleaner surfaces and slightly quicker polishing.

59 The properties to be desired in a polisher are to some extent contradictory in nature. In the first place the polisher must be susceptible of a certain amount of flow so that it can be formed by application of the flat or curved tool to give it the right shape. The rate of flow must, however, be slow so that the surface which has been imposed on it will in turn be imposed by it on the glass surfaces which are to be polished. If it is too soft, it will rapidly go out of shape during the polishing process, and thus frequently require re-forming. This re-forming takes time, although it is not necessary to heat the polisher for this purpose; it suffices to warm the forming tool slightly, or it may even be used cold. If the viscosity is too high, dust falling on the polisher will cause scratches before sinking into the surface of the polisher and thus becoming innocuous.

When considering polishers for use on machines, the problem of maintaining the shape of the polisher for long periods without frequent recourse to forming it, and without using a very hard grade of pitch, has led to the practice of " loading " the polishing pitch with various materials which, by increasing the viscosity and resistance to change of shape, allow the use of very soft pitch and thus make easier the production of high quality surfaces of impeccable accuracy.

There is a wide range of materials available which have been adopted as standard by one or other of the optical houses whose methods we have been privileged to examine.

One of the simplest methods is to add to the hot pitch dried willow wood flour, until the pitch will take up no more.

Cotton wool is used by some firms, and Ross Ltd of London made good use during the recent war of pulled yellow felt.

Polishers made of these materials are not stamped, it having been found that stamping reduced the efficacy of the polisher. Two or three annular grooves in each polisher are all that are required, and the polisher is applied to the block while still quite warm, and rubbed about by hand until it neither binds nor swings ; the pin of the machine is then lowered into place and the machine started.

Loaded polishers of this type will withstand polishing speeds and weights much in excess of the conditions normally associated with pitch polishers, and are much less likely to scratch.

Testing pitch for viscosity

60 The illustration (Fig. 28) shows a simple but quite satisfactory device for measuring viscosity of samples of pitch removed from the cauldron from time to time during the hardening process. It consists of a piece of steel $\frac{1}{4}$ in. diameter, with a truncated conical point of 14° included angle, terminating in a $\frac{1}{2}$ mm diameter flattened point. Attached to this is a weight. The rod with its weight is held loosely in a vertical position by the top of the rod passing through an eyelet in a wooden upright which itself is supported on a flat wooden stand and has a total weight of 1 kg. The point of the rod is allowed to bear on the top of the pitch, or other substance whose viscosity it is desired to control, which is immersed in water and thus kept at any desired temperature. The length of time taken for the rod to fall a given distance is determined.

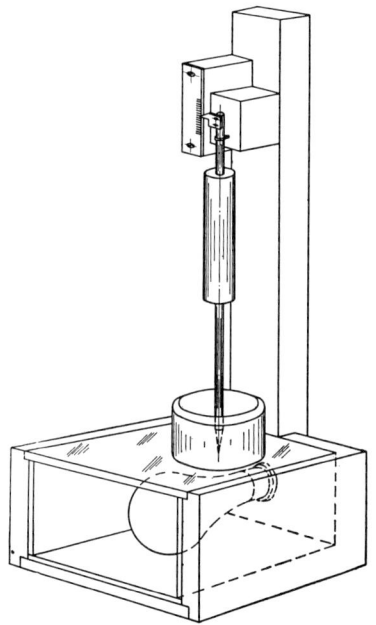

Fig. 28—**Pitch viscosity testing apparatus**

It will be found that the most useful grades of pitch to stock in temperate climates are two, in which the rod falls $1\frac{1}{2}$ and 3 mm in five minutes respectively at a temperature of 70° F, the former being used mainly during the summer months and the latter in the winter.

Although speculum metal can be polished on soft pitch the addition of one part of beeswax to sixteen of 1½-mm pitch makes it much easier to avoid scratches.

For polishing Iceland spar, and other soft crystals, a softer pitch is better—*e.g.* 4 mm. For further information concerning the polishing of materials other than glass see Chapter 6.

One may add here, however, that our opticians do not seem to make a practice of using pitch of differing viscosity in polishing the various

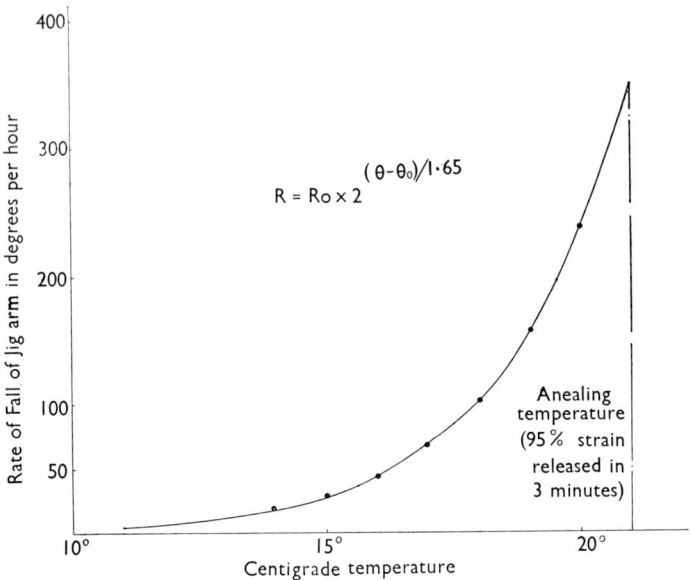

Fig. 29—The viscosity–temperature relationship for 1½ mm pitch
The viscosity is expressed as the rate of fall of the jig arm.

glasses and crystals. Of two men working on the same substance at the same time one may get better results with 1½-mm and the other with 3-mm pitch.

Doubtless other factors have to be taken into consideration such as pressure on the work, state of dryness of the polisher rate of traverse of the part being polished—indeed all those factors which add to the heating or chilling of the skin of the polisher. Thickness of the pitch polisher must also be of importance and Newton's practice is of interest in this connection (§11). With a thinner polisher a softer pitch, less likely to scratch, could doubtless be used without too great a distortion of its surface by flow.

As the polisher becomes thinner by reason of the constant recutting of the grooves as they close up, so it becomes harder. It is possible that reluctance to delay his work to make up a fresh polisher sometimes induces a man to go on with it too long. An old and good rule, whether in forming a polisher or in hand polishing, is to work with the tool and the prism being polished feeling neither warm nor cold to the hand.

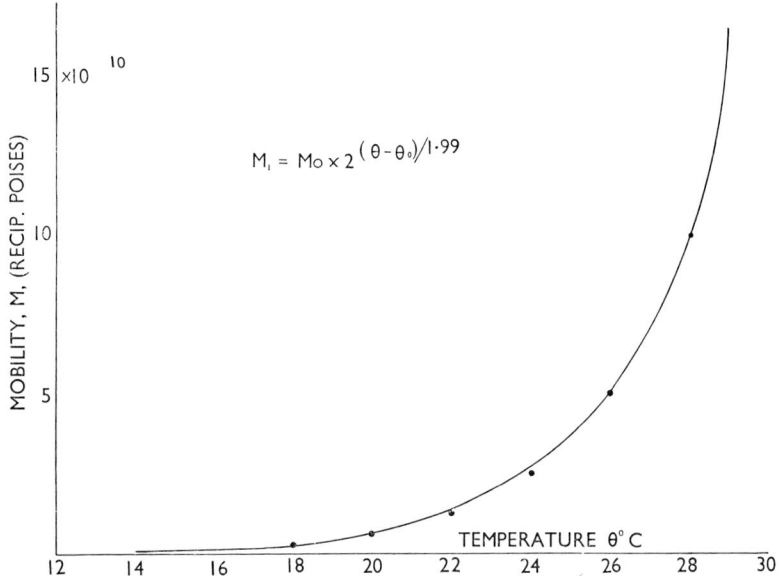

Fig. 30—The mobility–temperature relationship for 3 mm pitch
The values were determined by Searles cylinder method.

Viscosity–temperature relation for pitch

61 It is useful to know the viscosity-temperature relation of pitches whether for polishers or mallets. Experiments were made in the Hilger laboratory using the same technique as in determining the annealing temperature of glass (§295 *et seq*). Pure polishing pitch was used ($1\frac{1}{2}$ mm, 70°F). Fig. 29 shows the rate of fall of the jig arm in degrees of arc per hour for the temperature θ of the pitch specimen. From this the viscosity in poises can be found from the formula given below. Alternatively, the viscosity can be found by Searle's cylinder method (Goddard and Boulind, 1934): Fig. 30 shows the mobility M (reciprocal of the viscosity in poises) so determined for different temperatures of 3 mm pitch. The curves in Fig. 29 and 30 have been carefully drawn from observed values; the constants (1·65°C and

1·99°C respectively) in the formulae were determined by taking two points on each curve. To ascertain the degree of accord between the experimental observations and the formulae so deduced R and M have been determined from the formulae and the values plotted as heavy points. The closeness of these to the experimentally determined curves show to what a high degree of approximation the viscosity–temperature relation follows the exponential formula. The same accord (to a somewhat lesser extent) has been found with various specimens of glass (see Appendix 1).

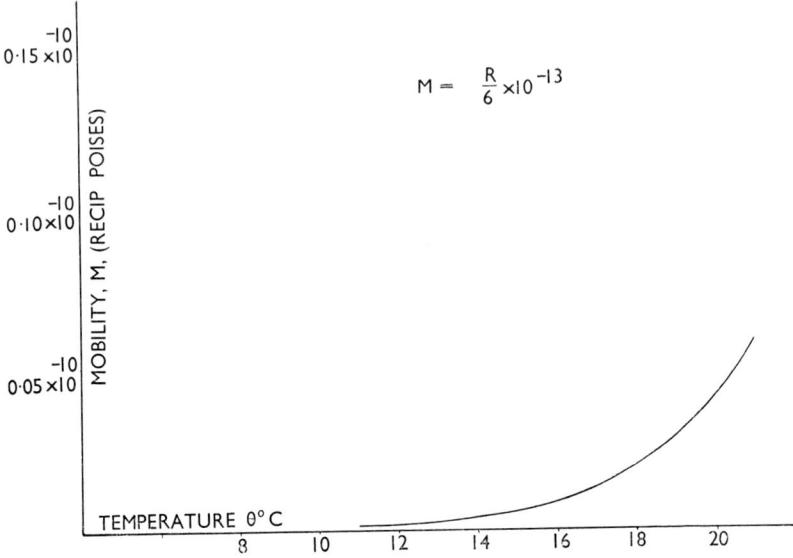

Fig. 31—Absolute mobility–temperature relationship for $1\frac{1}{2}$ mm pitch
The curve was determined from the curve of Fig. 29.

Let θ, R and $1/M$ represent the temperature of the specimen in degrees C, the corresponding rate of fall of the annealing jig arm in degrees per hour and the viscosity of the specimen in poises, respectively; θ_0, R_0 and $1/M_0$ representing the same quantities at the annealing temperature. From the rate of fall of the jig arm the absolute viscosity in poises can be found from the expression (Fig. 31)

$$1/M = (6 \times 10^{13})/R$$

while from Fig. 29 we obtain the expressions

$$R = R_0 \times 2^{(\theta - \theta_0)/1·65}$$

whence

$$M = (10^{-13} R_0/6) \times 2^{(\theta - \theta_0)/1·65}$$

William Taylor (1932) has pointed out that pitch is often not homogeneous and that this must be taken account of. He says—

It is essential that the pitch must have the right degree of viscosity and this must be uniform throughout the mass ; the means of ensuring these things, of mitigating the ill effects of their absence, was one of the secrets of the skilled artist craftsman. During the war Leicester had to produce lenses of the highest accuracy in considerable quantities and for many purposes. The supply of skilled craftsmen was totally inadequate and we had to find means by which unskilled help could do work at least as good as that previously done only by craftsmen of long experience. This was in its true and broad sense a mechanical engineering problem. By employing discs of glass with two polished flats at opposite sides of their perimeters, and viewing the glasses by polarised light transmitted through the flats with an apparatus known as a Babinet compensator (a kind of polariser by means of which one can measure the amount of double refraction) we were able to see the strain produced in the glass when one surface was heated by applying hot pitch. We saw the strain released as the pitch and glass cooled but if the pitch or portions of it were truly solid and not a viscous solid, some strain of the glass remained permanently. If on the other hand the pitch was not sufficiently solid the strain disappeared soon, thus a test of the quality of pitch was established and we learned that it must be self-annealed at ordinary temperatures and, for example, retain some strain for half an hour but no strain at all for an hour.

A makeshift substitute for wood pitch is gas pitch, obtainable from the Tar and Ammonia Products Works, The Gas Light & Coke Co., Beckton, East Ham, London, E.6. If the order specifies to them that the pitch should be " 60°C, K. & S. Grade " (as supplied to Adam Hilger Ltd.) it will be about our $1\frac{1}{2}$ mm standard viscosity. (See above.)

Owing to the rapid change of the viscosity of pitch with temperature (it is halved for each $1\frac{3}{4}$°C rise of temperature), a given pitch will only be at its best as a polisher over a range of $\pm 7\frac{1}{2}$°F, say ± 4°C. The thickness of the polisher should be about $\frac{1}{4}$ inch.

Many skilled workers also like to add various ingredients to the pitch ; for example I find in my notes of 1915 that at that time an approved mixture used by them was pitch—8 parts, rosin—1 part, beeswax—1 part ; later on pitch—10 parts, rosin—1 part, beeswax—$\frac{1}{2}$ part was preferred. However, as time went on I was unable to find any strong consensus of opinion as to the purpose served by such additions, although it is stated, and seems reasonable, that the addition of beeswax reduces the tendency to scratch. The general routine in our workshops is now to use pure Swedish pitch, of standardized viscosity.

During the Great War of 1914–18 when we trained many women to do glass polishing, we carried to extremes the attempt to remove foreign matter, in the hope that by so doing we could remove the tendency to scratches which is one of the main faults which occur in the polishing of glass by unskilled workers. With this aim we filtered the pitch through filter paper in an oven at from 200° to 300°C, while the air was fed into the workshop through a chamber screened by two separated filters of fine chiffon (100 mesh per inch).

Experiments extending over several months showed that pitch filtered in this way is less liable to cause sleeks than that which has only been passed through fine muslin. Fine sleeks however, were still produced even with pitch so treated, and with rouge ground to such a state of fineness, as not to deposit in slight alkaline distilled water after several hours.

Several years experience with this technique left us with the belief that little advantage was derived from this extreme caution. Scratches are not chiefly due to foreign matter in the pitch, but to dust (chiefly siliceous in nature) which falls on the polisher. Meticulous care to avoid this is not so important as to ensure that the pitch is so soft that such particles sink into the polisher rather than scratch the glass, or at worst cause the kind of light scratches known as " sleeks " which are distinguished by an abrupt and relatively severe head at one end, tailing away to invisibility at the other.

Importance of avoiding dust

62 The worst dust particles can be kept from falling on the polisher by enclosing the machine as far as possible with a shield made preferably from some transparent material such as " Windolite."

One well-tried method is to cover the polishing machine, top, back and sides, with sheets of "Windolite". This gives all the advantages of complete enclosure without the restriction of ventilation which is otherwise an objection.

The importance of correlating the workshop layout with other precautions against falling dust is underlined by some experiments on the rate of fall of dust particles and emery under various conditions. The method of experiment was to illuminate the air in which the dust was moving with a powerful beam of light, and to photograph the particles against a dark background.

The source of light was a high-pressure A.C. mercury vapour lamp, so that the illumination was intermittent and of a frequency of 100 cycles per second. Thus each falling particle appeared as a dotted line in which the separation of the dots was a measure of the speed.

TOOLS AND MATERIALS 85

A specimen photograph enlarged six diameters is shown in Fig. 32, and although the emery of 1·0 grain with which the exposure was made was what we should regard as well-graded, it will be noted that a great variety of grains or agglomerates are present, including some extremely small.

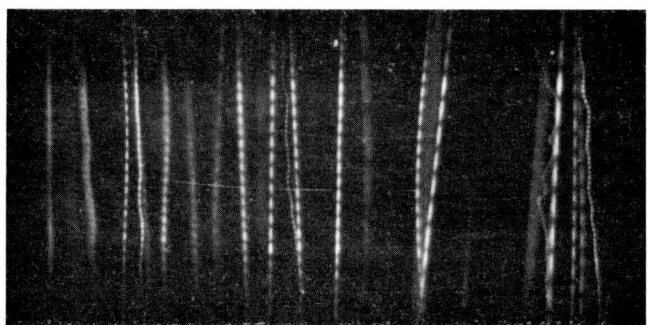

Fig. 32—Fall of emery grains

Photographs taken in a closed chamber with various powders gave the following series of results—

Material	Diameter (average) in 0·01 mm	Rate of fall in mm per sec
Lycopodium powder	—	28
Emery	1·0	31
,,	2·5	36
,,	3·7	50
,,	5·0	62
Carborundum	18	125
,,	80	250

The same method was then applied to an examination of the movement of particles in air currents caused by normal working conditions, that is with the machine still, with the machine running, and shortly after the workshop door had been slammed.

The chief source of draughts seemed to be the machine belts and pulleys, most of which have since been eliminated by motorization of the machines affected, but the currents of air caused by the slamming of the door were much greater, though of limited duration. The measurements were then repeated after the provision of a canopy as described, and this was found to be very effective in reducing the tendency for dust to be carried over the machine.

Following is a summary of the results obtained with lycopodium powder—

Conditions	Movement without screen		Movement with screen	
	Angle to horizontal	Speed mm per sec	Angle to horizontal	Speed mm per sec
Machine stopped	50°	167	70°	42
Machine running	30°	100	60°	63
After slamming door	20°	200	45°	83

It would appear that a refined method of grading the finer grains might be derived from these results—a uniform horizontal current of air distributing the grains according to their size in receptacles placed in line below a thin stream of the falling powder. After this idea had occurred to me I found that Sir Henry Bessemer described in his Autobiography, p. 69, a method of grading bronze powder by blowing it gently through a tunnel 40 feet long and 2½ feet wide; the powder was deposited slowly on a strip of varnished cloth running along the bottom of the tunnel. The author says—

It is difficult to imagine the beauty of this golden snowdrift of 40 ft in length, varying at every foot in appearance, and ranging from pieces too coarse for use, and which required further lamination, to the extremely minute particles which, between the fingers, felt like the dust of pure plumbago, or some other wonderfully smooth lubricant.

(Methods of grading which have been used with success are described in §77 *et seq.*)

Whatever the system of dust removal the ventilation required is at least 600 cubic feet per hour for each worker. The following figures are based on the supposition that the workshop contains 17 men requiring, therefore, not less than 10,000 cubic feet per hour. The three methods which come into consideration are—

(1) A large aperture for the entry of the air, extending the whole width of the workshop, so that the dust particles would have to be lifted over a partition.

(2) Filtering through thick cloth.

(3) Electrical filtration.

(1) Large aperture near the ceiling.

A corollary of this is a double self-closing door to enter the workshop, and the workers should be required to change their shoes on entering the clean room. The space above the partition should be so large that the velocity of the air should not be more than 1½ ft per minute. The air should be fed in near ground level and well distributed at the rate of 10,000 cubic feet per hour. The 1½ ft per minute corresponds with the settling velocity of particles less than 0·015 mm in diameter, so that the particles carried over the partition would be not greater than 0·0005 in. This corresponds with our fine smoothing emery and should, I think, be satisfactory. However, in the two following methods the air is filtered from particles larger than 0·0001 in.

(2) The second method is to use filtering screens of a closely woven thick cloth, of which several layers should be used. The material should be of sufficient total area and so arranged, for instance, by V formation that the velocity of air through the cloth is not greater than 20 ft per minute. For 17 occupants this means a filter of $8\frac{1}{2}$ sq. ft. total area.

A suitable filter cloth is made by Rylands of Manchester under the name of " Swansdown " and four to six layers of this cloth should prove satisfactory to remove all particles larger than 0·001 in. This cloth should be renewed frequently, and this is one of the main objections to filtration of this type.

(3) The third arrangement is the electrical filtration provided by the " Precipitron " electrical filter made by the Sturtevant Engineering Company under licence from the Westinghouse Company. In the " Precipitron " the air passes through an ionizer, in which metal wires, charged to approximately 13,000 volts D.C. in the neighbourhood of a number of earthed tubes, cause ionization of the air thus furnishing a source of positive and negative charges. These collect the dust particles on earthed plates to which they adhere until removed by cleaning. I think I am right in saying that the cleaning requires to be done once a fortnight.

The size of outfit needed for 17 workers would, of course, have to be specified as to deal with 10,000 cubic feet per hour. The " Precipitron " is used by Barr and Stroud for a balsaming room and found very satisfactory. Hilger & Watts also use it for a " Contacting " chamber. A description of the Precipitron is given by Penney (1937), and the makers issue a full account in booklet form.

An article by Carnall (1948) gives information concerning the dust problem as it affects the coating of photographic plates. There is much information there of interest to the optician. Although the optical glass worker is only concerned with *abrasive* material in the air, yet in preparing surfaces for cementing and for metallic or other coatings, fibres are as important as they are in the photographic coating room. For such purposes air filtration, even if perfect, would only partially solve the problem, and prevention of dust originating from the personnel becomes necessary, for every man and woman is surrounded by a dense cloud of dust made up of hair, epidermal debris, wool, cotton and silk fibres.

The article cited gives very full advice on this aspect of the subject.

Design and lay-out of an Optical Workshop
63   Col. Dévé in his book *Le Travail des Verres d'optique de Précision** says—

The risk of scratching glasses in the course of surfacing requires exemplary cleanliness and care to avoid the accidental mixture of abrasives of different grain size, or grains of dust falling on the tools or polishers.   The ground, the walls and the ceiling of the workshop ought to be sound, easily washable, and covered with varnish paint. The colour of the paint should preferably be an imitation of red marble up to a height of 2 metres, because this renders soiling by the polishing materials less unsightly.   Framed windows in the ceilings are absolutely to be forbidden, for dust accumulates in the angles and the least breath of wind blows it into the workshop.   One can cite as a model installation that of the polishing rooms of the ' Atelier de Construction de l'Artillerie ' at Puteaux.   There are no windows, properly so-called, but a fixed waterproof glazing. No air is admitted to the rooms except what is filtered through a special ventilator.   Large channels in the angles of the rooms bring in the air and evacuate the used air.   In this way smoke and dirt have no access to the optical shop.   The floor is covered with thick linoleum on which the glasses can fall without too much risk of breaking or scratching.   These precautions would be illusory if the workmen introduced with them the mud or dust of the streets.   Access to the workshops can only be obtained by passing through a vestibule, where the outdoor shoes are changed for those of the workshop and the overalls put on, as is customary in bakeries. A boiler suit is superior to the white blouse which is still in use in old-fashioned optical shops.   The wearing of flowing vestments in the neighbourhood of gears and belting is imprudent.   Along the workshop walls and away from the daylight are little recesses in which the workmen can go to examine by artificial light the glasses which they are polishing.   These recesses should be illuminated by approximately monochromatic light, preferably a mercury lamp.   Finally, it is necessary to maintain a constant temperature in the workshops which use pitch polishers.   If, for example, in winter one neglects one day to heat the workshops, pitch prepared to have a suitable viscosity at 18°C would be too hard at a temperature of 10°C and one would risk having numerous surface scratches.   The good optician should, as befits a very delicate operation, be careful, attentive, methodical and clean.   The good optician habituates himself never to soil his left hand, using only his right hand for handling

* An English translation was published by Adam Hilger Ltd. under the title *Optical Workshop Principles*.

the lenses, abrasives, greasy bodies, etc. He keeps his left hand always clean in order to handle a caliper or other instrument, some delicate piece or the handle of the optical tool. (*Free translation.*)

One may add that the hands and benches should frequently be washed, especially after the use of carborundum or emery, white cleaning rag and camel hair dusting brush should be kept in a covered enamelled jar like those used in the domestic larders for keeping sugar, rice, etc. For my own use I always keep an old linen handkerchief which has been boiled in distilled water, and a camel hair brush which has been cleaned in absolute alcohol, and use them for the final cleaning. If kept in a covered jar they will remain grease-free and very effective in cleaning optical work for some weeks, even when in daily use.

Workshop conditions in Germany also, naturally, receive special attention. In one workshop visited the walls and ceilings were covered with American cloth and regularly washed. Temperature and humidity controlled ventilation was installed and all air was filtered. Cleanliness and orderly arrangement of work were insisted on and workers were trained in this direction.

64 My own conclusion up till 1944* was, then, that while it is of great advantage to protect the machines by a canopy of the kind mentioned earlier, no form of filtration is worth while except the electrical filtration described above which is expensive to instal if ample ventilation is to be provided for a large workshop and in which the plates require cleaning once a fortnight. During the period when we filtered the air of our optical workshop through chiffon (§61) it was not found possible to ensure that the cleaners renewed the chiffon every fortnight as was stipulated. The consequence was, that it became clogged and ventilation was insufficient and then the workers would open the windows and the filtration system became entirely ineffective. Further, even when working properly it was only the coarser dust which was removed.

It should be insisted on that overhead gear of any kind should be avoided. This includes belts, pulleys, girders whose flat tops or flanges collect dust, and skylights.

Having come to this conclusion our efforts became directed rather to the finding of a polishing medium which would not produce sleeks even under rather unfavourable conditions, and we eventually found that such a medium could be obtained from mixtures of waxes, an account of which now follows.

* I was informed later (1944) by The Building Research Station (Garston, nr. Watford, Herts.) that four layers of closely woven thick cloth (*e.g.* that known as " Swansdown," made by Rylands of Manchester) will remove all particles bigger than 00·006 in., say between smoothing and fine smoothing emery.

Wax polishers

65 For a considerable amount of work done by unskilled workers in our workshops during the Great War of 1914–18 wax polishers were employed. A great variety of mixtures was tried, for example—beeswax $3\frac{1}{2}$ lb, black rosin 2 lb, is noted by me as exceptionally good for avoiding scratches. One competent foreman after much experience with polishers of this character prepared—beeswax $3\frac{1}{2}$ lb, black rosin 2 lb, pitch $\frac{1}{4}$ lb, and tallow " a small quantity." Beeswax and rosin melted together make a good polisher for Perspex (methyl methacrylate).

There is no doubt that wax polishers are far less liable to cause scratches and sleeks than are pitch polishers. They are, however, difficult to make flat. Wax mixtures such as those mentioned do not flow to the same extent that pitch does and they can only be flattened by scraping, at least until the stage is reached when almost the whole surface has been in this way made to fit the forming tool very closely. If from this stage onwards the tool is rubbed on the polisher with the customary rouge paste, eventually a satisfactory flat polisher is formed which will do excellent work and keep its shape very stubbornly.

Wax polishers polish more slowly than pitch and do not so readily produce accurate surfaces.

The difference of behaviour of pitch and wax as regards flow may be very simply demonstrated in the following way. If a ball of pitch such as is used for polishers, and one of wax mixture consisting of beeswax and rosin in equal proportions, are placed under equal weights and left for a week, the pitch will become entirely flattened, while the wax will remain unchanged in shape.

An amusing example will further illustrate the " fluidity " of pitch. Away back in the 1890's when there were only two or three in the Optical Shop, a barrel of Swedish pitch lasted a long time. The level surface of the pitch was broken when pieces were broken out with chisels or old files but the pitch soon flowed level again. Eventually the bottom of the barrel was reached and a number of tools, including chisels and files were found which had carelessly been left lying on top of the pitch from time to time.

The relative liability of pitch and wax polishers to cause scratches may be illustrated as follows. If two small flat polishers, one of pitch and one of wax, are placed on a piece of glass with a few grains of carborundum between the glass and polishers and the polishers are allowed to rest ten seconds and then pushed along the glass plate for an inch or two, one can distinctly feel and hear the carborundum under the pitch polisher scratch the glass, whereas the wax polisher moves smoothly and without sound over the glass. On examining the glass it is found that where the pitch polisher has passed there are numerous severe scratches, whereas the wax produces only extremely faint ones.

66 Another typical polishing wax is made by mixing equal parts of beeswax and putty powder; another which though slow is very " foolproof " is made by mixing three parts of rouge with four parts of paraffin wax. Although the writer has tried a great number of variants of the above materials for polishers obtained by various mixtures of pitch, beeswax, paraffin wax, rosin, putty powder, rouge and other materials, he has never been able to establish any certain superiority of any of them over the simpler ones already mentioned.

There is, however, an undoubted advantage in the addition to a chosen mixture of flour of willow wood, and at the present date (September, 1951) the standard polisher in our optical shops at Hilger & Watts (Hilger Division) is as follows : 16 oz $1\frac{1}{2}$ mm pitch, 2 oz rosin (colophonium), 1 oz beeswax, 12 oz willow-wood flour.

With this polisher and a weight of $8\frac{1}{4}$ lb in addition to the weight of the polisher holder itself, polishing normally takes from $2-2\frac{1}{2}$ hours for a 13 in. diameter block, with little tendency to cause scratches or sleeks.

The chief advantage of the addition of the wood flour is that considerably greater pressure can be used without distortion of the surfaces. Further, the stamping of the polisher is not necessary.

Wax-faced pitch polisher

67 In the manufacture of those graticules which are illuminated through the edge and viewed under magnification every blemish becomes a luminous point in focus. When made by standard pitch polishers many of the blemishes causing definite rejection are almost invisible by the usual transmission test.

For work of this kind we have found that a wax polisher is worth using, although it is much slower than pitch, taking about twice as long. To obviate the difficulty of forming up the polisher we have tried a wax-faced pitch polisher with success. It was prepared in the following way.

A 9-in. flat iron tool was coated with $1\frac{1}{2}$ mm Swedish pitch, and flattened as if for a pitch polisher, the tool being of such a heat that once pressed the polisher retained its shape without noticeable flow.

Hot beeswax and rosin (proportions 1 : 1) was then poured on, and allowed to run all over the pitch (and off the edge of the tool in places). One's finger wetted with cold water was found an efficient means for sealing off the overflow.

The polisher was then pressed on a wet 9-in. iron tool which already had a polished surface.

The polisher was then pressed twice (at right angles), on to a wet grooving tool and finally trued on the 9-in. flat tool and cooled off in water. Flattening on the warm tool was difficult until the latter was polished with fine emery paper.

Stamping with gauze was effected in the usual way, using a rather cool tool and heavier pressure than usual (about 1 cwt.). At the first sign of sticking, the gauze was stripped off, the tool cooled somewhat and cleaned up with petrol and methylated spirit.

Mallets

Mallet pitch

68 When a number of lenses are to be ground and polished together, they are held on to a curved tool by means of blobs of pitch known as mallets (in the U.S.A. as "buttons") and, as with the polisher holder, the radius of curvature should be such that all the mallets are of equal thickness.

The name "mallet" is not in general use, and seems to have been derived from a misapplication of the French "molette," the word used by Cherubin d'Orlèans (see §8) to describe the metal block holders. The composition and properties of mallet pitch are as variable as those of polishing pitch, but the mallet pitch should always be more viscous, but not too much so, than the polishing pitch; otherwise the simple method outlined in §131, for setting the lenses in the block would not be feasible, and the lenses would be strained, the release of that strain, after knocking them off the mallet, causing them to assume a bad figure.

Strain can be minimized by employing quite soft pitch, and reducing its mobility by the addition of a loading agent as in the case of polishing pitch, though there is no need in this case to restrict the loading agent to soft, non-scratching materials.

Wood ash, charcoal, wood dust, red and yellow ochre are all used by various manufacturing concerns, each mixture being supposed by the user to have a particular advantage over any other.

Additions of various resins and waxes are also made to the pitch base in order to modify the solidification characteristic; after all precautions have been taken we still find it advisable to leave the malletted lenses for 24 or 48 hours according to size, as the strain in pitch, which is a super-cooled liquid with melting point not far above room temperature, is gradually released even at room temperature.

69 One of the mixtures which has found approval for standard practice is a simple one, namely 5 lb of pitch (4-mm at 70°F, see §60). and 8 lb of red ochre; the resultant viscosity of the mixture should lie between ½ and 1 mm in 5 minutes at shop temperature when tested on the pitch jig.

If the lenses are of small diameter and the polishing is carried out at higher speeds, so that more heat is generated, a stiffer mallet may be used in which the fall of the standard jig is only ¼ mm. This is attained by using 7 lb of yellow ochre instead of 8 lb of red. Both the above will be quite suitable over a range within $\pm 2°C$ ($\pm 3·6°F$) of the temperature for which they give the above standard tests; and probably there will be no serious trouble within a range of $\pm 4°C$ ($\pm 7·2°F$).

If the lenses are thin, the ½-mm pitch will be likely to distort them due to strain; the only remedy is to use a softer pitch up to 2½-mm, though this will prove too soft for all but the slowest machine polishing. The addition of the more fibrous loading agents, such as charcoal and wood ash, seems to restrict the mobility of the pitch without increasing its tendency to strain.

Where continuous lens production is in progress, and large numbers of blocks are made and knocked off the mallets every day, specimens of mallet pitch should be taken daily, after vigorous stirring, tested for viscosity, and the mixture hardened or softened to the degree that experience has shown to yield the best result.

Some firms have made such a study of the increase in hardness of pitch in normal routine production that they are able to add a daily or weekly amount of soft pitch to their stock pot and automatically maintain the contents at the desired hardness.

Abrasives

70 As the actions of the various forms of diamond grinding tools do not strictly belong to the abrasive class at all, they are dealt with under the sections devoted to the production of prisms and lenses in quantity.

For roughing by hand, the fastest material is carborundum; for trueing and smoothing, aluminium oxide (Al_2O_3) in various forms, natural and artificial, is generally considered better.

Carborundum (silicon carbide, SiC)

When carborundum was first introduced into the Optical Shop of Adam Hilger Ltd as a fast roughing material, about 1911, it was much opposed, but the opposition was short-lived. A pan of the new abrasive was placed beside a pan of the old on the roughing bench, and operators were allowed to use which they preferred. It was not long before the emery was ignored and finally scrapped.

The finishers were afraid that an epidemic of scratches would break out owing to the particles which glistened on the floor and which imagination led them to believe would soon be floating around them. They were mistaken. When carborundum falls on a polisher, the carrier is not the air, but more likely the coat-sleeve or arm of the individual. When one is polishing plenty of scratches will occur if somebody is sweeping the floor at the time, even in shops that use no carborundum.

Emery

The impure natural forms known as emery have been used for grinding from time immemorial, and up till the end of the nineteenth century the emery from the Isle of Naxos had formed the principal source of supply for the optical industry. Purer and quicker cutting natural forms known as corundum are found in the United States of America, Canada, Madagascar, and elsewhere. We have used Naxos, Madagascar and Canadian corundums, and found the Canadian distinctly the best of the three, while the artificial forms sold under various names (aloxite, alundum, etc.) are better still. It is stated in Strong (1940) that the natural forms of corundum cut more quickly than the artificial types. This is not in accordance with our experience. The artificial corundum known as aloxite we find to be quicker than Naxos, Canadian or Madagascar corundum; it maintains its cut longer, is pure and free from " mud " and organic matter, and its wide use in industry ensures that it is always available. The freedom from the fibrous matter which we used to find in the natural kinds is particularly advantageous in elutriation, as such impurities act as " rafts " to float grains of all sizes. All these abrasives require grinding and grading by elutriation; a very fine and uniformly graded material being of the highest importance if speedy polishing is to be attained.

Boron carbide

About fifteen years ago a new abrasive became available, boron carbide. First produced at the Electro Schmelzwerke A. G., Bavaria, it is now made by the Norton Co., Worcester, Mass., and can be obtained from their agents Alfred Herbert Ltd, Coventry, England, under the trade name " Norbide."

Boron carbide (B_4C) is the hardest known material next to diamond, and although chiefly used for metallurgical work is also sold in grain sizes suitable for trueing and smoothing.

It is many times as expensive as carborundum or emery (£4 per lb for 320 grain size as against 1/5 per lb for the same size of Aloxite) but, if its efficiency in optical work is as great as its physical properties would lead one to expect, it may be very useful for working the harder kinds of crystal.

Grading of emery

71 The usual nomenclature for distinguishing the emeries of different grades is to name them 1 minute, 2 minutes, 40 minutes, 60 minutes, up to 240 minutes (although emeries of more than 120 minutes are very little used). These designations indicate the duration of the decantation by means of which they have been selected by the following process. The following passage is freely translated from Dévé (1936, pp. 34–36)—

> The process of grading is known as elutriation and is based on the time occupied by the grains in passing through a vessel of water 1 metre high and about 30 cm in diameter. The weight which causes the fall of a particular grain is proportional to its volume, that is to say to the cube of its dimensions. The force which resists that fall through the water is mainly a force of fluid friction proportional to the area, that is to say to the square of its dimensions. If then one doubles the linear dimensions the volume is multiplied by 8 and the surface by 4. If we imagine 8 little cubes of 1 mm side they will weigh as much as a single cube of the same material of 2 mm side. The surface of such a little cube will be 6 sq. mm whence the 8 small cubes will together have an area of 48 sq. mm while the surface of the large cube will be 24 sq. mm; thus when the large cube and the 8 little ones which weigh together as much as the large one are thrown into the water together, the force which restricts the fall of the little cubes is double that for the large cube which will therefore arrive at the bottom well before the little ones. The procedure, then, is that after the emeries are crushed the mixture of grains of various sizes of about 10 litres volume is put into the vessel, which is then filled with water. One then stirs up emeries of all kinds thoroughly from the bottom of the vessel with a forked tool. Water is then added to make it overflow and carry out all the floating impurities. One then lets the vessel rest for two hours, at the end of which time the water will only contain the finer grains of emery. The water with the floating emery is decanted by a tap half-way up the vessel or by a syphon, and the water thus drawn off together with the floating grains is emptied into another well-polished and very clean vessel where one lets it remain for several days. At the end of that time one throws away the water and dries the deposit, which is the 120 minute emery. The process of stirring, filling up and decanting is repeated after one hour, the resulting emery being then called 60 minute emery.

72 The method for washing of emery given in §71 is as recommended by Dévé; a somewhat different account is given in the little book by

Halle. According to the latter, a clean glass vessel, three or four times as high as it is in diameter, is used to shake up a quantity of emery about a sixth or an eighth of the volume of the vessel. It is then filled with water to the height of about a finger breadth from the top.

The mixture is stirred with a glass or iron rod (not too thick) drawing it in an upward direction. Through this process, stirring in various directions, the large grains are drawn temporarily to the top, where some of them are kept by the fatty foam or froth which forms. By slowly filling the vessel with water and stroking the patches of froth with the rod, they are caused to flow away over the sides. When the froth has been completely removed in this way, and the emery is uniformly distributed in the vessel, it is poured into a second carefully cleaned glass of approximately the same size, taking care, however, in the pouring that the largest grains at the bottom of the first vessel are not poured over with the rest.

With the first washing are obtained many sizes of grain which are purified and graded with repeated and slower washing, so that in the last pouring only the finest grains remain.

When the emery has settled to the bottom, the clear water on top is poured slowly away, the emery stirred once more, and finally poured into the vessel in which it is to be kept for use. The finer kinds are best kept in wide-mouthed glass vessels with ground stoppers ; the medium kinds in porcelain vessels with easily removable tops, while the larger grains can be kept in earthenware vessels, always kept closed. Carborundum and the freer cutting emeries—namely what is left after the first pouring—can be useful if they are carefully washed.

Other abrasives

73 Halle adds the following remarks on other abrasives—

For the final grinding of quartz (trueing and smoothing) the carborundum which has already been used is preferable if it is carefully washed.* Pumice powder is subject to the presence of harder grains which unfortunately cannot be got rid of by washing since both settle at the same rate. It is therefore recommended first to wash a small proportion and test it for quality of hardness by using it on a valueless piece of soft crystal on a ground glass plate. If fine scratches are produced on the crystal the material is unsuitable for use with soft crystals and metals.

* *Note by Author :* Messrs. Barr and Stroud use carborundum for all their glass grinding processes, including trueing and smoothing.

"*Graustein*" and "*Blaustein*"* are ground to powder in a mortar and washed in the same manner as emery. The finer the abrasive the more carefully must it be kept from pollution.

All these grinding materials must be kept in their vessels in a moist condition, with so much water in fact that a centimetre lies above the moist powder.† To get the grinding powder out one should use a slip of copper or brass about 2 mm thick with a spoon-shaped end, although for the medium and larger grains a strip of wood suffices.

It goes without saying that the vessels should all be numbered with the grade of abrasive contained therein and its nature.

Pumice powder is not suitable for grinding glass, and crystals of a like hardness, on account of its softness, but should be used for the softer crystals and metals after the finest emery. Two grades of the pumice are sufficient.

Graustein and *Blaustein* are still softer than pumice powder and find their use chiefly in the finest grade and for metals.

Tripoli powder can be used for glass surfaces in a dry condition. Various other polishing materials are available, such as tin oxide for the polishing of speculum metal, fluorspar, Iceland spar, gypsum, mica and other soft crystals. Diamantin yields on hard steel a deep black polish, particularly if it is used with a few drops of fine oil. Chromium oxide polishes fluorspar, arragonite and other crystals of like hardness, while ruby powder is good for the polishing of some crystals.

With a few exceptions, one should seek polishing materials of the purest kind from a chemical works. It is as important to wash polishing materials as abrasives.

Tripoli—A porous siliceous rock containing when pure about 98 per cent of silica and generally some alumina and oxide of iron, and known as tripoli, is found at a number of places in Illinois, Missouri, Oklahoma and Tennessee in the United States. It should

* Blaustein is Lazurite—$Na_4(NaS_3Al)$ $Al_2(SiO_4)_3$—which is the predominant constituent in Lapis Lazuli. Graustein is described in the dictionaries as " Greystone " or " volcanic rock " and is obviously a very general term in its ordinary usage, but I am informed by Dr. Walther Gerlach that among German opticians " Graustein " refers to a white grinding substance out of which tools 60 to 80 mm can be made. These have reticulations cut in them and the surface is then ground on them with pure water. This procedure is certainly 50 years old, although he tells me that in Germany it is no longer used. There can be no reasonable doubt that it is a very similar stone to the one known in this country as " Water of Ayr " stone. It was in use for smoothing metallic surfaces in the form of flat reticulated grinding tools when I came to Hilger's in 1898 and for many years afterwards. It is still occasionally used by us.

† *Note*—Some workers prefer to keep their abrasive dry, applying it in this condition to the wet tool. This must of course be done when smoothing materials soluble in water, when one uses oil as the moistening agent.

not be confused with tripolite, which is another name for diatomaceous earth.

The physical properties of tripoli differ greatly in accordance with the source. They may be freely divided into two qualities—
1. The Missouri-Oklahoma type.
2. The Illinois-Tennessee type.

The first is a very porous material either loosely coherent or compact, in either case easily crumbled to powder, consisting of particles 0·01 mm in diameter or finer, which will scratch steel, are doubly refracting and are probably chalcedony. Some beds are of such compact material that blocks of suitable size may be removed by special methods of working and used for filtering, which is one of its uses.

The second varies in structure from an entirely non-coherent powder of uniform grains of about 0·002 mm in diameter to dense material, cut with difficulty by a knife. This dense variety of tripoli is heavier, having a specific gravity of 2·2 to 2·5, and is much less absorbent than the Missouri variety. It grades imperceptibly into unaltered chert in places. None of the material is suitable for filtering.

Tripoli, diatomaceous earth, chalk and talc are among the relatively soft natural abrasives and will doubtless continue to be used, although they are encountering increasing competition from manufactured compounds.

At Hilger's we have occasionally used slate for smoothing. It is particularly useful for smoothing dense flint glass with a minimum danger of the marking to which such glass is prone in unskilled hands.

The Hilger classification of emeries

74 Until early in 1916 we were content to purchase our abrasives since they were found to be sufficiently well graded. Between November 1914 and July 1916, however, grave deteriorations began to take place in the purchased abrasives. Fig. 33 shows under a magnification of ×370 fine abrasives bought under the same designation and from the same supplier in November 1914 and July 1916 respectively; Fig. 34 exhibits a like comparison for an abrasive of finer grain. The progress of work in our optical shops was so much slowed up by this deterioration that the output was certainly halved. We therefore made a systematic study of the size of grains of various emeries, and decided to use the phrase " size of grain " as meaning the average diameter of grain in hundredths of a millimetre. It may seem hopeless at first sight to decide what the average size may be of grains which vary so much in shape and size in a single specimen. One finds, however, that an observer comes to much the same conclusion on examining a specimen under the microscope on different occasions, even with carborundum, the grains of which are usually long compared with their width. By size one understands here the diameter of a grain

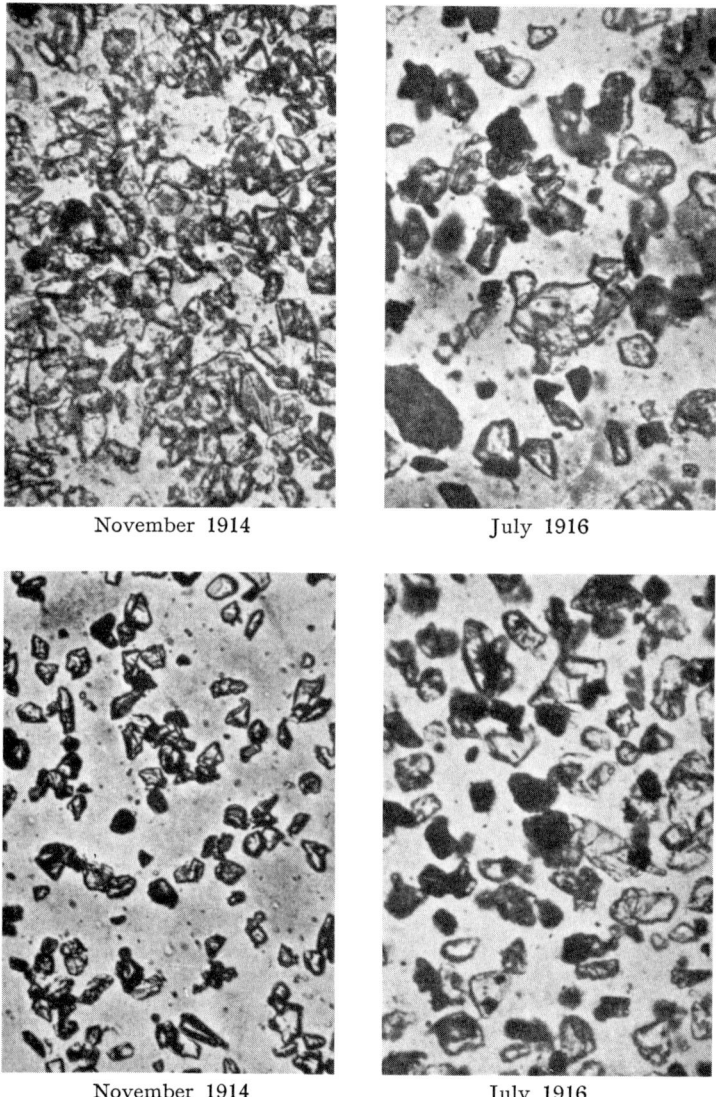

November 1914   July 1916

November 1914   July 1916

Fig. 33—Microphotographs of commercial abrasives in 1914 and 1916
Magnification × 370

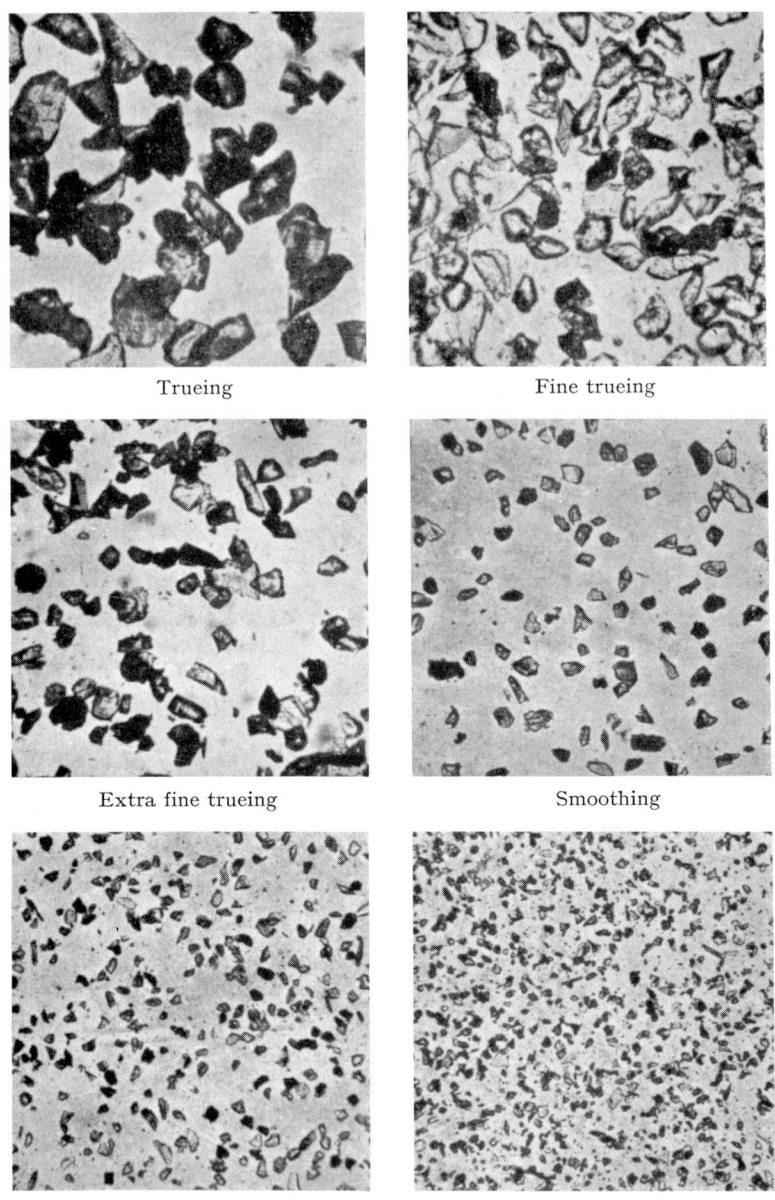

Trueing — Fine trueing

Extra fine trueing — Smoothing

Fine smoothing — Final smoothing

Fig. 34—Microphotographs of graded abrasives
Magnification × 100

if its bulk were reduced to a sphere. Even when the observation is made by different people the method of simple observation under the microscope enables us to class the fineness of given samples of emery in a way which meets with the agreement of the skilled workmen who use them. The next step was to define the size of grain, and a number of experiments showed that the following formed a complete and useful series—

Operation	Kind	Size of Grain ($\frac{1}{100}$ mm)
Roughing	Carborundum	26 to 40
Fine roughing	,,	11 ,, 20
Trueing	Corundum	10
Fine trueing	,,	5
Smoothing	,,	2·5
Fine smoothing	,,	1·25
Final smoothing	,,	0·6

A sufficiently comprehensive series would be provided by—

Roughing singly	Carborundum	40
Fine roughing singly	,,	20
Trueing singly	Corundum	10
Trueing, one emery in the block	,,	10
Smoothing in the block	,,	2·5
Fine smoothing in the block	,,	1·25

The following are the grain sizes (Hilger classification) of the Carborundum Co.'s emeries which we mostly use for smoothing—

No.	Grain size
320	3·7
500	2·5
600	1·0

A specimen of cerium oxide polishing powder had an average grain size of 0·2.

75 It is interesting to compare these with the figures given by Dévé (p. 82 of 1936 edition) for the average diameter of the " large " grains, as bought under the classification of " minutes " (§71).

From Dévé.			Specimens bought in London, 1942.
Emeri No. 1 (fin)	-	- 0·170 mm	
,, 3 minutes	-	- 0·112 ,,	
,, 5 ,,	-	- 0·068 ,,	
,, 10 ,,	-	- 0·048 ,,	
,, 20 ,,	-	- 0·042 ,,	0·030 mm
,, 30 ,,	-	- 0·038 ,,	0·018 ,,
,, 60 ,,	-	- 0·022 ,,	0·006 ,,

Quality of grading

76 The third factor, and a very important one, is the quality of grading, which we define in the following way. Let D_{max} be the diameter of the biggest grains (excepting monstrosities) and D_{av} that of the average grains, then imperfection of grading, or grading factor, is defined as—

$$10 \{(D_{max} - D_{av})/D_{av}\}$$

Here again the observer will at first be confused and undecided as to what he should consider a "monstrosity." By this is meant an exceptionally large grain present in such small numbers (a single specimen, for example) that it may be regarded as an accidental inclusion and not representative of the bulk. Obviously, if in a small sample more than one such occurs, the consignment is fitter for rejection than for grading. Perfectly graded material would therefore be marked 0 and less perfectly graded 1, 2, 3, and so on. Experience shows that an abrasive marked 4 or less is good; one marked 10 would be considered badly graded and is found to be gritty and scratchy in use.

Preparation of emery by Adam Hilger Ltd

77 Having found by about the middle of 1916 that it was impossible to buy good and well-graded emery from any of the various available sources, we built our own elutriating plant. We bought aloxite, ground it in a ball mill and found that the resulting mass contained grains of all useful sizes. This was elutriated in an apparatus similar in principle to what is used by geologists in grading soils, etc., and the resulting abrasives were of a better quality than we have ever been able to purchase before or since.

78 The elutriating apparatus was bulky and as soon as it became possible once more to purchase well-graded abrasives we scrapped the apparatus to use the space for more other urgent purposes, but we have often regretted since that we were forced to do so. Some of our workmen still treasure minute quantities of the finest of our aloxite abrasive of a grain size of about 0·8, and for special work prefer it to anything else which is available. It is worth while describing the apparatus here. Referring to Fig. 35, a number of metal tubular vessels made of thin sheet metal were arranged vertically in a row, they increased in diameter from about 2 inches up to about 8 inches. Water from the cistern (so that the pressure of water was always the same, therefore the flow of water constant) was allowed to flow through the vessels in succession. In every case the flow being from the bottom to the top. A charge of ungraded abrasive was put into the first vessel, that is the one of smallest diameter, the vessels connected in sequence and the tap to the water supply turned on. After some hours it was found that the grains had distributed themselves throughout the various cylinders naturally, so that the flow of the water was sufficient to maintain the grain of emery of a particular size without its either rising or falling. The lighter grains rose to the top and passed through the tube to the next cylinder, where those whose tendency to fall was just balanced by the slower upward flow of water

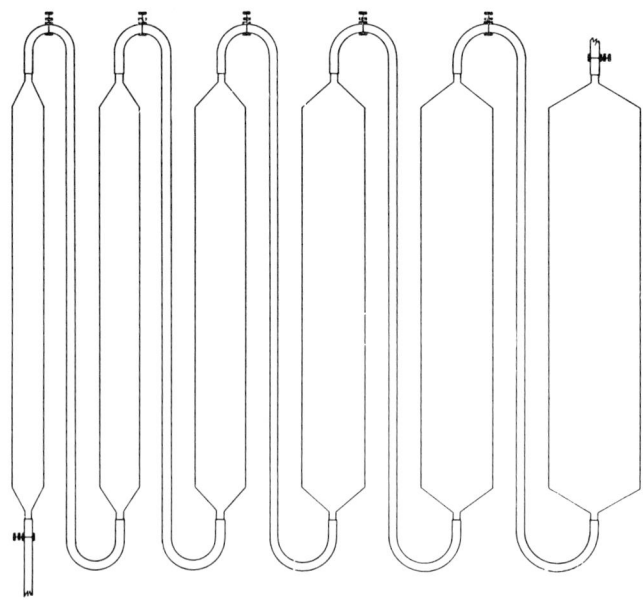

Fig. 35—Hilger elutriating equipment

through this larger cylinder remained, the still smaller particles passing on to the next. After a sufficient time had elapsed, which was found by experiment which did not need to be repeated, the tubes connecting cylinder to cylinder were disconnected and appropriate vessels placed below each of the cylinders and the water with its emery allowed to flow into these vessels. It was there allowed to settle, the water poured off and the remaining abrasive when dried was found to be extremely well graded.

The improved commercial abrasives

79 In 1923, however, a further form of Al_2O_3 (Sira abrasive) of great purity, hardness and uniformity of grading became available. It is sold by the United Kingdom Optical Co., Mill Hill, London. Owing to its sharp crystal edges, and to the fact that they maintain their quick-cutting properties, grinding with Sira abrasive is a rapid process. This abrasive is not so fine in grain as the finest of those graded by us, but on account of its uniformity and because of the shape of its grains, it gives a ground surface with shallow pits of even depth. Other excellent abrasives of all grades are described in §73.

80 An interesting table by G. Ritchey, cited by Dévé, p. 77, gives the pressure recommended as the best for the use of various emeries.

Pressure in grams per cm^2

Emery	Best pressure according to Ritchey	Note by Dévé
12 to 20 min	15 g	to 150 g for spectacle work
30 to 60 min	10 g	
120 to 140 min and pitch polishing	6 g for large tools. 9 g for small tools.	For surfaces of high precision
Paper Polishing	—	30 g (to 45 g)
Cloth Polishing	—	50 g (to 500 g for spectacle work)
Polishing on waxed cloth	—	25 g

It is a common mistake of opticians to use too large a charge of the finer grains in smoothing. If this is done the weight on the tool is shared by an unduly large number of grains and the pressure on the individual grains may be insufficient to cause the fracturing of the surface which constitutes the action of grinding (see §38).

The depth of grey caused by different grades of emery

81 It is desirable to know the depth of grey caused by various grades of emery and we have ascertained this in the following way. A plate of glass was greyed on one side with emeries of grain size 10, 5, 2·5 and 1·25 (see §74), the degree of working down the emery being in each case that which is used in applying those particular grades of emery. That is, each of them except the last two were worked down until the emery had begun to lose its cut, while the last two (which are used as smoothing emeries) were worked down fully so as to produce the fine smooth type of surface which polishes readily. The plate was then polished with a one-sided pressure so that while about half of it was left with the original grey the other half was polished at a slight angle to the grey, giving Newton's fringes about 5 or 6 to the inch when a proof plane was firmly held in contact with the grey half. In this way it was possible to see what was the depth of the grain by counting the number of Newton's rings from the point of full greyness to the point where the greyness disappeared. The following were the results—

Grade of emery	Depth of grey (inch)
Fine trueing, 5·0. Worked till "cut" reduced	0·00030 (a few deep pits were still left needing about 0·00020 to remove, making total depth of deepest pits 0·0005)
Smoothing, 2·5 worked right down	0·00011
Smoothing, 1·25 ,, ,,	0·00008

(*Note.*—The depth of grey does not include the under grey referred to in §37.)

Rayleigh (1901) found that a "very finely ground" surface was fully polished (except for very few small pits) by removing 0·00004 in. by polishing.

The emeries now obtainable are very well graded indeed, and the carborundum sufficiently so for the roughing purposes for which it is mostly used. The methods of manufacture and grading are described in §§70–2, 74–8, 84–7. The other grinding materials, however, (diamond, pumice, etc.) should be carefully graded by washing; the process is described in §§73, 118.

The powders appropriate for various materials other than glass, are mentioned in Chapter 5 (dioptric substances).

William Taylor (1920) has stated that experiments in the laboratory of Taylor Taylor & Hobson have shown that the layer of glass removed in polishing a lens surface with rouge is about six wavelengths in thickness.

Coarseness of grain

82 A good idea of the coarseness of grain on a glass surface may be obtained from the limiting angle at which regular reflection is obtained of a bright object. This depends on the wavelength: for example, with a particular trueing emery it was found that an image of a light emitting the green line of mercury 5461 was detected at all angles of incidence greater than the angle whose cosine is 0·077. For light of wavelength 4358 regular reflection was seen at cosine 0·063.

83 Occasionally in dry weather the rapid drying of the block during the smoothing process is annoying. In that case a modification of Plateau's solution may be used, made up as follows—

Fill a bottle $\frac{3}{4}$ full of distilled water, and add 1/40 by weight of sodium oleate. Shake well and allow to stand until dissolved. Then

106 PRISM AND LENS MAKING

fill up with pure glycerine, again shaking thoroughly. Allow to stand until the solution is clear (usually a week or ten days), then siphon off, avoiding the scum which rises to the surface, and add 3 drops of ammonia to each pint of solution.

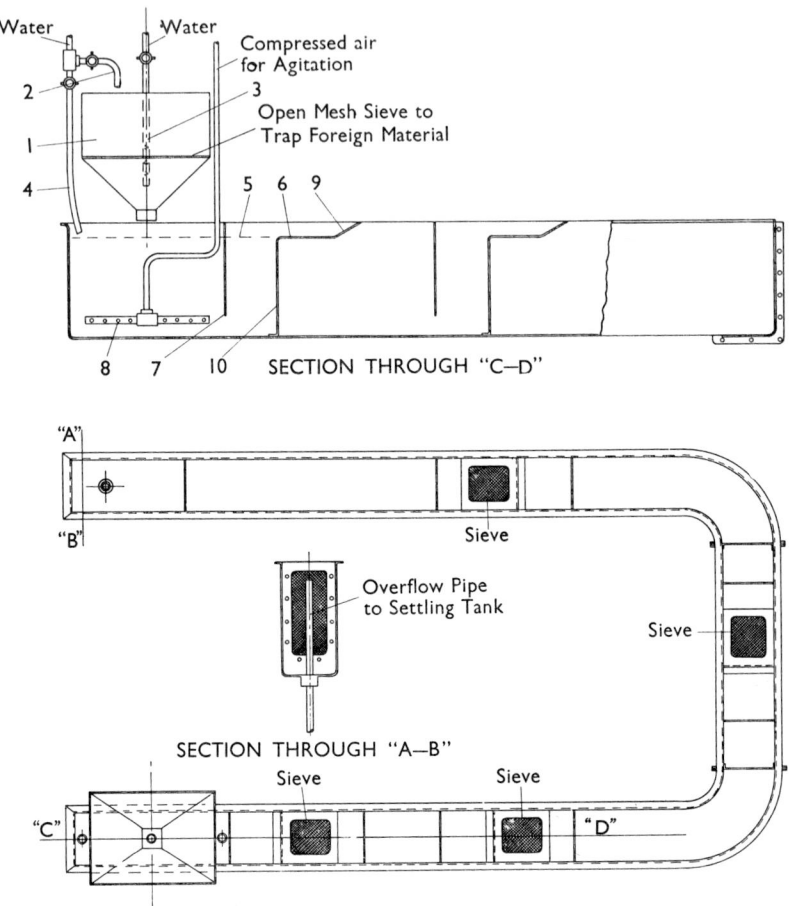

Fig. 36—Barr & Stroud recovery and grading tank for carborundum

Recovery tank for cleaning and grading carborundum

84 As has been stated in a footnote to §73, Messrs Barr & Stroud use carborundum for all optical abrasive purposes from roughing to smoothing.

I am indebted to Sir James French, Managing Director of Messrs Barr & Stroud for the following description of their Carborundum Recovery Tank—

Carborundum has a specific gravity not very different from the optical glasses customarily used, and it is therefore not easy to separate the two. It was found, however, that a mixture containing even as much as 20 per cent of glass seems to be actually an advantage, forming a soft bed which holds the carborundum. The effect of this can easily be felt if it is compared between two tools with pure carborundum.

After a rough breaking of the coarse carborundum (which is very cheap) it is taken to the tank.

The width of the tank which is shown in Fig. 36 is 12 in. and the water depth when empty about 2 ft. As the flow through the tank is not rapid, the side effect is small. The carborundum with floor sweepings is tipped into the hopper (1). To moisten the abrasive which is apt to cake, the water pipe (2) was provided. In addition, pipe (3) with a suitable tap passes down the centre into the throat. It is pierced with holes. If the downward movement of the consolidated carborundum is not good enough, a little central water can be admitted. The pipe (4) admits water to the tank. Its level (5) is controlled by the height of the overflow pipe shown in the section through A, B. This level (5) reaches just up to the level of the sieve such as (6).

The solid baffle (7) ensures that all material entering the first compartment becomes wetted. It can only proceed by passing under the baffle (7). At the bottom of the first compartment there is provided a compressed air pipe (8). The water and material in the first compartment can be maintained in a state of ebullition, thus keeping the carborundum in suspension. Between the baffle (7) and the sieve (6) there is a certain amount of wave motion produced. Each wave flows over (6) carrying a certain amount of water with the carborundum. From time to time the accumulated dirt that collects on the upper part of (6) can be brushed on to the sloping solid shelf (9). The wall (10) is, of course, solid. This description applies to the succeeding compartments with their sieves. As the water overlaps (6) and passes into the next compartment, the quantity being small, the sieve tends to regulate the currents. Although the length of the compartment seems small, the regularity of flow is amply sufficient. From the final compartment, as shown in the section through A, B, the water flows through the overflow pipe to a settling tank and thereafter to the second, and similarly the third stage, if required.

When the sediment in the bottom of the first compartment is about 1½ in. we stop the inflow of water and ladle out the water carefully until the bottom layer is exposed. Each compartment can be independently emptied.

All carborundum pots and vessels are sent to the Grading Department, where they are carefully washed, cleaned and refilled. This is done frequently and the service is, I think, much appreciated by the workers. All pots are, of course, glazed earthenware with covers.

The manufacture of carborundum and aloxite abrasives by The Carborundum Co. Ltd*

85 Carborundum (silicon carbide) and aloxite (fused alumina) are made in the works of the Carborundum Co. Ltd, at Niagara Falls, Canada, in the electric furnace, and are shipped to this country for making up. The two materials are complementary to each other, fused alumina being similar to emery and corundum in that its grinding efficiency increases with increasing tensile strength of the material to be ground, while carborundum or silicon carbide has exactly the reverse action, *i.e.* its efficiency becomes lower as the tensile strength of the material increases.

Preparation of the aloxite

86 The first crushing of the lumps of aloxite, which are for the most part of the order of size of cubes with 3 in. to 6 in. sides, is performed in a jaw crusher. The coarsly powdered material is then elevated to the top floor of the building by bucket elevators, and passes through two more sets of rollers which complete the pulverization. The entire product then passes over a pulley type magnetic separator, to remove tramp iron and some portion of the iron removed from the crushing machinery as well as of the iron derived from the original bauxite.

Grading

87 The material has next to be sifted. The first stage in this process is carried out by an eight-screen vibratory separator. The material is fed to a series of screens of successively finer mesh. The tailings from each screen are delivered to bins, and the fines which pass all the screens are passed over another similar separator with finer screens.

The screens are vibrated rapidly by a cam and at each stroke the material is thrown upwards and forward to the lower end of the

---

* Taken by permission from *The Industrial Chemist*, August, 1930.

screen, an action which corresponds as closely as possible to hand sifting and is extremely efficient, as it combines with that action the action of a conveyor.

These fines from the second set of screens are separated into their respective grades by cone washers. This simple and effective type of apparatus for the grading of finely-ground solids depends for its effect on the fact that different sized particles will settle out at different rates from a suspension in water. The material mixed with water is fed into the first of a series of inverted conical vessels from one side at the top, while water is admitted from the bottom at a carefully regulated rate. The fine particles overflow to the next cone through an opening near the top, while the coarser go to the bottom and are withdrawn at intervals through an opening provided for the purpose. At the bottom the rate of flow is such that the larger particles are kept in thorough agitation, while the decreased velocity at the top leaves them to fall almost undisturbed.

The muddy sediment of graded material thus produced is dried in copper bins over steam coils and passes over a six-decker screen to remove particles of dirt, etc.

The grading of the carborundum is carried out in a different way, which is fully described in the reference cited.

Polishing materials

88 There are three materials in general use for polishing optical work ; these are rouge, cerium oxide and putty powder. Others of occasional use are mentioned in §§73, 92, 118–9·2.

Rouge is a red oxide of iron (ferric oxide), Fe_2O_3, usually prepared in a fine state of division by calcination of ferrous sulphate. The best rouge for lens polishing should have a good red colour and should contain little or no free sulphate. Excess of sulphate is detrimental to good polishing as it causes the powder to aggregate in balls and thus gives rise to sleeks in the polished surfaces. It is also likely to cause enamelling, which is sometimes difficult to remove in the final stages of polishing. A light-red coloured, very fine-grained rouge generally implies that the calcination has not been complete or that the temperature of heating is too low. Such a rouge will usually have a high sulphate content. On the other hand, if the rouge has a very dark brown or purplish colour it indicates that the temperature of calcination has been too high, and a rouge of this colour will be too hard and gritty for successful lens polishing. (A rouge manufactured by Messrs Hopkins & Williams Ltd to the specification of the British Scientific Instrument Research Association and called " Sira " rouge has been specially prepared for the best optical work. It is a fine, uniformly grained powder and is free from any trace of sulphate.) To get the rouge in the form of a water paste in which it is suitable for use, it is mixed in a jar with about three times its bulk of water, allowed to stand for twenty seconds and the top poured off into a second clean jar. The grit and coarse particles, if any, will remain behind in the first jar. When all the rouge has settled in this second jar, the surplus water is carefully poured off. The remaining rouge paste is then ready for use and will remain so as long as it is not allowed to dry. This is suitable for pitch polishers. For cloth polishers a little fuller's earth (one-twelfth part) may be added to the rouge. Dévé states that extra-fine polishing rouge has a grain of 0·003 mm to 0·006 mm, while Tripoli powder for polishing on paper gives 0·002 mm to 0·003 mm.

White rouge

89 Ophthalmic opticians, both in the United States and here, have for many years been seeking for a white polishing material which would be as effective as red rouge. The matter is of very real importance in the ophthalmic industry, for in the United States there is actually trouble in renting premises from landlords for prescription work, owing to the very pervasive character of red rouge. Indeed, on many grounds, a white polishing material would be much appreciated by all optical glass workers.

A material known as *white rouge* has been on the market for a long while and has been carefully tried, but one of the foremost ophthalmic opticians in this country informs me that he tried it as long as 15 years ago and rejected it as very inferior in polishing qualities to red rouge.

Cerium oxide

90 During the recent war cerium oxide was introduced as a polishing powder on a commercial scale. It polishes faster than rouge, and its use effects a marked reduction in the soiling of the workman's hands, clothes and surroundings ; it is, however, more expensive.

The following information has been collected from the experience of other firms concerning cerium oxide—

1 It should not be used on a polisher already impregnated with rouge, owing to its smaller grain size.

2 Very little at a time should be used ; one application of fresh oxide is usually sufficient for a block, clean water being added when necessary.

3 It is less liable than rouge to cause sleeks and staining unless it is allowed to run dry.

4 Reports on the reduction of polishing time vary from 20 to 50 per cent.

5 A more perfect polish is said by some to be obtained with rouge.

Owing to the export of monasite sand being prohibited by the Indian Government supplies of cerium oxide failed in 1949. A substitute, called ceri-rouge G, made from zirconia, can be obtained, but is so far not generally found satisfactory.

When the observed image in an optical instrument is focused on one of the polished surfaces it is customary in Germany to use carefully graded thorium oxide for polishing that surface. It is claimed to be a fast polisher and above all cleaner than any other known abrasive (U.S.N. 1945).

91 It seems unlikely on present evidence that cerium oxide will displace rouge in the mass-production of low-cost optics ; in precision polishing, however, where labour costs are mostly far more important than working materials, the saving in polishing time becomes a major item, and this consideration alone may result in the optical workshops of precision instrument factories adopting the faster polishing material, although the opinion is held in some circles that a more perfect polish is obtained with a rouge than with either cerium or thorium oxides.

Polishing materials for substances other than glass

92 A carefully washed diamond powder sold in the U.S.A. under the name of " No. 10 Super High Polish ' Star Dust ' " is stated to have been used for lenses with a saving of 33 per cent of the time.

Putty powder is in fairly extensive use, and is of special utility when polishing soft materials such as Iceland spar and the synthetic halides which are assuming such importance in infra-red research instruments. It is also of service when using polishers of the wax class.

Rottenstone* is a light earthy-looking mass resulting from the decay of siliceous limestones, from which the lime has been removed by leaching. It is used as a polishing agent for various metals, celluloid and plate glass, and for the finishing of the surfaces of furniture and musical instruments, including pianos.

In England true rottenstone is found near Hull in the Yoredale rocks of Yorkshire, as well as in Derbyshire and South Wales. Belgium is also a producer.

Further notes on polishing and polishing powders are in §§33, 73, and 118–9·2.

Note—In Great Britain the use of putty powder (tin oxide) and chromium oxide in factories is permitted only if the Home Office regulations of exhaust ventilation, protective clothing and washing facilities are carried out.

* *The Mineral Industry of the British Empire and Foreign Countries* : *Abrasives* : 1929. Printed and published for the Imperial Institute by H.M. Stationery Office.

Cleaning liquids

93 The following cleaning liquids are found useful in the optical workshop, in addition, of course, to water for substances soluble therein, such as glue and the so-called " gums " sold as adhesives.

For cleaning off shellac, Canada balsam, de Khotinski cement ("Coates"), black wax, sealing wax, and many varnishes (dull black and others)—
> Methylated spirit
> Alcohol.

For pitch, rosin, beeswax, tallow and mixtures of the same, and some dull blacks—
> Benzene
> Turpentine
> Turpsad
> Petrol
> Paraffin.

For polymerized H.T. cement—
> Acetone

For metals—
> Nitric acid
> Aqua regia
> Strong sulphuric acid with chromic acid added
> Potassium cyanide (highly poisonous).

The addition of a drop of nitric acid will often assist cleaning with other liquids. For example, the optical contacting of surfaces is difficult even when the surfaces appear quite clean ; it may be successfully accomplished, however, if the surfaces are lightly wiped with cotton wool dipped in dilute nitric acid or ammonia, and then washed off with water, finishing with ammonia, wiping the latter off with an old linen handkerchief which has been boiled in distilled water.

Moisture condensed from breath is a useful cleaning medium, and distilled water is better than tap water.

Any rag or linen used for cleaning must itself be scrupulously clean in order to avoid depositing on the glass surfaces any substances dissolved by the cleaning fluids from the cloth. Excessive rubbing should be avoided as the electric charge produced tends to collect fluff on the surface of the glass.

Carbon tetrachloride is a volatile, non-inflammable and very powerful solvent for waxes (including paraffin wax), but a good ventilation for carrying away the vapour is obligatory (Section 47 of the Factory Act). Carbon tetrachloride is rather less toxic than chloroform. The use of tetrachlorethane, also a powerful solvent, is now considerably restricted owing to its strong narcotic and poisonous action, which causes jaundice, fatty degeneration of the organs, albuminuria and haemoglobinuria.

For cleaning dense barium crown, and indeed for cleaning glass generally where it is desired to remove every trace of grease or other foreign substance, the following has been found very effective. Treat for 30 minutes with a mixture of caustic soda 4 oz, water 10 fluid oz, methylated spirit 5 fluid oz. Then rinse off, and soak a further hour in water 10 fluid oz, ammonium di-chromate 2 oz, commercial HCl (conc) 2 fluid oz. After this treatment, followed by rinsing in tap water, the glass will be found perfectly clean. It is scarcely necessary to say that one must only touch glass which is cleaned in this way with something that is itself clean. Tweezers of brass are useful for handling smallish pieces.

The above cleaning method was the only one that was found suitable when preparing for the deposition of lead sulphide.

Stains on glass

Certain glasses are peculiarly sensitive to stains which may occur sometimes during the polishing process, and more usually in cleaning or in lying about in the open air. Dense barium crown is particularly sensitive in this respect.

If one leaves a polished piece of dense barium crown overnight in a glass jar which has once contained ammonia (although the jar may have been rinsed out) it will have a surface sheen on it the next morning.

Wrapping paper for avoiding tarnish

Some glasses, e.g. dense flints, and most metals are very liable to tarnish, particularly in the neighbourhood of substances containing sulphur, such as rubber tubing. An excellent wrapping paper, the use of which will entirely prevent tarnish for many months is the " Glass bleached tissue paper " used in the Cutlery trade for wrapping silver or silver plated articles. I was unable to obtain this when I last tried in 1946, but so useful a material will doubtless come on the market again.

Cements, plasters and varnishes

94 There is a good article on Laboratory Cements and Waxes by L. Walden (1936) *J.S.I.*, **13**, 345. *Modern Laboratory Practice* (Strong, 1940), p. 555 and *The Industrial Chemist,* Jan. 1943, p. 23 also contain useful information on the subject.

The cements which have been recommended throughout this book for various purposes are collected here together with some others which have also been found useful by Hilger & Watts.

1. Khotinski, a very strong and not brittle cement, useful for cementing plates on the end of tubes, is no longer obtainable. The best substitute is (2) below.

2. Cenco Seal stick (U.S.A.) is obtainable from Messrs Edwards, Kangley Bridge Road, Sydenham, London.

3. Edward's No. 3 is also a fair substitute. These three cements are liable to cause strain and, where this is to be avoided, rosin 4 parts, beeswax 1 part may be used, although it is less strong.

4. Shellac dissolved in methylated spirit makes a good varnish for protecting optical surfaces of parts of which other surfaces still need working. Parapan black semi-glossy, XE. 1319 paint, however, is better, and may be dissolved off by xylol.

5. Paraffin wax is of use for the purpose mentioned in Chapter 9. When melted it is very limpid, and will run between surfaces which are in contact and bind them together with very little alteration of the angle between them.

6. Plaster of Paris 2 parts thoroughly mixed with 1 part of hydrated lime is now used for making plaster blocks of prisms (§181), instead of the 4 parts Portland cement to 1 part Plaster of Paris recommended in the 1st Edition of this book. The latter is difficult to break away from the prisms when used for blocking.

7. *For joining the plates of cells.* Cells needed in large quantities now have the parts joined by heat, but a useful substitute, convenient for occasional use, is made by mixing equal parts of Portland cement and fine glass powder with sodium silicate (Tw. 140)* to a smooth paste. A thin coat is put on the edges to be cemented, which are applied to the surface to which they are to be joined, and the surplus paste wiped off. To set, stove at 120°C for three hours.

8. *For joining plates on to tubes.* Where a non-melting cement is needed No. 7 may be used, although in the Hilger workshop the silica bonded cement made by B.S.I.R.A. to their own formula is preferred.

9. The C3 cement developed by the British Scientific Instrument Research Laboratory (replacing shellac—Venetian turpentine cement), is a hard and tenacious cement, strong enough to cement a stack of plates *on a chuck for edging.* The chuck is heated till the cement, when applied, coats it thinly. The stack is then pressed firmly home on the chuck and adjusted central while still warm.†

---

* *i.e.* measuring 140 on the Twaddel hydrometer.

† This cement was originally prepared from shellac and Venetian turpentine, but owing to the wide variations in different samples of the latter and its scarcity during the war years, the formula was modified to that of C3, which is Angelo—Gaunna Shellac 45 parts by weight, Canada Balsam 55 parts by weight.

An alternative to this cement is the Montan wax mixture described in §177.

10. Beeswax 1 part and Rosin 4 parts is a medium hard cement, perhaps the most useful of all for optical purposes.

11. Hard Black Wax of low melting point—

Beeswax	40 parts
Paraffin wax	40 ,,
Hard balsam	16 ,,
Rosin	80 ,,
Vegetable black	10 ,,

An example of the use of this is the fixing in position of a prism in a recessed mount. The prism is set to the correct angular position and the melted wax is poured into the recess. The low melting point of the wax permits it to be poured thus without risk of breaking the prism.

12. Black wax for packing plates which have very delicate surfaces. This was used for packing Schumann plates for the ultra-violet, the sensitive surfaces of which can be damaged by the friction of soft wrapping paper. Very small pellets of the wax are put on the plate near each corner, so that the sensitive surface is not in contact with anything—

Paraffin wax	40 parts
Hard balsam	16 ,,
Beeswax	40 ,,
Vegetable black	5 ,,
Olive oil	10 ,,

13. Red Laboratory wax. I have not found any quite satisfactory substitute for the red wax which was obtainable from Germany prior to 1914 and known as "siegel Lac" (not "siegel wachs," which is sealing wax). The German wax, while pliable, had the property of holding an object indefinitely in any position into which one pushed it. A near substitute was made up at Hilger's from the following ingredients—

Castor oil	10 parts
Hard balsam	32·5 ,,
Paraffin wax	30·5 ,,
Beeswax	74·5 ,,
Lead oxide	5 ,,
Vermilion	10 ,,
Iodoform	0·5 ,,

The multitude of ingredients, and the doubtful utility of some of them, remind one of the cabalistic potions of Mediaeval times; but it is a good wax.

Other substitutes are " Sira " adhesive wax supplied by British Drug Houses.

14. " Araldite," a synthetic resin adhesive for bonding metals, glass, porcelain, mica, quartz and various other materials, has been found very useful : it sets under application of heat alone ; no pressure is necessary and no volatile substances are evolved during the setting process.

As a surface-coating resin it displays a remarkable degree of adhesion to metals and good resistance to chemical attack and abrasion, although it will not stand up to concentrated nitric acid for prolonged periods, and is of doubtful reliability with chloroform. It has been used widely for the coating of copper wire ; flexing and similar treatments do not impair its outstanding insulating properties.

" Araldite " was developed by Messrs Ciba Limited, Basle, and is marketed by Aero Research Limited, Duxford, Cambridge.

15. *Non-reflecting Black Varnish.* Berger's dead flat non-reflecting black is a very useful dead-black for metal. It is almost absolutely non-reflecting, although to retain this property fully it should not be rubbed. This black can be obtained from Lewis Berger & Sons Ltd., 12 Dartmouth Street, London, S.W.1.

Varnishes or paints which will reduce the interior reflection on ground glass surfaces are sometimes useful. Such reflections inside prisms can sometimes be objectionable. Two which have been found satisfactory are—

16. " Ebonide " manufactured by Messrs W. Canning & Company,
17. " Paint M " made up by B.S.I.R.A.

Both the above are air drying. Ebonide is of a nitro-cellulose type while paint M is basically a bituminous material.

18. Montan wax (see p. 302).
19. Keene's cement (see p. 311).

Dermatitis due to cleaning liquids

94.1 At least 350 substances common in industry involve a risk to the skin. It is not surprising therefore that dermatitis occasionally occurs as a result of using various cleaning liquids. Turpentine and turpentine substitutes have on one occasion at least caused quite a disturbing number of cases at the same time. The dermatitis was overcome within a week simply by washing the hands and arms in warm water prior to starting work and rubbing in, lightly, an ointment consisting of one part lanolin to two parts castor oil ; the procedure was repeated at the end of the day's work.

Should any cases of dermatitis arise it would probably be found worth while introducing into the workshop hygiene one of the barrier creams which have been designed to prevent various types of industrial dermatitis. Incidentally they are usually much appreciated by the female workers as face creams. Full particulars may be obtained from the Innoxa Laboratories, Balls Pond Road, London, N.1.

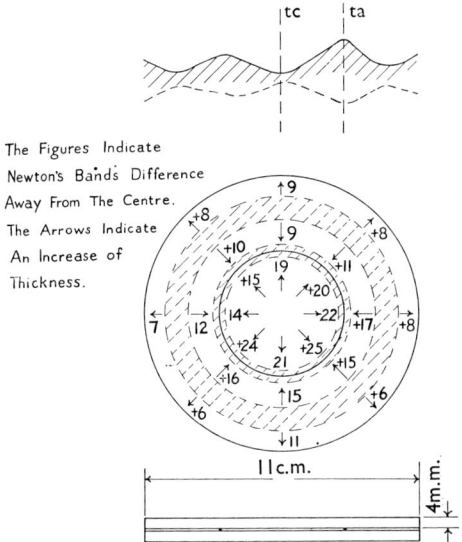

Fig. 37—Distortion caused by the setting of a cement

Experiments on the shrinkage of cements

One should remember that cements alter in shape on setting, for example : beeswax 5 parts, rosin 34 parts by weight (the customary beeswax and rosin cement of the workshops) shrinks considerably on setting, this was demonstrated in some experiments the ultimate intention of which was to find a convenient way of making certain non-spherical surfaces. Two discs with a circular piece of wire between them were cemented together with the beeswax rosin and allowed to cool down, the shapes to which the discs were bent are shown in Fig. 37 (the figures indicating bands and the arrows a rise in the height of the surface). It was found that the bending of the glass caused by beeswax and rosin cement was roughly proportional to the thickness of wire used.

The plates distorted as in the figure were worked flat uncemented and reversed and the procedure repeated ; under these circumstances there was a difference in thickness between t_a and t_c of about 50 bands. The thickness of the wire was $\frac{1}{2}$ mm. Using a considerably thicker disc as the base plate twice this difference was obtained.

CHAPTER 5

DIOPTRIC SUBSTANCES

Introduction

95 By a dioptric substance I mean a transparent material through which rays of light pass in straight lines.* Although for certain purposes coloured materials may be used, yet in the majority of cases the glass, or whatever the substance may be, should be colourless, clear, free from bubbles or opaque pieces, and free from any defect which will cause the deviation of rays of light passing through it.

At the time of Newton the optician required no more of the glass than that it should conform to the above-mentioned qualities, and although two main types were known (crown and flint) neither Newton nor any opticians made use of their different properties. Glass as made in different countries and for different purposes varied in chemical composition ; for example, window glass and mirror glasses would be soda glasses, Venetian and Bohemian were traditional potash glasses, glass from which the domestic glassware known as " crystal " was made owed its brilliancy to the presence of lead, but there was no preference for one type of glass over another for *optical* reasons, nor any use made of their various properties before the discovery, in the middle of the 18th Century, that lenses could be made free from " colour " by the use of two different kinds of glass (achromatic object glasses ; see below). The differing properties of crown and flint glasses which made them suitable for combination in this way were their *refractive indices* and their *dispersive powers*.

In the middle of the 18th century, flint glass, or " English crystal " as it was once called, was principally made in England. In making this, sand was replaced by powdered flint—hence the name. The denomination " crown " glass had its origin in the primitive method of its fabrication. With the aid of a tube about two yards long, about twenty pounds was taken from the furnace and blown into a bulb (Fig. 38). The tube was rotated to flatten the bulb, and then, by means of hot glass applied to the latter, an iron rod was attached to it. The tube being removed, the glass was put in the furnace again and a constant rotation given to it by means of the iron handle, whereby the centrifugal force caused the flattened sphere to open and finally to take the form of a crown, ∩, which was gradually

* It is not impossible that, to gain the effect of non-spherical surfaces, heterogeneity of a glass may be deliberately introduced by suitable heat-treatment, in which case some rays will deviate from straightness within the glass.

118

Fig. 38—Making a disc of crown glass in 1750
(*Reproduced by permission of Messrs Chance Bros*)

extended more and more to form a round disc. It was from the supposed resemblance of the form ⌒ to a crown that the glass gained the name of " crown " glass.

An excellent description of the method of manufacturing optical glass, together with much information concerning its faults and testing, will be found in Wright (1921). The plant and processes used in the manufacture of glass table-ware, bottles, tubing, rod, and sheet, are described in Angus-Butterworth (1948).

Chromatic aberration of lenses and its correction

96 Before considering the different properties of glass, and other glass-like substances, it is necessary, or at least desirable, to give those who are not acquainted with even the elements of lens calculations, an idea of the properties which are required to be known for such calculations.

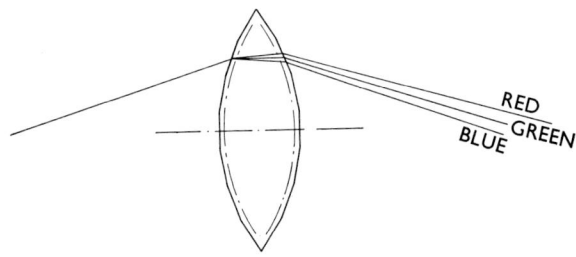

Fig. 39—A simple lens represented as an assembly of small prisms
The diagram explains the origin of chromatic aberration.

If we consider a simple lens, such as that shown in Fig. 39, we see that the lens may be looked upon as being made up of a large number of truncated prisms. We see further, that a beam of light passing through any one of these prisms is deviated, and owing to what is known as the " dispersion " of the glass, the violet rays are deviated more than the red ones ; the blue, green, and orange rays occupying intermediate positions.

Such behaviour causes the rays of different colours to come to a focus at different distances from the lens. One will thus get either a reddish image with a blue surround, or a blue image with a ruddy surround according to the focusing of the telescope, and in either case the definition will be defective.

Sir Isaac Newton, deeming this defect unavoidable in refracting instruments, proceeded to develop the reflecting telescope.

It was left to a gentleman of Essex, Chester Moor Hall, in the early eighteenth century, to show that the chromatic aberration, as it is called, could be corrected by making use of two different kinds of glass. This idea was perfected in 1758 by the English optician, John Dollond, who is frequently credited with the original discovery of the technique.

It may be interesting to add that Hall based his discovery on the erroneous opinion that the eye is achromatic, that his first instrument

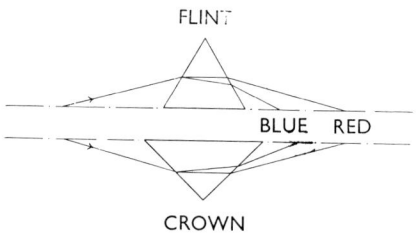

Fig. 40—Deviation and dispersion
The two prisms have equal deviation for the red, but unequal dispersion.

was made as early as 1733, and that telescopes of his were certainly in use as late as 1827. (*Enc. Brit.*, 11th Ed., article " Telescope.")

Optical glass may still be generally classified into two families, crown and flint. Crown glass has a dispersion considerably less than that of flint glass. If two prisms are taken, one crown and one flint, such that the *deviations* for the red light are the same, then the

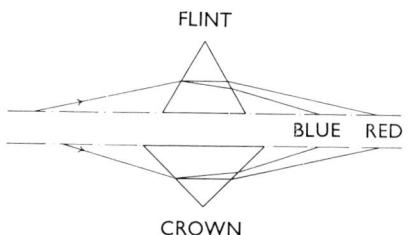

Fig. 41—Deviation and dispersion
The two prisms have equal dispersion, but unequal deviation.

dispersions will differ, as shown in Fig. 40. Alternatively, if the flint prism is made weaker so that the *dispersions* are the same, then the deviations will be different (Fig. 41). It will be easily understood that by putting together two prisms of equal dispersion in opposition, the dispersions can be neutralized while yet leaving a certain amount of deviation.

Reverting to the consideration of a simple lens, it now becomes clear that an assembly of compound prisms, each of zero dispersion,

E

would constitute a lens with zero dispersion also. A lens so formed is called achromatic, as the image it gives is, to a fair approximation, colourless.

Now the properties which will tell a lens-designer, at a single glance, whether two glasses are suitable for making a simple achromatic lens are n, the refractive index, and ν (the Greek letter pronounced "*nu*"), which is defined as the reciprocal of the dispersive power.

The optically important properties of dioptric substances

Refractive index

97 It was well known prior to 1617 that rays of light on passing to a different medium, say from air to water or from water to air, were deviated, but the exact law governing the degree of this refraction only became available to the world after its discovery by Snell shortly before 1617. It was first published and, indeed, first utilized by Descartes after the death of Snell.

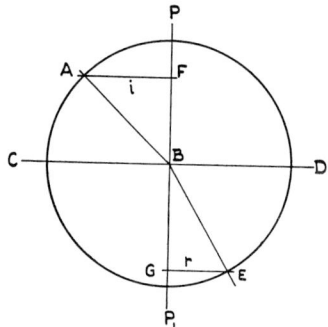

Fig. 42—Snell's law

Snell's law underlies the design of all optical instruments in which prisms or lenses are concerned. Referring to Fig. 42, if a ray of light AB falls on the surface CD of a piece of glass, or other transparent substance, it changes its direction to the course BE. Draw PP_1 perpendicular to the surface and passing through B and call i the angle ABP and r the angle P_1BE. Snell's law says that $\sin i/\sin r$ is a constant, no matter at what angle the ray of light falls on the piece of glass. This constant is called the refractive index of the glass and is usually denoted by the letter n. The refractive index varies for different rays of the spectrum and it is usual to state the refractive index of a glass for the yellow sodium lines (present as dark lines in the solar spectrum) and to indicate it as n_D. There are two of these lines close together, known as D_1 and D_2, but the wavelength taken to define the refractive index is the mean between the two, and is represented by D.*

The lens-computor requires to know not only n_D but what is known as the " dispersive power." In glass catalogues it is customary to give the inverse of this, and it is this which is usually designated by the Greek letter ν referred to above.

Spherical and chromatic aberrations

98 Everyone knows that a lens with convex surfaces will form an image of a distant object. A focus so formed, even with monochromatic light, can never be of good definition since, in a positive

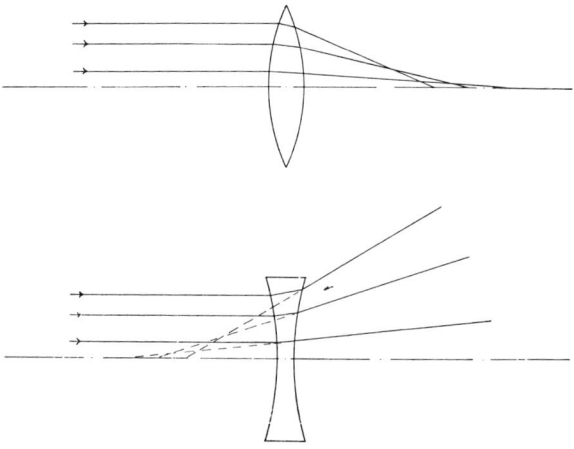

Fig. 43—Spherical aberration

lens, the rays which pass through the marginal part are brought to a shorter focus than those that go through the middle part of the lens, known as the paraxial rays, while in a negative lens the same is true of the virtual image (Fig. 43). This failure of a single lens with spherical surfaces to bring the rays from a point to a point focus is known as *spherical aberration*. As we have seen, there is another type of aberration, the *chromatic aberration*; since the refractive index of every dioptric substance is different for different colours, the rays which pass through the marginal part of a lens will not all be brought to a focus at the same point (Fig. 39), the blue rays being concentrated nearer to the lens than the red ones. Spherical as well

* The letter D was first used to distinguish this doublet by Fraunhofer. He had observed the lines in the spectrum of the sun.

as chromatic aberration can be corrected by combining a positive crown lens with a negative flint one.

The principle, first used to correct the *chromatic aberration* by Moor Hall and Dollond, is simple and has been adumbrated above. Since flint glass has a considerably greater dispersion than crown, it is possible by putting a crown and a flint lens together to annul the chromatic aberration while at the same time leaving the lens with sufficient refracting power to bring rays from a distant point to a focus. This invention was the starting point of the deliberate efforts to make glasses with different properties with a view to making more perfect lenses.

The two most important properties of a glass are, then, the refractive index and the ν value, where

$$\nu = \frac{n_D - 1}{n_F - n_C}$$

in which expression n_D, n_F and n_C are the refractive indices for D, F and C—three lines in the solar spectrum in the yellow, bluish green, and red respectively. The refractive index gives an indication of the refracting power of a material, while ν is a measure of the dispersing properties of the glass; a large value indicating a small dispersion.

If the ν values of the crown and flint glasses are nearly the same, the curves of the lens will be extremely deep, and this is a disadvantage not only in manufacture (since deep curves are more expensive to make than shallow ones), but also because deep curves can introduce other types of aberrations.

The first aim, therefore (in making simple achromatic lenses), is to choose ν values as different as possible in the two glasses.

By combining the methods of correcting these two types of aberration, it is possible to correct the chromatic and spherical aberrations simultaneously and tables are available giving data for such lenses. Reference to such tables will be found in Chapter 16.

Such was a state of the art when a further series of forward steps was initiated by Schott and Abbe about 1880. These and subsequent developments are described in §107 by R. Kingslake and P. F. DePaolis.

We must, however, first consider other important properties of dioptric substances.

Absorption

99 The colour of a transparent substance is due to the absorption of some of the visible rays, which for practical purposes may be considered as extending from 4000A in the violet to 9000A in the red, although under very special circumstances it is possible to see radiations of considerably shorter wavelengths than 4000A. The transparency of ordinary glass extends only a little beyond this range both in the longer and shorter wavelengths,* and since the ultra-violet and infra-red regions of the spectrum are of great importance nowadays, the former for the analysis of metals and alloys and the latter for the analysis and estimation of organic substances, the optician must know how to work those substances which are transparent to these regions.

One of these, Iceland spar, has an importance of a character different from that of the others. Although it is transparent a good way into the ultra-violet, its importance in optical instruments depends not so much on that as on its having been the first known means of producing fully polarized light. It is true that nowadays we have "polaroid" available in large sheets, but the perfection of polarization obtained with this material, although it has improved greatly in recent years, still falls short of that which can be obtained with Iceland spar, which therefore remains the only practical polarizing material for the manufacture of high-grade polarimeters.

Most of the other substances which are more transparent than glass in the ultra-violet and the infra-red were also originally obtained from natural sources, although they can now be produced in both larger and more perfect specimens in the laboratory.

Sections 116 and 117 give an account of some of the principal crystals which are optically useful, their regions of transparency, and the methods of synthetic preparation.

Durability

100 The following is abstracted, with permission, from Hampton (1942)—

Durability is a measure of the resistance of a polished surface of glass to atmospheric attack. The varying humidity of the atmosphere results in corresponding variations in the water content of the glass surface, accompanied by alternations of deposition and evaporation of moisture. These alternations result in the separation of soluble constituents from the glass and the production of a surface film. In obtaining desired optical properties, either by developing a new type of glass or modifying an existing one, the final choice of batch should be decided by the durability. Want of durability appears in three ways—

(1) The formation in course of time of a milky film on the polished surfaces. This is particularly liable to occur on enclosed surfaces such as those in the interior of binoculars. Even the slightest milkiness is very detrimental to visibility while the effect may sometimes be so bad as to make the instrument entirely useless. Although some optical glasses still tend to develop this milky film, ordinary glasses

* The transparency-wavelength curves for a number of glasses are given in Fig. 228.

from a reputable maker may be relied on. In America the milky effect is known as " dimming ".

(2) Fungoid growth sometimes appears in old telescopes as fine branching threads on the surface of the lens. The fungus doubtless lives on solutions of the more soluble constituents of the glass in water deposited on the surface by a change of temperature. It frequently causes deterioration of glass surfaces in instruments used in the tropics ; high humidity and condensation can result in etching and misting and these conditions encourage the growth of fungus, most species of which grow fastest when the relative humidity is above 90%. Manufacturers have tried various remedies, but without complete success. I suggest, as worth a trial, a device traditional in France for preventing the growth of mould on bottled fruit. Just before capping the bottle a pellet of burning sulphur is lowered on a spatula into the container for a second or two, the cap then being applied immediately. I have found this effective for several years in preventing the growth of mould on home-made jam.

(3) An iridescent film, usually blue, sometimes occurs in glasses containing lead. Unless very severe this does not impede transmission ; indeed Dennis Taylor has shown that it may decrease the reflection of the surface and increase transmission. This observation led to the technique, called " staining " in America, for reducing such reflection (see Chapter 13).

It is important that the manufacturer should be able to determine fairly rapidly the effect on durability of variations in batch compositions, and that the user should have some indication of the probable performance under service conditions. With long-established types of glass the effect of long exposure to atmospheric conditions is fairly well known, but there is no similar guide in forecasting what a new glass will do. Several laboratory methods have been put forward for determining the durability, but such accelerated attacks as a rule differ markedly from those to which the glass is exposed in use ; results often do not agree, and in the few cases where it has been possible to compare them with results under severe natural conditions there appears to be little correlation. Particulars of tests employed, and discussed in the literature, are given in Dr Hampton's paper from which this section is taken.

Some recent work on the durability of optical glass has been published in America (Jones, 1941). Observations were made on the behaviour of polished glass surfaces exposed to ordinary service conditions, and two types of surface change due to weathering were noted. Glasses subjected to humid atmosphere became covered with a hazy film of soluble alkaline salts—" dimming ". On the other hand, contact with condensed moisture led to the separation of lead or barium oxides

and the formation of a silica-rich film exhibiting interference colours—
" staining ". Various accelerated durability tests were investigated
and it was concluded that two tests were required, one to determine
the tendency to dimming and the other to determine the tendency to
staining. The dimming test involved exposure to a saturated atmosphere for 28 days at a constant temperature of 50° C. The best staining
test was found to be that developed by Roger (1936), using nitric acid
solution of known strength at 25° C, and observing the time required for
the formation of a surface film showing a dark blue interference colour.

The test described below was developed in this country over a period
of some years and, in fact, was being used as a routine method before
the recent American work was published.

In attempting to devise a new test for the determination of durability
it was laid down that the method should be capable of discriminating
between slight differences in durability, and the variables should be
capable of accurate control ; it was also considered essential that
the results should be comparable with those that are obtained under
natural conditions by exposing specimens of the same glasses to the
action of the atmosphere in the tropics for periods of one to two years.
In practice, the instruments are in many cases subjected to a relatively
high temperature during daylight and a low temperature during the
night. The result is that moisture condenses on the glass at night
and is evaporated off during the day. An apparatus was therefore
constructed in which cycles of temperature were introduced, and
the name " thermodyne " was coined to describe it,* in contradistinction to the thermostat. Fig. 44 shows the thermodyne heating and
cooling apparatus ; the heating and cooling control circuit ; specimens
in position under the bell jar and examples of the results.

The thermodyne method of determining the durability of glass
surfaces consists essentially of submitting either freshly broken or
optically polished surfaces (sealed off in a flask together with a quantity
of water) to a series of temperature cycles from room temperature to
60°C, in an atmosphere of air saturated with water vapour, for a period
of 12 days. Each cycle is of two hours' duration. The attacked
surfaces are subsequently examined visually and photographed for
record purposes. The type of attack which results is characteristic
of the type of glass. The degree of attack is recorded by an arbitrary
number, comparison always being made with the attack on specimens

* Both the principle of the Thermodyne and its name were originated by
myself. I am prone to sleep at committee meetings, and on one such occasion
when durability was under discussion, I found on waking that the subject did
not seem to have advanced since I closed my eyes. Thereupon I said " What
you want is a thermodyne," and when asked what a thermodyne was said it
was the opposite of a thermostat. From this remark was developed by Messrs.
Chance the apparatus described. (Author.)

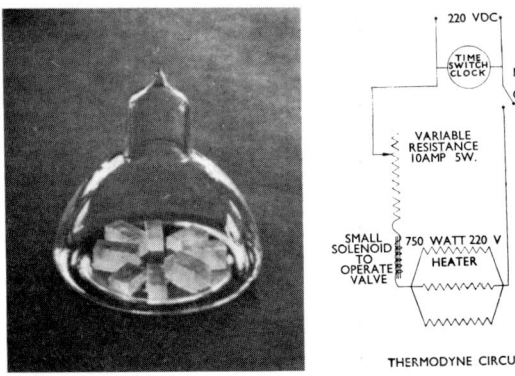

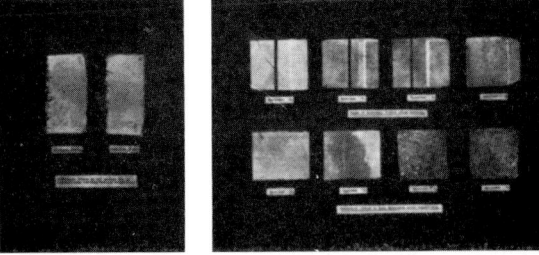

Fig. 44—The thermodyne

At the bottom are shown (*left*) thermodyne attack of two glasses having substantially the same optical properties. On the left is shown the original glass and on the right a modification of improved durability; (*right*) the thermodyne attack of (*top*) faces of binocular prisms after tropical service; (*bottom*) thermodyne attack on the same specimens after repolishing.

of similar standard type tested simultaneously. Control is possible on the following points: (*a*) Upper and lower temperature limits and the duration of the cycle; (*b*) humidity of the atmosphere in contact with the specimens; (*c*) composition of the atmosphere in contact with the specimens; (*d*) condition of the test surfaces.

The results obtained with the test were compared systematically with the results which had been obtained by exposure of the same glasses to tropical atmospheres. Without going into details of the experimental procedure, it may be said that this method of test has given valuable results.

It may be pointed out that both the conditions for producing the two types of weathering noted by the American experimenters are embodied in the thermodyne method, and while it is recognized that no single test is available for determining the durability of the wide range of optical glasses manufactured, comparison of the results of the thermodyne test with the results of natural-exposure tests shows that the accelerated test is fairly reliable. In typical tests on binocular prisms returned from service, the specimens appear in the same relative order when ranged according to their state after this test or after natural exposure; moreover, the two specimens which were badly stained naturally show heavy staining in the accelerated test.

It may be said generally that the thermodyne attack is rather more severe than that resulting from years of exposure in the tropics, but the more severe nature of the accelerated attack accentuates differences in durability. At the same time the test is not so severe as some of those described in Dr Hampton's paper, and it is perhaps most valuable in assessing the durability when modifications of an old type of glass are being developed.

It is possible that with the development of non-reflecting surface films, this question of glass durability will become of less importance, and it is worth noting that Jones and Homer (1941) have recently used the method employed for their accelerated weathering test to produce silica-rich films which have been stabilized by baking to form a durable film. (Here ends the abstract from Dr Hampton's paper.)

Halle (1921) gives a very simple test for durability which, however, I have not tried in comparison with other methods. He says (*translation*)—

> One cleans the polished surface carefully, and then puts the glass under a bell jar in a hydrochloric acid atmosphere. It is then taken out and left for 24 hours under a carefully sealed bell-jar which should be quite free from dust. It is important that there should be no trace of ammonia vapour therein. If after this treatment there is any deposit, even the slightest, the glass is not suitable

for optical work. To make the observation more severe, one looks at the light at a slanting angle, and draws the finger across the surface ; if there is no deposit at all, the surface remaining completely clear, the glass is good.

Pfund (1946) states in a paper on *The ageing of glass surfaces*, that the alteration of a glass surface with age lies in the formation of a thin film which eventually consists only of amorphous quartz. This is proved by the measurement of the refractive index ($n = 1\cdot46$) and the infra-red reflecting power at 8 to 10µ. The change can be artificially produced by heating the glass surface to about 70°C, the surface being in contact with the combustion gases of a flame. One day of this treatment corresponds approximately to a year of the natural alteration. It is possible to determine the refractive index of a thin film of this kind, without knowing its thickness, by measuring the wavelengths of the interference maxima for two different angles of incidence. The reader is referred to the paper for details.

Mechanical properties

Strength

101 The strength of glass is usually a matter of indifference to the optician ; due respect being paid to its brittle nature, there are few occasions when it need be put under dangerous strain. It is, however, important for the optician to remember that the tensile strength of the glass is only about $^1/_6$th of the compressive strength, and even this relation only holds for a short while after the tensile stress is applied : a glass which will carry a load for, say, 2 minutes may break under the same load at the end of a somewhat longer period.

Delayed fracture in glass

A paper by Gurney and Borysowski (1948) gives a good deal of information concerning the delayed fracture of glass and other materials under tension, torsion and compression ; this, the paper states, is usually associated with attack by the surrounding atmosphere on the substances under stress. The remarkable delayed fracture of glass under tension in air has been shown by Preston and his collaborators, in an important series of papers on the subject (1946), to be mainly due to atmospheric attack.

In reply to an enquiry addressed to the Research Laboratories of Messrs Pilkington Brothers Ltd, Glass Manufacturers, St. Helens, Lancs., England, I was told that, if zero risk of failure is to be attained, plate glass, which for a short time will withstand a tensile load between 4,000 and 8,000 lb per square inch, cannot be relied on to withstand an indefinitely sustained loading of more than 1,200 lb per square inch.

Hardness

102 Most of the ordinary glasses do not differ so much in hardness as to affect grinding and polishing methods very much, but when it comes to dealing with crystals, in which the differences may be very great, the hardness both of the crystal and of the abrasive will repay consideration. Materials and methods appropriate for the softer crystals will be found in §§92 and 119.

Hardness is also of prime importance when it comes to diamond engraving of glass. On a soft glass, the loading of the diamond must be considerably less than with one of the harder varieties if it is desired to produce a line of the same width and free from splinters. Information on this point will be found in §272.

A good account of hardness and its measurement will be found in O'Neill (1934).

There is a large number of different types of instruments for measuring hardness, but as far as the interest of opticians is concerned, some of these can be neglected, as for example the Scleroscope.

Four of the methods may be of interest to opticians—

(*a*) *The Mohs test.* In this test, introduced by Friedrich Mohs (1773-1839), the hardness of the material is classed according to which of a selected series of materials will scratch the substance or not. The Mohs test is discussed by Auerbach (Winkelmann, 1908) from which the following remarks are abstracted (*translation*).

In mineralogy the need for a means of indicating the property known in ordinary speech as "hardness," a property which is of value for the characterization and identification of minerals, has long been felt. This can be met by the action of a point of one substance on a flat surface of another substance whose relative hardness is to be compared, as for instance the scratching of the latter by the former. This is the Mohs test, according to which one calls one substance harder than another when the point of the first scratches a flat surface of the latter.

Mohs built up the following scale of hardness—

1. Talc   3. Iceland Spar   5. Apatite   7. Quartz   9. Ruby
2. Gypsum   4. Fluorspar   6. Felspar   8. Topaz   10. Diamond

[The numbers are called the hardness numbers and such numbers are given for 78 substances on page 860 of the reference.] The system is admittedly of a purely empirical character and depends on two tacit assumptions neither of which is, as we shall see, rigidly true. The first is that if a point of substance a scratches a flat surface of a substance b, then a point of substance b does not scratch a surface of substance a. The second assumption is that when, according to the definition, a is harder than b and b harder than c, then also *a* must be harder than *c*. In order to test these assumptions Auerbach submitted 14 various Jena glasses, which differed very considerably in hardness, to systematic scratching experiments.

The whole 182 (14 × 13) combinations were tried. Very surprisingly it proved that each of the glasses scratched each of the others, even the softest the hardest. In order then to come to a conclusion concerning the hardness one must abandon the original naive definition and call that one of two glasses the harder which scratches the other more strongly. But a decision still remains difficult, since under the microscope important qualitative differences between the scratches appear; in some cases there is flaking or splintering running sometimes lengthwise, sometimes perpendicular to the scratch.

Thus the two assumptions referred to above are not strictly justified. This is not to say that they may not be in the main right and they have, indeed, served a very useful purpose.

Impressed with the desirability of a less ambiguous and more precise measure of hardness Auerbach devised the apparatus described in (*c*) below.

(*b*) *The permanent indentation methods*, represented by the Brinel, the Vickers Pyramid Diamond test and the Rockwell diamond hardness-testing machines, class materials according to a scale of numbers called after the type of instrument—for example, the Brinel, Vickers Pyramid, or Rockwell numbers. They measure hardness in terms of the *permanent* indentation made by a ball (Brinel) or by a diamond of pyramidal or conical shape. What they measure, therefore, is the flow of material. In all of them the hardness is expressed in terms of the load in kg/mm^2 of the area of the permanent indentation. Such a test can be carried, as by Wilfred Taylor (1949), even to the measurement of such hard substances as glass.

In the reference cited Taylor says—

I described recently a form of hardness microtester which permits light loads to be applied to a diamond indenter of pyramidal form as a means of determining the hardness figures of minute individual crystals in a mosaic[1]. It is also possible with the same instrument to draw the loaded indenter across a surface and to determine the hardness figure by means of a scratch test.

It was intended by this latter means to attempt to assess the hardness figures of many of the known types of optical glass, and a beginning was made in this direction. In addition, the opportunity presented itself of observing the form of the resulting fracture when the loaded indenter was used in a static manner and brought into contact with a plane polished glass surface. It was then discovered

DIOPTRIC SUBSTANCES

that a lightly loaded pyramid diamond left a perfectly defined and lasting impression on the polished surface. Thus, in the case of borosilicate crown and extra dense flint glasses, perfect impressions were formed with loadings up to about 50 gm. ; with loadings of 100 gm., fractures appeared emanating generally from the corners of the square impression, and at some intermediate loadings fractures sometimes appeared after a short interval of time (Fig. 45).

With borosilicate crown glass 509/650, the diagonals of the impression under a load of 50 gm. measured about 12·3 μ, giving

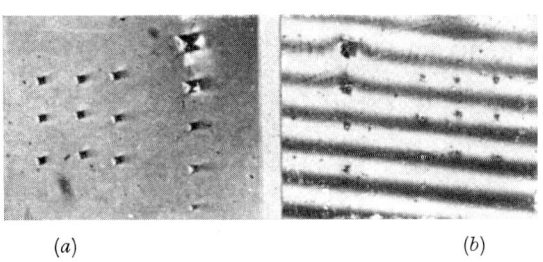

(a)   (b)

Fig. 45—Diamond impressions on glass
(a) Nine impressions under a load of 20 gm and single impressions under loads of 100, 50, 20, 10, and 5 gm.
(b) As (a) but with the addition of interference fringes to show the raised rim round each of the two larger impressions.
Wavelength 546 mμ, *Magnification* × 200
(*Reproduced by permission from* "*Nature*")

a hardness figure of 613 ; and under identical conditions the hardness figure for extra dense flint glass 650/337 was 433 with the diagonals measuring 14·6 μ.

The interesting fact is that the surface layer was found to be sufficiently plastic to take an impression. Thus, with a loading of 100 gm. the volume of the displaced borosilicate crown glass was 115 μ³, and an interference test showed a raised rim of glass around the depression (Fig. 45b). The volume of glass raised above the original surface was found to be about 90 μ³, but accurate measurement was difficult and there may well be perfect correspondence.

In view of the uncertainty which exists regarding the nature of a worked glass surface, similar tests were carried out on a newly fractured and on a fire-polished surface of the same borosilicate crown glass. No difference in behaviour was noted and no evidence was forthcoming to support the presence of a soft superficial layer of glass on the optically worked specimen.

1. Taylor (1948), *J. Inst. Metals*, **74**, 493.
2. Glazebrook (1923), " Dictionary of Applied Physics," **4**, 331 §15.

(*The passage above is quoted with the permission of the Author and of the Editor of* "*Nature*")

A micro-hardness tester was designed by Hanemann and made by Zeiss in the form of a set of accessories to their large metallurgical microscope *Neophot*. Whereas the Vickers, Rockwell and Brinel instruments as customarily used are limited to the determination of the general crystalline structure of an alloy, the Hanemann instrument permits one to select an individual crystal and determine its hardness by a microscopic indentation.

Brinel originally expressed hardness in terms of the pressure in kg/mm^2 of the *projected* area permanently indented; later, to obtain consistency, he used the *full* area of the indentation. Such a correction, necessitated by the fact that the depression altered in shape as it became deeper, is not needed in the Vickers and Rockwell systems

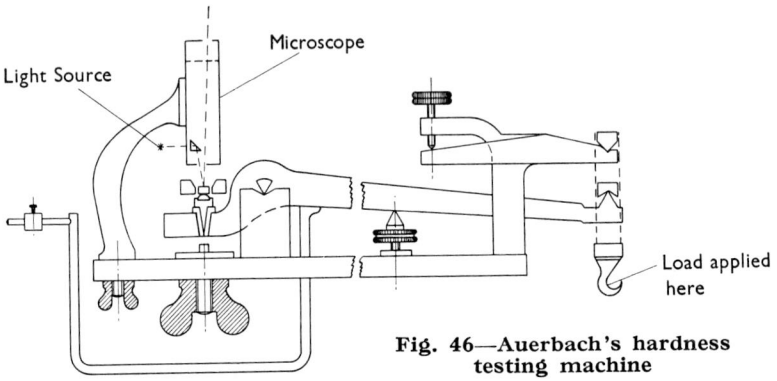

Fig. 46—Auerbach's hardness testing machine

in which, whatever the depth, the shapes of the indentations are geometrically similar.*

(c) *Auerbach* (*loc. cit.*) defines hardness in terms of the area of contact produced under pressure loading between a hemisphere and a plane surface, both of the material in question, *during the application of the load* just when the impressed material has reached the limit of elastic deformation.

In brittle bodies this becomes evident by the formation of a crack—in others by the beginning of plastic deformation. In the latter case the biggest elastic deformation is measured by applying increasing loads to the same spot until permanent deformation takes place; the area of contact can then, in opaque bodies, be determined either by photography sideways, or by flashing the impressing spherical surface with a thin deposit of silver which, when treated subsequently

---

* A Micro-hardness testing machine with 136° diamond pyramid is made by Hall Telephone Accessories Ltd, Dudden Hill Lane, London, N.W.10; with this instrument indentations can be made like those described by Taylor, and evaluated on the Vickers Hardness scale.

with sulphur vapour, shows by a mark the dimensions of the contact area.

In transparent bodies the area of contact under pressure is measured by observation through the flat transparent surface of the pointed indentation as shown in Fig. 46.

If R is the radius of the hemisphere and d_e the diameter of the indentation measured while under load L, then, from theoretical considerations, Hertz derived the formula—
$$H_A = 6L/\pi d_e^2$$
Auerbach in applying this formula found that a factor was required to correct for the curvature of the surface and used the expression $(6L/\pi d_e^2) \sqrt[3]{R}$, which he called the "absolute" hardness.

In his apparatus Auerbach used a hemisphere of 1 mm radius, L was in kilograms, d_e in mm ; thus H_A was given in kilograms per square millimetre. On p. 866 of the reference the Auerbach absolute hardness in kg/mm² is given for 44 substances varying from cast tin (11) to diamond (2,500).

(d) *The ruled line test.* By this test hardness is measured in terms of the width of a diamond line ruled under known load. An instrument for this purpose was developed by Bierbaum and called by him the "microcharacter." The instrument is made by the Spencer Lens Company, and a copy was also made by Zeiss. This instrument measures, therefore, the dimensions of a *cut* made by a diamond under known load.

The optician's interest in hardness

It is the Auerbach hardness which should be of interest to opticians concerned with roughing, trueing and smoothing, as will be evident from the nature of the grinding process as described in Chapter 3 ; the abrasive must be harder than the substance being ground in the " Auerbach " sense.

As stated above Zeiss made an instrument of the microcharacter style, and also another one of the Brinel (the Hanemann–Zeiss microhardness tester) type, but I can find no mention of any firm now making an instrument which will measure the Auerbach hardness, which seems to have lost whatever interest it formerly had for metallurgists.

On the hardness of *polishing* powders, there appears to be little written. This may be due, partly, to the difficulty of determining the hardness of a powder and, largely, to the very many possible different grades of a given powder ; different grades of rouge, for example, may be drawn from the same furnace simply because the contents were not stirred during firing.

Where, however, the powder also occurs naturally as a mineral, the hardness is usually known and I give a list of such minerals below.

Hardness (Mohs' scale)	Compound	Mineral Name
2	PbO	Litharge (red)
2	PbO	Massicot (yellow)
$5\frac{1}{2}$	MgO	Periclase
$3\frac{1}{2}$	CaO	Lime
$2\frac{1}{2}$	Pb_3O_4	Minium
5–6	Fe_2O_3	Haematite
6–$6\frac{1}{2}$ (crystals) 2–6 (massive)	MnO_2	Pyrolusite
6–7	SnO_2	Cassiterite
$5\frac{1}{2}$	PbO_2	Plattnerite
$5\frac{1}{2}$–$6\frac{1}{2}$	$Fe.Fe_2O_4$	Magnetite

(*A supplementary list is given in* §116.2)

Considering the most modern views on the nature of polishing (see Chapter 3) one would not expect much guidance as to polishing from such a table; further, Mohs' scale of hardness, used in the above table, is no more than a rough guide. For example, a number of very different glasses (two kinds of plate glass, high-expansion glass, hard crown, borosilicate crown, medium barium crown, extra-dense flint and double-extra-dense flint) all gave a Mohs hardness of 5; yet these glasses differ very much in hardness as judged by the optician's experience in handling them or their behaviour to a diamond in the ruling of fine lines. However, the Mohs scale is simple and inexpensive, and that it does permit of a certain amount of discrimination is shown by the textbook figures following—

Material	Hardness (Mohs)
Quartz	7
Fluorspar	4
Calcite	3
Rocksalt	$2\frac{1}{2}$

Information concerned with micro-hardness will be found in the *Fiat Review of German Science* 8 *Physik der festen Körper Teil* 1. "*The Hardness of solid bodies*" by Pöschl and Bückle.

Coefficient of expansion and variation of refractive index with temperature

103 These two properties are mentioned together since the optician is usually interested in both at the same time. For example, in correcting a large plane parallel plate, anyone working locally with vigour will find, if he examines the plate by reflection interferometrically immediately after the working, that he has apparently (if working with crown glass for example) *added* to the thickness of the glass. On the other hand, if he does the same with a piece of D.E.D.F. (double-extra-dense flint), he will find very little change. This is because, working with the Fizeau bands*, the change in optical path length is $2(t \cdot \triangle \mu + \mu \triangle t)$ and, with an increase in temperature, t increases while μ diminishes. With crown glass $t \cdot \triangle \mu$ is less than $\mu \triangle t$, and negative in sign, while with D.E.D.F., $t \cdot \triangle \mu$, while still negative is almost equal to $\mu \triangle t$.

This combined effect is also, of course, operative in the case of prisms or other objects immediately after they have been worked on and, for this reason, ample time must be allowed for such objects to acquire a uniform temperature before a reliable test can be made. Matters can be accelerated by standing the prism on a flat brass plate and covering it as far as possible with an envelope of sheet brass, leaving therein only such apertures as are necessary for the testing.

The expansion of glass comes to the notice of the optician most palpably during the working of a single prism or the like. At the beginning of the work the prism is probably colder than the hands which are holding it and the surface in contact with the polisher therefore becomes concave, which results in the marginal regions being polished more than the inner regions. When polishing has proceeded for some little while, however, the surface in contact with the polisher becomes warmer than the remainder, with the result that the middle part of the surface receives preferential polishing. When the temperature has become uniform, these two effects combine to make the surface concave in the middle and convex at the margin, an effect characteristic of hand polishing. The skilled optician is very sensitive to the feel of the work and can distinguish very precisely whether the work is being polished in the middle, in which case it " swings," or on the margin, in which case it " binds ", and he instinctively regulates the speed and pressure of polishing accordingly.

The coefficients of expansion of glasses differ considerably, varying from 7×10^{-6} per 1°C up to 10^{-5} for Chance's high-expansion glass, in both cases over a range of about 20°C to 500°C.

Very many examples of coefficients of expansion, refractive indices, and variations of refractive indices with temperature, are given by Morey (1938).

Crystalline quartz is very different from glass in its behaviour owing to its high heat-conductivity, which along the axis is $13\frac{1}{2}$ times and, perpendicular to the axis, $8\frac{1}{2}$ times that of ordinary crown. Fused silica has a heat-conductivity $2\frac{1}{2}$ times as great as crown glass.

* See §222

Coefficient of expansion of quartz glass (fused silica)

Although for most purposes one can assume that fused silica has a zero coefficient of expansion there may be occasions when its expansion cannot be ignored. In such case one may take the expansion between room temperature and 100°C as being given by the formula—

$$l_1 = l_2(1 + 0.332 \times 10^{-6} . t + 0.00146 \times 10^{-6} . t^2)$$

This is the mean of the two almost identical formulae given by Scheel (1912).

The nature of glass

104 It is not easy to define the material which we all know as " glass." Dr F. W. Preston has described it as " that which, if there were none, we should have to make bottles of something else." The nature of its physical structure can, however, be fairly definitely stated.

The following passage is abstracted by permission from the article by Dr K. L. Loewenstein (Works Chemist at Nazeing Glass Works Ltd, Broxbourne, Herts.) published in *Science News*, No. 8 (Penguin Books), 1948.

The art of glass manufacture depends on melting together a number of substances, which, on cooling rapidly, will set to give a clear, transparent solid. From a physical point of view this corresponds more to the liquid than to the solid state. For example, the change from liquid water to solid water (ice) takes place at a definite temperature, and as long as both water and ice are present the temperature of the mixture will be 0°C or 32°F ; with glasses this is not so ; on cooling, the viscosity increases continuously and there is no definite temperature at which the glass can be said to solidify ; similarly, on heating glass from room temperature, it will soften more and more without any break in the temperature curve.

For this reason the glassy state has been termed " vitreous," and glasses are often described as " supercooled liquids." The decomposition of glass into solid constituents can take place, especially when the glass is cooled very slowly ; this is called " devitrification." This often takes place years after a glass article has been made, and there is no known method by which it can be arrested.

The constituents of glasses

(1) *Glass-formers*—These are pure substances which can exist as glasses by themselves, and without which no stable glass can be made. For example, silica (sand) can be made into a glass, but lime, present in most glasses, or lead oxide, present in many, cannot form a glass.

By far the most important glass-former is silica ; others are germanium dioxide, beryllium fluoride, and phosphorus pentoxide ; the last, in particular, has now been extensively investigated for special electrical and optical purposes.

What makes a substance a glass-former is a question of relative atomic size of the atoms making up the substance. Firstly, the cation (silicon, phosphorus, etc.) must be large enough to space around itself a number of anions (oxygen, fluorine, etc.), to permit the formation of a three-dimensional network structure ; and secondly, the anion must be small enough in relation to the cation to fit into such a structure. For example, beryllium fluoride is a glass-former ; on substituting a smaller atom, *e.g.* lithium, for beryllium, or a larger atom, *e.g.* bromine, for fluorine, we find that the substitutes are too small and too big respectively to fit into the network structure.

It has been found that the ratio of the atomic radii of the cation to the anion must be approximately 0·3.

The high stability of silica glasses compared with other glass-formers is ascribed to the fact that the ratio of the number of cations to anions is 1 : 2, so that each silicon atom can group around itself four oxygen atoms in the shape of a regular tetrahedron, and that each oxygen can hold two silicon atoms ; furthermore, as no oxygen can hold more than two silicon atoms there is a certain flexibility in the joining together of SiO_4 tetrahedra. Hence the energy in the supercooled state is not very different from that in the crystalline state, and the tendency to devitrify is small.

If, however, other groups besides the SiO_4 tetrahedra are incorporated in the glass structure, the tendency to devitrify is increased.

(2) *Other constituents*—The breaking-up of the three-dimensional SiO_4 network by other constituents will lower the melting point of the mixture, and good glasses can be obtained at temperatures as low as 1,250°C, compared with 1,750°C for the melting point of silica.

This description may be supplemented by the following information kindly sent me by Mr R. E. Bastick of Messrs Chance Brothers Ltd.

The atomic or molecular structure in glasses is characterized by an extended network which lacks symmetry and periodicity but in which the forces between the atoms are comparable with those which occur in the corresponding crystal. This follows from the fact that certain substances (*e.g.* silica) which can exist in the vitreous state also exist in the crystalline state, but when they are in the vitreous state there is little tendency to crystallization. From a knowledge of the crystal structure of glass-forming oxides it is known that the oxides most suitable for glass formation are those of elements which provide small and highly-charged positive ions. In the crystalline oxides, the oxygen ions form polyhedra around the positive ions and if the ions are small these polyhedra are triangles or tetrahedra. It can be shown on geometrical grounds that such polyhedra can form an open network lacking symmetry and periodicity, but still maintaining electro-neutrality in the system, with the angles of valency of the oxygen ions differing only slightly from those in the corresponding crystal lattice.

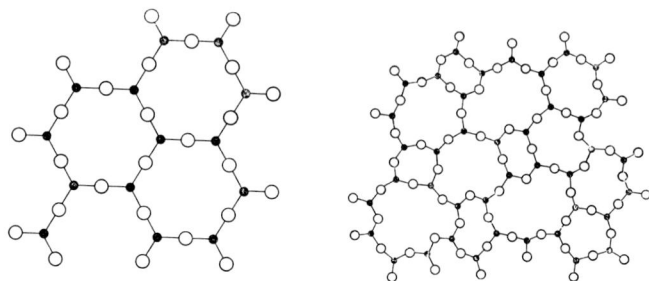

Fig. 47—Crystal and glass
The difference between the structure of crystal (*left*) and of glass (*right*) consisting of the same constituents.

Zachariasen (1932) has laid down the requirements which must be satisfied by crystalline oxides in order that they should form glass, and these requirements are as follows—
1. Each oxygen ion is linked to no more than two cations.
2. The oxygen polyhedra share corners with each other and not edges or faces.
3. The number of oxygen atoms surrounding the positive ions must be small (three or four).
4. At least three corners of each oxygen polyhedron must be shared with another polyhedron.

According to these conditions the glass-formers are confined to the oxides of the following elements: silicon, boron, phosphorus, vanadium, germanium, arsenic, antimony, niobium, and tantalum; and, of these, the oxide of niobium is the only one of which no examples of glasses are known. For various reasons, practical glasses are confined almost entirely to the glass-formers silica, boric oxide, and phosphoric oxide, silica of course being the network-former of practically all commercial glasses, while phosphoric oxide and boric oxide are used in special cases.

Fig. 47 illustrates the difference between the structure of crystal (left) and of glass (right) consisting of the same constituents.

DIOPTRIC SUBSTANCES   141

The Zachariasen conditions also apply to compounds related to the oxides, as for example, beryllium fluoride which is known to occur in the vitreous state. Obviously therefore, so far as inorganic constituents are concerned, the number of substances which are capable of forming a glass is very limited. Of course, once a glass-former is present, it is possible to modify the structure by what are know as network modifiers, a term which covers such substances as the alkalies, lime, barium, lead, zinc, and the many other constituents which are used, for example, in optical glasses.

105 The surface structure of glass has been studied by electron diffraction with which method, as in the case of X-rays, a pattern results consisting of diffused scatter through which broad rings are more or less distinctly visible. With glass and similar vitreous substances the diameter of these rings corresponds to the atomic separations within chaotically distributed molecules, and this pattern is now widely accepted as evidence of amorphous as distinct from crystalline structure.

The atoms in the glass form an extended three-dimensional network but, as is also shown by X-ray diffraction experiments, this network is not periodic as in crystals. On the other hand it is not entirely random; since atomic distances in the molecules of the constituents are in themselves fixed, it is only the orientation of the molecules which is random.

The X-ray diffraction rings of glass are described in an article on the structure of glass by Ackermann (1948). The paper contains a useful bibliography on the subject.

The development of optical glasses since 1880

106 Mention has been made in §98 of the initiation by Schott and Abbe, about 1880, of a series of improvements in optical glasses. These made possible the construction of lenses with wider angle, larger aperture, better colour correction and other advantages.

These improvements culminated in the introduction recently of quite new types of optical glass by the Eastman Kodak Company. The major forward steps since 1880 are shown in Figs. 48–50. The following description (§§107–11) by R. Kingslake and P. F. DePaolis of the Kodak Laboratory is included here by permission of Dr C. E. K. Mees (Head of the Laboratory and Vice-President of the Company), and of the Editor of *Nature*, in which journal it first appeared.

New optical glasses
by R. KINGSLAKE *and* P. F. DEPAOLIS

107 Before 1880 the only types of optical glass available to lens designers were the flint–crown series, in which the addition of progressively increasing amounts of lead oxide led to a progressive increase in both refractive index and dispersive power, so that in effect there was a fixed relation between the dispersive power and refractive index of all available glasses (Fig. 48).

In order to achromatize any lens, the positive elements must be of lower dispersive power than the negative elements, and in those days this meant that the refractive index of the positive elements had to be low, and that of the negative elements, high. This had two very adverse effects : On the one hand, it tended to make the Petzval

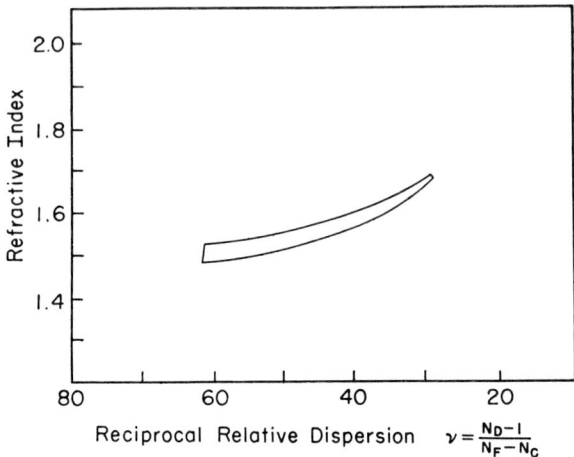

Fig. 48—**Optical glasses prior to 1880**

sum large,* giving a strongly inward-curving field ; and, on the other hand, it made the surfaces of the positive elements strong and those of the negative elements relatively weak. As the lens on the whole is positive, this resulted in considerable amounts of zonal spherical aberration, and also indirectly in large residuals of all the other aberrations.

* *The Petzval sum*
A stigmatic lens system, that is, one which images point objects as point images, will not, in general, image a plane object as a plane image. The image surfaces of nearly all such lenses (apart from photographic anastigmats), will be curved concave towards the lens. When the object is placed at infinity, the image surface shape—in the neighbourhood of the axis—will be determined approximately, solely by the focal lengths and the refractive indices of the component lenses in the system. The reciprocal of the radius of curvature of the surface, at the vertex, is called the Petzval curvature (or the Petzval sum), after Joseph Petzval, a Hungarian, who investigated this error about the year 1840. T. Smith has suggested that the English mathematician Airy has a claim to priority here, and that this aberration could justifiably be called the Airy curvature.

For two reasons, then, a crown glass was needed with low dispersion and high refractive index and also, if possible, a flint glass of high dispersion and low index. The invention of barium crown glass in the 1880's by Abbe and Schott helped enormously to meet the first requirement, but the problem of the low-index flint was not solved at that time. (Incidentally, plastics, crystals, and liquids mostly tend to have higher dispersion for their index than glasses, but these materials are generally undesirable for other reasons).

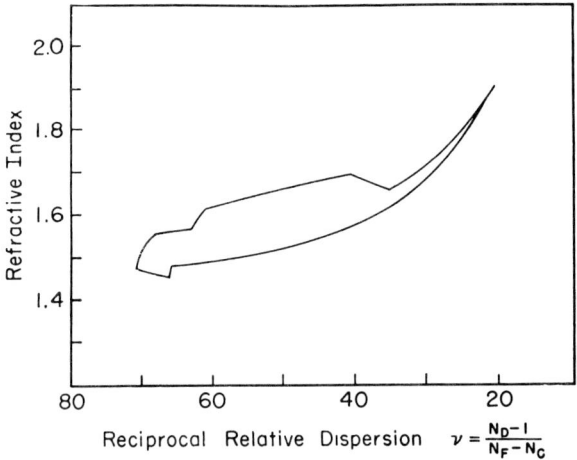

Fig. 49—Optical glasses prior to 1934

With the introduction of barium crown glass, many new types of photographic lens became possible, the first to be developed being cemented triplets of the Dagor type, in which the three glasses are common crown, light flint, and dense barium crown in that order. Barium crown glass also greatly improved all lenses of the air-spaced type (the Cooke Triplet and the Celor, for example) by weakening the surfaces of the crown elements and enabling the designer to shorten the lens without adversely affecting the Petzval sum. To be sure, some successful lenses of the meniscus type; *e.g.*, the Omnar, were designed without using barium crown, but these were the exceptions, and they did not become very popular.

The refractive index of barium crown glasses was raised slowly and steadily by Schott until the early 1930's (Fig. 49). At that time they introduced SK–16 and SK–18, which marked the upper limit of the ordinary barium-soda-lime-silica types. These glasses are chemically unstable and present difficulties in lens manufacture because of their susceptibility to acid staining and their brittleness.

108 Following the first World War, G. W. Morey, who had done work of great value in connection with the production of optical glass for military purposes, discussed the advances which could be made in optical glasses with C. W. Frederick, chief of the lens designing bureau of the Eastman Kodak Company. Frederick suggested that what was wanted was a very high refractive index with low dispersive power and that even a small production on a laboratory scale would be useful. Morey undertook to study the optical properties of glasses in relation to chemical composition, especially with a view to the production of high-index crown glasses. To this end all high atomic number cations were chosen for systematic study in silicate, borate, and phosphate glasses. Small melts of twenty to forty grams were made to indicate the field of glass formation. The more promising ones were repeated in larger melts of fifty to one hundred grams.

By 1933, the work had progressed to the point where silicon and phosphorus were discarded as glass-forming elements. Boric oxide had by now proved to be by far the best fluxing agent. Oxides of elements such as lanthanum and thorium found in the rare earths, and columbium, tantalum, tungsten, titanium, zirconium, and strontium were used in major portions up to 80 per cent by weight, with or without the usual barium, zinc, magnesium, and aluminium.

109 In 1934, samples of unusual glasses in the region with an n_D of about 1·85 and a ν of 43·0 were in existence and their properties well measured[1]. About this time, the work was expanded to a larger scale by the Kodak Research Laboratories. Under the direction of S. E. Sheppard, L. W. Eberlin and P. F. DePaolis made a systematic study of the solubility of the rare elements in boric acid and the limits of glass formation. The results of the combined work were revealed in patents by Morey and by Eberlin and DePaolis[2].

The solubility of lanthanum in boric acid is remarkable, and its contribution to higher refractivity without increase of dispersion is a revelation. The oxides of tantalum, thorium, and tungsten are soluble in the lanthanum borate base glass in amounts up to 35 per cent. These new borate glasses are very stable and fairly hard. They are harder than flints, suitably stable to the atmosphere, and amenable to optical shop practices of moulding, grinding and polishing.

Early in the development work it was found that the new rare element borate glasses were extremely corrosive to all known pot-refractories. A decision was therefore made to use platinum for the actual production of these glasses. This was justified on the basis that no platinum would be lost by contamination and that the glass, once homogenized, could be poured in its entirety, free from striae, into a single slab or into cast shapes without striae, seed, bubbles, or other defects usually attending a glass made in a refractory pot.

The first of the new glasses to be made had a refractive index of 1·7445 and a ν-value, or reciprocal dispersive power, of 45·8. The glass was slightly yellow, but further work established the origin of the yellow colour, and finally glasses were produced as colourless and homogeneous as any other ordinary optical glass.

Pilot plant operation began September 1937, and the first commercial glass was delivered in June 1939. Production increased rapidly, and more than 125,000 pounds of rare-element glass were produced during the second World War (1942–1945). Much of the success of the enterprise was due not only to the platinum equipment but to the Kodak method of using all-electric heating and a small-pot, "multi-pot-multi-stage" process.

Under sustained production conditions during the war, the yield of finished usable glass in a cast form was 95 per cent of the theoretical glass available in the batch.

Electric heating was retained in a plant that was erected during the war and operated, from December 1942 to September 1945, on a continuous basis of approximately 5,000 pounds of finished glass a month. The process consisted of feeding thirty-two ten-pound platinum-lined pots every twenty-four hours, starting one every three-quarters of an hour and progressively moving these pots through the various stages of the glass making process.

110 Since 1940 the types of these glasses in production have been extended, and at the present time the following seven types are being made—

	n_D	ν
EK–110	1·69680	56·2
–210	1·73400	51·2
–310	1·74500	46·4
–320	1·74450	45·8
–330	1·75510	47·2
–450	1·80370	41·8
–448	1·88040	41·1

All these glasses contain thorium, and in the case of folding cameras, in which the lens may rest for a long time in close proximity to film, the radioactivity of the thorium may be a disadvantage. A thorium-free equivalent of EK–320 is available.

The new glasses were first used in lens design in 1934, actual production of lenses began a few years later, and to-day many of the Eastman Kodak "*Ektar*" lenses contain the high index glass. The versatility of these glasses is evidenced by the large number of lens patents which specify such glasses.

111 After the transfer of the production of the new glasses to the factory, the Kodak Research Laboratories continued their study of optical glasses. Theoretical considerations by M. L. Huggins and K. H. Sun[3] predicted the possibilities of further new glasses, and their predictions have proved correct. Of the glass systems of this type which were worked out by Sun, the most useful were flint glasses containing titanium oxide and using fluorine in addition to silica.

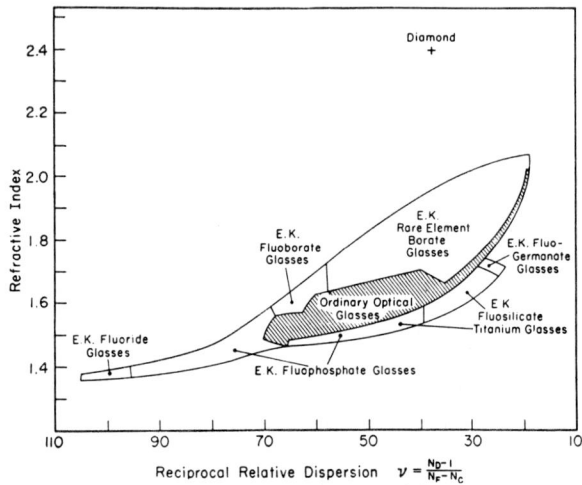

Fig. 50—The new Kodak glasses

These unusual glasses lie well below the ordinary flint line (Fig. 50). For a given index, the dispersion is appreciably greater than that of ordinary flints. In this sense the glasses may be termed super-flints. The best glasses lie in the region 45 per cent SiO_2, 28 per cent TiO_2, 27 per cent NaF ranging in optical properties from $n_D = 1.65/\nu = 29$ to $n_D = 1.58/\nu = 36.6$. The glasses are mouldable, very resistant to tarnish, and easily fabricated by usual optical methods.

The fluosilicate flints are almost as useful to the optical designer as the high-index glasses, since they extend the possible difference between crown and flint index and between crown and flint dispersion. With these glasses three-element lenses have been designed to give even better performance than the usual four-element types. K. H. Sun succeeded in producing novel mixtures in some twenty quite different glass fields including fluo-borate, fluo-germanate[4] and fluo-phosphate systems.

A most interesting group of glasses suggested by Sun are those containing no oxides and composed entirely of fluorides. These glasses show the characteristic low refractive index and extremely low dispersion previously available only in fluoride minerals. The refractive index in most of the glasses approximates 1·38–1·39 and the ν-value, 100. Moreover, the glasses are transparent to below 300 $m\mu$ in the ultraviolet and to 5μ in the infra-red, so that they may be very useful in the making of instruments requiring optical transparency over a wide range of wavelengths. Difficulties have been met in the production of these all-fluoride glasses, but it is possible that these difficulties may be overcome in the near future.

The extension of the frontiers of optical glass by this work is illustrated in Fig. 50. In this figure the range of optical glasses known before 1934 is shown crosshatched, while the larger area shows the glasses that can be made at the present time*.

Bibliography
1. British patent 462,304, March 3, 1937 ; United States Re-issue patent 21, 175, August 15, 1939.
2. United States Patents 2,206,081 ; 2,241,249 ; 2,434,146 ; 2,434,147 ; 2,434,148 ; 2,434,149.
3. Huggins, M. L. and Sun, K. H., *Amer. Ceram. Soc.*, 1944, **27** (1), 10-12, 13-7.
4. United States patent 2,425,403.

Problems arising in the manufacture of optical glass

112 The properties and constitution of glasses are very fully dealt with in Morey's well-known book on the subject (Morey, 1938).

In this and the following sections I shall deal only with the problems which concern the user in respect of quality.

If a glass is not durable, clear, homogeneous, free from specks and bubbles and well annealed, the rejection of a finished piece of work may result, however perfectly the optician may have used his skill. Durability is dealt with in §100, the other problems in §113.

113 Hampton (1942) dealt with the following problems—
1. The difficulty of the supply of raw materials of the necessary quality.
2. The effect of heat treatment on the optical properties of the glass made.
3. The rather more general question of durability.

---

* Messrs Chance Bros. have developed modified forms of rare earth glasses in which a higher proportion of silica is included than in the Eastman Kodak glasses. For notes on annealing of these modern glasses see §291 *et seq.* and Appendix I.

This section and §114 are mainly extracted verbatim from the paper cited above with the permission of the authors and publishers.

Raw materials—Sand

Before the 1939–1945 war, sand supplies for the manufacture of optical glass came almost exclusively from Germany. The sand deposits at Lippe and Hohenbockaer were of the highest quality known and a serious problem would have arisen had we not been able to find a native source. It had been known for some time (Turner, 1940) that in the Western Highlands at Loch Aline there existed a deposit of sand of fairly high purity as regards iron content, but it was in a very inaccessible position so the deposit had never been developed before. Arrangements were made for taking advantage of this deposit and purifying it and, since 1941, optical glass in this country has been made from it. After treatment the iron content does not exceed 0·007 per cent, a lower average than had previously been achieved with the foreign material. Recently a new method of treatment has been developed, and sand is now available in sufficient quantities for optical glass manufacture with an iron content not exceeding about 0·005 per cent. Thus, at the present time, the British optical-glass industry has a source of supply of sand which is of higher quality then any which has previously been available.

Heat treatment and refractive index

114 There was an intensive period of research on annealing during the 1914–18 war and up till 1926. This was all directed towards a shortened time-schedule for annealing, and the sole criterion of success then applied was the absence of double refraction. Glass at high temperature is a viscous liquid, and there is no evidence of any sharp change to a definite solid state as it is cooled to room temperature. The viscosity-temperature curve is continuous and, at room temperature, glass can be considered as a liquid of enormous viscosity. In cooling it from a high temperature, therefore, any temperature gradients which exist are, as it were, frozen into the glass, with the result that instead of being optically homogeneous and isotropic it becomes optically anisotropic, behaving as a uniaxial crystal. As a result of elastic deformation, variations in refractive index are present, and on examination in polarized light it is found to have different optical properties in different directions. Sufficiently, slow cooling reduces these birefringent effects to a negligible amount and the old test of an annealing schedule was that it should reduce the double refraction to something of the order of 0·01 λ per cm. It was, however, found on occasion that glass which appeared satisfactory from the point of view of double refraction, and which was known to be free from striations, still did not give a satisfactory optical image, and it was possible to find two pieces of glass which, while appearing equally satisfactory in polarized light, gave substantially different pictures when examined on an interferometer (Fig. 51). At the same time it was clear that glass could be used satisfactorily although showing much double refraction in a strain

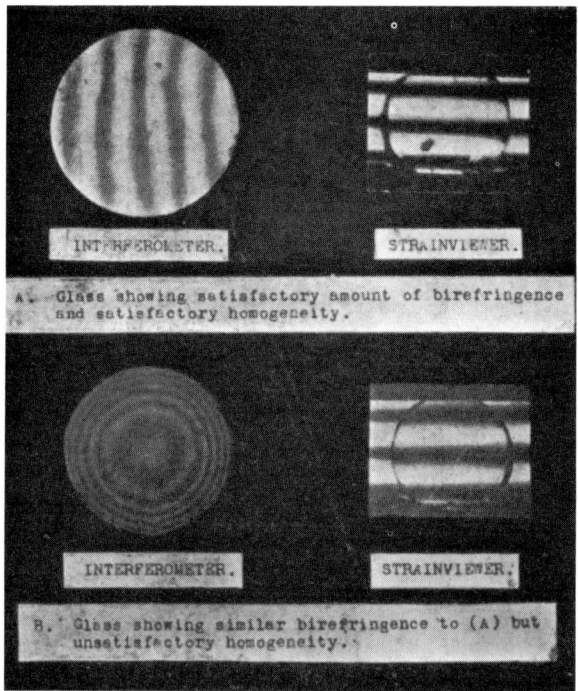

Fig. 51—Results of heat-treating glass
Interferometer and strain-viewer photographs.

viewer. The strain viewer is so sensitive to the effect of temperature changes that it is necessary to maintain a block at a constant temperature for a considerable time and to handle it with gloves, wool or other insulating material during examination, the warmth of the hand holding the block being sufficient to change its appearance at once in the strain viewer : yet this same glass when used, say, in a submarine periscope, which might rise suddenly from the temperature of the sea into 20° of frost or sunshine, may give an entirely satisfactory performance, so that it is clear its optical performance is not directly related to its strain-viewer appearance. Occasions have been known where lenses have been found unsatisfactory and have been returned for re-annealing, and although they looked in the strain viewer rather worse after re-annealing than they did before, the optical difficulties had disappeared and the glass was satisfactory. It is clear from all this that there is something other than double refraction which can be affected by the annealing process.

It has long been known that glass which is cooled rapidly has a lower refractive index than glass which has been cooled slowly from the

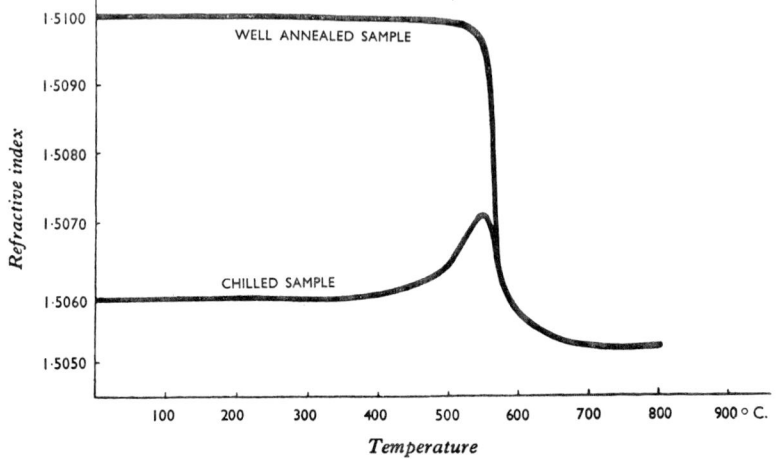

Fig. 52—Effect on refractive index of heating annealed and chilled samples of BSC glass

same high temperature. These differences are relatively large, amounting to, say, six or more units in the third decimal place.* It can be shown by simple calculation that the plus and minus variations due to compression or tension in unannealed glass which is stressed to the tensile limit will not exceed more than one or two units in the *fourth* decimal place. These differences due to tension or compression are, therefore, almost completely masked by the general lowering of refractive index, due to cooling rapidly from a high temperature. There has been a great deal of research on this question, as, for instance, on the effect of heating specimens from the same melting, some of which have been chilled and some of which have previously been annealed.

Fig. 52 shows the result of one such experiment (Lebedeff, 1926). Small specimens of the glass were heated at a constant rate, extracted

* There is also a small, but measurable, effect on the absorption, F.T.

from the furnace at various temperatures, and allowed to cool rapidly in the air. The refractive index of the cold glass was then measured. Up to a temperature of some 500° or 600° C there was no change in the refractive index of either sample. As the temperature was raised, however, there was a marked change in the refractive index, which, in the case of the well-annealed samples, tended to fall until a temperature of about 800°C was reached, beyond which it remained constant. In the case of the chilled specimens, the index rose a little

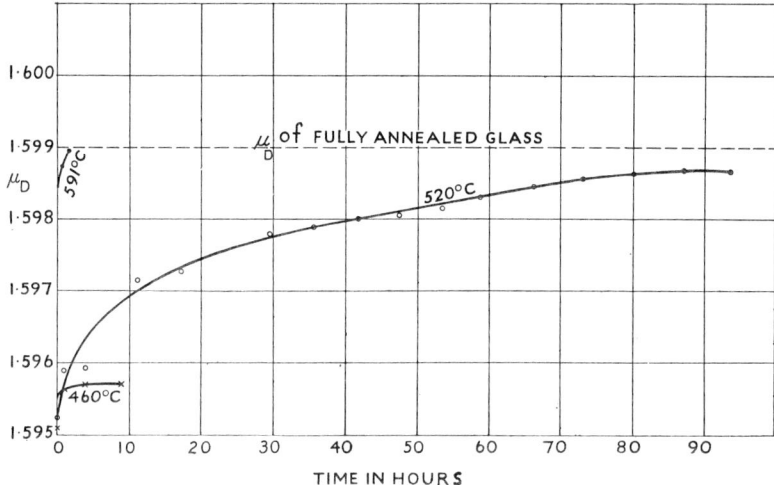

Fig. 53—Effect on refractive index of heating, for various lengths of time, chilled samples at constant temperatures.

and then, at about 800°C fell to the same curve as the well-annealed glass.

In another series of experiments (Twyman and Simeon, 1923) chilled samples of a DBC (melt 3271) were found to have a refractive index of 1·595. Three specimens were kept at temperatures of 460°, 520°, and 591°C, and after soaking for various times at these temperatures were cooled and the refractive index remeasured. Fig. 53 shows the rate of approximation to the refractive index of the completely annealed glass.

Hampton (ref. cit.) shows that an equilibrium index seems to exist for each temperature and shows how this can be determined.

The cooling should be rapid so as to avoid the refractive indices proper to 460°, 520°, and 591°C respectively changing in the direction of those proper to the lower temperatures through which the glass has to pass in cooling. According to various authorities the curve is approximately linear over the range where, in the first experiments, the index was changing rapidly. The velocity with which the specimens of glass approach the equilibrium index is also a function of temperature, and this velocity increases very rapidly as the temperature rises. It is found that the rate of change of index with time at a constant temperature can be represented by a relation similar to that found by Adams and Williamson (1920), for the rate of change of double refraction. It was shown by Hampton (1925–6) that the velocity constant was inversely proportional to the viscosity of the glass. Approximate calculations of the velocity constant for refractive index, based on such meagre information as is available, suggests that a similar proportionality exists in the case of velocity constants for the approach of refractive index to the equilibrium index and, although the matter cannot be considered as proved, it is almost certain that this velocity also is inversely proportional to the viscosity of the glass. Since the viscosity of glass falls very rapidly with increase of temperature, the velocity of approach also changes very rapidly, and may vary in the ratio 100 to 1 for a change of, say, 50°C. This fact makes the determination of the equilibrium index–temperature curve very difficult, and results at both ends of the annealing range are of doubtful value.

Without at the moment going into the question of what causes this change in the equilibrium index, it is clear that we have here an explanation of curves given by Hampton (*loc. cit.*, p. 398.) A well-annealed specimen is heated to a high temperature and samples removed from the furnace. At the lower temperature, the viscosity is so high that the velocity of approach to the equilibrium index remains what it was at room temperature. Above about 500°C, the viscosity of the glass is beginning to fall to a reasonable value and the refractive index of the glass turns towards the equilibrium-index line during the heating. It does, in fact, rather overshoot this line, and approaches it asymptotically. It will be noted that the experimental curve shows that the refractive index as measured on cold samples is constant above a certain temperature, whereas extrapolation of the equilibrium-index–temperature curve suggests that the refractive index should continue to fall. It may be pointed out that the experimental evidence does not disprove this implication. The velocity of approach to equilibrium at this high temperature is so large that even during the high rate of cooling which is implied by the word " chilling " it is probable that the glass index can readjust itself to the equilibrium value corresponding to the temperature. Thus, it is to be expected that the refractive index of a sample chilled from, say, 1000°C will follow the equilibrium line for some distance until the viscosity becomes too high for the instantaneous adjustment to take place. The refractive index measured on the chilled sample, therefore, taken at 1000°C, would be recorded as the index corresponding to some lower temperature, and it follows also that, for a given rate of chilling, this recorded index will be independent of the temperature from which the sample was chilled, provided this temperature is sufficiently high. It may be shown experimentally, that the limiting index is lowered with increase in the chilling rate.

These results have an important bearing on the annealing of slabs of glass. If a piece of glass is uniform in temperature while being held prior to the annealing operation, the only index changes introduced into it will be the plus or minus variations due to the temperature gradient during cooling, and these will probably be small, perhaps not more than a few units in the fifth decimal place. If a small piece of glass is held at, say, 550°C until the index has approached the equilibrium and is then cooled slowly from this temperature, it will have a general refractive index slightly higher than the equilibrium index for 550°C, together with the plus or minus variations due to the gradient. If it is held at some other temperature, say 530°C, it will have a final index slightly higher than the equilibrium index for 540°C, together with the plus or minus variations; but the two specimens of glass will have different refractive indices, although their double-refraction pattern may be the same. If, therefore, both these temperatures exist during the annealing operation—that is, if a temperature difference exists during the soaking period—then different parts of the finished plate will have different refractive indices. Without the most elaborate precautions it is very difficult to get any sort of annealing kiln where the temperature differences are so small as not to give a risk of departures across a slab of optical glass as high as 10°C. After much experimental work kilns have been designed and are in use where the temperature differences over a length of 2 feet or more does not exceed 3 or 4°C. These kilns are capable of yielding glass which is of the highest standard when viewed on the interferometer, although the strain-viewer pattern may not be appreciably different from that given by less uniform kilns. Experience has shown that the glass which is satisfactory on the interferometer, due to the absolute uniformity of index, is capable of being used for the very highest types of optical instruments, and it is becoming generally accepted in the optical industry that the interferometer, and not the strain viewer, is the instrument to be used. It is now a routine operation for samples from all kinds of optical glass to be examined on the interferometer, and for the strain viewer to be used to a considerably less extent.

There was for many years an objection, which was based on painful experience, that the highest quality prisms could only be obtained by

F

sawing from large blocks and not by moulding. Since moulding is much more economical in the use of optical glass than cutting up large pieces, recent developments in annealing, which have proved beyond doubt that moulded prisms can give results comparable in quality to those produced by any other method, have been of the utmost importance in amplifying the supply under the difficult conditions of the present time.

115 There is a lesson in the above for the glass worker. When he finds trouble in getting good definition with, say, a prism, even although the surfaces are flat, he should examine the piece for veins by the knife-edge test or interferometer (Chapters 11 and 12). If there is no fault in the way of veins or other gross heterogeneity he should return the piece to the glass maker. It is highly probable that a re-annealing may make the glass good. An example of this is given in §287.

Note on the production of optical glass in Germany during the 1939–1945 war

It is stated (" Production of Optical Glass in Germany and France," Item No. 9. File 29/41 *Report by Combined Intelligence Objectives Sub-Committee*, London, H.M. Stationery Office) that during the year 1944 Schott produced 1,700 metric tons of clear optical glass, together with 28 metric tons of coloured filter glasses.

It is added (*loc. cit.*) that, instead of breaking up the glass from the pot and moulding the broken pieces after selection, a process had been developed in recent years in Germany (Schott and Gen) in which the optical glass is cast into slabs which are then broken and sawn into blocks. The process is said to be cheaper and to produce a greater quantity of large pieces of glass than the old process of allowing the glass to cool in the pot. It may be remembered that Sir Charles Parsons in his experiments with the large-scale manufacture of optical glass during the first world war, allowed the pot to cool and then sawed it into pieces of the size required.

Artificial crystalline materials

116 I am indebted for this section to Mr J. Skinner, Research Graduate working under Dr A. C. Menzies in the Laboratory of Hilger & Watts Ltd.

Introduction

The transparency of ordinary optical glass extends very little beyond the visible region of the spectrum, and therefore components designed for ultra-violet or infra-red wavelengths must be made of some other material. Some small success has been achieved in the development of special glasses for this purpose, but there are available certain crystalline materials whose transmission and dispersion are suitable. Crystalline materials are also frequently used as components in achromatic lens combinations, and calcite is used for the best polarizing optics.

In addition to calcite, crystals of quartz, rocksalt, sylvine and fluorite have been used. The principal source of good rocksalt and sylvine was at Stassfurt, 30 km due south of Magdeburg.

Optically clear natural crystals of sufficient size have always been scarce, and are becoming increasingly difficult to find. For this reason, work on the artificial production of these crystals was started, and with the exception of quartz and calcite, considerable success has been achieved. Rocksalt (sodium chloride), sylvine (potassium chloride) and fluorite (calcium fluoride) can all be produced artificially, and the technique of the methods developed for their growth has been successfully applied to other substances which do not occur in nature in the form of large crystals, with the result that there is now available a wider variety of substances from which to choose. Potassium bromide, lithium fluoride, sodium fluoride, and silver chloride, all of which have desirable optical properties, can be grown successfully and are now in use. It is usual to refer to such crystals (incorrectly), as " synthetic."

By using mixtures of pure salts in known proportions a considerable number of " mixed crystals " or " solid solutions " have been grown as homogeneous single crystals (§117). In this manner attempts have been made to produce materials with improved properties such as reduction of deliquescence, which in the case of crystals is often aggravated by the presence of impurities. The natural crystals most liable to deliquescence can be distinguished by their feeling " soapy " to the touch. Single crystals of thallium bromo-iodide, a solid solution of thallium bromide and thallium iodide, are now becoming available; this material is sufficiently transparent for use as a prism material at infra-red wavelengths between 28 μ (the useful limit of potassium bromide) and 40 μ.

It is generally supposed that the crystals which are found in nature have been formed either by solidification from the molten state during the cooling of the earth's crust, or from aqueous solution. In the latter case the crystals frequently contain inclusions of the mother liquor which render them less suitable for optical purposes, for such crystals are no longer homogeneous and have undesirable absorption bands in the infra-red region. These inclusions are often to be seen in natural rocksalt.

Artificial crystals also may be grown either from the molten material or from solution, but although some beautiful specimens of piezoelectric crystals are now being grown from solution in water, most of the crystals for optical purposes are being grown from the melt. In optical properties these synthetic crystals compare favourably with natural crystals, and this is hardly surprising when one considers that, in the laboratory, it is possible to start with very pure materials, and that such factors as the rate of growth can be carefully controlled, whereas in the natural process these were matters of chance.

Single crystals of rocksalt up to 35 lb in weight have been grown having the same transparency as the natural crystals, and showing none of the flaws and inclusions so common in large natural crystals. Artificial sodium chloride crystals show the same properties of cleavage as natural rocksalt and are less affected by a humid atmosphere. Potassium bromide also crystallizes in the cubic system and cleaves like rocksalt, but is rather more deliquescent. Lithium fluoride, another cubic crystal developed for its ultra-violet transmission can in some instances be used instead of fluorite. It has a low solubility in water and is consequently little affected by atmospheric humidity.

Silver chloride crystallizes in the cubic system but does not cleave along the usual cleavage plane. It is transparent in the infra-red region as far as 20 μ, and being insoluble in water it is stable in a damp atmosphere. It suffers a photo-chemical change, however, when exposed to ultra-violet radiation; it darkens in direct sunlight and slowly in diffuse daylight. Suitable coatings can be applied to protect it from ultra-violet light. The crystal has remarkable plasticity; it can be turned on a lathe, sawn or subjected to any of the operations common to plastic materials. When hot it can be pressed into any desirable shape or rolled into sheet (Kremers, 1947).

The methods employed for crystal growth have developed along two lines. In the process due to Stockbarger, 1936, which has been further developed and is used extensively by the Harshaw Chemical Company in the United States (Kremers, 1940), and in this country by Taylor, Taylor & Hobson, a platinum crucible with a conical bottom is loaded with the very pure salt. The temperature is raised until the contents of the crucible are molten, and then the crucible is slowly lowered through a sharp temperature gradient into a region where the temperature is somewhat lower than the melting point of the salt. Since the apex of the conical bottom is the first point to enter the region of lower temperature, crystallization begins here and gradually extends upwards until the whole contents of the crucible are transformed into a single crystal. The lowering of the crucible takes about one week and a further ten days are required to cool the finished crystal slowly to room temperature. Very careful control of temperature in both the upper and lower parts of the furnace. and of speed of lowering, are required during the whole process,

Employing this technique the Harshaw Chemical Co. have, in regular production, synthetic crystals of sodium chloride, potassium bromide and lithium fluoride of between 30 and 35 lb in weight. Fluorite is grown up to $3\frac{1}{2}$ lb, and silver chloride crystals weighing 10 lb are also being produced, the latter being grown in Pyrex crucibles.

The second method which has been used extensively in Germany, is due to Kyropoulos and Pohl and was largely developed in the 1st Physical Laboratory in the University of Göttingen. In this process a seed crystal, which is held in a water-cooled chuck, is lowered so that it dips into the pure molten salt contained in a suitable crucible. The temperature of the melt, which is initially some 30° or 40°C above its melting point, is then allowed to fall to a few degrees above the melting point, where it is held steady. The flow of water through the chuck holding the seed crystal is then slowly increased and growth occurs on to the seed. The crystal slowly increases in thickness and diameter until it approaches the sides of the crucible; the chuck from which it is suspended is then slowly raised. By this means a cylindrical shaped crystal is gradually drawn upwards out of the melt. The rate of lifting is of the order of 3 to 4 mm per hour. Again careful control of temperature and rate of lift are necessary. When the crystal has grown to the desired size, it is lifted clear of the melt and transferred to a preheated furnace for slow cooling to room temperature. Crystals weighing about 4 lb can be grown in approximately two days, and a further time is needed for cooling. Larger crystals take a longer time. The resulting crystal is single, having the same principal axes as the original seed.

Most of the alkali halides including sodium chloride, potassium bromide, potassium chloride, potassium iodide and lithium fluoride have been grown by this technique. Sodium chloride crystals up to 40 lb in weight have been grown during the war by Dr Körber of I. G. Farben. in Ludwigshafen. Employing the same method, crystals of sodium chloride and potassium bromide are now being grown in England in the laboratories of Hilger & Watts Ltd.

The growing scarcity of calcite, needed for high-quality polarizing prisms, etc., and the difficulty of producing this crystal artificially, has stimulated research for the production of large crystals of sodium nitrate, which crystallize in the same habit as calcite and also exhibit strong double refraction. Owing to its low melting point (308°C) which makes it difficult to maintain a high temperature gradient, and its anisotropic character, it is not an easy crystal to grow. A method which has been used successfully has been described by Stöber (1925).

Pure sodium nitrate is placed in a roughly hemispherical glass vessel which stands on a metal base plate, to the under side of which water cooling can be applied. Above the glass vessel is a heavy metal disc containing heating elements. The whole apparatus is surrounded by heat lagging material. Initially, the top heater is turned on, but no cooling is applied to the lower plate. The temperature is allowed to rise until all the sodium nitrate is molten. The water flow to the lower plate is then gradually turned on, the temperature of the top heater being kept constant. The cooling of the lower plate slowly reduces the temperature at the bottom of the glass vessel and at some stage the lowest point of the vessel reaches the melting point, and crystallization commences here. Heat is continuously drawn away from below as the water flow is increased, and so the contents of the vessel are slowly transformed into the solid state, the solid-liquid boundary slowly moving upwards through the melt. It may be necessary in the later stages to allow the temperature of the top heater to fall. When the whole contents of the vessel are solid, the temperature at the top and bottom are evened out at some point below the melting point, and the crystal is then slowly cooled to room temperature. Provided the lines of heat flow are vertical, and all parallel, a single crystal results. Sodium nitrate is soluble in water and deliquescent. Its birefringence is greater than that of calcite.

The table on pp. 162-3 gives a list of optically useful crystals and their properties. Where the hardness is expressed as a plain digit it is on the Mohs scale as given in Dana's "System of Mineralogy."

116.1 In addition to the crystals mentioned in §116 and in the table, caesium bromide is now available from the Dow Chemical Company. A purity of 99·6 per cent is claimed, impurities being—

Rb 0·05 to 0·01 per cent
Na 0·005 to 0·05 per cent
K 0·005 to 0·01 per cent
Ca 0·005 to 0·01 per cent

It can be further purified from water solution, adding two to three times the value of isoprophyl alcohol. The melting point is 636°C approximately. The residual ray reflection is at about 110 to 120μ. The limit of transmission is unknown but from the general properties of alkali halides one might expect it to be beyond 40μ. It is not particularly hygroscopic. In spite of the high price it may be an important material for the infra-red.

Artificial sapphire in sizes useful for optical elements and of acceptable optical qualities is now obtainable from Linde Air Products in sizes up to 20 mm diameter and 10 mm thick. Interferometry tests show that the samples are sufficiently uniform in refractive index for most optical requirements. Calculation shows that the birefringence of the sapphire is not objectionable in lenses of moderate field view if the crystal is cut perpendicular to the optical axis of the crystal and it is said that methods have been devised to compensate the effects of double refraction if it becomes too great. A 1 in. focal length f/1·5 Petzval lens has been made with two sapphire elements. The refractive index of sapphire for D is about 1·769 for the ordinary, and 1·760 for the extraordinary ray. The artificial sapphire requires to be annealed.

Polishing is achieved with diamond dust using a lap of tin-lead alloy. The ratio of tin to lead is varied during the process, softer and softer being used as the finishing proceeds. The quantity of diamond required is trifling—indeed it is thought that the polishing is effected by sub-visible particles of sapphire. Up to eight hours polishing on the machine is required. *From Hopkins and O'Brien, 1949.*

Composition and hardness of various polishing and abrasive powders

116.2 (This list was compiled by Mr. S. J. Underhill, Adviser on optical techniques, Hilger & Watts Ltd.)

Material	Chemical composition	Hardness (Mohs)	Origin	Remarks and sources of information
Polishing powders				
Rouge	Fe_2O_3	4—7	Synthetic	Anhydrous ferric oxide (1) is used for polishing glass and gemstones (5).
Cerium oxide	CeO_2	—	,,	Now much used for polishing optical glass.
Zirconium oxide	ZrO_2	5·5—6·5	,,	Also used for optical glass polishing.
Titanium oxide	TiO_2	3·8—4·2	,,	Used for optical glass polishing.
Chromium oxide	Cr_2O_3	7—7·5	,,	Extensively used for polishing metals.
Putty powder	SnO_2	6—7	,,	Tin dioxide, used in polishing precious stones (5).
Glassite	Fe_3O_4	5·5—6·5	,,	Principle ingredient, black oxide of iron.
Diamantin	Al_2O_3	9	—	—

Abrasive Powders

Material	Chemical composition	Hardness (Mohs)	Origin	Remarks and sources of information
Diamond	C	10	Natural	—
Corundum	Al_2O_3	9	,,	Mined in the Transvaal (2).
Ruby powder	Al_2O_3	9	,,	Red coloured Corundum.
Emery	Mainly Al_2O_3	7—9	,,	Variety of Corundum containing much admixed magnetite and haematite. Mined chiefly in Naxos.
Aloxite	Al_2O_3	9	Synthetic	—
Alundum	Al_2O_3	9	,,	—
"Sira" abrasive	Al_2O_3	9	,,	A uniform fine grained powder, used for the final smoothing of glass before polishing.
Carborundum (silicon carbide)	SiC	9	,,	—
Boron carbide	B_4C	9	,,	Produced in electric furnace (2). Used as a lapping abrasive.
Silex powdered quartz	SiO_2	7	Natural	—
Jasper powder	SiO_2	7	,,	An impure opaque form of cryptocrystalline silica (2).
Pumice	—	6	,,	Vesicular lava (8) an igneous rock, cooled rapidly; often contains harder impurities which cause scratches (1). Consists largely of aluminium silicate (11).
Argillaceous earth	Mainly aluminium silicate	—	,,	Argillaceous or clayey rocks consist of the finest fragments worn from older rocks; examples :— slate, shale, etc. (2).

Material	Chemical composition	Hardness (Mohs)	Origin	Remarks and sources of information
Water-of-Ayr stone	—	5—8	Natural	A fined grained sand stone.
Vienna lime	CaO. MgO	—	Synthetic	Vienna lime produced in Bavaria is similar in composition to " Sheffield Lime " ; an analysis of a sample of the latter made at Imperial Institute showed lime 57 ; magnesia 40; and less than 1% of silica, alumina and ferric oxide (8).
Tripoli	Mainly SiO_2	6—7	Natural	A porous siliceous rock when pure contains about 98% silica and generally some alumina and oxide of iron, not to be confused with Tripolite which is another name for diatomaceous earth (8). The variety found in Derbyshire is called Rottenstone (10).

References

1. *Abrasive materials* by A. B. Searle.
2. *Rutley's Elements of Mineralogy* by H. H. Read.
3. Harmsworth *Self-Educator*.
4. *Handbook of Chemistry and Physics* by C. D. Hodgman.
5. *Silica and the Silicates* by J. A. Audley.
6. *Engineering Materials* by Judge.
7. *International Critical Tables*.
8. Imperial Mineral Sources Bureau Booklet *Abrasives*.
9. *Materials Handbook* by G. S. Brady.
10. *Chemical Synonyms and Trade Names* by W. Gardner.
11. *Kingzutts Chemical Encyclopaedia* by R. K. Strang.

E.B. *Encyclopaedia Britannica* 11th Edition.

Optically Useful Crystals

Name of material	Formula	Mol. Wt.	M.P., °C	Hardness	Solubility in cold water grams/100 cc	If produced artificially	Special properties of natural crystals	Special properties of artificial crystals	Notes, uses, etc. Transmission limits are only approximate. At the red end they are the points at which the transmission of a 2 mm thickness falls to about 50%
Aluminium Sesquioxide (Corundum, Ruby, Sapphire Alumina)	Al_2O_3	101·94	2050	9	0·000098	Yes	Emery is bluish grey or black corundum containing magnetite (Fe_3O_4)	Grown for jewel bearings	—
Calcium Carbonate (Calcite, Calc-spar, Iceland spar)	$CaCO_3$	100·08	Sublimes at 898·6°C	3	0·0014	—	—	—	Doubly refracting crystal used for polarising optics Transmission 2100A
Calcium Fluoride (Fluorite, Fluorspar)	CaF_2	78·08	1360	4	0·0016	Yes Harshaw Chemical Co. U.S.A.	Natural Fluorite usually coloured	"Vacuum grown" crystals are very perfect	1200A (best specimen) to 9μ. Used with quartz for achromatic lens combinations
Lithium Fluoride	LiF	25·94	870	136 Kg/mm²	0·27	Harshaw Chemical Co. U.S.A.	—	Grown in large sizes of high quality (the Harshaw vacuum grown LiF is free from absorption bands due to hydrolysis)	1200A...7μ. Achromatic lens combinations
Potassium Bromide	KBr	119·02	730	13 Kg/mm²	53·48	Harshaw, Hilger & Watts Taylor Taylor Hobson	—	Grown up to 9 in. diam., 6 in. thick	Infra Red optics. Hygroscopic 2100A...28μ
Potassium Chloride (Sylvine)	KCl	74·56	776	14 Kg/mm²	34·72	Grown artificially Harshaw Hilger & Watts	—	Artificial crystals grown from very pure KCl show no absorption at 3·2 or 7·1μ	2000A...21μ with narrow absorption bands at 3·2μ and 7·1μ
Silica (Quartz)	SiO_2	60·06	1470	7	insoluble	Yes Silica Syndicate Very rarely sufficiently homogeneous for best work	—	Not grown in sufficient size or quality for optical purposes	Grown artificially for use as oscillator plates. Lens combinations, ultraviolet optics. Transmission 1850A...3·5μ

DIOPTRIC SUBSTANCES

Substance	Formula	Mol. Wt.	M.P.	Hardness	Solubility	Commercial	Quality	Notes	Remarks
Silver Chloride (Cerargyrite, Horn Silver)	AgCl	143·34	455	1·5-2 (Dana) 1·1-5 (Handbook of Chemistry and Physics) 14 Kg/mm² (Zeiss microhardness tester)	0·000089	Yes Harshaw Chemical Co. Hilger & Watts Ltd.	Grey or coloured (not suitable for optical use)	—	Infra-red window for use with aqueous solutions. Must be protected from blue or ultra-violet light which causes it to blacken. Material can be rolled into thin sheets or pressed into shapes, lenses, etc. . . . 22μ
Sodium Chloride (Rock-salt, Halite)	NaCl	58·45	804	2, 26 Kg/mm²	35-70	Yes Harshaw Hilger & Watts Taylor Taylor Hobson	Often cloudy and with water inclusions	Less hygroscopic than natural crystals believed to be due to greater purity (absence of magnesium and calcium salts). Very large clear and perfect crystals are grown	Infra-Red optics. Transmission 2000A . . . 17μ
Sodium Fluoride (Villiaumite)	NaF	42·00	980	3·5, 84 Kg/mm²	4·0	Yes	—	—	—
Sodium Nitrate (Soda Nitre, Chile Saltpetre)	NaNO₃	85·01	308	1·5-2	73·0	Harshaw Chemical Co.	—	—	Birefringent—can be used to replace calcite in many polarizing optics. Hygroscopic. Transmits to about 2500A
Thallium Bromide	TlBr	284·31	460	13 Kg/mm²	0·0525	—	—	—	—
Thallium Chloride	TlCl	239·85	430	10 Kg/mm²	0·3220	—	—	—	—
Thallium Iodide	TlI	331·3	440	—	0·0064	—	—	—	—
Binary Mixed Crystals									
Silver Chloride / Silver Bromide (KRS 13)	35% AgCl / 65% AgBr	— / 4	455 / —	— / 17 Kg/mm²	=6×10⁻⁵ / —	—	—	—	Described by Koops as yellow to olive green. Surfaces turn dark on exposure to light. Almost unusable. Transmission 14μ
Thallium Bromo-Iodide (KRS 5)	44% TlBr / 56% TlI	— / —	— / —	30 Kg/mm²	0·02 / —	Harshaw Chemical Co.	—	—	Most useful of the Binary Mixed Crystals. May be used in open air. Transmission . . . 40μ
Thallium Chloride / Thallium Bromide (KRS 6)	60% TlCl / 40% TlBr	— / —	— / —	35 Kg/mm²	— / 0·1	—	—	—	Used where blue absorption of Thallium Bromo-Iodide is troublesome. Transmission . . . 30μ

Barium and strontium fluorides are said to be of value for the infra-red.

Decolorizing coloured crystals

Blue fluorite (Blue-John) has been examined in order to ascertain its usefulness if the blue colouration could be removed. The raw material fluoresces a bright blue; plates (0·25 in. thick) transmit to $\lambda 2500$. The colour can be successfully removed by heating the crystals to 595°C (heating above this temperature produces milkiness), and this temperature lowers the limit of transmission to $\lambda 1750$ but does not diminish the fierceness of the fluorescence. In the infra-red the decolorized material has several absorption bands which reduce the transmission to 20 per cent after passing through a plate only 0·25 in. thick. It would seem that on account of fluorescence in the U.V. and absorption bands in the I.R. that this material could not be used in those regions. Thus it appears that the only use for it would be in the visible (possibly in microscope objectives).

Iceland spar of a " yellow " colour can be decolorized; but it is very doubtful whether it becomes more transparent in the ultra-violet. A piece $1 \times 1 \times \frac{1}{2}$ in. was heated to 342°C and decolorized to a large extent. A larger piece ($1 \times 1 \times 1$ in.) from the same crystal was heated to 420°C and, at first sight, appeared to have clarified but, on closer examination, an extremely faint neutral shade was seen in one part of the piece and, on observing in the arc light, this part was found to be slightly misty. The misty part divided from the clear part by a vein.

The decolorization was found to be permanent for four months during which it was exposed to the air and light on the roof of building in London.

It is noteworthy that the temperature required " permanently " to remove the " smoke " from quartz and the colour from Iceland spar seems to be the same, viz. not less than 400°C.

Binary mixed crystals as optical materials

117 This section is translated from Koops (1948).

Optical elements made from crystal are, in comparison with those made of glass, considerably more expensive both in production and in the raw material. On that account they are only used in optical work when the prescribed conditions cannot be fulfilled with glasses. This is the case:

(1) In polarization optics, where double refracting crystals are used.

(2) In optical work of the highest quality where, through a combination of crystals and glasses, particularly good colour correction can be reached.

(3) In the optics for ultra-violet and infra-red radiations which are normally absorbed by glasses.

Originally only those crystals were used which were found in nature in sufficiently large individual pieces. In the course of time crystals were added which could be artificially produced from solutions or melts (Stober, 1924, and Kyropoulos, 1926). With these one could replace the rare crystals provided in nature. For example ; lithium fluoride replaced fluorspar, sodium chloride and potassium chloride replaced rocksalt and sylvine. Further optical materials were obtained for spectral regions for which no dispersion optics were previously available, as for example, potassium bromide and potassium iodide for the longer infra-red.

The available optical materials so obtained satisfied requirements for a good while. Nearly all crystals for the infra-red have indeed the unpleasant property of being soluble in water, so that their optical surfaces are attacked by the water vapour in the atmosphere. Since, however, measurement in the long wave infra-red is in itself laborious and requires great care (and sensitive instruments because of the slight amount of energy to be measured) a careful handling of the optical work introduces no great increase of difficulty in the work. One works in thermostatically controlled and dry rooms and takes care to put the optic into a desiccator in the intervals between experiments. To extend the available dispersive materials into yet longer wavelengths was, for certain work, desirable, yet not absolutely necessary (Barnes, 1938).

The supply of optical materials in this region, first became once more a serious problem when it became necessary to make measurements in the open atmosphere in the regions for which rocksalt, sylvine or potassium bromide optics were otherwise satisfactory. All materials which up till then had given good results in the laboratory failed in this respect.

Experiments to protect those materials with films had little result. Nearly all such sealing films absorbed radiation, and even if they did not do so they had other unfavourable properties ; for example, considerable loss of light by reflection since they consist of substances of high refractive index, or in other cases sensitiveness to short-wave radiations as is the case with films of silver salts. An entirely unobjectionable sealing film has not yet been found.

A solution of the problem can only be expected from crystals of new materials which can be produced by the well-known method of cultivation of single crystals. These materials must fulfil the following conditions—

(1) They must be transparent far into the infra-red. Since the transparency depends on the characteristic frequencies of the crystal lattice or one or more groups of atoms, only such materials come into question whose characteristic frequencies

lie far in the infra-red, *i.e.* materials with large ions and correspondingly small lattice energy.

(2) They must be isotropic, since on account of the small amount of energy available in the infra-red, optical systems of wide aperture must be used.

(3) They must not be hygroscopic.

(4) It must be possible to polish them by the methods customary in the optical industry.

Since the time of Rubens (1915) a small group of salts has been known which fulfil conditions 1 to 3 and, out of these, large single crystals could be produced. These are the thallium and silver halides. The chloride and bromide are cubic, the iodide dimorphic. Thallium iodide has above 170 to 180° a dark red cubic modification, which at lower temperatures changes to a yellow rhombic form (Barth, 1926). Silver iodide is in the α–modification, cubic at temperatures above 146° (Strukturberichte, 1913–28). None of these materials, however, fulfils condition (4).

If one compares conditions (1) and (4) it is obvious that they are to a great extent antagonistic. Only those materials are polishable by ordinary optical methods which have a certain degree of hardness, but the hardness is directly dependent on the lattice energy (Joos, 1943), and only those materials can be polished which have a lattice energy exceeding a certain minimum. Since condition (1), however, is only possible with materials of small lattice energy, the fulfilment of the one condition to a determined point excludes the fulfilment of the other condition. This happens in the above-mentioned group of substances, they are sometimes so soft that they can be cut with a knife and damaged or deformed by a blow.

(*Note.*—One may mention that this great plasticity can be used for finishing. By pressing thallium chloride at a high temperature in the direction of the normal to the 111 plane between highly-polished stamps, windows can be produced which although not of optical quality, have been used satisfactorily to protect a small number of bolometers for a considerable time under the most severe conditions.)

These considerations point the way to further development. It is improbable that salts can be found to fulfil all four conditions. Fulfilment of condition (4) must be achieved by the thallium and silver halides, but the hardness of these materials must be raised so that one can polish them.

Through experiments directed towards this end one must determine the degree of hardness needed to make polishing possible under the ordinary conditions of optical finishing (pitch-polishing). For measuring the hardness the micro-hardness tester of the firm Zeiss* was used on which the hardness is given in kilograms per square millimetre. With the micro-hardness tester reliable measurements can be made even on small single crystals, The values are plotted on " guides " in order to get a better comparison. It was found that a hardness of at least 15 kilograms per square millimetre is to be sought.

Since the difficulty of polishing depends not only on the hardness of the material but also, for example, on the solubility in water, the regions overlap. Potassium chloride is relatively easy to work; potassium bromide, in case of need, is also possible; on the other hand the silver chloride and tellurium bromide which have the same hardness, could not be polished in the ordinary way even by the most skilled opticians in Jena.

To raise the hardness the method well known in metallurgy is to add other components. It is difficult not to spoil the homogeneity of the materials in combining foreign materials with them, since the optical test is a very severe one.

The hardness is obtained—

(1) Through the addition of small quantities of non-isomorphic substances. In order to attain approximate optical homogeneity, one of the substances transparent to the infra-red must be chosen which has optical properties similar to the basic substance, on account of the unavoidable segregation.

A crystal from a melting of thallium chloride was combined with 1 per cent silver bromide; this increased the micro-hardness from 10 kilograms to 17 kilograms per square millimetre, and windows of this material were polishable.

This method was soon given up however, since the following method led to the required goal considerably more easily and with fewer disadvantages.

(2) The combination of a larger quantity of isomorphic material, that is by the production of binary mixed crystals.

It is known that with metals mixed crystals are sometimes considerably harder than the pure components (Sauerwald, 1929). Fig. 55 shows the usual melting diagram of a complete mixed crystal diagram. The abscissae show the percentage composition, the ordinates the temperature. L is the liquid curve, S the solid curve; H gives the variation of hardness for the various compositions of a mixed crystal in the case of metals. The same effect takes place in ionic crystal mixtures. If one adds silver chloride to sodium chloride one gets a crystal which is considerably harder, this is shown in the third row in Fig. 54. The hardness increases, for example from 26 kilo-

* The Zeiss *Mikrohärteprüfer nach Hanemann* see §102.

grams per square millimetre (the hardness of pure sodium chloride), to 39 kilograms per square millimetre by the addition of 7 per cent silver chloride, and 45 kilograms per square millimetre by the addition of 10 per cent silver chloride. Segregation makes these crystal mixtures, however, so heterogeneous that they cannot be used in optics. The dependence of hardness on silver chloride content is determined by measurement. The point of the crystal whose hardness is to be determined is bored with a fine drill and the silver proportion

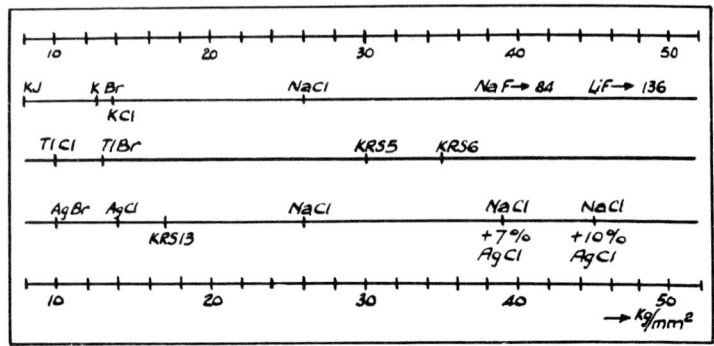

Fig. 54—Change of hardness on addition of silver chloride to sodium chloride

of the powder determined by titration. Since the growth of the crystal takes place very regularly, the alteration of refractive index which is produced by segregation increases very constantly. A slice taken through the crystal perpendicular to the axis has a definitely negative focal length.

If one limits oneself always to the mixed crystals which show a minimum in their melting diagrams, then one has, in the combination which corresponds with the minimum of the liquid curve, a material that behaves like an element (Einstoffsystem) which therefore has a definite melting point and in solidifying exhibits no segregation. This material, corresponding to the minimum, must not be confused with a chemical combination, for it does not require a simple definite " Stochiometric " proportion of the components.

Fortunately the mixed crystals of thallium and silver halides belong to those systems which have a minimum in the liquid curve. The melting diagrams are shown in Fig. 55a.

Symbol and composition of the mixed crystals
KRS 5   44 per cent TlBr and 56 per cent TlI
KRS 6   60 per cent TlCl and 40 per cent TlBr
KRS 13   35 per cent AgCl and 65 per cent AgBr

DIOPTRIC SUBSTANCES

Some properties of these crystals

Symbol	Micro-hardness kg/mm³	Specific gravity gm/cm³	Coefficient of expansion °C⁻¹	Solubility gr/100 gr
KRS 5	30	7·2	$45 \cdot 10^{-6}$	ca. 0·02
KRS 6	35	—	$50 \cdot 10^{-6}$	ca. 0·1
KRS 13	17	—	$39 \cdot 10^{-6}$	ca. $6 \cdot 10^{-5}$

From Fig. 54 and the Table one gets the hardness. All these mixed crystals are easily polishable; they are distinguished by the number

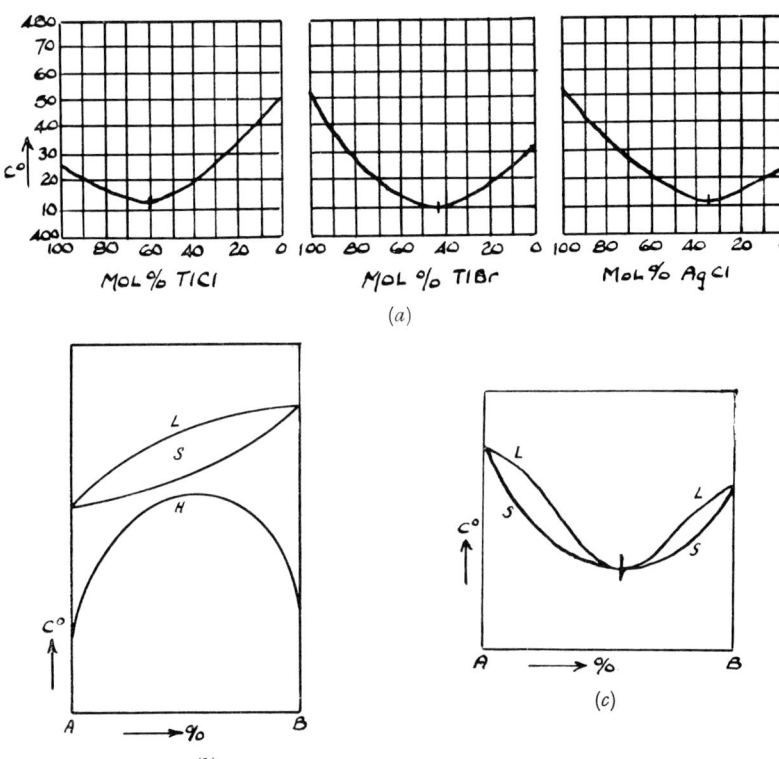

Fig. 55—Melting diagrams of mixed crystals

of the experiment with the letters KRS (*Kristalle aus dem Schmelz-fluss*), an abbreviation for " crystals from the melt."

The mixed crystal of thallium bromide and thallium iodide is interesting on other grounds. Small additions of thallium bromide to thallium iodide makes the cubic lattice spacing pertaining to the temperatures above 170 to 180°C stable for room temperature (Barth, 1926). The crystals then show a dark red colour, so that it might be thought that they could be used for infra-red filters, absorbing the visible light, but the optical heterogeneity resulting from segregation prevents such a use.

Of the three materials so discovered KRS 5 is the most useful. In order still further to increase its optical homogeneity the minimum of the melting diagram was again measured and a small shift found as compared with that given by Morksmeyer (44 per cent instead of 42 per cent thallium bromide). KRS 5 crystals were produced early in 1941 by the firm Carl Zeiss and used as windows, cells, lenses and compound objectives and correction lenses for mirror systems.

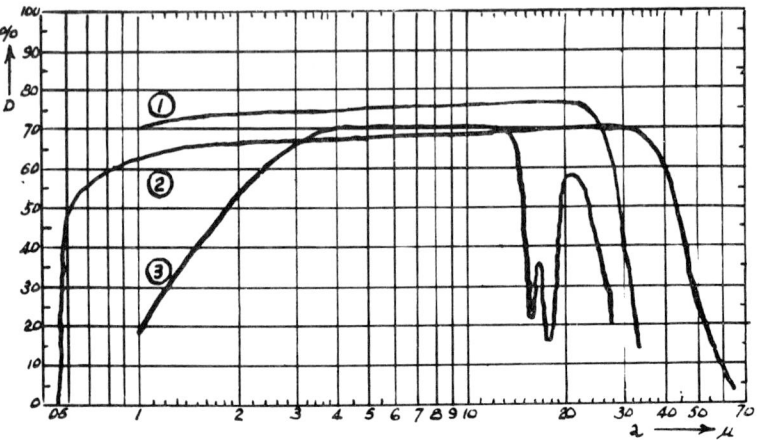

Fig. 55A—**Transmittance of mixed crystals**
(1) KRS 6, thickness 3·5 mm.   (2) KRS 5, thickness 2·4 mm.
(3) KRS 13, thickness 9·5 mm.

KRS 6 was only used in certain cases where the blue absorption of KRS 5 was injurious. KRS 6 is almost white compared with the red KRS 5.

KRS 13 shows the disagreeable property of the silver salts; mechanically worked surfaces turn dark in the light, particularly in the presence of moisture. In spite of the very favourable extremely low solubility KRS 13 is practically unusable; its colour is yellow to olive green.

The optical data of these materials were determined outside the Zeiss works, as follows :—

Transparency by Herr. Prof. Czerny, Frankfurt. This is shown in Fig. 55A. The diminished transparency of 70 to 80μ in the spectral region where there is no absorption is occasioned by loss by reflection at the two air-crystal surfaces which results from the large refractive indices of the latter. This loss can be reduced to about 5 per cent for each surface in the infra-red up to about 4μ by reflection-diminishing layers. For the long wave infra-red, there is no substance available for making suitable films, all substances which fulfil the optical conditions absorbing strongly.

The reflection-diminishing films were produced in the Zeiss works and were used in making triplets from KRS 5. The increase in transparency was considerable.

The dispersion of KRS 5 and KRS 6 were measured on various prisms by Herr Professor Hettner and his pupil Fraulein Leisegang in Jena (Hettner and Leisegang, 1948).

The high refractive index of these materials makes possible the computation of optical systems with extreme apertures; for example, in the Jena triplets from KRS 5 a focal aperture of 1/0·7 was arrived at, with 35 mm focus and an image angle of $\pm 15°$, with a resolving power in the image plane of about 30μ (this objective did not need to be corrected chromatically since the dispersion curve KRS 5 in the region that was used is very flat and since the receiver in the combination was very selective).

Summary

By changing from single systems to binary systems in the production of synthetic crystals for optical work, optical media for the infra-red can be produced which, as compared with those hitherto available, are very insensitive to the influence of the atmosphere. They possess far-reaching transparency in the infra-red and exhibit also high refractive indices.

Cutting, grinding and polishing crystals, plastics and metals

Trueing and smoothing

118 *Selection of quartz crystals*—One of the most important of the operations in working crystals in the works of Hilger & Watts Ltd is the selection of quartz from which it is intended to make prisms and lenses. Quartz is extremely expensive and the labour of cutting it and getting it " to axis " is a considerable part of the whole labour in making the prism. A great deal of work can be saved by careful examination for internal flaws and other defects. If the crystal is immersed in methyl-salicylate the refractive index of this liquid is sufficiently close to that of quartz to enable even small defects in the interior of the crystal to be located and avoided. The Brazilian Government, which grades and levies export duty on quartz according to the supposedly usable volume, uses this method of examination.

Cloudy and coloured quartz can be improved in appearance by heating to 400°C and cooling at a rate of about 1° per minute. Transmission in the ultra-violet is improved thereby but does not equal that of naturally clear crystal.

The properties of greatest importance to the optician in working crystals are their hardness and degree of hygroscopy.

Starting first with the soft crystals, in Germany Wienerkalk (Vienna lime) or the very finest argillaceous earth is used for grinding very soft materials, especially sodium and potassium chlorides. It might also be worth while trying grinding and polishing soft crystals with chromium oxide on steel plates. For polishing, many use a glass plate with silk stretched thereon, using first chromium oxide for grinding and for the final polishing only the damp silk. Pumice powder is recommended by Bernard Halle (1913), for finishing and smoothing soft crystals, but special care must be taken in the washing; often it is contaminated with grains of hard material which unfortunately cannot be washed away since they settle at the same rate as the pumice itself. Before starting an important piece of work, therefore, it is desirable to test a small amount on a soft crystal which is of no importance, grinding it with the pumice on a grey glass plate. If fine scratches are produced on the crystal the pumice is unsuitable for the grinding or smoothing of soft crystals and metals. *" Blaustein "* and *" Graustein "* are even softer than pumice and therefore suitable for grinding still softer crystals. The former is excellent as a smoothing material in the form of a solid tool for smoothing speculum and other metals. Emery and carborundum for smoothing materials of a hardness comparable with glass, have already been described. For grinding and smoothing diamond, corundum, ruby, jasper, and other hard crystals, diamond powder is suitable on account of its great hardness. To prepare the diamond it should be broken up in a hardened-steel cup using a hardened-steel plunger and must, like other grinding materials, be carefully washed.

Polishing

Rouge is satisfactory for polishing most crystals. Polishing can often be hastened by using two kinds of rouge, a hard and sharp one to begin and a softer and less sharp kind for the final polishing. The quality of the rouge depends upon the temperature at which the iron sulphate is roasted, and different qualities can be drawn from the same furnace if the material is not stirred during heating. For softer materials such as Iceland spar and most metals, tin oxide (putty powder) is useful although at Hilger's rouge is on the whole preferred (see also §92). Chromium oxide polishes fluorspar, aragonite and other crystals of similar hardness, while diamantin gives a good deep black polish on hard steel, particularly when it is mixed with a few drops of fine oil, but in every case it is important that all such materials are obtained in the purest possible state from a reliable chemical firm. Purity and uniformity of grain are essential and, of course, all polishing materials should be washed before use and kept in covered receptacles. The small quantities required by the opticians for their daily use should be kept in covered porcelain jars similar to those sold by druggists for ointments. The grinding material for bench use should always be kept moist in its jar.

A method of dry polishing used in France up to fairly recent times, when rightly carried out, gives very good results ; it does not give such complete control over the shape of the surface, but might be useful where water cannot be used. Good letter paper is cemented on a plane plate and another plane plate pressed on it, so that the superfluous cement is pressed out. The paper is then sprinkled (after the drying of the cement) with powdered Tripoli, with which one then polishes exactly as with rouge, but keeping the material dry. The fine scratches which are likely to result can be reduced by cleaning the superfluous polishing material from the tool, breathing on the latter and drawing the glass across it with a heavy pressure. Suitable paper made for this purpose can be purchased in Paris under the name of " Berzelius paper ".

For taking the fine grinding material out of its jar one should use a copper or brass rod about 2 mm diameter with a spoon-shaped end, while for trueing emeries a wooden strip is suitable.

Instructions for polishing ruby are given in the *Industrial Diamond Review*, October 1946, page 314. The material used for grinding is diamond powder broken down in the manner described and washed to appropriate grain size. Laps of pear wood well seasoned with the grain of the wood perpendicular to the face of the lap are used to form a lap which runs at 300 revolutions per minute. This is said to produce a good polish. It is scarcely likely to be of optical quality yet doubtless the opticians by using some ingenuity might form a flat polisher. Suggested alternatives for pear wood are poplar or alder. There are various other substances of great hardness which might be worth trying for trueing and smoothing hard crystals. Boron carbide (B_4C) has a micro-hardness 7000 kg/mm^2, silicon carbide (SiC) 4500 kg/mm^2, while corundum (emery) has a micro-hardness of only 3000 kg/mm^2.

In the optical shops of Hilger & Watts it is customary to polish ruby on natural quartz using rouge. In polishing metals, which are on the whole softer than crystals, it is found difficult to keep the surfaces flat. As a remedy for this it is found that if the final grey is done on a piece of greyed quartz, without any loose abrasive, a surface is produced which rapidly polishes flat. We have used this for stainless steel, copper and brass. No emery or other hard grains should be used because they become embedded in the metal and it is impossible to polish them out; these are particularly objectionable when it is proposed to rule a grating on the surface, since they will damage the diamond. For Iceland spar or copper we use putty powder.

Finally magnesia (fine and, of course, washed) is used for polishing ivory and might be worth trying for the very softest crystals.

I am indebted to Madame la Comtesse de Maigret for the following method of working ruby. In edging this material, the usual laps get out of order very quickly so boron carbide (B_4C) is used. For sawing, a diamond-charged saw; for smoothing, a flat copper tool charged with diamond; and for polishing, diamond powder used on a paper polisher. The diamond powder for polishing is ground in a mortar, and the grains are probably less than 1 micron in diameter.

In *Machinery*, 21st August, 1947, page 208, there is the following note on *flame* polishing—

Flame polishing, a recently developed technique, has been employed with considerable success on synthetic sapphire and ruby. The rod is rotated and an oxy-acetylene flame traversed along the length, a smoother finish being obtained in this way than is possible with a diamond polishing technique.

For cutting hygroscopic crystals such as sodium nitrate or rock salt a wet cotton thread forms a very good method. It is slow, but is practically without risk of splitting or damaging the crystal—and the large crystals of these materials are very expensive. For roughing, a cloth polisher with water alone may be used, while for smoothing and polishing the usual abrasive and rouge mixed with paraffin is recommended. One of the very difficult soft crystals to polish is thallium bromo-iodide. This material is useful owing to its transparency in the infra-red, particularly over the region for which infra-red rays can traverse considerable distances in air, the region which is sometimes referred to as the " atmospheric window." When polished in the usual way it is found that this substance becomes covered in scratches. Various polishers have been tried by us, such as mixtures of beeswax and pitch, ordinary beeswax alone, and cloth. The method we finally adopted is to polish the material with cloth and putty powder (tin oxide).

Further notes on the treatment of rock salt and other difficult crystals

119 (1) *Polishing rock salt*

Any good method of polishing is used up to the last few strokes.

For these, a flat plate (good selected plate glass) is used, on which is stretched a piece of real silk, and the rock salt is given two or three strokes only using a little rouge and concentrated salt solution. The silk should be only just damp. This is more a cleaning process than a polishing one.

(2) *Permanence of surfaces of synthetic rock salt*

Polished surfaces of synthetic rock salt are reasonably permanent provided that they are never touched and have been finished off in the way described above. Any foreign contamination, such as bits of polishing material left on the surface, leads to its deterioration through moisture. Optical work dealt with in this way does not need protection subsequently except as a precaution against extremely unfavourable atmospheric conditions. It is of great help to have pure materials, and the purity needed should be specified to the suppliers.

(3) *Lithium and calcium fluorides, sodium bromide*

For polishing these materials, at no part of the process must one use carborundum. One must use a soft material, *e.g.* alundum, and preferably *worn* alundum. Scratches must be avoided, because they may start a cleavage. The best polishing agent is Linde's Alumina. This is obtainable in two grades, A and B. It is very fine powder, and some opticians get on better with A, others with B. The correct way of using it is to use a thick cream and go on polishing until almost dry. This is the fastest as well as the best material for polishing this substance, and more encouraging to the optician than any other. It can be obtained from Linde Air Products, Ltd., Tonawanda, New York

(4) *Lithium and calcium fluoride—Temperature*

It is important to handle fluorite at all times with great care to avoid possible temperature differences. Even a few degrees difference may shatter the crystal, especially when in a large piece.

(5) *Sawing; fluorides*

The usual process is to use a hacksaw with plain copper blade, and allow water and alundum to drip into the cut, but diamond impregnated saws, when available, are much quicker and better (see §19).

(6) *Sawing; soluble crystals*

These are always sawn with a thread saw; the thread is of "six cord" three-strand crochet cotton and is made into a belt by:
- (a) scraping the ends with a knife;
- (b) moistening the frayed fibres with waterproof mending cement such as "Duco household cement."

The string should run fast (two or more feet per second) over smoothly running pulleys and should bear lightly on the crystal so that the speed of actual cut depends only on the travel of the workpiece which may be an inch or so per hour for a one-inch thickness. With screw-feed lathe type operation, the cutting requires only occasional attention.

The string should be barely damp when passing through the cut; a " sopping wet " string would dissolve sides as well as the bottom of a cut. A method can easily be devised of getting rid of the saturated salt solution and re-damping the string with fresh water without interrupting the running.

(7) *Sawing; sodium nitrate*

A string saw is used, but in this case hot water is needed because of the relatively high negative heat of solution and brittleness. The best way is to have a tiny gas flame playing on the string just before it enters the crystal. There is no danger in the string getting too hot so long as it is moving.

(8) *Silver chloride; die casting*

Warm the material and squash between plates of highly polished chromium sheet steel at about 150°C. To make it thinner it is then passed through a rolling mill having chromium-plated rollers, also warmed. Sheets of about 1/16 in. thickness can be cut with scissors.

(9) Polished surfaces of sodium nitrate usually tarnish badly in a matter of hours, whereas cleft surfaces have remained as bright and transparent as crown glass for at least 12 years protected only by a glass show case.

Sodium nitrate is likely to be opaque at wavelengths below about 2500A.

(10) *Purity of materials*

It is doubtful whether Analar material is pure enough for artificial crystals to get the best results. It is believed that very slight traces of of iron, aluminium, copper, magnesium, etc., are to be avoided, but small quantities of foreign alkali ions can often be tolerated.

(11) *Annealing of crystals*

The principal requirement is that cooling should take place in a gradient-free box. For example, fluorite is cooled in a graphite box with wall $\frac{1}{2}$ in. thick. The material is soaked at an elevated temperature from $\frac{1}{2}$ hour up to maybe 8 hours, depending on the size, and is cooled reasonably slowly. The important point here being not to cool the outside faster than the inside. If the material conducts heat well, one may cool it faster than a badly conducting material. Fluorite with melting point 1378°C is annealed at 800°C.

Crystals occasionally turn cloudy when annealing. The reason is that there must have been some impurity which comes out on heating. There is nothing to be done but start again with purer material.

(12) *Strain*

By the correct arrangement of the temperature gradients when growing crystals, one may ensure that they have little strain. The best practice is to use no thermocouples and no regulators, but to make the furnace so that they are stable. This requires a large amount of power.

(13) *Anion impurity*

Calcium fluoride is ruined by the presence of even a small amount of chlorine. For example, it is ruinous to clean dirty fluorspar with hydrochloric acid.

The following publications by Dr Stockbarger concerning artificial crystals contain many useful references*—

The Production of Large Artificial Fluorite Crystals (*Discussion of Faraday Society on Crystal Growth*, **5** 294).

Improved Crystallization of Lithium Fluoride of Optical Quality, *Ibid*, 299.

Artificial Fluorite, *J.O.S.A.*, **39** 731.

119.1 Arthur G. Hall, of the Massachusetts Institute of Technology Optical Shop finds it preferable to use garnet of about fourteen micron grain size for the soft and difficult crystals. Attempts at using very fine abrasive invite seizing of the surfaces and resultant serious scratching. Furthermore, polishing, when proceeding correctly, is so rapid that there is not much point in striving to secure an extremely finely-ground surface upon which to start polishing.

He prefers to use a beeswax lap, with which more rapid polishing may be performed with less tendency toward scratching the surface. A small amount of moisture will, by its solvent action, speed up the process but will produce bad pitting of the surface unless used

---

* (See also A. C. Menzies and J. Skinner, 1949, *The growing of crystals by the methods of Kyropoulos and of Stober*; Discussions of the Faraday Society. No. **5**, 306).

cautiously. The lap he says may be used almost dry, in fact he has found that to finish up with an unstained, clean surface, the last polishing spell should preferably be performed on a *completely* dry lap. While a pitch lap will nearly always scratch if used in this way, the beeswax lap will not, if a sufficient quantity of polishing material is present. If too little polishing agent is used with the beeswax lap, the work is sometimes prone to pick up particles of wax, which build up on the surface, necessitating removal with xylene or other appropriate solvent before polishing is resumed. With reference to this last point, it has been found that while more than a very small amount of polishing agent is detrimental with a pitch lap, a fairly liberal amount works better on the wax lap. A more or less burnished surface then develops on the beeswax lap which will produce a good polish.

Linde "A" powder (Linde Air Products Company, Tonawanda, New York) has proved effective with a beeswax lap, giving a rapid polish. He also finds this combination very useful in the polishing of titanium dioxide, a substance difficult to polish with rouge or other such agents. A fairly heavy, creamy mixture of Linde "A" powder and water can be used here as in the polishing of the sparingly soluble alkali halides. It is stated by the Linde Co. that their A powder is "alpha Al_2O_3" with a Mohs hardness of 9 and a grain size of 0·3 micron. They also supply a B powder, "gamma Al_2O_3", with a Mohs hardness of 8 and a grain size less than 0·1 micron. The first has a hexagonal, the latter a cubic, structure. (See also §118.)

Professor Rudolph Koops points out that he finds the polishing of crystals can be greatly eased if one selects as the surface to be worked, the one as nearly as possible parallel to the cubic surface (001) of the crystal. The most unfavourable orientation is that of the octahedral surface (111). A very simple method of determining the orientation of those crystals which have no natural surfaces which he introduced in the Jena factory has recently been described in detail by Smakula and Klein (1949).

Silver chloride is ordinarily worked into optical shapes by rolling and pressing. This entails expensive, accurate and highly polished hard moulds and powerful presses. Such equipment is not available to the ordinary laboratory, so that while this technique is successful for production, it is not appropriate for experimental optical pieces*. Therefore, some way to work lenses, mirrors, windows, etc., by more conventional laboratory and shop methods is highly desirable. The

---

* There appears, however, no reason why, to avoid the need of powerful presses, the operation should not be carried out at a raised temperature according to the principle illustrated in application to "Perspex" in §123; silver chloride has quite a high melting point—460°C.

following technique has been successful in producing acceptable lenses for the infra-red.

The silver chloride is ground on iron using soap and Aloxite. If flats are to be made, the use of two laps with the work being moved between them by a septum produces nice flatness and parallelism with little strain. The lapping of silver chloride takes over an hour for three thousandths of stock removed. The process should not be rushed or the surface will become badly charged; of course one must work in darkness to prevent reduction of the silver chloride, and its consequent darkening.

For the polishing operation, prepare a beeswax lap as described in the above section on KRS–5 and press it flat or to the curve desired. Score the lap in quarter-inch squares. Prepare a one-half-saturated solution of photographic " hypo " ($Na_2S_2O_3 \cdot 5H_2O$) and apply two drops of hypo to the wax lap. Use a standard polishing stroke. Rub for a few strokes and then add two more drops of hypo. There is always a crust which is difficult to remove, but it *can* be removed. When this crust is removed the lap will probably have deposits of silver on its edges. Remove this by washing with hypo and scraping if necessary. If this is not removed the polishing action is much retarded and scratches will result. Apparently what happens is that the surface is dissolved at the same time as the mechanical motion on the wax lap scrapes it away. If the dissolving action seems too rapid for the size of the piece being used, reduce the concentration of the hypo by dilution. If the piece being polished is thin the solution may creep around the edges and get on the top. Hence if the top must be free of stain and pits, it should be covered with wax or lens varnish.

No abrasive is needed in this polishing action, which is especially true as the finishing touches are being put on. The crust may be more speedily removed by using Linde " A " abrasive, but the entire operation may be carried out by a combination of the dissolving action and the mechanical rubbing on the wax lap.

After a " wet " or rub, don't take the work from the lap and look at it: stains will *immediately* result, and so will pits. Plunge the work into clear water to stop the action of the hypo, and then wipe it lightly. One must always remember to work in an area of low light level or in complete darkness, using dull light for inspecting purposes. Needless to say, ordinary tungsten filament lamps are much safer than fluorescent lighting which has strong ultra-violet intensity.

Using the above-noted technique, Mr Frank Cooke has made lenses and flats which are accurate to three fringes per inch. The surfaces have shown almost no hair lines and no scratching. It is believed, therefore, that given time and correct working conditions a skilled optician can make optical flats of silver chloride by following the above-outlined technique.

The refractive index of silver chloride for the visible and infra-red has been published by Tilton, Plyler, and Stephens (1950), of the National Bureau of Standards. They give the following values—

Wavelength	Refractive Index
5780A	2·06932
2·02μ	2·00603
4·62	1·99837
8·22	1·98766
10·34	1·97877
11·86	1·97104
14·98	1·95127
17·80	1·92842
19·80	1·90083
20·56	1·90083

Notes on the idiosyncrasies of crystals from the point of view of the working optician (see also §§118 and 119)

by H. W. YATES, Manager of the Optical Shop at Hilger & Watts Ltd. (Hilger Division).

119.2 *Calcite* has a very strong cleavage ; great care is needed when smoothing so as to avoid scratches ; 3·7 emery is the coarsest used. The polishing is usually done with putty powder on 3 mm pitch or on a cloth polisher, depending on the orientation of the cleavage planes ; *i.e.* if the part to be polished is nearly parallel to the cleavage plane then a cloth polisher is used. There is no great difficulty in the working of this crystal provided care is used.

Calcium fluoride is subject to local hardness ; much care must be taken in maintaining an even temperature ; any rapid change will cause it to fracture. It can be ground with fine carborundum and finally smoothed with 304 emery.

Polishing is done on pitch, the hardness being varied to suit ; it is usual to commence with 3 mms. The polishing material can be either putty powder or rouge although it is more usual to use the former.

Lithium fluoride : the procedure is exactly the same as for calcium fluoride.

Potassium bromide is very liable to cleave ; much care must therefore be taken in working. Mineral oil is used when smoothing, as this crystal is soluble in water. It is polished on raw pitch with rouge or putty powder and the minimum of moisture. Surface brilliance is obtained by rubbing finally with clean chamois leather. A pitch polisher coated with wax is sometimes used as the crystal very easily scratches. Protection from direct contact with the hands is most essential.

For *rock salt* the procedure is the same as for potassium bromide, but is slightly easier to polish free of marks.

Sodium nitrate can be treated in the same way as potassium bromide, but it is not quite so hygroscopic.

With thallium bromo-iodide there are no signs of local hardness or any danger of cleavage. Polishing is done on cloth with putty powder, but it is extremely difficult to polish clean.

Quartz is one of the best crystals for optical working, it can be treated exactly the same as for glass polishing, but it should be remembered that it is much harder than glass. There is very little tendency to cleave and local hardness is only apparent when polishing a twinned surface.

Sundry notes on artificial crystals

120 The following mode of packing rock salt prisms has been used successfully for many years.

The prisms, clamped in their mounts, are wrapped in tissue paper thoroughly dried, then in thoroughly-dried cotton wool.

The whole is then placed within a tin, together with some silica gel, and the lid waxed on.

For export the following notice is put outside the tin—

NOTICE TO CUSTOMS OFFICIALS

This canister contains a rock salt prism which is liable to be destroyed by exposure to the atmosphere. It must not be opened in a moist atmosphere and should be kept for two hours at a temperature of $10°C$ above that of the surrounding air before opening. If it is intended to repack the prism, the packing must be kept very dry.

Stability of rock salt surfaces

It is stated that, in Germany, polished rock salt surfaces have been stabilized by heating the material gently after it has been polished, up to $500°C$, holding it there for two hours and then bringing it slowly down. A French manufacturer of table salt is supposed to employ a similar process.

Apparently, deliquescing of the surface is at least partly due to minute cracks; it has been observed that a natural face of rock salt is frequently stable. The heat treatment is sufficient to bring about sufficient flow in the surface to seal up the cracks, so it is said.

Similar stabilizing of potassium bromide might be tried. Judging by the melting points there does not seem any reason for departing from the $500°$ used for rock salt, at any rate for a first trial, although it may turn out that a somewhat lower temperature will suffice.

Optical plastics

by Dr D. STARKIE, Imperial Chemical Industries, Plastics Division

121 During the last ten years lenses made from plastics have become of industrial importance.

The optical plastic materials in commercial use at the present time are polymethyl methacrylate and polystyrene (sold under the trade names of " Transpex I " and " Transpex II " respectively) in Great Britain, and polycyclohexyl methacrylate and polystyrene in the U.S.A. The methacrylate resins correspond, as far as their optical constants are concerned, with normal " crown " glass, and polystyrene with normal " flint " glass.

The materials as now supplied are said to be as free from inhomogeneity and strain as optical glass.

Optical plastics have a higher coefficient of thermal expansion and a lower surface scratch resistance than optical glass, and these properties should be taken into account when designing an instrument with plastic optical components. Processes for producing hardened surfaces have, however, been developed up to the laboratory scale.

Much attention has been paid to increasing the softening temperature of 120°C as determined by normal laboratory methods.

The optical plastics have a remarkably high transparency to light which is greater than that for optical glass. In general they are highly transparent to short infra-red rays, and cut off at a lower wavelength in the ultra-violet than glass. Since optical plastics are very pure chemicals made under accurately controlled conditions, there is very little change in refractive index from batch to batch, and an accuracy in refractive index to $\pm$ 0·0001 is usually guaranteed. The materials have a very much higher impact strength than glass and have a low density and, in many cases, have been used where these properties are of importance.

Established methods of manufacture of lenses from optical plastics are (*a*) moulding under heat and pressure in optically-worked stainless steel moulds, and (*b*) grinding and polishing using the normal glass-working technique, in Great Britain, and (*c*) casting the lens from monomer in optically-worked glass moulds, and (*d*) grinding and polishing, in America.

For pressing in stainless steel moulds, a preform is first machined from a sheet of optical plastic and buffed. This preform is heated to the moulding temperature and, while hot, is introduced into the mould, when heat and pressure are applied. The moulded lens is cooled down under pressure and no further treatment is required after the lens is taken from the mould.

In casting from monomer in a mould, it is usual to add a catalyst to the monomer and to carry out a preliminary partial polymerization before introducing the liquid into the mould. Various devices are employed for overcoming the distortion of the lens which would normally arise from shrinkage of the material during polymerization and, in fact, cyclohexyl methacrylate is preferred to methyl methacrylate in the U.S.A. where the casting method is used owing to the smaller shrinkage of the former during polymerization.

In grinding and polishing lenses in optical plastics, a lens blank is accurately machined from a sheet of material and this is ground, smoothed and polished in the normal way.* Slow polishing speeds are used and polishing is done on pitch laps.

It is stated that no emery must be used in preparing these plastics for polishing—they must be polished straight from the machined surface and the best material is putty powder. [The proprietary polishing material " Silvo " is also used. F.T.]

Advantages of optical plastics over glass for lenses

The advantages of high impact strength and lightness in weight have already been mentioned. There remain the advantages which can be obtained from the methods of fabrication employed for plastic lenses. Optical plastics require less time for grinding and polishing than glass ones. Further reproduction from a master surface, either by moulding or casting, allows lenses with aspheric surfaces to be made cheaply once the mould has been prepared, and the low temperatures used in either method of reproduction allow practically no deterioration of the mould surface. Reproduction of large-diameter lenses from a mould is very little more difficult than for small-diameter lenses and, as optical plastic materials are readily available in sheets and blocks of large area, relatively cheap large-diameter lenses are a possibility.

Use of optical plastic lenses

During wartime, plastic lenses were used in binoculars, sighting instruments, cameras, projection systems and magnifying systems. There have been many applications where accurate plastic aspheric lens systems have been used.

An outstanding peace-time use for large-diameter plastic lenses has been reported from America. This is a spherical mirror combined with a correcting lens of the Schmidt type to project an enlarged

---

* Taylor, Taylor & Hobson have no difficulty in polishing Perspex or Polystyrene, but they have not been able to get homogeneous Polystyrene. (Note by F. T.: Polystyrene readily forms long chain molecules, particularly in conditions such as drawing it into threads, the heterogeneity is probably due to this phenomenon. Achromatic O.G.'s made of Perspex and Polystyrene have a 1 per cent change in focus for 30°C).

image on to a screen of the picture on the end of a cathode-ray tube of a television receiver. Systems up to 12 in. diameter have been made and used with success, and it is contemplated making still larger diameter systems for television projection in cinemas.

Considerable numbers of plastic spectacles lenses are in use and have proved particularly valuable to people who are obliged to wear spectacles while playing games.

U.V. transmission and optical constants of methyl methacrylate

121.1 *Transmission*—Two samples of methyl methacrylate were examined in the laboratory of Adam Hilger Ltd. The first was slightly yellow, while the second was sunlight-polymerized and obtained from colourless starting material. A 5 mm thickness showed no colour.

In order to compare the transmission with that of Vita glass 2·3 mm thick the transmission of a thickness 2·3 mm was calculated from the light absorption curve, a correction being made for the reflection of the two surfaces.

Wave length λ	Sample 1	Sample 2	Vita glass 2·3 mm
3,400		approx. 88–92%	
3,300		78·2	84·0
3,200	47·0	75·5	76·0
3,100	32·0	70·4	63·0
3,000	11·0	61·6	43·0
2,900		44·4	34·0
2,850		30·6	26·0
2,800		13·4	19·0

The transmission of the resin is much better than common window glass which does not transmit beyond 3,100 A at all. It compares well with Vita glass and in the spectral region below 3,100 A to 2,920 A, which has special therapeutic properties, the transmission of the resin is distinctly greater. The transmission does not compare with quartz which is 92 per cent approximately at 2,600 A.

It is unlikely that such a resin can supplant quartz for optical instruments; it is on the other hand very possible that it could be used instead of Vita glass. From this point of view it would be interesting to determine whether the resin undergoes solarization or decrease of U.V. transmission after exposure to ultra-violet light or daylight—a disadvantage of Vita glass.

Refractive indices

Refractive indices were measured with a spectrometer for a 60° prism of the material polymerized in sunlight, for lithium and sodium lines and for three lines of the mercury arc spectrum. The results are given below and from them the refractive indices for the C and F lines (to which dispersion in the visual region is usually referred) have been calculated by the contracted formula—

$$n = A + B/\lambda^2$$

Line	λ	n.Obd	n. Calcd.
Lithium	6,708	1·5030	—
Sodium	5,893	1·5057	—
Mercury	5,460	1·5081	—
,,	4,350	1·5171	—
,,	4,050	1·5214	—
C	6,563	—	1·5035
F	4,861	1·5123	—
F	4,861	—	1·5123

The medium dispersion $n_F - n_C$ is therefore 0·0088 and the ν value $(N_D - 1)/(n_F - n_C)$ is 57·5
With these may be compared the corresponding figure 60·6, for a silicate crown glass, of refractive index 1·511.

Combined Optical Industries Ltd : Report on technical progress from 1939 to date

122 (Combined Optical Industries Ltd, who have kindly provided the report set forth in this section are the largest manufacturers of optical work in plastics in the United Kingdom.)

For the purpose of this report, progress has been divided under the following headings—

      (a) Raw material   (d) Hardening
      (b) Design   (e) Aspheric machine
      (c) Process   (f) Other plastics

(a) Raw material

During the years 1939–1943 the Research and Development staff, with the full co-operation of the Plastic Optics Research staff at I.C.I., developed test equipment such that the new material could be improved with regard to the following properties—

Resistance to temperature—In this respect the softening point was increased from 40°C to 60°C.

Optical homogeneity—In this respect Transpex I and Transpex II were developed to give an optical homogeneity of one wavelength of light per inch of thickness.

During this period, test equipment was also developed to measure the coefficient of linear expansion and the coefficient of refractive index with temperature.

G

(b) Design

The development of the two different optical raw materials, *i.e.* Transpex I and Transpex II, with corresponding differences in optical properties, *i.e.* refractive index and dispersion, together with the development of the technique of cementing, made possible the design of optical systems which could be corrected for colour and spherical aberration. In this respect the following list gives a few of the systems which have been designed from 1940 onwards—

- 3 × Galilean binocular
- 6 × Galilean binocular
- 6 × Mirror erecting binocular
- F 6·3 Anastigmat 3½ in. and 9 in. focus
- F 6·3 40 in. Telephoto camera lens (as used by R.A.F. in reconnaisance work)
- F 4·5 Wide-angle anastigmat, 0·3 focus to 3·5 focus
- F 1·3 Television projector

Besides these a number of special purpose instruments and magnifiers have been designed for Government Departments, and recent designs which are being completed include a 5× lens erecting binocular and a F 0·7 reflecting projection system to be used for television projection.

(c) Process

1939–1940.—A production process was developed for the manufacture of spectacle dies, and a full range of dies was manufactured such that a comprehensive range of ophthalmic prescriptions was available. A process for the moulding of these ophthalmic lenses was also developed.

1940–1943.—During this time the moulding process was adapted for the manufacture of "precision" lenses. This involved more accurate control on the temperature of the preheat, on the moulding temperature and on the moulding pressure. A new technique known as "skin moulding" was used, and from the experience gained on the mass production of lenses by this method an automatic press for the moulding of precision lenses was developed. On this press the temperature and time cycles involved were set by electrical mechanism and did not depend upon the judgment of the operator. Simultaneously with this development, better methods of preforming and buffing were used; for example, the buffing machines were equipped with a vacuum chuck instead of the previous spring-collet chuck, and vacuum chucks were also supplied to the polishers instead of the lens being held by hand.

(d) Hardening

1939–1940.—The Research Laboratory was engaged in the development of the process of chemical deposition of silica for hardening

Perspex, and this process was developed to a production method which was applied to all spectacle lenses manufactured by the Company. The process, however, was rather unsatisfactory in that hardened films of only one-fiftieth of a thousandth of an inch were possible. In 1943 a new hardening process was developed, in which the lens was coated during the moulding process with a hard plastic film of a cross-linked resin. With this method, hard films of one-half of a thousandth of an inch in thickness were possible, and these films gave a very much greater resistance to abrasives. It is also to be noted that the cross-linked resin is impervious to all organic solvents and the new hardened lens can be cleaned by any of the proprietary brands of cleaning fluids on the market. At the present time this process is giving good results on a laboratory scale and a semi-automatic machine is being constructed which will manufacture on a commercial scale.

(e) *Aspheric machine*

In 1942 a machine for the automatic generation of non-spherical surfaces was developed, and mechanical improvements have been going on up to the present time. The machine consists essentially of a moving template which moves across the die surface and, by the addition of grinding or polishing abrasives, the die surface is ground and polished. By manipulation of the pivot about which the template is free to swing, surfaces can be formed which, to the second order of approximation, are the surfaces formed by the rotation of the various conic sections, *i.e.* circle, hyperbola, ellipse and parabola. At present the machine will generate only the prolate form of these surfaces of revolution, but modifications are now in hand such that the oblate form also can be generated.

Other plastics

A number of other plastics are appropriate for making lenses; among materials which can be given the desired shape by moulding or casting are the three classes benzyl-cellulose materials, pheno-plastics and amino-plastics. A nitro-cellulose material such as the product known under the trade name " Trolyte F " is also available. Further materials which can be cast to form a correcting element are : synthetic resins consisting of polymerized vinyl-compounds such as polystyrol, mixed polymerizates, resins, such as the phenol resins, and methacrylic acid esters. Of these, however, I have no direct personal experience. (*Here ends the combined Optical Industries Report.*)

Operative data concerning the moulding of Perspex

123 The following results are obtained from experiments made by me in 1943 and demonstrate the kind of use that can be made of a knowledge of the variation of mobility with temperature. The annealing temperature was found by Twyman's method (§§293 *et seq*) to be 68·3°C. It will be remembered that the annealing temperature is defined as that at which 95 per cent of the stress disappears in 3 minutes.

It was found that throughout the range from 64° to 86°C the viscosity at temperature θ relative to that at the annealing temperature was

$$10^{-0\cdot125(\theta - 68\cdot3)}$$

i.e. the mobility increases ten times for each 8° rise of temperature. We may assume that the formula continues to hold for somewhat higher temperatures and it is known by experience that with a pressure of 4 tons per sq. inch prisms could be moulded from the Perspex in a few seconds at 90°C. Let us suppose that we wish to mould with a pressure of 5 lb per square inch ; we must then raise the temperature of the Perspex until it has a viscosity of 1/1,800 of that at 90°C.

At 90°C the mobility from the above formula is $515M_0$, the mobility at the annealing temperature 68·3°C. Let t be the increase of temperature required to increase the mobility in this proportion then

$$10^{t/8} = 1,800$$
$$t = 26\cdot1°C$$

Hence one should be able to mould prisms with a pressure of 5 lb per square inch at 116°C. Cooling could be anything up to 10 times the fastest cooling schedule quoted in §301 with the pressure on, and the prisms should be removed from the mould at 60°C.

Moulded lenses of glass

123.1 The progress that has been made in recent years in moulding plastics has been described in preceding sections and it must have occurred to many that there should be no reason why what has been done with low temperature plastics should not, with suitable modifications in procedure, be done with glass, which is in its mechanical behaviour a plastic following much the same law of variation of viscosity with temperature as methyl methacrylate and the other transparent plastics which have been used for this purpose.

The nearest approach to making usable lenses by a moulding process seems to have been attained by the Eastman Kodak Company experiments under U.S.A. Government Project AC-11, Section 16.1 of NDRC. The descriptions of the experiments are contained in the British Board of Trade Technical Information and Documents Unit photostat copy O.R.R.151/49.* This document formerly " Restricted " has now been declassified to " Open " and the following abstract is published by permission.

---

* In America copies would be more conveniently obtainable from the Office of Technical Services, Department of Commerce, Washington 25, D.C., U.S.A., by quoting the reference PB. 28634.

Several different moulding presses were tried of which the third only will be described. The size of the lens which could be accommodated on this press was 50 mm. The press was constructed according to Fig. 56a. It consisted of a 16 inch steel channel column 70 inches in height. A pneumatic piston was attached to the upper part of the column and connected with a moveable head which slid between accurately machined gibs attached to the column. A fixed block was clamped to the column to carry the supporting tube for the lower mould. The tube was accurately aligned with the tube which carried the upper mould. These tubes were adjustable in the vertical direction and were clamped between the channels in the blocks and the cover plates. A bracket on the lower block supported an annular brass plate through which the tube passed. This brass plate supported the rubber gasket and the glass moulding chamber. A similar brass plate served as a cover to the chamber and it was clamped by three tie rods with thumb nuts to the lower plate in such a manner as to make gas-tight seals at the gaskets. A section of rubber tube was slipped over the tubes and the brass ports in the plates to provide a gas tight seal which permitted the tubes to move freely along their axes. Copper tubes soldered to the rims of the brass plates formed a water cooling circuit to keep the temperature of the plates low enough to avoid damage to the rubber.

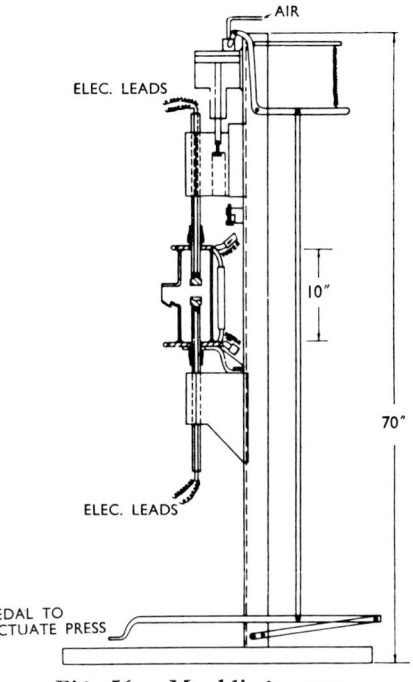

Fig. 56a—Moulding press

The heating of the moulds was by high frequency induction coils within the chamber and immediately surrounding the moulds. This arrangement is not shown in Fig. 56a which is actually of the model 2 press, but this differs from model 3 only in being smaller and having resistance instead of induction heating.

A representative sample of the better surfaces obtained with this press is shown by the Newton's rings of Fig. 56b. The surface which was intended to be flat departed from the surface of the plane quartz test plate by about 65 rings.

The final conclusion was that a high yield of small lenses can be produced quite satisfactorily, but as the size of the lens increases much beyond 35 mm in diameter the effects of the many uncontrolled variables render the rate of yield of impractical value. These variables include the amount and shape of the glass inserted between the moulds, the temperature of the glass at the instant of pressing, the

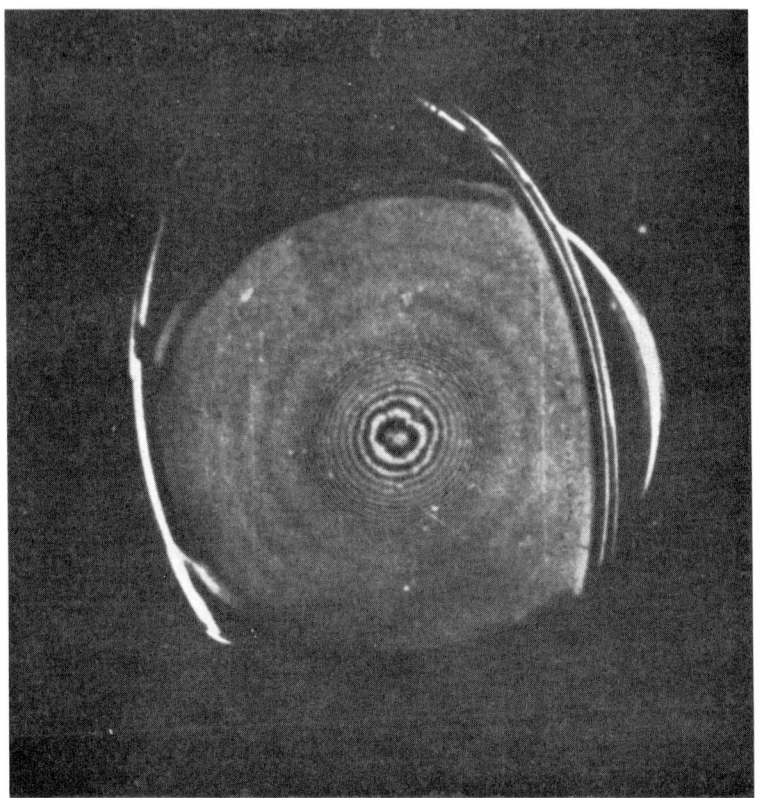

Fig. 56b—Moulded surface

duration of pressing, etc. If a greater degree of success is to be attained, therefore, these variables must be governed by mechanical or other devices and the perfected press must be constructed to perform the necessary actions in automatic sequence. (*Abstract ends.*)

It appears to me that these experiments demonstrate the feasibility of good moulded lenses being produced by developing the process on the lines suggested in the document in question.

CHAPTER 6

PRODUCTION OF LENSES IN QUANTITY

Introduction

124 The skilled optician should be practised in producing single prisms and lenses of high quality, a process which, like glass blowing, although very simple requires much skill and experience. The elements of the process have already been described in Chapter 2.

The great bulk of manufacturing work, however, consists of making numbers of prisms and lenses all of the same size and shape.

In the mass production of lenses, polishing processes have developed in two directions. The first aims at increasing the number of polishing spindles which can be attended by a single workman. The way in which this aim is achieved is described in Chapter 7. In the second, in use in other factories, particularly in some making large numbers of lenses (such as camera lenses) of high quality, carefully designed machines are used to secure that the best conditions for good polishing are automatically maintained. For instance, in Taylor's polishing machines (W. Taylor, 1918) the rising and falling motion of the member which actuates the polisher is eliminated, and reactions, due to inertia, which are a cause of the polisher becoming deformed, are thus abolished. At the same time the liquid is automatically applied to the polisher when the adhesion between the work and the polisher increases owing to the drying of the latter (W. Taylor, 1922). Similar machines are used for lens smoothing.

In roughing glass to shape, a departure from the traditional processes was made by Carl Zeiss by the use of copper laps charged with diamond dust and rotating at high speed (1500 revs or more per minute) with an ample supply of lubricant.

Somewhat similar processes have been developed and extended by Taylor to the accurate shaping of lenses, using bonded abrasive wheels. A lens-forming machine described by him (W. Taylor, 1925), built on the turret principle, has three lens-holding spindles, and two abrasive wheels operating simultaneously on two lenses at once, one roughing, the other trueing. With this machine, owing to the principle of grinding adopted, spherical surfaces may be ground indefinitely. Correct curvature is maintained by the grinding wheels being fed forward to compensate for their wear.

The optical workshops of Hilger & Watts Ltd. do not, in normal times, carry out work on " production " lines ; the instruments in which they specialize are rarely put in hand in batches of more than three dozen. Yet, since more than once during my management of the firm we were called upon to assist other firms by the supply of considerable quantities of high-quality lenses and prisms, we have also studied appropriate methods of production. These methods were to a great extent worked out by ourselves and for that reason they were often different from anything in vogue elsewhere. Although doubtless other firms which had the problem of quantity production always before them must have developed methods ahead of us in many directions, there is plenty of evidence that in this country many processes developed at Hilger's in the last forty years have now become standard practice in the trade.

The optical tools can of course be made to radius by the methods described in §46 *et seq.*, but for production work the curve generating machine described in §128 is greatly more speedy and accurate.

Moulding, trepanning and roughing by hand ; machine roughing by abrasive wheels and diamond laps

125 In Chapter 2 we considered in some detail the processes involved in the making of a single optical component, using only the simplest apparatus and a minimum of special equipment. In this chapter we shall deal with the manufacture of large numbers of lenses of the same kind : under these conditions, equipment of limited functional application is required to perform each operation with the highest possible speed and efficiency.

Mouldings

Until the last few years there was a considerable prejudice against moulded lenses or prisms owing to the heterogeneity which often resulted from the chilling incidental to the process. Now that this effect is better understood it can be avoided by efficient annealing, and there should be no need for a continuance of the objection.*

Where large numbers of identical parts are required, the use of pressings or mouldings offers many advantages ; less glass is removed in the preliminary shaping operations, and of course less time is required. The saving of glass enables the glass-maker to offer mouldings at about the same price as a slab of glass from which the same number of lenses could be sawn, and the labour of sawing, rough edging and roughing to curve is saved. Successive pressings from the same mould are consistent for diameter and thickness within close limits, which permits collets and other mechanical fittings to be prepared for long production runs without alteration.

* The reader should however refer to the end of §180 for our preference at the moment of going to press.

A firm which has installed equipment for moulding blanks considers that moulding is worth while if the rate of production amounts to only a few hundred per month, or even less. A method is in use whereby a ribbon of the molten raw glass is moulded as it is poured into the orifice of the moulding machine, so that a strip emerges from which the lenses are stamped out like cakes from a ribbon of dough.

In 1945 it was still considered in Germany that moulded lens blanks were insufficiently homogeneous for the highest class of optical instruments, and it was therefore customary, for the best work, to cut the blanks from slab glass. The optical parts for binoculars, however, were generally made from mouldings. In the United States moulded lenses were used up to 6 ins. in diameter even for lenses of the highest quality. (U.S.N., 1945)

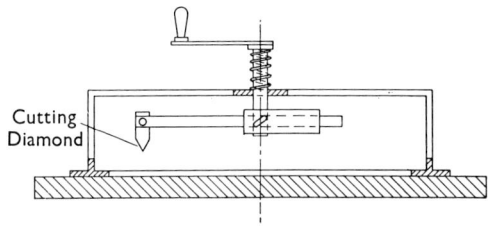

Fig. 57—Apparatus for diamond-cutting discs from plate

Lens moulding specifications

Specifications for lens mouldings are prepared from the drawing of the finished lens, but with the following allowances on the final dimensions—

	Up to 1½ in. dia.	1½ to 3 in. dia.	3 to 4 in dia.
Centre thickness ...	+0·05 in.	+0·15 in.	+0·25 in.
Radii { convex ...	+0·05 in.	+0·07 in.	+0·12 in.
{ concave ...	−0·05 in.	−0·07 in.	−0·12 in.
Diameter	+0·12 in.	+0·2 in.	+0·3 in.

In the case of deep convex lenses, it is advisable to check by scale drawing the maximum possible diameter of blank consistent with the stipulated centre thickness and diameter.

Trepanning

Glass for lenses is frequently available in plate form. If so it can sometimes be cut with a diamond in the simple apparatus indicated in Fig. 57, the handle being turned and pressed down at the same time. The best and quickest method is by trepanning. I do not know of any machine on the market that is specifically designed for this process ; it appears to be customary, with others besides Hilger & Watts, to convert drilling machines, the chief alteration required being to see that the motor has ample horse-power to prevent the slowing down of the tool (particularly important when the disc to be cut is of large diameter) and to provide an ample flow of lubricating liquid through the spindle. The manual adjustment of the machine is similar to that of a normal drilling machine, although it is convenient for the table movement to have micrometer adjustment with locking screws to facilitate setting for any particular work.

The glass plate is stuck to a sheet of plate glass, the diamond tool affixed to the spindle, and the table then adjusted so that the cutting edge of the tool is quite near the edge of the plate, thus effecting the maximum economy of glass.

Lubrication for the cutter must be provided through the hollow spindle, the best lubricant for this purpose being the Honilo cutting oil.*

The machine in the Hilger roughing shop is lubricated with Tecalmite grease by a single pump from which pipes run to all bearings and slides. This is desirable, as the speed used for trepanning is much greater than that for the drilling machine of like size. It is advisable to relieve the pressure on the cutter frequently to clear the swarf. The speed of cutting is high, a 1 in. diameter disc $\frac{1}{4}$ in. thick being trepanned in 100 seconds, its diameter being accurate to within 0·002 in.

The trepanning tool is known as a " copper-bonded Neven drill " and is made by I.D.P. (Impregnated Diamond Products) Ltd who will supply a shank to the trepanning cylinder to suit any drilling machine. The trepanning tool (Fig. 58) consists of a steel cylinder, the front edge of which bears a ring of impregnated diamond, thicker than the steel cylinder so as to provide ample clearance.

There is no apparent limit to the thickness which can be trepanned other than that imposed by the internal length of the steel cylinder. For a $2\frac{1}{2}$-in. diameter disc a speed of 1200 r.p.m. is recommended, although it is no use attempting to drive at this speed unless the horse-power of the motor is sufficient. Naturally, unless the drilling machine is a powerful one, which is unnecessary for this glass cutting, the horse-power of the motor supplied to drive it is likely to be insufficient for the glass cutting.

I am told that the machine in the Hilger roughing shop has a motor of only one-third h.p., whereas I should consider $1\frac{1}{2}$ h.p. desirable.

* This is one of a series of oils made by W. B. Dick & Co. Ltd, which they recommend for cylinder boring.

Possibly this is the reason why the trepanning tool is being run at only 600 r.p.m. instead of the 1200 recommended by the makers. If the latter were used the cutting time would probably be much shorter than that mentioned above.

Fig. 58—Trepanning drill in position

Roughing by hand

Roughing is the traditional name given to the generation of a curved surface by the use of a coarse abrasive in a rotating iron dish or tool. The speed of rotation can be as high as practicable provided always that the abrasive mud is not thrown off, and speeds up to 4000 r.p.m. are sometimes encountered in the roughing of small lenses of short radius.

For flat roughing the simplest arrangement is to screw the roughing tool face upwards on a post around which the worker can gradually step sideways, thus changing the direction of the stroke relative to the tool. Such a device, though in use at Hilger's during the author's memory, is now obsolete, but is not to be scorned as a means of educat-

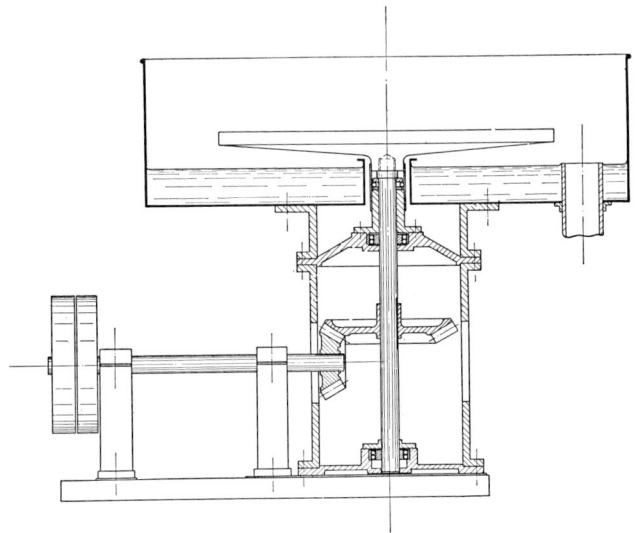

Fig. 59—Hilger roughing machine

ing the beginner in the initial processes. It will certainly avoid the embarrassment and annoyance occasioned to his foreman by the work on which a learner is practising being rapidly ground too thin.

Next in order of simplicity is the same arrangement, but so made that the tool is supported on a vertically rotated spindle. The speed depends on the diameter of the tool—the quicker the better so long as the carborundum is not thrown off. A roughing machine of this type for flat work is shown in Fig. 59, designed by Mr Dowell of the Hilger Company to my instructions in 1915. The diameter of the roughing tool is 3 feet. In re-turning the tool we usually compensate for the wear which causes the flat to assume the shape shown in Fig. 60a by turning it to the shape indicated in Fig. 60b. The life of the tool is then greatly lengthened before the departure from flatness is sufficient

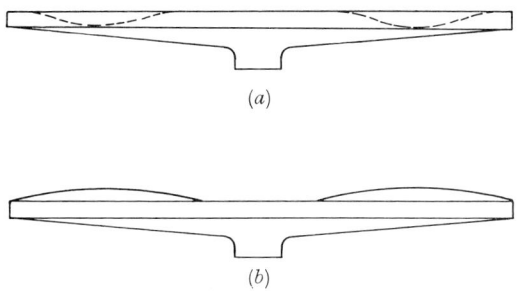

Fig. 60—Shape of rougher to compensate for wear

to necessitate re-turning. The flat tool needs re-turning every few months.

Fig. 61 shows a 4-spindle roughing and trueing machine which takes tools up to 12 in. diameter. Each spindle is independently driven by electric motor and each spindle is in a separate compartment to avoid

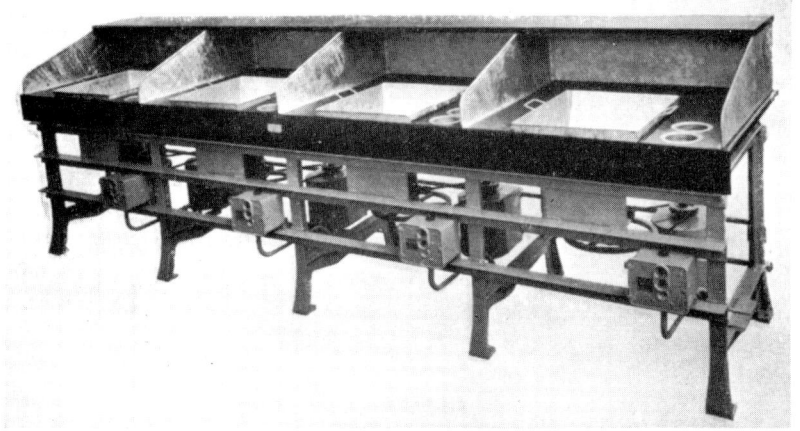

Fig. 61—Four-spindle roughing and trueing machine
(Bryant Symons & Co., London)

the accidental mixing of abrasives. This machine is made by Messrs Bryant Symons & Co., of Northumberland Park, Tottenham, London, N. 17, England.

Machines of the type detailed above are of the greatest utility where a large variety of components is made under conditions of small quantity production, as the vast majority of optical parts possess surfaces that are either flat or spherical.

Roughing lenses by hand is still very common, particularly in the smaller establishments, but there is little doubt that in time it will be replaced by one or other of the machine methods now to be detailed.

Machine roughing by abrasive wheels and diamond laps

A new, even revolutionary, development in roughing technique was the insertion of diamond into copper laps, first described by Carl Zeiss, Jena (U.K. Patent No. 14126, 1907). When I was in the U.S.A. in 1913 my friend Dr C. E. K. Mees showed me an edging machine in use at the Kodak works on which such laps were used, and with his permission I had a machine on a similar principle made by Hilger's on my return to England.

This machine is described in §143, and is still in use.

Once the principle of the diamond lap had been adopted, the step to making other machines based on its use was an easy one, and a number of machines were designed and adapted by Mr J. H. Dowell (Chief Designer at Hilger's) and myself for slab milling glass blocks, roughing lenses, and for other purposes.

The slab milling of glass was effected by replacing the cutter of a standard milling machine with a diamond lap, and increasing the spindle speed to give a linear cutting speed of 1600 f.p.m.

Ample lubricant of the " water-soluble " type must be used and the machine driven with adequate power, as if the spindle slows down the lap is soon destroyed. For a cut 4 inches wide, removing $\frac{1}{4}$-inch of glass at one cut, a motor of $1\frac{1}{2}$ h.p. is advisable ; under these conditions a table feed of about 1–3 in./min is normal. Glazing of the lap will not occur if the speed is high enough and the lubricant ample.

Early lens roughing machines

126 An early lens-roughing machine made by us is shown in the Fig. 62. The diamond lap A is mounted on a vertical spindle and the work to be ground is carried in a headstock B, with a horizontal spindle. The position of the headstock can be adjusted along a bar C mounted on a table D, which can be slowly rotated about a vertical axis E.

The grinding tool A is adjustably mounted so that the distance from its axis of rotation to the centre E of the rotating table D can be varied to suit the curvature it is desired to grind on the work, the position of the headstock B being adjusted along the bar C to give the desired thickness to the work.

When the headstock is set in this position the table D slowly rotates and feeds the work into the grinding tool A, the work also rotating on the horizontal axis of the headstock. A curve equal in radius to the distance from the centre E of the table D to the face of the grinding tool A is thus generated on the work as it is fed into the grinding tool. In 1917 we had ten of these machines in use.

Machines operating on the principle shown in Fig. 62 suffer from their inability to generate a radius longer than the maximum distance from lap to table centre, and the impossibility of producing a concave surface of smaller radius than that of the lap in use. It is also true that the direction of work feed, although not strictly parallel to the direction of cut, is nearly enough so for grooves in the lap to generate corresponding grooves in the work. Small though these may be, they

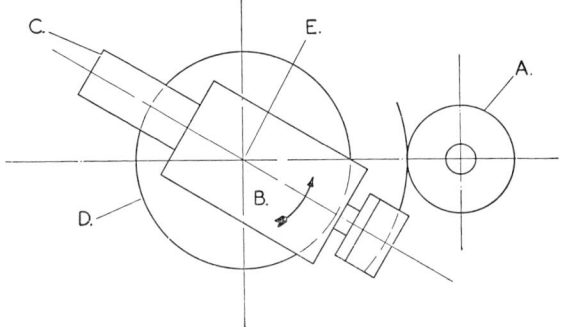

Fig. 62—Early lens roughing machine with diamond lap

are very deep in terms of the optician's unit of measurement, namely, a wavelength of light, and milling glass under these conditions must be expected to give an imperfect result.

A cylindrical grinder was therefore adapted with an angle setting to the cupped abrasive wheel. By altering the angular setting of the cupped wheel any radius, convex or concave, could be generated. Geometrically the principle was the same as that of the Adcock and Shipley machine described in §128. Our machine was used both with diamond-charged and corundum wheels. The direction of cut being at right-angles to the direction of feed, the fineness of finish attainable was then limited only by the fracturing action of the grinding tool used.

Most of the grinding tools used in machining glass owe their effect to the tendency of glass to conchoidal fracturing; the fractures usually extend below the apparent surface for two or three thousandths of an inch, and must be removed by fine grinding before polishing can commence.

A bonded abrasive wheel has been developed which, it is claimed, produces a surface fine enough to pass straight into polishing without the necessity for fine grinding. An abrasive agent as fine as this cannot be expected to remove material as quickly as a diamond lap, and it is therefore better, for maximum life, to restrict its use to the final stage of grinding. Since, to avoid waste of time, it is essential that laps should be interchanged as seldom as possible, the use of a pre-polish machining operation would seem to assume the availability of more than one machine.

Except for the sub-surface fracturing associated with the diamond tool, the surface produced is as good as the machine slides and rigidity allow, and only the minimum of material has to be removed in the later grinding with loose abrasive. Indeed, given careful setting of the machine (which must be amply rigid for the job in hand) a block of glasswork 9 in. in diameter can be machined with surface irregularities no greater than 0·0001 in.

Diamond laps

127 For grinding a material as hard as glass with an abrasive wheel the bond must be made very soft in order to prevent glazing. This, however, inevitably increases the rate at which the wheel will wear, and although there are some applications in which the use of a bonded stone is essential, the very much longer life of diamond-charged laps has led to their increasing adoption.

The method detailed in Chapter 2 for notching and charging saw blades can be applied to the manufacture of cupped wheels, but although laps made in this way will give good results, the concentration of diamond obtainable is not high and, if the lap should slow down or the coolant fail for any reason, the diamond is soon scraped out and the whole lap must be re-turned and re-charged.

Sintered diamond tools

The development of the sintering process has provided the optical trade with diamond tools of which the life is to be reckoned in years of continuous service. The process of holding diamonds in a metallic binder and manufacturing tools from the same was invented by P. Neven, after whom they are called " Neven " tools, and all original patents regarding metal-bonded wheels are in his name.

Mr Neven commenced manufacture of this type of metal-bonded tools in 1922 at his factory in Belgium ; they are now made by Impregnated Diamond Products Ltd. (I.D.P. for short), Gloucester.

The initial cost of these tools may seem high (upwards of £30 for cutting discs of 12 in. diameter), but those who have in the past made their own laps will appreciate the saving in time and money which

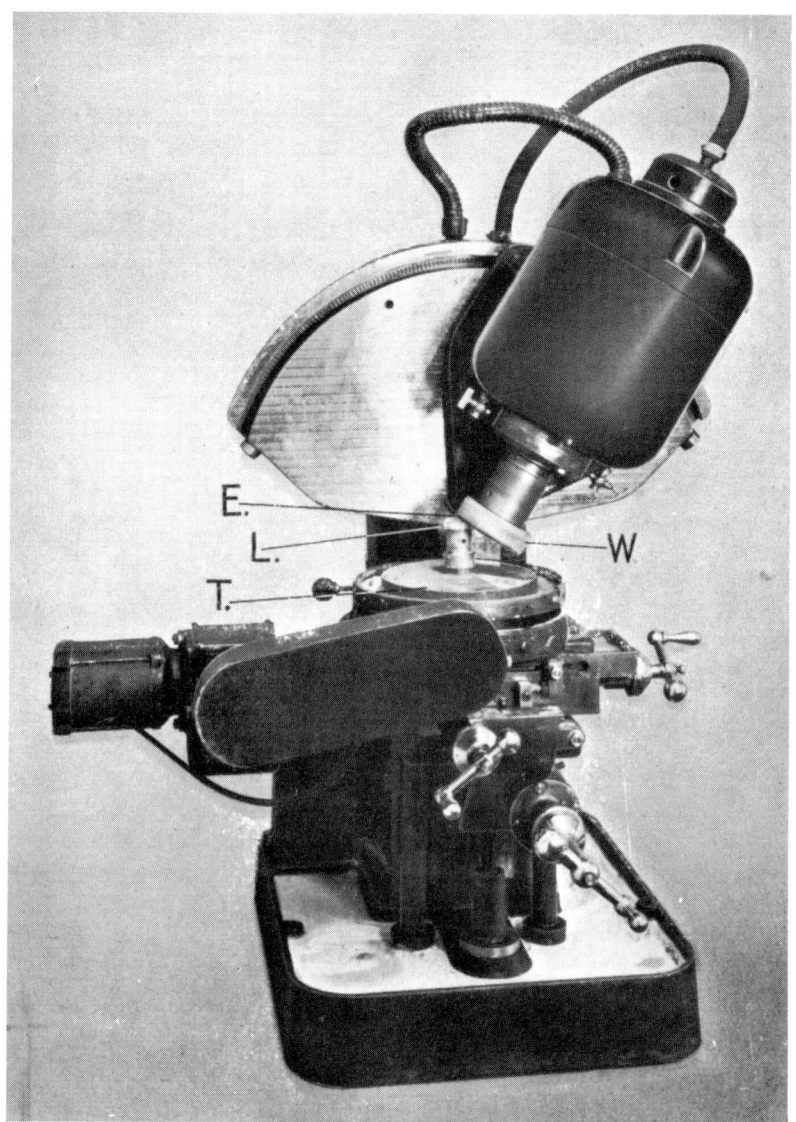

Fig. 63—Adcock and Shipley lens-roughing machine, OVS

accrues from the use of tools which are virtually permanent. One such blade now in use at the Hilger Division of Hilger and Watts Limited has cut over 10,000 sq. in. of glass and quartz and is still in service.

Modern lens roughing machines

128 A parallel development has taken place in the provision of machine tools specially designed for machining glass and quartz during and since the recent war, and a number of these are shown in these pages.

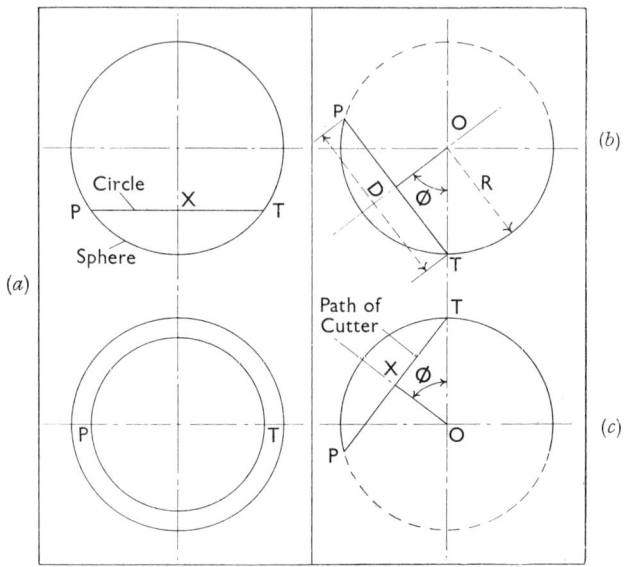

Fig. 64—Adcock and Shipley lens-grinding machine
Diagrammatic illustration of its basic principle

Two lens and prism grinding machines are manufactured by Adcock and Shipley, Ash Street, Leicester, England. The first of these, the OVS, is shown in the illustration (Fig. 63). The restriction on radii that can be ground by the earlier lens grinding machines is removed by the adoption of the spherical grinding principle described below.*

If a circle is scribed on the inside or outside of a sphere, as shown in Fig. 64*a*, then the circle will touch the sphere around the circumference of the circle.

---

* The description is taken by permission from the article by A. H. Shipley in *Machinery* for 1st March, 1945.

If the sphere and circle be now tilted into the position shown in Fig. 64b the truth of the preceding paragraph is not altered. Accordingly, if the circle is the locus of a cutter point, then it is merely necessary to rotate the sphere around the axis OT while the cutter is revolving, in order to generate a cap of a true geometric sphere as shown by the full line in Fig. 64b. This illustration shows the arrangement for a concave spherical surface. The principle for milling a convex surface is the same, as can be seen from Fig. 64c. In this case the cutter circle is on the outside of the sphere.

The angle to which the cutter spindle must be tilted in order to generate the radius required can be obtained from the triangle TOX, Fig. 64b and 64c, where :—

D = the diameter of the locus of the cutter point or of the cutting edge of the cupped grinding wheel ;
R = the required radius of the sphere ;
and
α = the angle to which the cutter is to be tilted.
Then it will be seen that :—
$$\sin \alpha = D/2R$$

The diamond lap W (Fig. 63) is direct driven by a motor on the same shaft. Its cutting edge E must be exactly over the centre of the automatically rotating table T, and in these circumstances the lens L has a sphere generated upon it which depends on the diameter of the cutting edge of the lap and the inclination of the axis of rotation of the lap. The machine yields a very good surface and lenses roughed on it can be put straight into the block and immediately worked with trueing emery. This machine can also be used with an emery or carborundum wheel or a flying cutter for grinding or turning lens tools to radius (see Fig. 65).

For machining lenses, the rotating table carries a collet in which the trepanned or moulded lenses can be clamped directly ; thus no cementing is needed.

The cutting edge must pass through the centre of the lens, and be accurately centered in the cross direction (Fig. 66) ; spherical surfaces are then generated with radius in accordance with the above formula.

The second machine, the Model 2VS, is dealt with in detail under the production of prisms in quantity, §178, but is also suitable for curve generation on lenses of very large diameter, say up to 16 in. ; otherwise its operation is identical in principle with that detailed above for the OVS.

A comparable machine, for curve generation on lenses up to 4 in. diameter is marketed by Impregnated Diamond Products, and differs from the Adcock and Shipley in that the work is rotated more rapidly, and fed into the cutter under spring pressure instead of a positive feed.

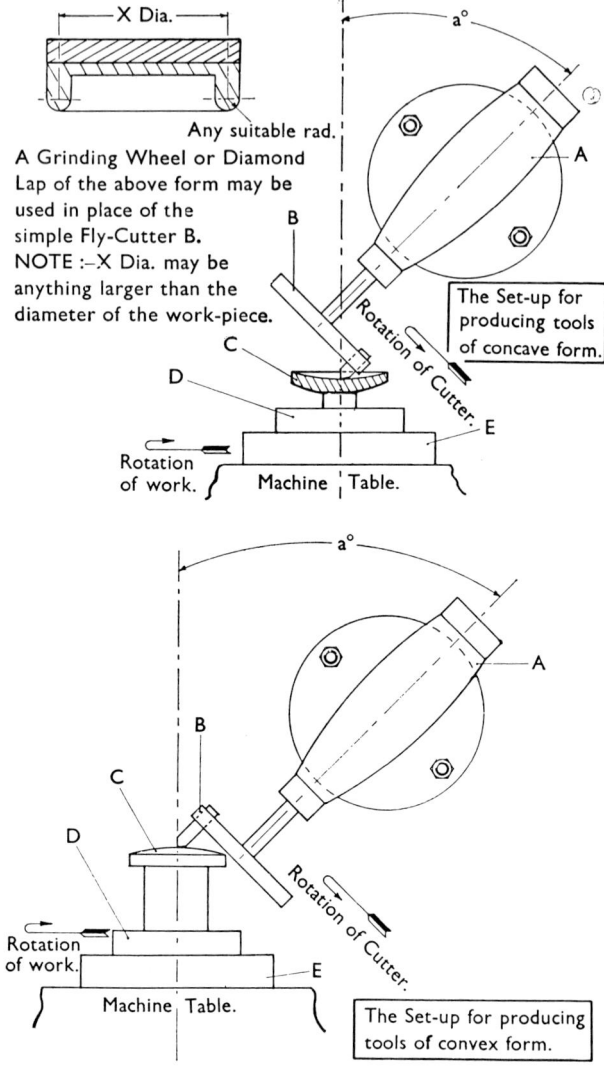

Fig. 65—Adcock and Shipley lens and tool curve-generating machine
The machine is shown with a single-point cutter for producing optical tools.

The development of this form of lens grinder has been carried still further in the U.S.A., where a machine has been constructed so that, once the lens moulding has been placed on the chuck and the " start " button pressed, the whole sequence of operations is automatic. The lens is held on the chuck by vacuum, the splash guard lowered, the spindle and table-drive motors started, the table advanced rapidly until near the cutter, then slowly fed into a pre-set stop ; two revolutions of the table are allowed, then the table is quickly run down, and all the motors shut off. The only human intervention needed in the operation of such machines is in changing the lens mouldings with a minimum of delay, and re-starting the automatic cycle, and one operator can attend to several such machines.

Process layouts for lenses

129 The process of making a lens can be broken down into seven basic operations, some of which are performed twice, once for each surface. They are as follows—
1. Machine both surfaces to radius and thickness. True both surfaces in trueing emery and chamfer both sides,
2. Stick on mallets (buttons),
3. Block up as many as the tool will hold,
4. Smooth the block,   ⎫ Repeat for other side
5. Polish the block,
6. Knock-off and clean up,
7. Edge and chamfer.

Inspection between the various operations has not been indicated, as the amount of inspection required varies with the grade of labour used, and most competent foremen rightly have their own conception of the process inspection needed on a given job.

Compound lenses, that is doublets, triplets, etc., are cemented after edging, but the components from which they are built up are produced by the above routine, or one of its variants.

One of the more important variations on the basic layout above is the recessed tool (or " spot black ") method which has come into vogue in recent years. The block-holding tool for a convex block is of slightly shorter radius than the true tool, and has the correct number of recesses machined into its curved face, all to exactly the same depth, and all exactly square to the radius passing through its centre.

The lens mouldings are mounted in these recesses with pitch or wax, and the block is *machined* to curve, smoothed and polished without the lenses being removed from the block holder. Given sufficient tools and a very large number of lenses, very high production rates can be achieved by this method ; it is, however, only worth the tooling-up if the numbers involved run into thousands.

We will now consider the basic operation steps in more detail.

Sequence of operations

Machine or rough to radius and thickness

130 The spherical grinder is set to the angle given by the formula $\sin \alpha = D/2R$, and the longitudinal traverse adjusted so that the cut passes through the middle of the lens. A lens moulding is placed in the collet the right way up, tightened, and a light cut taken, say 0·015 in. After the table has made a full revolution, the lens is removed and examined through a magnifier. One of two patterns will probably be observed (see Fig. 66).

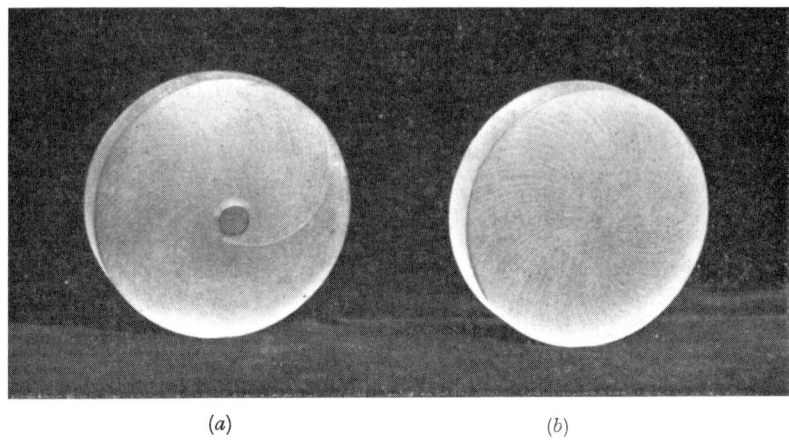

(a) (b)

Fig. 66—Appearance of surfaces roughed on the lens-grinding machine
(a) Incorrect   (b) Correct

Fig. 66a, in which the spiral marks of the cutter surround the central pip, indicates that the cut is passing to the left of the centre ; if the marks remain outside the pip the cut is passing to the right of the centre.

The longitudinal traverse is moved in the appropriate direction by half the diameter of the " pip," and a further light cut taken. Adjustment is thus made until the appearance is as in Fig. 66b. The machine surface can now be tested for radius.

This can be done by the conventional spherometer, or by a dial-gauge derivative of it, shown in Fig. 27, which we call the " Spheroscope." It is easier still to grind the lens very briefly in the trueing tool ; if this brief grinding is seen to be removing the grinding pattern evenly all over the surface, then the setting for radius is finished, and it remains only to measure the thickness of the half-finished blank after trueing, and to make any necessary adjustments to the setting of the machine stop.

The whole batch is now machined by the machine operator; it is usually found that where one machine only is worked per operator, the machining time of $1\frac{1}{2}$–3 minutes for a lens of $1\frac{1}{2}$ inches diameter is sufficient to enable the surface previously machine to be trued up and chamfered at the same time.

The other side is then machined, trued and chamfered in a similar manner, either on another machine (to avoid re-setting to a new radius), or on the same one if only one is available.

Sticking on

131 The next step is to stick on the back of each lens a blob of mallet pitch (see Chapter 4). Following is a well-tried routine for carrying this out.

The worker must be provided with—
(1) A bench to which are screwed two iron nose pieces to which the tools can be screwed when occasion requires (see Fig. 7),
(2) A block holder (see §68),
(3) A pair of optical tools of the required radius of curvature, accurately ground together and a wooden handle, Fig. 7, to screw to either of them as required,
(4) A gas ring,
(5) A cauldron of mallet pitch.

The cauldron of hot mallet pitch (about 1 ft in diameter) is placed on the bench, the hot mixture being a little stiffer than treacle. It is very important not to overheat the mixture, otherwise the pitch will harden. The mixture must be very well stirred immediately before use by means of a round stick 1 ft long and 1 in. diameter.

We will suppose that the lenses to be blocked are 2 in. in diameter. A sufficient number to form a block are placed on paper on a cold, flat, 1 in. thick iron plate, which is then gently heated with a Bunsen burner or gas ring to a temperature slightly hotter than can be comfortably borne by the hand. A blob of the melted mallet pitch is collected on the end of the stick, the stick being rotated so as to prevent the pitch dropping off; the said blob is then laid on a cold tool where it is turned over and over to cool it. As it cools it is formed with the hand into a flexible, round-shaped piece about $1\frac{1}{4}$ in. diameter and of the consistency of putty. Still hanging on the stick, its lower end is allowed to rest on the back of one of the lenses and about $1\frac{1}{4}$ in. snipped off with scissors.

The process is repeated until there is a lump of mallet pitch on the back of each of the lenses, care being taken that the lumps are all of equal size. The lenses are then taken up in the hand one by one and the mallet at the back of each pressed on to the lens and worked up into a nearly hemispherical shape (see Fig. 67), placed in a wooden tray and allowed to cool. It will usually be found that some of the mallets fall out of shape and one must go rapidly through them, re-shaping them with the hand until they are all cold enough to retain their shape. They are then allowed to cool until all are of about the same temperature. Before allowing them to cool make sure that they are of equal thickness.

In the United States the mallets (or buttons as they are called) are pre-cast; all that is required then is to warm in a small flame the bottom of the button, and press it on to the warm lens. Some Continental manufacturers place the lens in a steel mould and cast the mallet in this way, but it is not apparent where the advantage of this method lies, as it takes longer to carry out and more equipment to handle.

Fig. 67—Making a block of lenses
A typical pitch " mallet "

The optical tool which is to be used for smoothing the lenses is now screwed to one of the noses and thoroughly cleaned. It is extremely important that every particle of dirt should be removed, both from this tool and from the surfaces of the lenses which are to be put in contact with it. There is always a possibility of a small spot of pitch finding its way to the surface of the lens or the tool; if this occurs it must be cleaned off immediately before putting the lens on to the tool; a rag dipped in turpentine will serve for this.

If the lens is placed on the tool while the turpentine is still wet on the surface, it will stay in position, or if no turpentine has been used, the tool should be slightly moistened. The lenses should be arranged not less than $\frac{1}{8}$ in. apart (though there is no harm in their being farther apart up to $\frac{1}{4}$ in. or more) and in sufficient number comfortably to fill the tool. The following numbers and arrangements make good workable blocks, but of course the more lenses there are the better.*

5 in a ring.
6 round 1 = 7.
9 ,, 3 = 12 (see Fig. 69).
10 ,, 4 = 14.
12 ,, 6 round 1 = 19.

The block holder is now heated on the ring burner until it is just so hot that one cannot touch it without discomfort for more than a fraction of

* See also pp. 213-4.

a second. It must not be so hot that water sizzles on it. It is important that it should be heated uniformly, and for that reason the final heating should be carried out by holding it with the box-wood handle and waving it about in the flame to distribute the heat uniformly.

It is scarcely necessary to say here that under practical conditions of work it is not feasible to measure the temperature, nor is there any necessity for doing so.

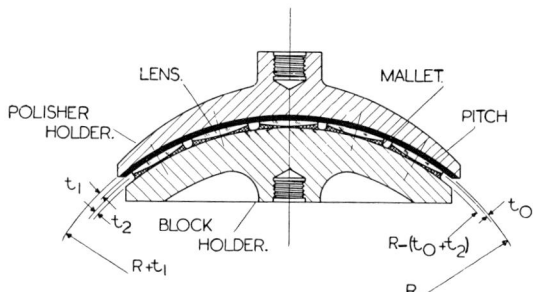

Fig. 68—Making a block of lenses
Sectional elevation of lens block and working tool

The block holder must now be gently placed centrally on the lenses which are arranged in the underneath tool and allowed to settle by its own weight. It will gradually soften the mallets and it should be allowed to do so until their thickness is between $\frac{1}{4}$ in. and $\frac{1}{2}$ in., say

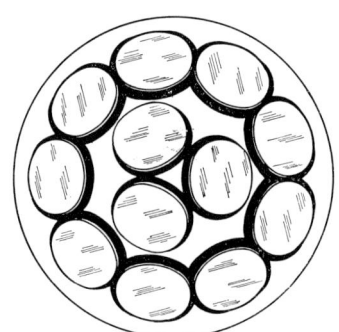

Fig. 69—Making a block of lenses
Plan of completed block

about $\frac{3}{8}$ in. Meanwhile, keep a watch to see that the mallets are settling down equally all round the block and, when they have settled down to the thickness stated above, cool off the block holder by sponging with cold water. Work is usually commenced before the block is quite cold.

210   PRISM AND LENS MAKING

Fig. 70—Smoothing a block of lenses

The block of lenses is shown in section in Fig. 68, lenses uppermost, with a polisher in position. Fig. 69 is a view of the block from above.

The block should not be removed from the forming tool until one is ready to smooth it, as otherwise, even if it is cold, the lenses will shift and the smoothing will take longer. Supposing the smoothing tool is

Fig. 71—Block of lenses being checked on the polishing machine for accuracy of curvature

ready, the block is pulled off and smoothing can at once commence. If the work has been carefully done it will be found that the whole surface becomes uniformly ground by the smoothing emery within a minute's rubbing.

Figs. 70a and 70b show blocks of concave and convex lenses being smoothed on the machine.

The foregoing description applies to the case where convex surfaces are to be polished, so that the lenses with their mallets are arranged face downwards on a concave tool. If a concave surface is to be worked they are, of course, arranged on a convex tool, the concave blockholder being lowered on the mallets. The subsequent procedure is exactly the same, except that as soon as the tool begins to settle down on the mallets, the pair of tools with the lenses between them may be turned upside-down, since then it is more easily seen whether the operation is being performed satisfactorily—it is important that the block should be central and the mallets of nearly equal thickness all round the outside of the block.

One aid in this direction is to fix three ball bearings or other suitable spacers to the true tool with wax, the thickness of the spacers being the difference in radius between the true tool and the block holder. When the blockholder is touching all three the block can be cooled off.

If the radius of the tool is one inch or less, and the lenses of such a diameter that only three can be accommodated in the tool, the mallets must be tapered to a point in order to get the lenses close enough together, and this is liable to reduce the strength of the block. In this case it is preferable to use the " pudding block " as follows.

On the back of each lens is stuck a very small " key " of soft black pitch, say $1\frac{1}{2}$ mm hardness (see §60). The three lenses are then " breathed down " into place in the tool, and a blob of mallet pitch of about two-thirds the volume of the tool, rolled into a ball. The tool and lenses are briefly passed through a Bunsen flame, as is the ball of mallet pitch, the latter being then pressed into the tool so that the lenses sink right into it when the hot blockholder is pressed into it. The whole is then cooled off in cold water. If the pitch is pressed right up to the lens surface, it is scraped or dissolved away so as to leave the lens surfaces slightly projecting.

Other methods of blocking lenses

132 A quicker method of blocking, appropriate when the number of lenses of one kind is very large, is described in the sections on ophthalmic lenses.

Still a further system of blocking is coming into use for large numbers of lenses. This is shown in Fig. 72. The block holder has a number of support blocks screwed to it, these being machined on the outside with curves to fit exactly the surface of the lens which is to be attached to them. The machine work for the blockholder and support blocks has to be very accurate, so that the thickness of all the lenses is precisely the same. The lenses are held to these special supports by means of thin tissue paper soaked in pitch or a pitch mixture. With this type of tool, very heavy pressure can be used and the distortion after unblocking is negligible. It is stated that this method is the only one found to produce lenses combining high surface accuracy and speed of working, while maintaining a high accuracy of finished thickness.

With this method of blocking very considerable pressures can be used and thus very hard pitch can be used for the polisher. The temperature of the pitch during the working can rise as high as 60°C and at this temperature even hard pitch will not scratch, provided it is quite free from grit. It should be mentioned, in connection with the choice of pitch, that it should be tested with the utmost care to be sure that there is no grit in it, and faulty specimens should be returned to the supplier until examination shows that the pitch is free from gritty material.

The advantage of the tissue paper is that it will lie on a concave or convex tool without buckling.

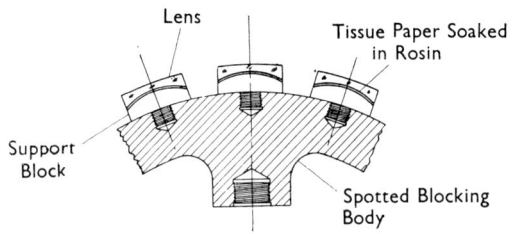

Fig. 72—Spot-blocking of lenses

Number of lenses per block

133 Normally, the number of lenses to be included in a block is the maximum number that can be arranged therein without touching one another.

This number can be determined easily enough by trial with the actual tool and a few lenses, but it is often necessary for the estimating section to know how many lenses will compose a block for each surface of a lens system before a price can be quoted.

The number of lenses which will fit into any tool is controlled by two factors, the depth of curve of the lens and that of the tool, both reckoned as solid angles. Perhaps the easiest way to define depth of curve of the lens is by the ratio R/D, R being the radius and D the diameter. Limiting values of R/D can be found for each successive ring of lenses. Lens blocks can be built up in one of two ways, either with one or three lenses in the centre, so that two series of values will be derived.

It is necessary also to define the depth of the tool in similar terms, and in this case it is of great assistance if the depth of optical tools is standardized by regulation, for example—

(a) for short radii, up to $1\frac{1}{2}$ in.—hemispherical,

(b) for longer radii to have a diameter equal to $\sqrt{3}$ times the radius, subject to a prescribed maximum of say, 9 inches.

Lenses under consideration would then be grouped into three divisions—those under $1\frac{1}{2}$ in. radius, for which hemispherical tables would be prepared, those between $1\frac{1}{2}$ in. and $4\frac{1}{2}$ in. radius, for which 120° tables would be used. Those over $4\frac{1}{2}$ in. radius are worked on 9 in. tools and the actual radius is irrelevant, for the number per block is the same as in plano work. If plotted on a graph the points representing the number of surfaces which can be worked on a block can then be found from the empirical formula—

$$N = 42 \cdot 5/D^{1 \cdot 85}$$

The following tables have been prepared taking the above assumptions for granted ; if other conditions apply they may serve as a model for similar tables to suit.

Hemispherical blocks
up to $1\frac{1}{2}$ ins. rad.

R/D	...	...	3·1	2·1	1·5	1·2	0·8	0·5
Lenses per block			40	19	11	7	3	1

120° blocks (Tool diam. $\sqrt{3} \times$ Radius)
$1\frac{1}{2}$ up to $4\frac{1}{2}$ ins. rad.

R/D	...	...	3·8	2·5	1·9	1·5	0·9	0·6
Lenses per block			40	19	11	7	3	1

Shallow blocks (Tool diameter 9 ins.)
over $4\frac{1}{2}$ ins. rad.

D in.	...	0·5	1·0	1·5	2·0	2·5	3·0	3·5	4·0	>4·2
Lenses per block		150	40	20	11	8	6	4	3	1

The following have been found by experience to be the most easily worked lens block numbers. It is true to say that in the hands of a skilled hand-polisher, almost any number and arrangement of lenses can be persuaded to take up a spherical shape of the desired radius, but for repetition machine polishing it is better to adhere to those arrangements which give the best results with a minimum of trouble.

		One in centre			Three in centre		
Centre	...	1			3		
1st ring	...	6	Total	7	8	Total	11
2nd	...	12		19	12		23
3rd	...	19		38	16		39
4th	...	34		72	21		60
—		—		—	—		—
10th	...	—		200	—		220

Two other arrangements which do not conform to the above table, but which may prove useful with certain lenses are four and five in a ring, with no centre. Although apparently very "bare" in the middle, they will be found to polish well, provided the block is on the deep side.

For best working, the following arrangements are recommended—

One ring of 3   =3
One ring of 6 round 1   =7
One ring of 8 round one ring of 3   =11
One ring of 12 round one ring of 6 round 1   =19

The following are references to papers on lens blocking : Redding (1916–7), Lee (1917–8).

Smoothing

134 Once the blocking has been completed, the block should pass as soon as possible to the smoothing machine as, if left for long out of the true tool the mallets are inclined to sink unequally ; the rate of sinking can be much reduced by storing the blocks under the surface of a tank of water.

Smoothing and polishing machines vary considerably in detail of construction and capacity, but most of them have this in common :* that they provide for one tool to be rotated around a vertical axis, while the other is moved to and fro in an approximately straight line in harmonic motion by means of a crank. Some or all of the following adjustments will be provided ; variation of crank speed, variation of spindle speed and variation of the ratio between crank and spindle speeds. §§ 134 to 138 refer particularly to this generic type of machine.

Contrary to standard practice in most optical workshops, it should be made a general rule that *the tool goes underneath whether convex or concave,* unless the worker is consciously changing the shape of the tool nearer to the required radius.

The reason for insistence on this rule is the fundamental principle that when two discs of equal size are rubbed together with emery on a conventional polishing machine, the upper element will always become more hollow, and the lower more convex.

Reversing their relative positions will reverse the drift, and if the stroke offset (i.e., the amount by which the reciprocating pin is off centre when in the middle of its stroke) is not too severe, the shape will return to flat.

The foregoing remarks are true also of curved tools and blocks. If a convex block is screwed to the machine spindle, and a concave

---

* But see §140.

tool smoothed overhand, then, as time goes on, the tool will assume a shorter radius, until finally the lenses produced lie outside the permitted departure from radius, and the tool is sent back to the tool-room for re-turning. If instead the block is made slightly smaller than the true tool, and the tool screwed to the machine spindle, then the change of tool shape may be controlled merely by centralizing or decentralizing the mid-point of the crank stroke. A central stroke will deepen the tool, an eccentric stroke will shallow it.

If the block be shallow in included angle, say of radius twice the diameter or more, then the reversal brings no complications. If, however the block is deep, then it will be found desirable to take the pin of the machine well outside the centre of curvature of the tool.

The block is smoothed twice in each of the grades selected, with the usual precautions when changing grades to avoid contamination and the attendant scratching of the block. For smoothing the block the American Optical Co.'s $302\frac{1}{2}$ grade followed by 303 is fine enough. The duration of smoothing is a matter of experience, depending as it does on the amount of emery on the tool, the speed and loading of the crank-pin, and the material of which the lenses are made. Fifteen minutes for a 9 in. block of crown glass lenses is a representative figure to form a basis for experiment.

Polishing

135 The method of charging the polisher with rouge and keeping it moist is the same for a block as for a single surface (see Chapter 2). In polishing deep blocks of lenses, the reversal technique is not so necessary as control over the shape of the block is achieved partly by alteration of stroke, and partly by careful grooving of the polisher to encourage the required change of shape, an expedient which is not available during the smoothing process.

For repetition work, one or other of the loaded types of polisher (§§59 and 66) should be employed. The polisher is warmed, rubbed on the true tool if it has stood long since its last use, and rubbed on the block by hand to ensure that the contact is good all over the block. The pin is then lowered into the socket, and polishing commenced with a medium length of stroke and a moderate offset of the crank arm. With the appearance of the block, after polishing has been proceeding for ten minutes or so, for a guide, alterations in the length and offset are made as follows (the remarks apply to a convex block)—

(*a*) If the centre lens is quite bright, but the edge lenses hardly polishing at all (a fairly common occurrence in the early stages) the grooves in the centre of the polisher should be opened out a little, and the stroke moved more over the edge. If this does not, after a further ten minutes or so, appear to be effecting the alteration required, the machine pin can also be moved towards the front of the block.

While taking these steps to achieve an even polish, it is highly desirable to check the shape of the block with the test-plate to ensure that the correction for polish will also be correct for shape. There is not much to choose between a block well polished but of the wrong radius, and one of the correct radius but only partly polished.

(b) If the edge lenses are brightened, the centre remaining grey, the edge grooves of the polisher should be opened up, and the stroke moved back nearer the centre.

(c) It may sometimes happen that the block appears to be taking evenly, but that closer inspection reveals that the edge of each lens is nearly polished while the centre remains almost grey. This can only mean that the individual lenses have moved since polishing commenced to the deeper curve possessed by the polisher, and indicates that the mallet pitch is too soft. (The opposite effect occurs if the polisher is too shallow). The only real cure is to knock off and re-block with harder mallet pitch. If for any reason this cannot be done, then the edge of the polisher should be cut back, and the stroke set quite central.

Principles governing correct hardness of mallets

136 In using the unstamped loaded polishers which are recommended the basic principle which dictates the hardness of the mallets is that the latter should be hard enough to withstand working for ten or fifteen minutes without the lenses moving. The polisher is by then warm enough to assume the shape of the block without recourse to the forming tool. In some factories, indeed, it is not the practice to make true tools in pairs other than for cemented surfaces, the polishers being formed on the smoothed block, the radius of the (brass) true tools being maintained either by reversal, or hand scraping, or both.

Since the principle just enunciated is contrary to that in vogue when the first edition of this book was published I shall run the risk of boring the reader by a recapitulation.

1. The polisher is loaded with wood flour, etc., so that at room temperature it is almost entirely free from flow.

2. The mallet pitch is hard, say about $\frac{1}{2}$ mm (§60), and time must be allowed for any strain to anneal out.

3. The polisher receives its general shape from the forming tool, but after cooling and grooving it is warmed up again and rubbed on the *block* until good contact is felt.

If the polisher is warmed and rubbed on the warm forming tool this is more to take care of dust on the surface than to impart the final

H

shape. Of course, where the tools are a good fit (and why need they be otherwise?) no harm is done by giving the polisher a good rubbing; but the point to be stressed is that the shape of the block should be decided by the figure of the smoothing tool, which is itself under control by the reversed smoothing method, and that first quality surfaces can be produced with a forming tool which is a poor fit, or even with no forming tool at all.

Load on the tools

137 The weight that should be used on the pin of the polishing machine is the subject of much difference of opinion. There seems to be no doubt that increased weight gives increased speed of polishing, but there is no evidence to show whether the speed increases indefinitely with the load, or whether it reaches a limit beyond which further weight has little effect. At Hilger's we have pushed the polishing weight up to 75 lb on a 10 in. diameter polisher, and thereby reduced the polishing time from 4 to $1\frac{1}{2}$ hours. The increase in efficiency was, however, largely offset by the difficulty of achieving

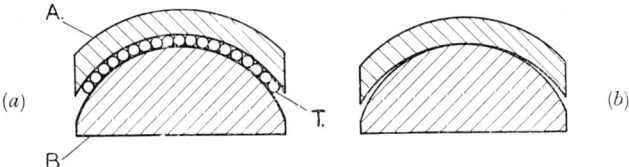

Fig. 73—Effect of thickness of abrasive when grinding deep curves

the desired shape as, under this magnitude of load, iron tools and plaster blocks flex an appreciable amount. Our general practice now is to use a load of 20 lb (including the tool) for an 11 in. diameter block. This amounts to about 0·4 lb. per sq. in. (29g/cm²). This may be compared with §§80 and 330. Since with the heavy pressure polishing takes place much more quickly but the surfaces are not so good, it has been suggested, and it seems very likely to be successful, that one should start with a heavy initial pressure and gradually reduce the pressure in one or two stages until the desired accuracy of surface is attained. Recent efforts to reduce the polishing time have been directed rather along the line of producing a finer grey and a better shape, so that as soon as the grey is removed the block is finished.

Polishing deep curves

138 At this stage it is necessary to recognize that in polishing deep surfaces certain phenomena become obvious that are not appreciable with shallow ones. If tools A and B (Fig. 73a) are ground together with an appreciable thickness of abrasive, their surfaces may both be spherical but the radii of curvature will differ by the thickness T of the abrasive, and if they are cleaned and put into contact they will bear in the middle (Fig. 73b). The thickness of a well-rubbed-down smoothing emery is about 0·0002 in. and this is quite sufficient to produce a very noticeable effect in a pair of tools subtending an angle of 120° and with a radius of curvature of less than, say, 4 in.

If two such tools are put together, they will be found to swing around the middle. If, however, the convex tool is now used to form a polisher, the said polisher will be of the same radius of curvature as the smoothed block of lenses ; or rather, very slightly less, since in forming the polisher the forming tool is itself polished to a slight extent and thus slightly reduced in radius.

It should be found, therefore, that the polisher thus formed fits the surface of the smoothed lenses almost exactly, and that polishing should commence uniformly over the whole surface.

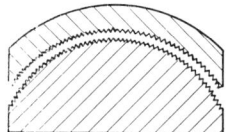

Fig. 74—Grain of surface of tools

The novice, then, should not suppose that it is necessary or desirable that when he has trued his tools they should fit over the whole surface. It is normal and correct—if they are deep and of short radius—that they should swing about the middle.

There is, however, another point to consider ; the tools after smoothing have a grain which has a not entirely negligible depth (see Fig. 74). As pointed out above, in forming the polisher this grain is partially polished away and the radius of curvature somewhat reduced. The polisher, therefore, may be expected to commence polishing the outside of the block of lenses first. In so doing it soon begins to accommodate itself to the shape of the block, becoming slightly less deep ; simultaneously with this, however, the outer portion of the lens block polishes before the middle and thus the lens block gets a little deeper. Thus polisher and block rapidly come into good contact and the polishing takes place uniformly.

In the routine prescribed by me for the manufacture of very large numbers of small lenses with deep curves during the 1914–18 war in which girls were employed, we had separate sets of tools of slightly different radius for the trueing, smoothing and fine smoothing, in which the average thickness of the emery, plus the depth of grey, was allowed for. It was found that this was very effective and had also the result that the tools did not require correcting so frequently as was otherwise the case.

Thus the worker must regard it as perfectly normal for there to be a slight tendency in deep blocks of lenses or in deep single lenses for

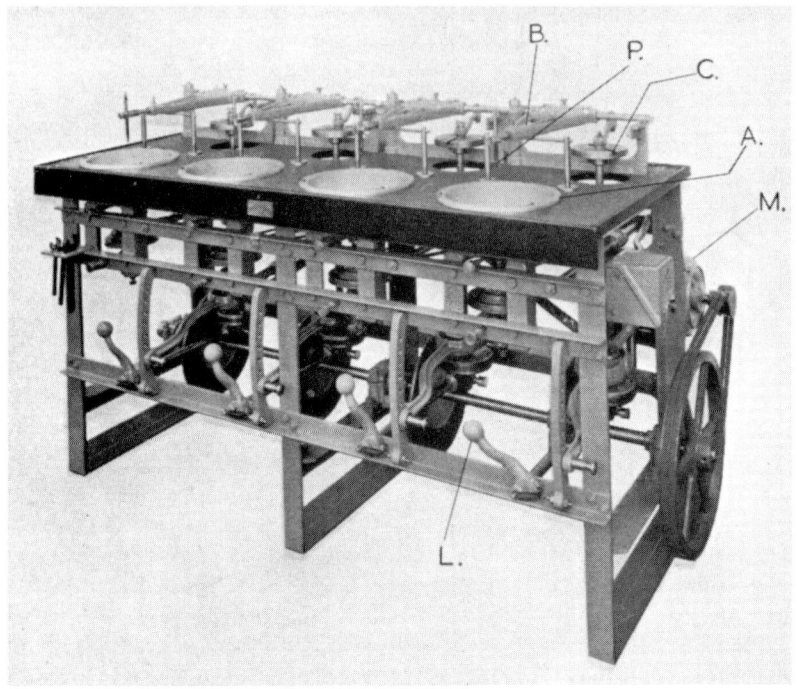

Fig. 75—4-spindle polishing machine
Bryant Symons, London

polishing to commence on the outside. Since the effect depends very much indeed on the depth of the surface, it requires a good deal of experience to recognize when one is dealing with the normal and indeed desirable condition, or when on the other hand one is in the presence of defective processes.

The above illustrations show the surfaces central with each other. Actually in polishing deep surfaces the polisher is always off-set and the effects just mentioned are less obtrusive.

Polishing machines

139 Fig. 75 shows a 4-spindle grinding and polishing machine of a type which has been in use for very many years. The tool screws into position at the top of the spindle which lies within the dishes A. When the top tool or polisher is in position the arm B is raised and the ball at the end of the pin P (see Fig. 75) lowered into the socket which is screwed into the boss of the upper tool. The spindle in question is then started by the lever L (the four sections of the machine run independently of each other) and the arm reciprocates with a stroke which can be varied by the adjustment provided at the crank disc C. The machine is independently driven by the motor M.

The makers are Messrs Bryant Symons Ltd, Northumberland Park, Tottenham, London, N. 17.

We find that with polishing machines of this type the following relations between rate of cross stroke to that of rotation of the spindle give good results—

Diameter of block	Spindle rotation revs. per minute	Cross stroke (forth and back) number per minute.
13"	24	48
9"	30	60
$6\frac{1}{2}$"	40–50	80–100
$3\frac{1}{2}$"	60–70	80–100
Single lenses, 1–$1\frac{1}{2}$ in. diameter.	120	90–120

The power required for such machines is very small. For example in a small shop running about 18 spindles of the type shown in Fig. 75, together with a like number of slowly rotating bench spindles and two edging machines, the following power sufficed.

Load	Power (kva.)
Motor only … … … … … …	0·18
Motor and main shaft … … … …	0·35
All shafting and bench spindles … … …	0·44
All the above and about 6 spindles actually polishing … … … … … …	0·90
All the above and the 2 edgers … … …	1·32

Figs. 76 and 77 show parts of a 24-spindle machine for lenses up to $1\frac{1}{2}$ in. diameter and a 6-spindle machine for blocks up to 6 in. diameter, both designed and made by Adam Hilger Ltd. The design of these machines aimed at combining all the essential movements with the greatest simplicity of construction, owing to the urgency of installing them. For this reason the frames were made of wood. The machines differ from the traditional type in which there is a separate reciprocating arm for each spindle, for all the driving pins in each row are driven from one shaft.

Other types of polishing machines

140 The types of polishing machines described in the previous sections are derived from Lord Rosse's machine for polishing large astronomical reflectors (§14). They all reciprocate the polisher on the revolving work. I have little doubt, however, that the round-stroke machine originated about 1913 is more efficient. The following

Fig. 76—24-spindle polishing machine
Hilger

extract is taken from William Taylor's Presidential address to the Institute of Mechanical Engineers (Taylor, 1932).

Prior to the War, lens-polishing machinery of the kind already referred to, comprising one or two cranks and connecting-rods to move the tool and weights to apply the necessary pressure, were in general use. But the eccentricity of the cranks, which determined the size of orbit of the upper tool, the eccentricity of that orbit with respect to the work, that is, the size of the epicycloid described by the upper tool on the lower one (commonly the work), the respective rates of movement of the two members, and the pressures between them, were all varied according to such judgment as the operator possessed. The product varied greatly with the skill of the operator, and long experience was necessary to get good results. Much evil resulted from the inertia of the weights, which moved up and down as the attachment to the upper tool moved to and from the zenith of its vertical orbit.

During the War, in order especially that hundreds of thousands of binocular telescope objectives of the highest quality could be

Fig. 77—One spindle of Hilger 6-spindle polishing machine

made by inexperienced girls, I reduced these variables by designing a machine whose geometry will be seen in Fig. 78. Here the lower member is a block of twelve lenses held by pitch to one metal holder c, and this is slowly rotated on its vertical axis. The upper member e, the polisher faced with a wax composition, has a central depression in which is seated the end of a crank-pin a carried on an inclined crank-shaft b. The axes of this shaft and of the lens holder c intersect in the centre d of the sphere of lens surfaces, and thus the crank-pin at every point of its normal circular orbit would be at a constant distance from the lens surfaces without guidance by the polishing tool e.

The crank arm *f* is carried by a pivot *g* and urged by a spring *h* so that the pressure between the polisher and work may be adapted to requirements. The lens holders for all sizes of sphere within the capacity of the machine are so constructed that the centres of curvature lie at the fixed point *d* where the two shaft axes intersect, and the crank-pin is adjustable by sliding in the crank arm to

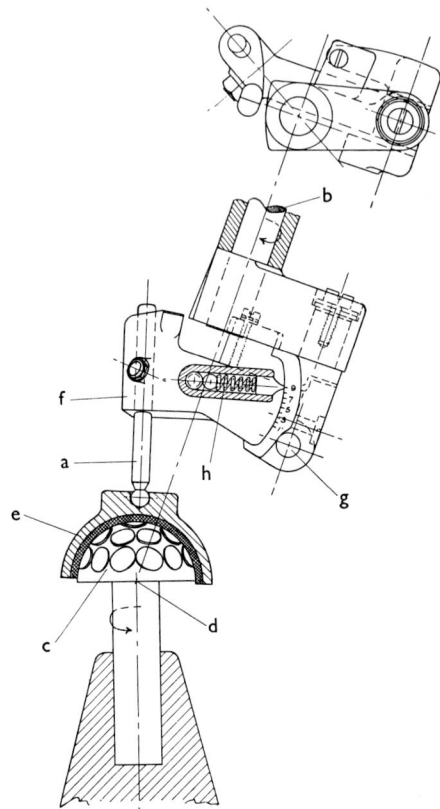

Fig. 78—William Taylor's type of polishing spindles

accommodate work of any curvature. Being set at an angle to the crank-shaft and in line with the point *d*, the crank-pin, when so adjusted, automatically varies the size of its orbit in strict proportion to the radius of lens curvature. At the same time the eccentricity of the orbit with respect to the work axis is varied automatically and precisely as it should be varied.

This machine entirely eliminated the inertia effects of previous machines, and it eliminated all the variables but one, the pressure between the tool and the work. But for this it provided a measure of value, being graduated with a scale of pressure by which the machine could be set.

On this machine was another device which, though stimulated into being by the War, I had conceived a generation before. The operation of feeding rouge and water to the lens-polishing tool had been performed with a hand brush. If the film of water between the polisher and work exceeds in thickness the diameter of the grains of rouge, no polishing can occur. If the film dries the polisher heats and adheres to the lenses, and may tear them from the tool. Thus the operator has to exercise ceaseless care in watching his work and keeping it sufficiently moist, and the number of spindles he can attend is very limited. Also, the work is very messy.

The traction between polisher and work rises in value continuously as the polisher dries after being wetted. Obviously the traction is a function of the rate of polishing, and the object should be to maintain this as high as possible short of the danger point.

The rule must therefore be to apply the rouge and water continuously or frequently when, and only when, the traction is a predetermined value short of the danger point. This cannot be done manually with certainty, but it is done by a device in which a supply of rouge and water is kept in circulation in a closed circuit of pipes and provided with pneumatically closed valves. One of each pair of machine spindles is mounted with slight lateral freedom so that it may be moved by the tractive effort between the polisher and work. Normally, by means of a spring whose strength may be varied, this motion is resisted and a vent in the pneumatic system is closed and the water supply thereby cut off. When, however, the traction rises to any value predetermined by the setting of the traction spring, the vent is opened and water and rouge are supplied to the polisher. Thus the whole process becomes automatic, the operator is relieved from much strain, his work is more cleanly, and he produces more and better work ; and this is the normal result of applying to an art the science of mechanical engineering.

Polishing machines with controlled relative rotations of work and tools

" *Valseurs* "—In all the grinding and polishing machines described here that one of the work or the tool which is uppermost is free to rotate as may be determined by the system of frictional forces between them. Since the friction is greatest at the places where there is a raised place on either the work or the tool the tendency will be to diminish their relative movement at such a place. It would appear that this is the very reverse of what is desirable, and devices (termed " valseurs ") have been invented in France for controlling this tendency.

The " valseur " is a special device for retarding the natural speed of rotation of the upper member of the pair so as to establish a more favourable relationship between the speeds of the two.

These devices were originally described in the *Revue d'Optique* for 1930 and 1931 (**9**, 481 and **10**, 201). Dévé in Chapter V of his book *Le Travail des Verres d'Optiques de Precision* (a translation entitled " Optical Workshop Principles " was published by Adam Hilger Ltd) analyses in great detail the movement of the work and polisher under varying conditions of their relative sizes and rotations. His analysis leads him to a number of theorems concerning the distribution of wear of the glass, and how it can supposedly be made more uniform by using a machine with adjustable relative rotations and " valseurs."

In my opinion this work—completely neglected, as far as I am aware, in every country except that of its origin, and not in wide use even there—deserves serious consideration by those responsible for the installation of grinding and polishing machines and methods.

Procedure after polishing

Detaching lenses from the block

141 A quick way of detaching lenses from the block is to plunge the block into a bucket of iced water. A better method is to put the block for a few minutes into a refrigerated space. A simple way of doing this is to have " Drikold "—or solid CO_2—in a wooden box which is placed within a larger wooden box with sawdust in between. The lenses fall off leaving very little pitch adhering to the surfaces. " Drikold " requires renewing every two days. Alternatively, where large quantities are concerned, or where supplies of " Drikold " are not easily obtainable, an industrial refrigerator may be used.

Before passing to the refrigerating box, the polished block is checked by proof sphere for radius and sphericity, and examined under an intense light for surface marks. If satisfactory, it is placed in the freezing chamber until the pitch has contracted and pulled loose from both the lens and the block holder. The lenses are then detached, left on the bench until the frost film has melted, and soaked for an hour in turpentine. They are then wiped with best cotton wool, laid on clean paper and wiped with a clean, soft cloth. If the second side remains to be done, they are returned to the sticking-on bench ; finished **lenses** are bathed in methylated spirit and cleaned for final inspection.

Centring lenses

142 Having polished the lens on both sides to the radius and accuracy called for by the blueprint we have now to edge it, that is, to reduce the blank to the finished diameter and tolerance required, and at the same time to ensure that the edge so generated is co-axial with the lens axis.

There are several variations of the method of centring, but all the methods fall into one of three classes, which will be discussed in chronological order.

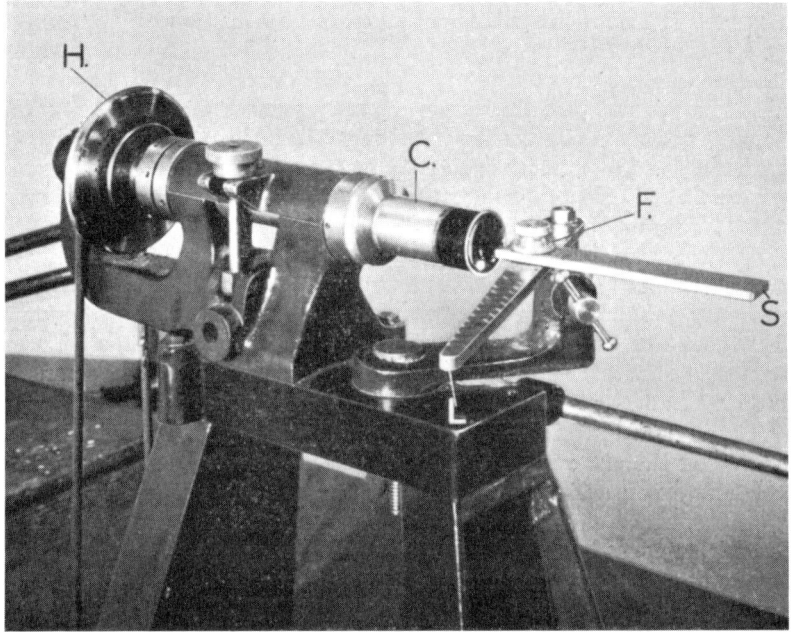

Fig. 79—Adcock and Shipley centring lathe with Hilger centring device

The oldest method, and one still giving valuable service, is the one detailed in Chapter 2; this consists of turning a brass tube true in a lathe, sticking the lens on the end with pitch, and observing the reflections of a lamp in the lens, which is adjusted until both the reflections appear stationary.

Pitch suitable for "sticking on," when tried on the standard jig (§60), should give 3 mm fall of the rod in five minutes at 21°C. Such pitch is suitable for sticking on lenses at that temperature $\pm 7\frac{1}{2}$°F (preferably within 5°).

Sticking-on pitch should be stocked for several ranges of temperature ; namely—

42–57°F for cold weather
52–67°F for mild weather
62–77°F for warm weather
72–87°F for hot weather

One method of speeding the process of centring is shown in Fig. 79, fitted to a centring bench of the type supplied with the Adcock and Shipley Edging machine shown in Fig. 82.

The device is constructed as follows : A strip of hard wood S is mounted so that it can rotate with friction, the friction being supplied by a spring through the felt washer F. This is brought gently against the face of the lens while the pitch is still warm. Two projections on the boxwood strip are thus pressed gently against the lens and, if the latter is rotating eccentrically, the longer end of the strip is caused to oscillate. If the pressure is slightly increased by means of the lever L the oscillation dies away almost immediately, the reaction of the strip with its friction being such as to set the lens central.

If the lens continues to rotate eccentrically, further increase in the pressure applied through the lever L will not improve the centring, and may well cause scratches on the lens surface if kept in contact therewith too long. The chuck should be warmed with a Bunsen burner to soften the pitch slightly, when the lens should move central without further trouble.

A final glance should be given to the two reflections in the lens before cooling off the chuck, as this method, like its parent method of using a wooden peg on a T-rest, can only be responsible for the truth of the exterior surface of the lens ; if between the back lens surface and the chuck there should be a piece of dirt, the back surface will inevitably be running out of truth, and the same effect will be produced by an untrue chuck.

Some degree of compensation for these last named difficulties is introduced if the adjustment for centring is performed not by reflections but by transmission through the lens.

Fig. 80 illustrates the set-up. The lens to be edged is mounted on a chuck screwed to a hollow mandrel, through the end of which the lamp illuminates the graticule G.

The lens shown at B is to compensate the convergence of the mounted lens, and transmits approximately parallel light to the viewing telescope T, which for convenience is provided with a small further focusing adjustment. The lens is warmed with a Bunsen burner until the pitch is slightly softened, moved until the image in the telescope is stationary and the chuck cooled off.

If the chuck is slightly out of true, the lens will none the less be truly centred but, in effect, oval to the extent of the amount of tilt involved. For the small angular errors encountered in the trueing of edging chucks, this defect is small enough to be neglected, unless the lens is a deep bi-concave, in which case particular attention must be paid to the trueing of the chuck, otherwise the effect may become undesirably great.

The third method of centring, which was used by both British and German manufacturers during the recent war is known as the " pressure chuck " system. Two chucks are used, very accurately lined up to be co-axial; usually the left-hand chuck is fixed while the second is pressed against it with a powerful spring, and can be withdrawn an

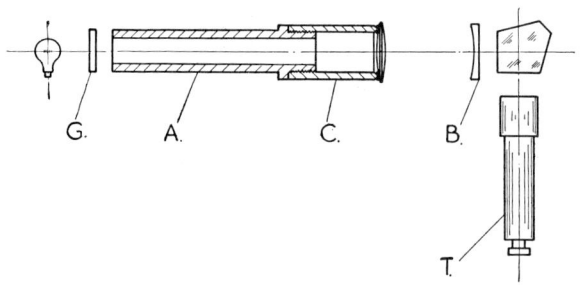

Fig. 80—Centring a lens by transmitted light

inch or so by means of a lever. No adhesive is used, the lens being placed between the chucks and the lever released; the two curves locating on the chuck rims automatically centre the lens.

This system is very efficient for deeply-curved lenses, and for shallower lenses where the centring tolerance required is not of the highest order. Representative accuracies of centring claimed for deep lenses are of the order of 0·0005 in. to 0·001 in., and this is amply close enough for almost any purpose. For checking the centring of shallow lenses (say up to 6 ft focal length) there is in my opinion no better way than that of transmission through the lens.

Edging lenses

143 Although the simple method of hand edging detailed in Chapter 2 is still in use and, in the hands of a skilled man, very safe and reliable, it is also very slow, and edging machines for quantity production are in general use. Without considering the machines in use for edging the non-circular shapes required by the makers of spectacle lenses, we may describe two machines specially built for the repetition edging of lenses.

A double spindle edging machine made by Adam Hilger Ltd is shown in Fig. 81. A machine of this type was shown to me at the

Eastman Kodak Works in Rochester, U.S.A., by my friend Dr C. E. K. Mees, now Vice-President of the Company, who very kindly gave me permission to copy it, and the copy made by Adam Hilger Ltd from my sketches has been in use from 1914 to the present time. The lenses are carried on two chucks A and B which can be interchanged by rotating them about a common axis a-a. When a lens is in the forward position a pin P on its bearing engages with a pin on the driving plate D which rotates it at a speed of about 120 r.p.m. It is in this position that the centring is effected in the manner which was shown in Fig. 79.

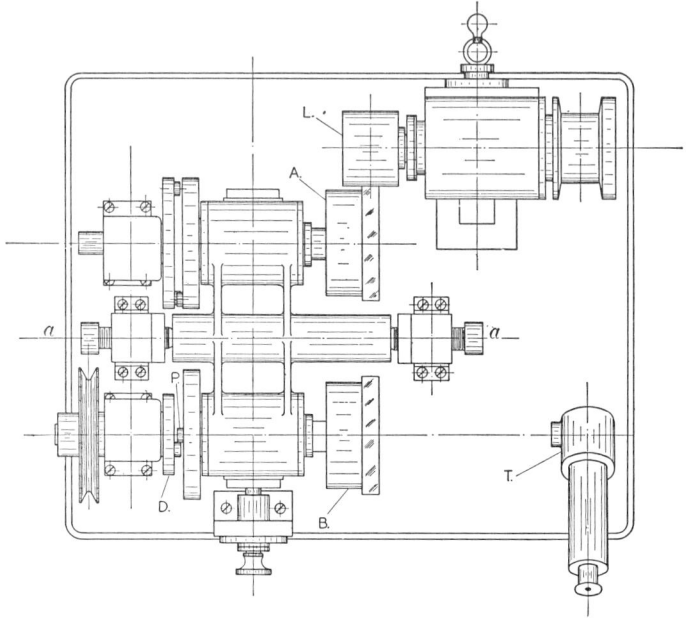

Fig. 81—Hilger edging machine

When the lens has been centred and cooled on its chuck, the lens positions are interchanged and the one which has just been centred is brought into position for edging by the diamond lap L. It is rotated slowly and continues to edge while the operator affixes and centres a second lens. It is essential that the linear speed of the periphery of the lens during edging should not be too great.

The Adcock and Shipley edging machine

This machine (Fig. 82) operates by the use of separate carriers, each containing its own bearings, which are adjusted to yield an identical diameter of workpiece.

If the lenses under process are large and thick, they may need more time to cool than the edging cycle of the machine will allow, and unless extra carriers are available the cooling time would control the time of edging.

The grinding agent in this case is a carborundum wheel of suitable grit and grade (180 to 220, P. or Q.), or diamond wheel 200 grit,

Fig. 82—Edging machine
Adcock & Shipley

rotating at 3500 r.p.m., across whose edge the centred lens oscillates along its own axis, while revolving. The complete grinding head pivots on the back shaft B, which also serves to bring the wheel into the correct position (see later) by a screw adjustment. On the front of the grinding head is a guide bar G, which, in the working position, bears on the right hand (feeder) stop point, which is itself brought slowly closer to the axis of the lens by means of the ratchet feed operated by the traverse crank T through the adjustable linkage L.

When the guide bar comes into contact with the fixed stop point attached to M the grinding ceases, since the forward motion of the wheel is arrested. The feed on the point F continues and retracts the bearing point clear of the bar. This serves as an indication that grinding has ceased.

The ratchet is then thrown out of action, and the feeder stop point rotated (clockwise) until the guide bar is clear of the fixed point M, by about 0·1 in., the whole machine turned off, the splash cover raised as in the illustration and the grinding head swung back until the bar Q catches in the restraining hook R. The lens is then measured for diameter, and if any reduction is necessary, the fixed point M is released from its clamping screw and the adjustment made.

When the lens is to size, the carrier spindle is removed and replaced by another on which is a lens ready centred, this having been prepared on the separate centring lathe during the edging of the first lens. Before replacing the splash guard and switching on the machine, care should be taken to see that when the grinding head is released from the restraining hook R, and gently returned to the working position, the lens should clear the stone by a small amount. The splash guard is now replaced, the machine restarted and the lens left to edge.

When the carborundum wheel is in use it is desirable that the lens should traverse right across the stone, clearing it at both extremities of the stroke. This adjustment is obtained by setting the stroke of the crank T and moving the grinding head traverse B.

The edging machine preferred in the Hilger optical works at the present time is the OVS lens-grinding machine with impregnated diamond wheel.

It will be found that the use of an impregnated diamond wheel for edging calls for modification to the speeds and feeds to obtain the best results. Good edging calls for two qualities in the finished lens, a smooth edge and freedom from chips or splinters where the polished surface meets the ground edge, and it is this latter condition which will not easily be achieved if a diamond lap is substituted for the carborundum stone without other alteration.

In general it may be said that the impregnated diamond tool is essentially a milling device, in that it performs most efficiently and gives the finest finish if a heavy cut is taken slowly.

The operating conditions can be altered part way towards the ideal by running the drive belt for the lens carrier from the countershaft instead of from the pulley provided. The traverse should be adjusted so that the lens does not quite leave the lap on either side.

A further step, giving even better results, is to tilt the grinding head slightly, to run the work carrier from the shaft as above, and to slow the crank stroke down to the point where the linear speed in the mid-stroke position is about 1 in./min. The whole amount to be edged off is now removed in one cut, and the feeder stop ratchet mechanism is not used at all.

If the operation times for the various arrangements detailed are compared there will be very little difference, but the finish of the edge will be found progressively better.

Fig. 83—Taylor, Taylor and Hobson edging machine

It should hardly be necessary to add that, whichever type is used, the grinding wheel should be trued as accurately as possible before work is commenced and, once fitted, a diamond tool should be left in place until worn out (generally some years) and not taken off unless absolutely essential. After the most careful trueing of the lap, the small residual lack of truth (measured in fractions of a ten-thou.) will gradually disappear until, when the lap is running perfectly true, as a result of some few months of use, the cutting sound will be a smooth hiss, with no trace of knock at spindle speed.

Taylor, Taylor and Hobson edging machine

For many years now Taylor, Taylor & Hobson Ltd, of Leicester, have used for the edge-grinding of lens elements a machine of their own make.

The operative portion is shown in Fig. 83. The procedure is unconventional in that a cup wheel is employed, and the lens is so presented to the wheel that the lens axis is at right-angles to the axis of the wheel, the outer diameter of the cup wheel being in the same horizontal plane as the axis of the lens. In this machine the work is rotated slowly once, or at most twice, while in contact with the wheel.

The machine is also provided with adjustable means for grinding chamfers on lenses at any required angle to the axis. The means for centring the optical axes of lenses on their shafts prior to grinding is not embodied in the machine, but is done separately beforehand. A special lathe is also provided for enabling operators to true the brass chucks on the work spindles when required. The procedure for centring and grinding a lens is therefore as follows—

Having mounted the brass chuck on the spindle and turned its annular edge true, the spindle is mounted in the crutch (Fig. 84) and is rotated by hand as illustrated. A gentle gas jet is directed on to the brass chuck to warm it. A stick of wax is then applied to the edge while the chuck is being rotated, the lens placed on the chuck, pressed down and adjusted until the reflected images from its upper and lower surfaces cease to move relatively while the shaft is being rotated. The shaft is then placed in a suitable rack with its axis vertical while the wax hardens, after which it is transferred to the machine (Fig. 83) in which it is driven, during grinding, by a tangential wire wrapped round a drum. This wire passes upwards out of the illustration and over a pulley, and then down to a weight. The other end of the wire (not shown in the illustration) is attached to a dashpot so that, when the shaft has been loaded into the machine, the operator pulls down the wire, which is of sufficient length in relation to the diameter of the drum to give one or two complete turns as required, and the rate of rotation of the drum is governed by the weight and the restriction to movement imposed by the dashpot.

When the wire has been pulled down the work is fed into contact with the wheel and rotates slowly during the cutting operation.

The foregoing is the procedure for grinding the cylindrical edge of the lens. If it is desired to grind a chamfer at any desired angle on the edge of the lens, the workshaft is removed from the edging station and placed into the crutch illustrated to the right and slightly above the grinding wheel. In this position the workshaft is rotated slowly by hand while the wheel grinds the chamfer.

Jorgan edging machines

As far as I can judge from the literature and photographs which have been sent me by the Jorgan Company of 100-108 Seneca Avenue, Rochester 5, N.Y., U.S.A., their 4 S lens multiple-spindle centring and

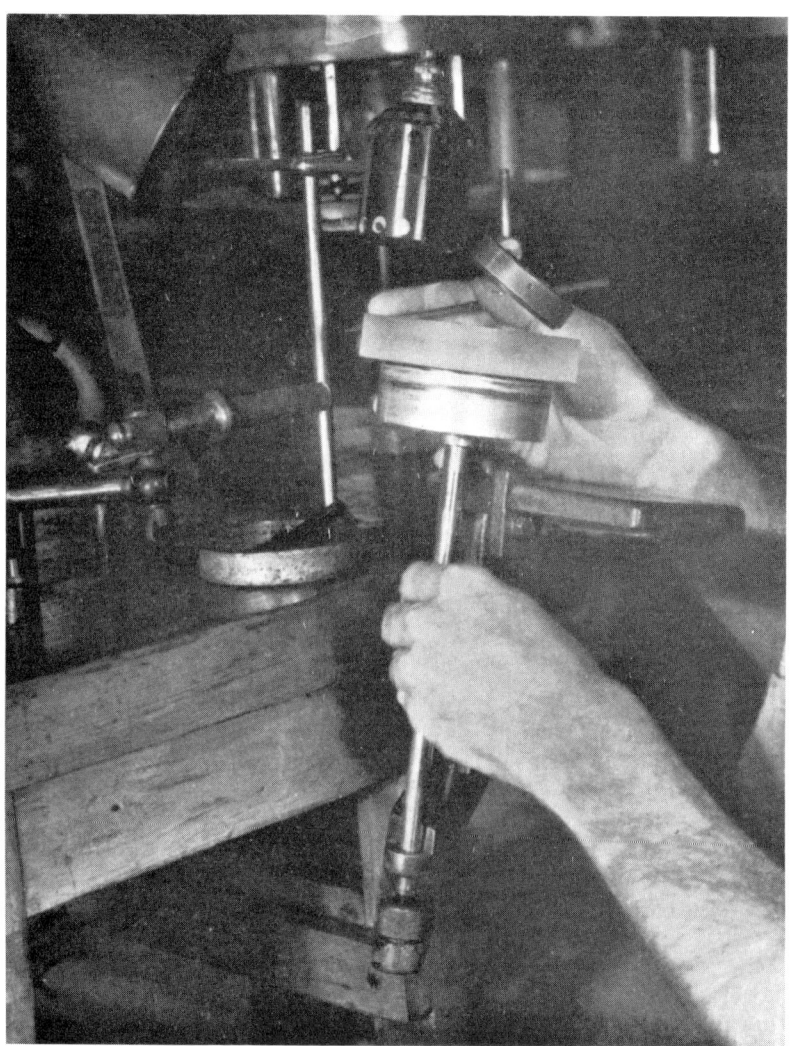

Fig. 84—Mounting the spindle in the Taylor, Taylor and Hobson machine

edge-grinding machine may be the best edger yet developed for mass production.

Chamfering

After edging has been completed on one of the machines described above, and before the lens is warmed off the chuck, it is usual to chamfer the intersection of polish and edge with a curved tool of appropriate radius and 400 abrasive.

In the case of the Hilger machine this is done on the machine itself with the work spindle in the nearer position, while on the Adcock & Shipley it is convenient to use the centring bench.

When the lens has been removed from the chuck, the other side is usually chamfered in a similar manner, and in this stage very annoying damage to the polished surfaces can be caused by imprudent handling. The defects produced are not so much scratches as a form of bruise, *i.e.* the damage does not extend through the polish layer. They can only be removed by polishing, a most undesirable retrogression now that the lens is finished, and to avoid trouble in this operation we have found the following small tool most effective—

A piece of ebonite or, better, " Tufnol " tubing* is turned at one end to accommodate the lens with say 0·002 in. clearance, and the tube chamfered away to clear the chamfering tool. The tube is then sawn part way down its length to allow of some spring, and a screw added to clamp it to the lens. The split chuck thus formed holds the lens firmly by its edge and no damage can occur due to the fingers slipping over the polished surface while contaminated with emery.

After the chamfer has been made, the collet and lens are rinsed in a tank of clean water before removing the lens and cleaning-up in the usual way.

While it is traditional at Hilgers to use very small chamfers on optical work, it does not appear that there is any valid reason for this except that it is evidence of careful work. As a reasonable chamfer for optical work intended for work of a more industrial character, the following specification has been adopted by us—

> Edges of all first quality optical work to have a chamfer whose width is 1/60th part of the longest surface terminating in that edge, with a minimum of $\frac{1}{2}$ mm.

Accuracy of centring

It has been found that each of the component lenses of an achromatic object glass of $1\frac{1}{4}$ in. diameter and 11 in. focal length should be centred

---

* Tufnol is a laminated synthetic resinoid material available from Tufnol Ltd, 121 Victoria Street, London, S.W.1, England, in the form of sheet, **tube**, **rod**, angle channels, and specially moulded shapes.

to within one thousandth of an inch if the very finest definition is to be attained. If this limit is much exceeded the image of a star becomes perceptibly affected. A further point that requires attention is that the balsam layer must not be wedge-shaped. A variation in thickness of balsam from edge to edge of more than one thousandth of an inch should not be tolerated. It may be mentioned, however, that the method of balsaming in the oven described below ensures that this accuracy is attained automatically ; indeed, only gross carelessness could result in such an error, whatever method of balsaming is adopted.

Finally, in testing components of objectives of other focal lengths but of the same types, the tolerance of centring must remain the same if the full resolving power is to be maintained, namely the two components should have their optical axes coincident to within one-thousandth of an inch. This is most easily attained by centring each lens accurately in itself and then ensuring that their edges are exactly registered.

Not only must each individual lens be edged centrally but they must be " set up " central with each other to a like accuracy (0·001 in.). For eyepiece lenses, or components of the same, an accuracy of centring to 0·003 in. suffices. If each component has been centred within the limit stated, a good way for testing their setting is to rotate the object glass so that its edge runs true ; the shift of the image should then be not more than 0·004 in.

Occasionally it is required to centre the components of a balsamed doublet lens to an even higher accuracy than that imposed by considerations of definition. For instance, it may be necessary to remove a lens from its cell for cleaning and to replace it without rotation of the image exceeding say ten seconds of arc ; such lenses are normally of fairly long focal length.

To comply with such a condition we should adopt an expedient which we call " jig centring." The flint component is edged by transmission as truly as a $20 \times$ telescope will allow, and the crown is deliberately edged two or three thousandths of an inch smaller.

The lens is placed in an accurate cell, which the flint component just fits, and rotated, while the image transmitted is viewed in the telescope. If the rotation of the image in the telescope exceeds the permitted amount, the lens is placed in a thermostatic oven, adjusted to the temperature at which the balsam can just be moved by finger pressure, until it has acquired the same temperature, which usually takes about 15 minutes. The lens is then replaced in the cell and the crown component moved until the image stays within the permitted amount on rotation.

When cool, the lens is tested for definition on the Hilger lens interferometer, and if any sign of astigmatism appears to be present, the lens, or batch of lenses, is annealed at the correct temperature until the strain is removed. If the lenses are supported on a flat glass plate in a levelled oven during annealing, and the temperature is correct, no movement should occur ; if it does the temperature is too high.

Balsaming

144 Doubtless every reader of this book knows that if a single lens is used as the object glass for a telescope, the definition is very poor (see §98). One can get a reddish image of a star, but it is surrounded by a blueish haze. To avoid this " chromatic aberration " as it is called, one must use a positive lens of crown with a negative lens of flint. The second surface of the crown and the first surface of the flint usually fit each other, and are then frequently cemented together, the compound lens being called an " achromatic " objective, or for short, achro. O.G.

Most achromatic objectives are cemented together by Canada balsam. Canada balsam, an oleo-resinous exudation of the so-called Balsam Fir, *Abies balsamea* Miller*, has as its chief constituents two separate resins and turpentine. The turpentine and other volatile constituents can be evaporated off by heating so as to obtain balsam of varying degrees of hardness. We use three kinds which we distinguish as hard, medium and soft. These degrees of hardness are defined as follows—

Hard : The standard test jig (see §60) sinks 5 mm in $3\frac{1}{2}$ minutes at 20°C. The test is made by the pot containing the balsam being entirely immersed in water of this temperature for about an hour before test. This balsam was so standardized at Hilger's from 1914.

Medium : On this the standard jig falls 25 mm in $3\frac{1}{2}$ minutes at 20°C (standardized at Hilger's since 1915).

Soft : This is obtained by passing the raw unclarified balsam of commerce through filter paper at 100°C and heating it until it loses about 7 per cent of its weight. The filtration is carried out in a steam-jacketed filter and the subsequent heating to produce the balsam of the various degrees of hardness is carried out as follows—

An old-fashioned pair of scales is used with the scale pan tinned (since iron discolours the balsam). The filtered balsam is put in the scale pan which is then heated from below to such a temperature that the balsam boils fairly vigorously. The temperature gradually rises as the volatile constituents boil off but it should never be allowed to exceed 200°C. The loss of weight for the various degrees of hardness is as follows : soft balsam, 7 per cent ; medium balsam, 14 per cent ; hard balsam, 25·8 per cent.

---

* And not as is frequently stated (indeed it was so in the first edition of this book) the Douglas Pine. The Douglas Fir—*Fseuda-tsuga trifolia* (or *Douglasii*) —also known as Oregon Pine, although neither a Fir or a Pine, gives Oregon balsam which has been offered as a substitute for the more expensive Canada balsam but is not so good.

If the hardening is carried out rapidly and in an atmosphere reasonably free from dust, the balsam remains clear and of a good light colour. On theoretical grounds it might be preferred to adopt a procedure which has been recommended to me, namely, to distil off the necessary amount of solvent by heating the balsam in a distilling flask under reduced pressure (7–10 mm of mercury, which can be obtained with an ordinary filter pump) at a temperature of about 130°C.

Interesting particulars about Canada balsam and its use are given by French (1918).

The method of measuring relative viscosity by means of the jig described in §60 is, as may be supposed, not very accurate; further, balsam kept in an ordinary container—say in a bottle with a cork—becomes, even in a few days, appreciably harder at the top than at the bottom, and at the sides than in the middle. The test is, however, a practical one and amply suffices for the purposes of the optician. If anyone wishes to measure the viscosity of specimens of balsam more accurately, the best method is by the immersion of a cylinder. By using this method it has been found by Dr K. M. Greenland of the British Scientific Instrument Research Association that the measured viscosity depends on the rate of shear. By plotting log shear stress γ against log rate of shear $d\gamma/dt$ he finds that

$$\gamma^{2.25} = \frac{d\gamma}{dt} \times \text{const.}$$

The use of soft balsam

Perhaps the most usual process is to use soft balsam. This can now be purchased of excellent clear quality and very pale in colour so that there is little need for the optician to deal with the crude balsam as described above unless the consumption of the works is very large. A thick iron plate on three legs is heated from below by a ring burner. On the plate is placed a piece of paper and on this the lenses are placed, crown and flint side by side. The lenses being perfectly clean, they are brushed on the contact surfaces with a camel hair brush to remove dust, and the crown lens placed on the flint. Newton's rings should be seen, indicating that the lenses are nearly in contact. The crown lenses are lifted one by one by the finger and thumb of the left hand while with the right hand the worker puts a brass wire into the pot of balsam which is also on the hot plate, and with it transfers a drop of balsam to the upper surface of the flint lens. The top lens is then replaced and gently pressed down with a slight rotary movement by means of a cork, until the excess balsam is pressed out. The lenses are then allowed to cool down slowly and as they cool are pressed central with one another by the fingers so that when they are cold and the surplus balsam has been cleaned away the two edges are coincident.

The balsam, however, is still quite soft and the lenses can easily be shifted relative to each other. It is necessary therefore to bake them.

For this purpose they are put in an oven fitted with a thermometer and baked for sixty hours continuously at a temperature of 77°C. The oven is then allowed to cool before removing the lenses. The effect of this process is to harden the edge of the balsam and it will be found that, even after a number of years, if the lenses are taken apart all but a thin rim of the balsam round the edge is quite soft and sticky.

For certain purposes it is important that the lenses should be permanent in the sense that even if warmed they cannot readily be shifted relative to each other. This requires balsam so hard that any sudden shock may cause the lenses to split apart at the balsam layer. At this point it is worth mentioning that where conditions of great cold have to be met, such as in aircraft, all balsam is liable to split and it is generally considered therefore that it should be avoided. None the less, for certain purposes medium or even hard balsam is preferable to soft, and may be manipulated in the following way—

Medium balsam.—The procedure is the same as with soft balsam except that the temperature of the hot plate requires to be higher so that the balsam is liquid enough to press out. No prolonged baking, however, is required, but in order to get the balsamed object glass free from strain it should be annealed in an oven for two hours at 40°C and cooled not more than 3° every quarter of an hour. This procedure—which is on the careful side—is suitable for achromatic object glasses up to 3 in. diameter. Lenses of $1\frac{1}{2}$ in. diameter and smaller can be dealt with much more rapidly.

Hard balsam.—The procedure is the same as for medium balsam, but annealing should be carried out at 60°C for two hours. Again the cooling for lenses of $1\frac{1}{2}$–3 in. should not exceed 3° each quarter hour, and again smaller lenses can be dealt with more rapidly. In Germany, Canada balsam is obtained in sticks, filtered, and concentrated to a melting point of 78°C. This is applied to the hot lens and after the usual pressing out and centring the cemented lens is annealed at 60°C for two hours. (*U.S.N.*, 1945)

The respective merits of hard, medium and soft balsam

145 To sum up, one uses hard balsam when one wants to reduce the amount of shifting as much as possible. Unless the cemented components are stout the work should be annealed, or the parts will be distorted. One uses medium balsam where the same effect is desired but where the temperature must not be raised too high, as in cementing Polaroid or gelatine filters between glass plates. Soft balsam is used where strain must be avoided, as in protecting half-wave plates of mica. Medium balsam is the most generally useful.

Hard balsam is in general to be preferred for small, medium for larger, and soft for large areas, since differential expansion of larger pieces introduces greater mechanical strain. If a soft balsam must be used, we prefer a hard balsam dissolved in Xylol; it can be used without heat and baked at a low temperature. Although in general it has been regarded as dangerous to use cemented surfaces for very low temperatures, prisms (of the same glass and small in size) cemented with medium balsam and baked at 70°C for $8\frac{1}{2}$ hours have stood a chilling test of –50°C.

Balsam will eventually harden without heat and, if cementing in which the hardness is of critical importance is performed only occasionally, the appropriate cement may be kept at constant hardness for a very long time by storing in conical flasks with flat-ground tops to which glass disks are cemented.

It is said that, in Germany, Canada balsam has been superseded by a satisfactory balsam extracted from the larch tree. Whether this was produced as an *ersatz* substance or not I cannot say.

Special precautions to facilitate good balsaming
145a Instructions I prepared at the beginning of the 1914–18 war for employing unskilled girls on high quality optical work prescribed that cleaning liquids should be filtered, cleaning rags washed in distilled water, and dusting brushes washed in absolute alcohol and stored in tissue paper. I pointed out that lenses should never be pressed together, except very lightly with the end of a boxwood stick, that really clean lenses slip over each other without the slightest friction, and that absolute alcohol may be used (sparingly) for final wiping. Skilled workers may think such particulars unduly meticulous but I consider they would result in an overall saving of time.

Assembly of lenses without cement
145b The desirability of a high light transmission in such instruments as submarine periscopes in which the number of surfaces is very great, is said to have resulted in a German technique for assembling lenses in optical contact, thus avoiding balsaming. The contact curves should match to half a ring or less and it was stated that the technique was fairly easy, the lenses being assembled by semi-skilled female labour; there was practically no scratching of surfaces and the assembly was stable in use.

In any case lenses over about 5 in. diameter should be mounted without balsam—either with oil (*e.g.* medicinal paraffin) or dry. It is almost impossible to balsam lenses of this size without distortion.

Halle (1921, pp. 23 *et seq.*) gives a number of recipes for soft cements under the generic title of " *Oelkitten*," while an article by König and Rautenfeld (1942) offers interesting suggestions on stress-free cementing.

Plastic cement as a substitute for balsam

146 The contradictory properties required in an optical cement, namely resistance to large temperature changes, freedom from strain, and high mechanical strength at all temperatures, have stimulated research with a view to making an artificial substitute. In particular the rapid introduction of air photography subjected the optical parts of aircraft instruments to the cold of the stratosphere and the heat of a tropical airfield in quick succession. One of the most successful synthetic substitutes for Canada balsam is n-butyl methacrylate, manufactured under the trade name of H.T. Cement ; it is a good general purpose cement obtainable from Hopkins & Williams.

Unlike balsam, this material is used cold when laying the lenses together, although the lenses themselves should be warm (50° to 60°C) ; baking in the oven for 5 hours at 75°C or for 16 hours at 60°C polymerizes the cement, that is a new substance is formed which has a melting point in the region of 200°C, and which, being elastic but immobile, withstands temperatures below –40°C. The 75°C baking should be used if the lens is to stand a temperature of 75°C ; the 60°C baking is preferred if the lens is only to stand 60°C.

It also differs from balsam in that once the baking has been completed, the cement cannot be liquified except by raising to the very high melting point of the polymer. The routine for employing this cement is as follows—

The lens components are first paired for diameter to a tolerance fixed by the designer ; this tolerance is also the amount by which the components may move relative to each other during baking.

In the case of lenses whose centring is of critical importance, it is advisable to measure the residual errors on each component and pair them again for deviation, marking the " thick edges " which have to be opposed in assembly.

The pairs are cleaned and laid together cold, and the cement inserted as for Canada balsam (§144). The combination is now placed on a balsaming jig, which consists of a steel plate bearing a number of sets of three small locating pegs, two in each set being fixed and the third adjustable to allow for variations in diameter (Fig. 85). For small lenses the two pegs are replaced by an equivalent made from sheet brass.

Lenses with a convex exterior to the flint component are laid on a thin parallel ring of metal to avoid scratching the vertex, and to ensure that the lens sits truly level.

When the jig is full, it is transferred to the oven and baked according to the time and temperature stated by the maker.

It is advisable to chamfer the contact surfaces as well as the exteriors when using H.T. cement, as a slight contraction takes place during polymerization which is made good by the rim of cement held in the chamfer; if this is not present there is a risk that air pockets will creep in round the edges.

H.T. cement is readily soluble in acetone, much less so in petrol and methylated spirit, and where the cementing of a combination has not been successful, an overnight soaking in acetone is much to be preferred to heating the part to 200°C to soften the polymer.

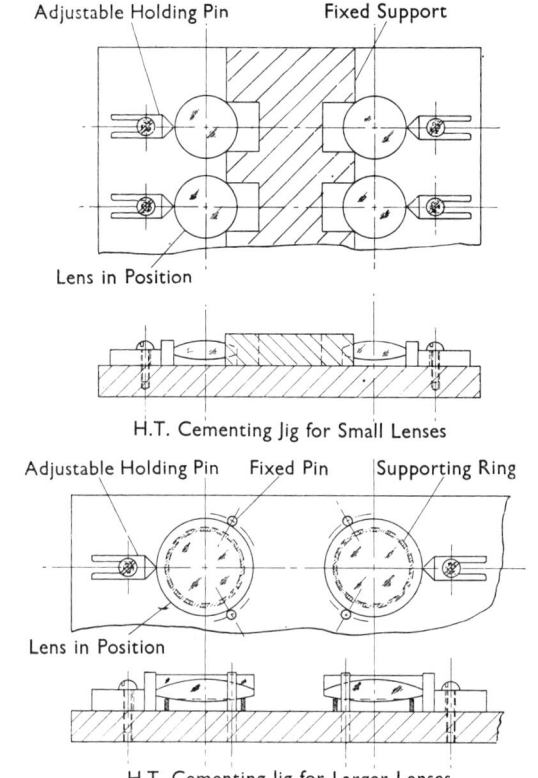

Fig. 85—Hilger & Watts jigs for cementing object glasses with H.T. cement

Precautions recommended for the use of H.T. cement
(1) The cement should be stored in an amber bottle, preferably covered with a screw-on bakelite cap. Direct contact of the cement with a cork or rubber bung must be avoided. If a cork or rubber bung is employed as a stopper, it must be protected with tin-foil or cellophane.

(2) H.T. cement is evacuated before it is sent out. It is very important that no unnecessary agitation or stirring should be employed, as dissolved oxygen often causes failures in the cement film. The dissolved oxygen may give rise to bubbles or interfere with the process of polymerization itself. The bubbles can be prevented if the polymerization is done in an atmosphere of nitrogen.

(3) The H.T. cement now available, unlike the original material, is stabilized for storage at ordinary room temperatures, although it is always desirable to keep it in the dark in a cold, dry place. Under these conditions the cement has an average shelf-life of two to three months.

Refractive index of H.T. cement

H.T. cement has a much lower refractive index than Canada balsam—

Refractive index (n_D) of monomer	1·42
,, ,, ,, of partly polymerized material supplied for use as an optical cement	about 1·44
,, ,, ,, of completely polymerized material	1·485

Use of H.T. cement for polarizing prisms

In cementing certain types of polarizing prisms, for example those known as the half-Lippich type which are used for high-quality polarimeters, it was customary to use linseed oil which has a refractive index of approximately 1·48.

As will be seen from the above notes, the refractive index of H.T. cement when it has been completely polymerized by heat treatment is almost identical with this and it is found to make a very good substitute for the linseed oil and is free from some of its disadvantages.

At the Watts Division of Hilger & Watts Ltd the following additional precautions are taken in the use of H.T. cement—

(1) In spite of the cement being stabilized, store it in a refrigerator at about 0°C whenever not in use.
(2) Use bottles not larger than 8 oz. capacity.
(3) Take the cement from the refrigerator about two hours before use, for it to assume room temperature, and replace it in the refrigerator immediately after use.
(4) Only order quantities sufficient for about 1 month and in any case, even if the cement is kept constantly in a refrigerator, do not use more than three months after its manufacture.

(5) Apply the cement with a solid glass rod and take care to avoid the formations of bubbles in the bulk as well as in that applied to the component. Be sparing of the cement and use it gently. Do not press out too much. The final film of cement should not be less than 0·001 in. to 0·0015 in. ; 0·002 in. is found most suitable.

(6) Do not use pressure or weight to hold the components together during baking ; the jig should control the components by the edges and the cement film should be horizontal.

(7) Use a water oven in order to secure uniform heating and bake at 75°C for five hours.

Close observance of these precautions is said to obviate the occurrence of bubbles and separation of cement from the optical surfaces. It must be admitted, however, that workshops which have used H.T. cement for long periods without trouble have sometimes suffered a recrudescence of such faults, whether because the procedure has become lax or not I cannot say.

It is also believed in some shops that H.T. cement renders uncertain the attainment of perfect definition in lenses for which it is used. I have not been successful in tracking down an actual instance of this ; if it is true, the work of Lyot and Françon on the scatter due to balsam would suggest that an explanation should be sought by tests for heterogeneity in the film of cement before and after polymerizing.

There is another peculiarity of the cement which might be suspect. It is an advantage of Canada balsam that strain occurring in cementing can be easily and completely removed by annealing. Since the polymerized H.T. cement has a jelly or rubber-like consistency no such process is possible, and it is easy to imagine strain being set up, during polymerization, which would be permanent.

Application

Two grades are used—
(1) *Thin*—about the consistency of golden syrup at room temperature (thick enough to gather on a glass rod).
(2)' *Medium*—about twice as thick as grade (1). Larger components (dia about ½ in.–2 in.) are joined with the thin grade and small components with the thick (dia up to approx. ½ in.).

Lenses

(1) Surfaces to be joined should have pronounced chamfers to provide a small ring of cement around the lens but not big enough to tear out in the cleaning process. The size of the

chamfer should be not less than 0·01 in. and not more than 0·025 in. Crown lenses at 45°, Flint lenses 90°. Inspect all lenses carefully for quality of chamfer which should be smooth having no chips or raggedness at the join with the polished surface.

(2) Select for diameter and pair to the following rules.
Up to 1 in. diameter, variation between crown and flint—Nil. Over 1 in. diameter, variation between crown and flint 0·001 in., the crown being smaller.

(3) Very carefully clean the batch of lenses and lay them together. Contact at the edge is preferable.

(4) Cement individually and fit into the jig (see below) which should be on a level surface.

(5) When the batch is completed the jig is left at room temperature for about $\frac{1}{4}$ hour and then placed on a level shelf in the water oven and baked at 75°C for 5 hours for small components and 6 hours for larger components (over about 1 in. diam).

(6) Remove jig from oven at the end of the bake and allow it to cool, avoiding draughts. Dismantle the jig whilst still warm.

(7) Clean up the surfaces using as little cleaning fluid as possible.

(8) When surfaces are clean, cut away the surplus cement from the edge with a sharp blade. Avoid breaking the small ring of cement in the chamfer. Don't allow any cleaning fluid to contact the edge after the cement is cleaned off.

(9) Inspect lenses under bright light for blemishes.

Prisms and flat work

(1) Inspection for chamfers as for lenses.
(2) Components should be flat to 5 rings per 1 in. with edge contact.
(3) The top component should be about 0·001 in. smaller in diameter.
(4) Arrange the jigs so that the cement film is level.
(5) Proceed as for lenses, with special emphasis on the need to avoid bubbles in the cement film rather than pressing them out.

To describe in greater detail the two types of jig shown in Fig. 85—Type 1 jig consists of mild steel plates with screw holes to accommodate the vertical guides for the lens. These guides are right-angled brackets, one of each set having an elongated slot in the base to accommodate the thread of the locking screw. The vertical part is machined so that a 90° angle is presented to the lens (with a slight flat 0·01 in. for protection) and must be square in both directions with the plate of the jig. A thin parallel ring is made up about 0·002 in. smaller than the diameter of the lens which protects the under-side. For small lenses up to about ¾ in. diameter, the third upright is fitted into position with light finger pressure to hold the combination properly whilst the screw is locked. For larger lenses, a light spring is placed between the upright and screw head which is not locked.

In cases of the smaller lenses the support ring is replaced by a piece of paper to protect the outside curve of the flint.

Type 2 jig is used for small doublets. A carefully machined central portion replaces 4 of the uprights previously mentioned. The edges must be square in both directions and smooth. A strip of paper protects the lenses.

All parts of jigs are made in mild steel, chromium plated on nickel. The base plate must be flat and have a smooth finish.

Cleaning
> (a) Methylated spirit is found to be suitable for cleaning the lenses, especially if the cement has been used sparingly and some of the surplus removed with tissue paper before baking.
> (b) Carbon tetrachloride, in which the jigs can be soaked.

Synthetic cement of high refractive index

147 Another useful synthetic cement is that known in the trade as "Sirax," and prepared by Stafford Allen & Sons Ltd, 20 Wharf Road, London, N.1, England. Two grades are made, which, after the solvent has been volatilized by baking, have indices of 1·65 and 1·8 respectively. These cements are very useful in the manufacture of certain types of polarizing prisms.

Thickness of cement layer

148 Since the refractive index of Canada balsam is close to that of glass the reflections from the two contact surfaces are slight. It is, however, sufficient for Newton's interference fringes to be visible in light which consists of monochromatic radiations such as that from a low-pressure mercury vapour lamp. With H.T. cement they are still more visible. Observation of these fringes affords a convenient means of ascertaining the thickness of the cemented film. This is measured by observing the number of fringes which cross the patch of reflected monochromatic light when the object glass is tilted from a position of almost perpendicular reflection to a position at 45° to this. If m be the number of fringes the thickness is $6m \times 10^{-5}$ inches.

In a number of object glasses 1½ in. diameter which were observed the thickness was observed to be 0·00025 in. and in some of 2·35 in. diameter it was found to average 0·00030 in.

Balsaming in quantity

When numbers of object glasses require to be balsamed, it is convenient to adopt the following procedure. The lenses and medium balsam are placed on a sloping board drilled with holes, one hole being under each object glass. Metal pegs or nails are driven into the board near the holes in such positions that the lenses resting against them are fairly central with respect to the holes. The object glasses, together with the balsam, are put into the oven and raised to a sufficient temperature to make the balsam of the consistency of treacle, the object glasses having been first carefully cleaned, dusted and put in contact.

The board is then carefully removed from the oven and placed on a plate sufficiently warm to prevent the object glasses cooling down while the top lenses are lifted one by one, a drop of balsam put in the middle of the lower lens and the top lens replaced. This procedure can be carried out very quickly, since it is not necessary with this routine to press the balsam out.

They are then replaced in the oven which is brought to a temperature of 100°C and the heat then turned off. The lenses will pass through the annealing range round about 40°C slowly enough for the annealing to be good.

Method of determining at what temperature to anneal balsamed objects

At the same time that the objects are balsamed, balsam together two plates of glass about $\frac{1}{16}$ in. thick and the same diameter as the objects. Put these in water and heat slowly (say 3° every quarter of an hour) and try every few minutes whether one can be perceptibly slipped on the other by sideways movement of the fingers. The temperature at which this can just be done is the temperature at which annealing may be carried out.

Balsaming prisms and combinations of prisms

From what has been said above it will be realised that no balsam can be relied upon to hold cemented prisms in position relative to each other indefinitely, except at a temperature so low that the balsam is liable to split. The best that can be done if one must depend on the balsam alone is to use hard balsam and risk the splitting, which is not likely to occur in laboratory instruments that are not subject to rough usage.

Where possible, however, the cemented prism system should be held in metallic mounts which would geometrically hold the prisms in position even if there were no cement. An example is afforded by the method of balsaming direct vision prisms where it is desired that the refracting edges of all the prisms should be strictly parallel, a desideratum which is important in, for example, the compensating prisms of Abbe Refractometers.

In this instrument two identical direct vision prisms are used in series and rotated in opposite directions. It is essential for the accuracy of the reading that the prisms should possess no deviation in a plane perpendicular to their dispersion. This condition is easily and accurately achieved by the use of the jig shown in Fig. 86. In this the prisms lie so as to fulfil the necessary condition accurately. If a small drop of balsam is placed between each of the surfaces, a small weight put on the top and the compound prism put in the oven as usual and heat treated as described above, the result will be satisfactory. The jig itself must be accurately machined so that the two upper edges are parallel to each other and to the lower surface on which the centre prism rests.

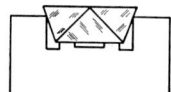

Fig. 86—Jig for balsaming compound prisms

The fabrication of prism assemblies consisting of more than one component is made much easier by the use of H.T. cement; one pair of components having been joined and baked in a suitable jig, the third part can be cemented on without fear that the second baking will cause the first cement film to shift. Each part-assembly can be tested for alignment before the next stage is commenced.

Balsaming gelatine filters

Here the ruling condition is that the temperature should not be high enough to destroy the gelatine. It was customary with makers of high-quality colour filters, consisting of a coloured gelatine cemented between glass flats, to use soft balsam and bake at a low temperature for five or six weeks. We have found, however, that in cases where it is desirable to use a harder balsam the following procedure is successful. A special balsam was used which allowed the standard jig to fall 11 mm in 1 minute at 20°C. The flats, after balsaming, were then heated to 84°C in the oven for forty-five minutes and cooled at not more than 3°C every quarter of an hour.

Except where high optical quality is needed this method is being replaced by a lamination process developed by Messrs Ilford Ltd, similar to that used during the War for the manufacture of goggles from a splinter-proof sandwich of glass and plastic, hot-pressed together. The duration and degree of heating is small enough to work successfully with most filters.

Cylindrical lenses

148·1   Deep concave cylindrical lenses may be ground and smoothed by hand with loose abrasive on a bar rotated in a lathe and polished by replacing the bar with a cylindrical polisher made up on a bar of appropriately smaller diameter.   With deep convex lenses the process

Fig. 87—Mechanism for working shallow cylindrical lens

is the same *mutatis mutandis*.   Analogy with spherical surface working renders further details superfluous.

When, however, the surfaces are shallow mechanical means are necessary to maintain polisher and polished co-axial. Such means are provided by the parallel motion illustrated in Fig. 87 which allows perfect freedom of surface movement only limited by the constrained orientation of the axis. This mechanism can be used on the ordinary type of smoothing and polishing machine.

CHAPTER 7

THE MANUFACTURE OF SPECTACLE LENSES

by A. A. S. MOORE, of the Technical Staff, United Kingdom Optical
Co. Ltd, Mill Hill, England

149 Although the manufacturer of ophthalmic lenses is not faced with the need for such a high accuracy of curvature and figure as is the instrument maker, there are a number of processes outside the normal methods of lens working which present a variety of technical problems.

The manufacture of spherical surfaces for spectacle lenses is very similar in principle to the methods of the instrument maker, but in addition to these there are toric lenses and fused and solid bifocals which require quite different manufacturing techniques.

Preparation of glass

The glass used for single vision lenses is a standard white crown with a refractive index of 1·523 and a ν-value 61 and is normally obtained as rolled sheet with a thickness of from 2 to 12 millimetres. A range of tinted glasses, such as Crookes' A1, A2, B and B2, and several other types are used for special prescriptions. In addition to the standard crown and tinted glass there are two flint glasses used for the manufacture of fused bifocals. These have indices of 1·625 and 1·654 and have to be selected for expansion.

When the glass is received in the glass store every individual piece of flint is tested for refractive index, samples are selected from each batch and checked for expansion, and each piece of flint glass is stamped with an identifying mark which will remain indelibly printed on the surface after moulding. Suitable marking inks are made by mixing various coloured oxides with gum and applying them to the surface of the glass with a rubber stamp. The glass is then put into stock until required.

The glass-cutting room draws suitable glass from stock to prepare and examine it ready for moulding.

To reduce the wastage of glass, and to keep the machine grinding times to a minimum, the glass is moulded to approximately the same curvature and shape as the finished lens.

As there are a large number of different mouldings, charts are kept which show the weight of glass required for each type of moulding and, by cutting up the sheet into pieces of a definite weight, mouldings of the required thickness can be produced irrespective of variation in thickness of the sheet. At this stage all faulty pieces of glass are rejected. The thickness of finished mouldings has to be correct within 0·2 mm.

Moulding

150 The moulding shop is set out in a number of separate units, each consisting of a lehr with one or two moulding furnaces built on each side (Fig. 88).

Fig. 88—Front view of moulding unit
One rotary hearth furnace is shown with the entrance to the lehr

At the front of the lehr is a large rotary-hearth furnace used for moulding larger work. Two smaller furnaces are situated at the middle of the lehr and these are used mainly for producing small flint mouldings used for fused bifocals. The large furnaces are about three feet in diameter with an opening at the front. The hearth is mounted on a large casting with a central vertical spindle which projects through the bottom of the furnace and is driven by a motor. The furnace is fired by gas through air-blast burners situated round its periphery.

A massive press into which suitable dies and plungers can be fitted is mounted slightly to one side of the mouth of the furnace, and is operated by a pedal. Another pedal operates an ejecting mechanism and is also connected to the belt striker which starts and stops the rotating hearth.

The furnace operator stacks the pieces of rough glass at the side of his furnace, and all subsequent handling is done with a pair of foils. These are pieces of $\frac{1}{4}$-in. rod flattened at one end and fitted in shielded handles. A piece of glass is picked up with the foils and laid on the hearth, the ejector pedal touched, the piece of glass carried just inside the furnace, and the hearth stopped. Another piece is placed on the hearth and the operation repeated until the first piece appears on the other side of the furnace. By this time the first piece is in a plastic condition, and is drawn by the foils down a metal chute into the metal die. The plunger is brought down by the pedal and the glass pressed to the shape of the die and plunger. The plunger is then released, the moulding ejected and slid down a chute into the lehr. Another piece of glass is then put into the vacant spot on the hearth and the cycle of operations repeated.

With the rotary-hearth furnace a continuous cycle can be achieved by choosing a suitable number of pieces and adjusting the temperature of the furnace so that the glass is in a suitable condition for moulding after its circular journey.

When the moulding lands on the travelling mat of the lehr it is carried into the hot zone, which is thermostatically controlled to a suitable temperature for the type of glass being used. As the mat travels through the lehr the temperature gradually drops until the mouldings emerge at the end after a period of $1\frac{1}{2}$ hours at a temperature of 100-200°C. They are then removed and packed into a box ready for examination.

The use of the two smaller furnaces down the lehr enables softer glasses to be fed into a part of the lehr which is operating at a lower temperature.

When the mouldings are received in the issue department they are sorted and put into stock ready for issuing to the various lens-working shops as required.

151 Spectacle lens forms fall into several distinct groups.
   (1) Flat spherical lenses which include plano-convex, plano-concave, bi-convex and bi-concave.

(2) Flat sphero-cylinders, one surface of which is a convex or concave sphere and the other a convex or concave cylinder.
(3) Meniscus : one surface a concave sphere and the other a convex sphere.
(4) Torics : one surface a concave sphere and the other a convex toroidal sphere. This is the most common form, although some with a concave toric and a convex sphere are manufactured.
(5) Fused bifocals.
(6) Solid bifocals.

A variety of shaped mouldings of different curvatures are required to produce these different lenses and Fig. 89 shows representative samples of various types.

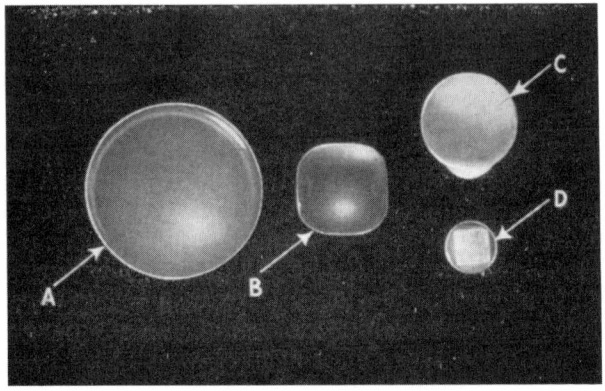

Fig. 89—Four types of mouldings
A 90-mm solid bifocal   c Fused bifocal crown
B Toric   D Fused bifocal flint button

At this stage it would perhaps be advisable to give a brief explanation of the curvature and focal lengths used in spectacle lens work. The international standard of power is the diopter. The power of a lens in diopters is the reciprocal of the focal length in metres. Thus a lens of 500 mm focal length has a power of 2 diopters and one of 100 mm, 10 diopters.

This is a very convenient system because the power of a lens or combination of lenses is obtained by simple summation of the power of each surface.

The radius of curvature required to produce dioptric curves is obtained by the use of the formula $R = (\mu - 1)/D$, R being the radius of curvature, μ the refractive index of the glass, and D the power in diopters.

Glass of $\mu = 1\cdot523$ being a universal standard, curvatures derived from the above formula with $\mu = 1\cdot523$ are rather loosely called " dioptric tools."

It is necessary to make a modification to this formula in the case of meniscus and toric lenses. If a piece of glass is worked with a concave surface on one side and a convex surface of the same radius on the other, the resulting lens has a slight convex power owing to the thickness of the glass. A reasonable compromise is achieved by substituting $\mu = 1\cdot53$ in the formula in calculating the radius of convex curves used on meniscus and toric lenses.

All concave curves are referred to as minus, and convex as plus. The contribution of a surface to the power of the lens is called " the surface power " and equals $(\mu - 1)/R$.

Spherical surfaces

152 With very few exceptions spectacle lenses have one spherical surface, the other being either spherical, cylindrical or toric.

The principles of grinding and polishing spherical surfaces have been adequately dealt with in the earlier part of this book, but as the requirements of spectacle lenses are not the same there are some differences in technique mainly in blocking and in smoothing and polishing machinery.

Blocking spherical lenses

Mouldings used for spherical lenses are generally diamond-ground singly on both sides to a thickness about 0·4 mm above the thickness of the finished lens. The mouldings are accurate enough not to require edging before grinding. This method ensures that each lens is free from prism.

To locate the lenses for making a block of the correct curvature, a cell mould is used. The cell mould is machined from a casting, or made of cement or low melting-point alloys, and consists of a number of cells located on a curve equal to the required curve of the block. The lenses are heated and put in the cells, and a lump of hot pitch is placed on top. A hot metal block holder (or runner, as it is called by the ophthalmic lens manufacturers) is then placed on the pitch and pressed downwards to squeeze out the surplus pitch and the whole then left to cool. When the pitch is sufficiently hard, the runner with the lenses attached can be removed from the mould and the finished block is ready for smooth grinding and polishing. As soon as the lens-working is complete, the lenses are taken off the runner and blocked in a similar manner ready to work the second side.

By using several moulds, an operator can block up large quantities of lenses very quickly.

Blocking cylinders

Cylindrical blocks are produced by a process known as printing. A cylindrical metal mould is used, a lump of hot pitch put on it and a hot runner placed on top of the pitch. A duplicate of the mould, but of opposite curvature, is thus produced on the pitch and, when the latter is hard, the lenses are heated and pressed on to its surface. The hot glass softens the pitch and sticks firmly in position when it is cold.

When the cylindrical surfaces have been worked the lenses are taken off the block and are transferred to a spherical cell-mould so that the spherical surface can be worked.

Grinding and polishing spherical surfaces

As spectacle lenses have to be produced in very large quantities the machines used differ from those described earlier. Lenses with surface powers less than 3 diopters are worked in blocks 14 in. in diameter. The machines used for working these large blocks are of the simple hanging-spindle type (illustrated in Fig. 89A) which are hand fed with emery for smoothing, and rouge for polishing. The number of spindles which an operator can attend is limited.

The problem of reducing the number of operators required to look after a given number of machines by automatically supplying a continuous flow of abrasive occupied the attention of lens makers for many years before a satisfactory method was devised. The main difficulty lies in the fact that moving parts of conventional forms of pumps have a very short life in the presence of emery or rouge.

One of the best types of automatic feed is that of the bowl machine. Fig. 90 demonstrates the principle. The bowl is fixed on a vertical spindle which can be run at a speed of 250-300 r.p.m. and a small stationary tube, usually called a scraper, is fixed on an adjustable mounting near the inside surface of the bowl. At the bottom of the bowl is an adapter which can hold either a block of lenses, a grinding tool or a polisher. At the back of the bowl (with a cylindrical weight in it) there is a boat-shaped cradle mounted on a horizontal spindle so that it can swing up and down, and underneath there is a ball-ended peg which fits into the socket of a block or tool. When the machine is out of use the weight is rolled to the back of the boat which will then remain clear of the bowl. When the machine is used the block or tool is placed in position, the boat lowered so that the peg fits into the socket and the weight rolled forward to apply pressure to

THE MANUFACTURE OF SPECTACLE LENSES

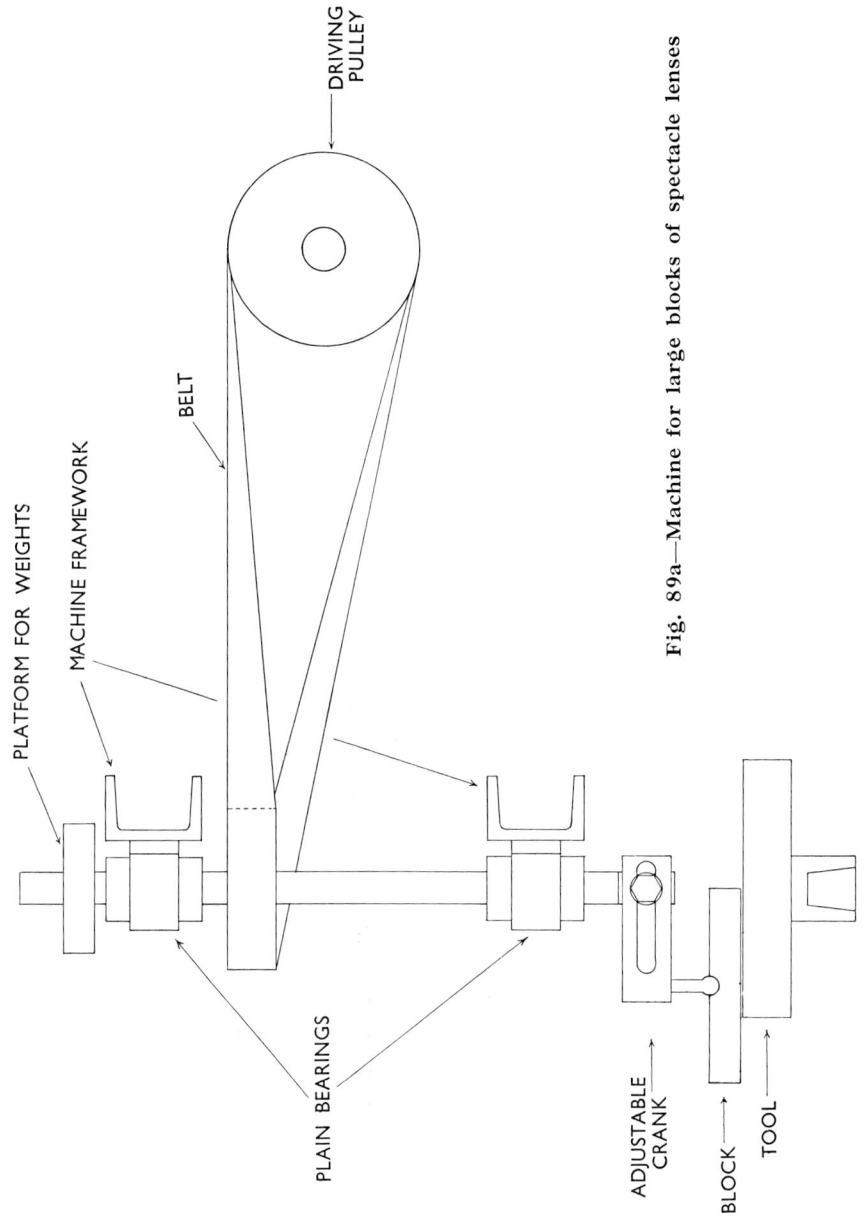

Fig. 89a—Machine for large blocks of spectacle lenses

the work. Then some water and emery, or in the case of polishing water and rouge, are put into the bowl and the machine started up. As the bowl rotates, the grinding tool or polisher rotates with it and the emery or rouge is thrown up to the maximum diameter of the bowl, where it is deflected by the scraper in a steady stream on to the block.

The scraper is the only part of the machine which is worn by the abrasive, and this wear can be taken up by adjustment until the scraper is finally worn out. A scraper, however, lasts several months and can be replaced at the cost of a few pence.

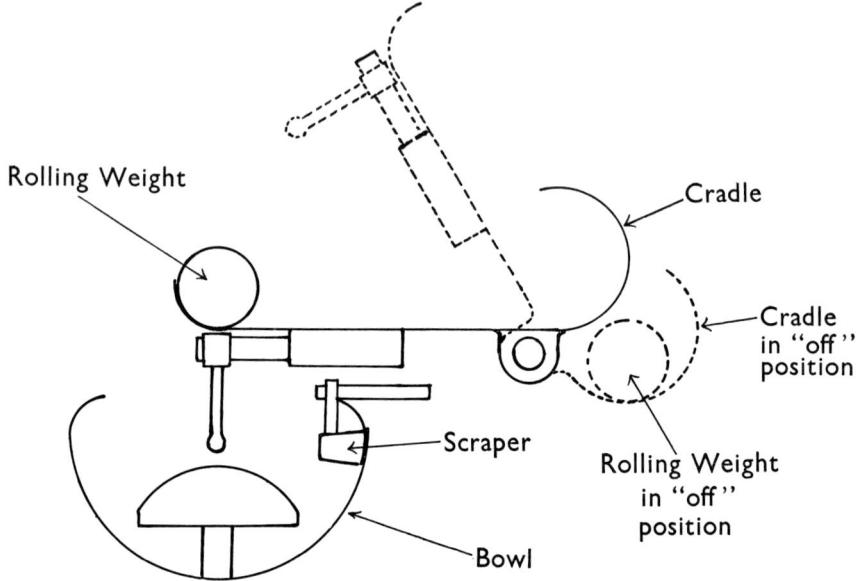

Fig. 90—Diagram of bowl machine

Another advantage of the bowl machine lies in economy in use of material, because it is continuously returned from the block into the bottom of the bowl to be thrown up and used again, and it is only necessary to make good the material used which clings to the blocks as they are taken out of the machine. This economy is of considerable importance when the amount of grinding and polishing materials used is approximately 1 ton per day. The bowl machine requires very little attention while it is running and this makes it possible for one operator to look after many machines. The machine is usually fitted with a mechanism which imparts a small sideways motion to the block so that, in addition to the rotation, the block is also swept across the tool. The very highest accuracy of figure cannot be obtained because the ratio of speed of the sideways movement to the rotation is of necessity rather low, but 2 in. diameter lenses in blocks up to seven have been polished to within 2 wavelengths.

The use of this machine is limited to blocks up to 7 in. in diameter, the peripheral speed of the bowl being too high with sizes in excess of this. Machines working larger blocks can be fed automatically by a centrifugal pump, which can be situated below the machine and the abrasive or polishing material pumped through tubes to the block and then returned to the pump through suitable channels to be used again.*

Toric lenses

Toric surfaces are sections of the periphery of a ring of circular section as illustrated in Fig. 91. R_1 is the radius of the base curve and R_2 is the radius of the cylindrical curve. The difference in dioptric power of these two curves gives the power of the cylinder in the finished lens. Stock is mass-produced with base curves ranging from 4D to 10D in 1D steps. The actual curves vary somewhat between various manufacturers, but in most cases low-power toric lenses are produced on base curves of approximately 6D.

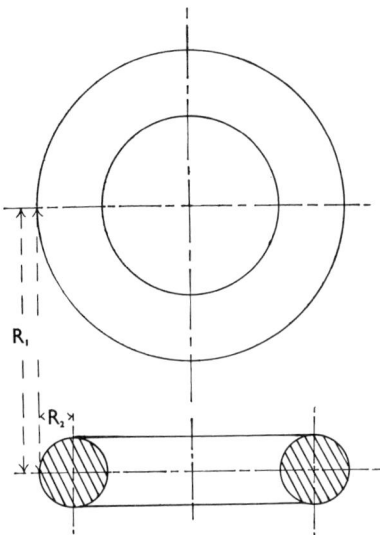

Fig. 91—Section of a toroidal sphere

The lenses are stuck round the periphery of a circular metal runner to form a block, which, owing to its similarity to a motor-car wheel, is called a tyre block.

Blocking toric lenses

153 The base curve is controlled by the diameter of the block and consequently the number of standard-size lenses which a block contains depends upon the radius of the base curve. The runner is a cast-iron

---

* Grinding and polishing machines possessing similar features, together with some novel ones, are made by Autoflow Engineering Ltd, 118-120 Clifton Road, Birmingham, 12. The speeds of grinding and polishing claimed are very high.

plate, slightly thicker than the diameter or width of the lenses, with a 2-in. hole in the middle for fitting the block to the machine. The diameter of the runner is chosen to suit the various base curvatures used. The runner is heated and fitted into an elaborate mould which

Fig. 92—Block of toric lenses ready for roughing

has a number of interchangeable and adjustable curved dies fitted round the circumference. Hot pitch is then poured into the space between the runner and the dies. When the pitch has hardened the dies are withdrawn and the runner taken from the mould It

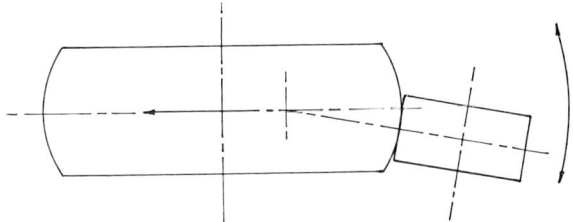

Fig. 93—Geometry of a toric roughing machine

now has a film of pitch with a number of convex spherical facets moulded round the periphery. The radius of curve is produced by the die, and the position of the die controls the thickness of the finished lens. Mouldings with the spherical concave side polished are then

THE MANUFACTURE OF SPECTACLE LENSES

heated and pressed into contact with the surface of the pitch so that they stick firmly when they are cold (Fig. 92).

The blocks are then ready for roughing. Diamond tools are now almost universally used for roughing and, although the design of toric grinding machines varies considerably in detail, the basic

Fig. 94—Toric smoothing machine
A Driving pulley   D Adjustable crank
B Weights   E Tool
C Guiding forks   F Block
G Can containing emery and water

principles are similar. Fig. 93 shows the geometry of a toric rougher. The block is rotated and a high-speed diamond tool is swung on a radius. The position of the diamond tool determines the base curve, and adjusting the distance between the centre and the tool varies the cylindrical curve.

After the block has been ground to a diameter slightly in excess of the final diameter of the block, it is transferred to the smoothing machine (Fig. 94). This machine has a vertical hanging spindle with an adjustable crank and peg, and the block is fixed to a horizontal rotating spindle. The smoothing tool is made of cast-iron with a concave toric curve machined on the one side, to suit the curve of the block, and a socket on the other. Two metal arms are fitted on opposite sides of the tool in line with the axis of the toric. These arms fit in forks on the machine to keep the axis of the tool in line

Fig. 95—Toric polishing machine

with the block. The peg on the vertical spindle is fitted into the socket and imparts a circular motion to the tool. The machine is started and is fed with fine emery and water until roughing marks have been removed and the block is of the correct diameter.

The polishing machine has two spring-loaded spindles working a polisher on each side of the block to increase the polishing speed. Fig. 95 shows a bank of toric polishing machines. The production of a toric surface is something of a compromise, and a study of a drawing of a toroidal ring shows the reason for this (Fig. 91). The rather exaggerated diagram shows that the base curve is convex on the outside, but gradually changing to concave on the inside. It can be appreciated that a toroidal surface could, strictly speaking, be generated only by a point or a line contact. This, of course, is impracticable except in roughing; therefore it is essential to use as small a smoothing tool or polisher as is possible, consistent with a reasonable rate and smoothness of polishing.

Smoothing tools and abrasives
154 Many materials have been tried for grinding tools but cast-iron has proved to be the best from all points of view. It is cheap and can be cast to the required shape, thus obviating unnecessary machining. Higher grinding speeds are obtained with cast-iron than with other metals, no doubt due to its porosity.

Abrasives are dealt with in Chapter 4. For ophthalmic lenses aloxite is very satisfactory for all stages of grinding. It is obtained in 80, 120 and 220 grades and after use for preliminary roughing operations is regraded by elutriation or settling (see §§ 77 to 87). Continual use of the abrasive reduces the grain size and makes it possible to obtain finer grades for finishing.

" Sira " abrasive mixed with very fine grades of emery is a useful finishing abrasive. It grinds quickly and leaves a surface which polishes rapidly.

Polishing materials (see also §§ 88 to 91) *and polishers*
155 Rouge is a word which is perhaps rather loosely applied by ophthalmic lens makers to any polishing materials, and of these ferric oxide is the most widely used. There are many methods used for its manufacture and consequently the properties of the finished products vary and rouges have to be selected to suit the requirements of each job. Some rouges can be used with certain polishers only, and some have a tendency to " blind " or cover up surface faults, thereby causing difficulty in later stages.

The introduction of cerium oxide during the war was an event of major importance to the optical industry, because of its superiority over rouge for polishing speed. It is very expensive, but its rapidity and cleanliness are important advantages. At the time of writing, supplies in the quantities required by the ophthalmic industry are limited in this country, but it is hoped that deliveries will improve.

Titanium dioxide, mainly used as a paint pigment, is another useful " rouge," sold by British Titan Products. It is white, its polishing speed compares with red rouge, and it has no tendency to " blind."

Other polishing materials, such as chromic oxide and putty powder, are very little used in the ophthalmic industry.

Polishers

Polishers are made of felt impregnated with wax. A piece of 4 mm or 6 mm hard felt is cut to the size of the polishing tool and, if it is required for a short radius curve, it has to be " ironed." This is done by soaking it in water and pressing it between two hot curved tools which mould it to the required curve and prevent it from buckling. The polishing tool is then heated and smeared with pitch and the felt pressed on to it and some molten wax brushed on the surface of the felt. When cold the polisher is trimmed on a lathe and is then ready for use.

Fused bifocals

156 Fused bifocals consist of a main lens of crown glass with a small area of flint glass fused to it, thus producing two different powers.

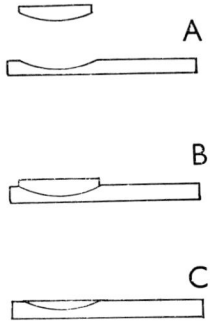

Fig. 96—Processes in manufacturing a fused bifocal
 A Flint and crown before fusing
 B Blank after crown and flint have been fused together
 C Finished fused bifocal

This result is achieved in the following manner. If we take a flat piece of crown glass and grind a small depression on a radius of 32·68 mm the power of this curve in crown glass will be $-16·00D$—see Fig. 96. If now a piece of flint glass is worked to the same radius the resultant power will be $+20·0D$. By fusing the flint on the depression, as described later, and grinding and polishing both surfaces to a continuous flat surface we have no power in the major part of the lens, but in the small circle of flint glass the power is $-16·00D +20·00D$, or in other words $+4·00D$. If the other surface is then worked to any specified curve it gives a power in the main lens, or distance portion, and an addition to the 4·00D in the small segment, or reading portion.

Manufacture of fused bifocals

Working the crown and flint

The crown mouldings are blocked up four at a time in cell moulds so that a depression is ground near the bottom of each moulding. The block is first roughed to open out the depression to 26 mm diameter, and then smoothed on a bowl machine with very fine emery until the depression is 28 mm. Inadequate smoothing causes serious trouble in the later stages.

The block is then polished on a bowl machine and it is at this stage that all precautions have to be taken to ensure that the surface is in the right condition to stand the high temperatures used in fusing. It is essential to use a rouge which does not " blind," and a plain hard felt polisher is used to assist in keeping the surface " open." This term refers to the type of surface produced by a plain felt polisher, compared with the surface polished on a hard wax or pitch polisher. In the early stages of polishing the surface appears entirely different with the two types of polisher. A lens which has apparently been completely polished on a pitch or hard wax polisher will break out into large holes if rubbed on a soft polisher, or if it is heated to a fairly high temperature.

The polishing cycle should be 50 per cent longer than the time taken to remove the last traces of grey. If all these points are observed the surface should be satisfactory.

The flint glasses D.F., $\mu = 1\cdot625$, used for reading additions below 2·00D, and E.D.F., $\mu = 1\cdot654$, for additions of 2·25D to 4·00D, are moulded to 28 mm diameter and one side ground flat and semi-polished. They are then made up into blocks of up to 18 and ground and polished in the same way as the crown.

The curves of the depressions vary from $+7\cdot00$D on a deep meniscus moulding to $-16\cdot00$D on flat mouldings. The opposite curvature is worked on the flint, but with a difference of 0·25D or 0·50D so that when placed together contact is made in the centre.

Fusing the flint and crown

When the flints and crowns have been polished and knocked off the runners, they are delivered to the fusing department, where they are cleaned and passed into a room which is kept as free from dust as possible by blowing filtered air into it. In this room operators put the flint and crown into contact so that there is no dust or dirt between the two surfaces. This is done by laying the crown on the bench, picking up the flint, brushing the surface of each component with a soft brush and then placing the flint carefully on the depression.

Owing to the difference between the two curvatures a small interference spot is visible in the middle.

The operator now presses the top surface of the flint close to the edge with a pair of tweezers and, if all is well, the spot moves in a straight line to the edge. If, however, there is some dust between the surfaces the spot is deflected and travels in a curved path. The flint then has to be picked up and brushed again until this test, applied in a number of positions round the flint, shows that no dust is present. Two short pieces of wire are then dipped in gum and placed $\frac{1}{2}$ in. apart under the flint at the bottom of the moulding. The wire has to be sufficiently thick to cause the interference spot to disappear at the top edge of the flint.

The assembled crown and flint are then placed on a carborundum slab, moulded to fit the underside curve of the crown, and placed on the travelling belt of a lehr. The hot spot of the lehr is thermostatically controlled at 640°C at which temperature the flint becomes quite soft and sinks by its own weight, gradually expelling the air from the top of the flint until the whole surface is in contact with the exception of two very small areas round the two supporting wires. As the mat travels onwards the temperature of the furnace gradually drops until the fused blanks emerge at the other end at a temperature of approximately 100°C. They are then allowed to cool and the two supporting wires pulled out.

The blanks are then checked for strain. Matching the crown and flint for expansion, as mentioned in an earlier section, is very important because any serious difference will cause the crown to flaw. Some tolerance can be allowed if the expansions are such that the flint is in compression and not in tension.

The blanks then undergo an inspection for faults in the segment. If by any chance some dirt has been left between the surfaces, a bubble is formed which may result in the rejection of the blank.

Very great care is necessary in the surfacing because the high temperature can cause the surfaces to break open, leaving a mass of tiny bubbles or scratches visible in the segment. It is rather a peculiar effect, because marks such as sleeks, which are visible before fusing, seem to flow and disappear during the process, whereas two surfaces which appear to be perfect under the most rigorous examination may be so bad after fusion that they are fit for nothing but scrap. The fault can only be minimized by paying great attention to polishing and smoothing materials and by giving ample machine cycles to both operations.

When the blanks have passed inspection they are sorted according to a code number, which is marked with a diamond on the crown, and are then put into stock ready for sale, and a certain percentage are ground and polished on the segment side to fulfil orders for the more common ranges of prescriptions. These are known as semi-finished blanks.

The code number identifies the curves moulded on the crown, the curve of the depression and the index of the flint. A chart is issued to each user so that the correct blanks can be ordered by code number to fit almost any individual prescription. The combination of prescriptions in bifocals is so enormous that they are sold in blank or semi-finished form so that the Prescription Houses can select and work blanks into finished lenses to suit each prescription.

Fused bifocals have the advantage in that the segment is almost invisible and they are comparatively cheap. The segments cannot be made over 22 mm diameter without introducing serious chromatic aberration

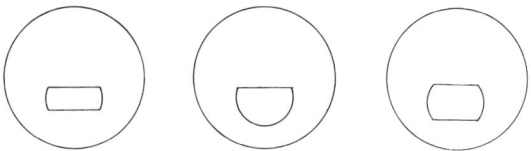

Fig. 97—Three most popular "Univis" shapes
Types B, D, and R

The Univis fused bifocals

157 The segment of the normal fused bifocal is round, but the United Kingdom Optical Co. Ltd secured the patent rights for the manufacture of fused bifocals with a variety of shaped segments.

A 32 mm depression is worked on the crown by methods the same as those used for the round segments.

The shape of the final segment is produced by a composite flint button consisting of two or more pieces of glass. Three of the stock shapes " B," " D " and " R " are illustrated in Fig. 97.

The " B " shape button consists of a narrow rectangular moulding of flint glass which is ground and smoothed on the two longer edges and two similar shaped pieces of crown glass ground and smoothed on one edge. These three pieces of glass are carefully cleaned and clamped together with the flint between the crowns and the smoothed surfaces in contact. In this condition it is passed through the lehr and fused into a solid piece of glass.

The " R " shape is made in the same way with a wide flint, and the " D " has one " D " shape flint and a rectangular strip of crown.

When the button has been fused it is first ground flat and semi-polished on one side, and then ground and polished in the usual way to the required curve.

Subsequent operations are carried out as previously described, but a temperature of 680°C is required to fuse the button owing to the presence of the crown.

When the segment side is ground and polished to a continuous curve the crown component of the button becomes invisible because it has the same refractive index as the main lens and the shaped flint segment only remains visible.

Solid bifocals

158 Solid bifocals are made by grinding and polishing two different curvatures on a single piece of crown glass, one curve being used for the distance portion (D.P.) and the other for the reading portion (R.P.).

Solid bifocals are made in toric form with standard D.P. curves of $-4\cdot00D$, $-6\cdot00D$, $-8\cdot00D$ and $-10\cdot00D$. The R.P. curve is varied to give the required addition and therefore has a longer radius of curvature than the D.P. It is thus quite impossible to use normal lens-working methods because the working tool must not be allowed to run over the dividing line between the two curvatures. Owing to these circumstances an entirely different method of working was devised by Bentzon and Emerson and patented in 1904 and 1910. This method depends entirely upon the use of precision machinery to achieve the necessary accuracy and quality.

Until 1939 as many as 10 different segment sizes were made as stock items, ranging from 19 mm to 50 mm diameter, but these have now been condensed to the three most useful and popular sizes 22 mm, 38 mm and 45 mm.

Sticking on

The mouldings have a D.P. curvature as mentioned above, the diameters being 73 mm for a 22 mm segment, 90 mm for a 38 mm, and 96 mm for a 45 mm.

The mouldings are first diamond-ground with a special tool which grinds the D.P. and R.P. together, and are then stuck singly, with wax, on holders which can be fitted to the machine.

The holder is an iron casting which is machined on the top to a curve to suit the various mouldings in use. Underneath, a female taper is bored in a boss, and it is by this taper that the lenses are located on the grinding, smoothing and polishing machines. It is essential that all the holders have the tapers machined to very fine limits so that there is no loss of truth when the holders are transferred from one machine to another.

Roughing

The roughing machine, in principle, is the same as the machine illustrated in Fig. 63 with a bottom rotating head and a top grinding spindle. A small precision lathe head is mounted, at the bottom, on a vertical slide which is held against an adjustable stop by two springs. The holder fits on a male taper on the lathe head. The top high-speed spindle can be set to any angle with the aid of a vernier and the whole of the top assembly can be moved laterally on a horizontal slide. A cast-iron tool is screwed on the top spindle, and the angle set to produce the required D.P. curve. The lateral position of the tool is then adjusted to a position to one side of the axis of the bottom head. Water and medium-grade emery is fed on the lens,

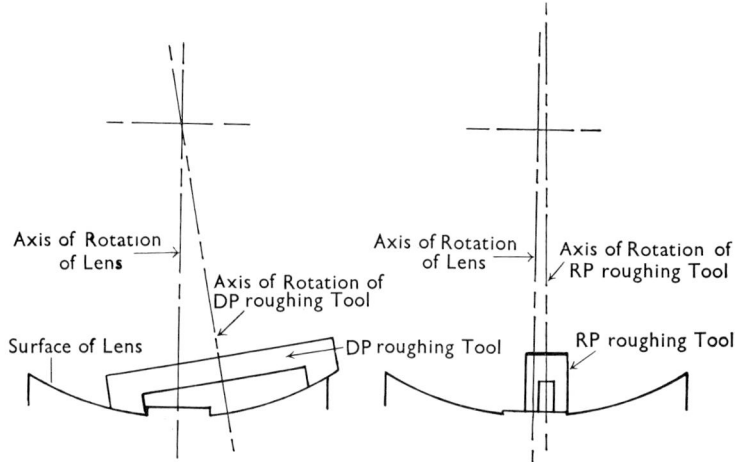

Fig. 98—Principle of D.P. and R.P. roughing

and the machine started. The stop is adjusted to allow the machine to grind about 0·3 mm off the surface.

Owing to the offset of the tool a circular area in the centre of the lens is left unground, but the outside band is ground true to the holder and correct to curvature.

The portion in the centre which becomes the R.P. is now left standing above the level of the R.P. This is ground down to the level of the D.P. by a similar machine, but this time by a tool of such a diameter that it covers the centre of the R.P. when the edge of the tool is set to the ridge or dividing line between the two surfaces. In this way a continuous spherical surface is ground on the R.P.

Fig. 98 illustrates the principle of D.P. and R.P. roughing.

If the holder is a bad fit on the bottom head, or the bottom spindle is out of truth, the R.P. will be generated on an axis different from the D.P. The effect is shown, greatly exaggerated, in Fig. 99. This is known as a lop-sided bifocal. Great attention to the condition of the taper and truth of the bottom head is necessary to reduce this defect to the minimum. Weekly inspection for truth is made and if the error exceeds 0·00015 in. the head is overhauled.

Smoothing

All subsequent operations are carried out on one type of machine. Fig. 100 shows the general principle. The bottom head is bolted on the front of a cast-iron frame and at the back there is an arm which carries the top spindle in two bearings. The arm is supported by two pivots, the top one allows the arms to be lifted up or down, and the adjusting screw moves the whole assembly backwards and

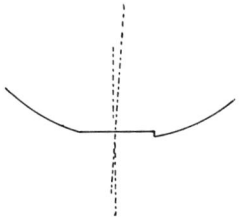

Fig. 99—A " lop-sided " solid bifocal
The diagram is exaggerated to show how the defect is caused by an untrue spindle or a badly fitted holder

forwards on the lower pivot. The top spindle has a small ball point and a collar with two driving pins. This engages with the ball hole and pegs on the tool as shown in the photograph. The angle of the spindle can be adjusted by loosening a simple clamping mechanism, and both the top and bottom spindles are driven by belts.

The D.P. smoothing tool is a ring tool made of brass and of the same dimensions as the roughing tools. The tool is placed on the lens and the arm lowered to engage the ball and socket and driving pins, and the position is adjusted so that the inside edge of the tool coincides with the ridge, then the angle of the spindle is adjusted, if necessary, so that it is at right angles to the tool. The machine is started and fed with fine emery and water. Referring to Fig. 100 it will be appreciated that the roughing machine will only generate a true sphere if the axes of the two spindles are in precisely the same plane. If they are not, a toric will be produced. In production it is not possible to keep the roughing machine in sufficiently close adjustment,

THE MANUFACTURE OF SPECTACLE LENSES   271

hence the necessity for the different type of drive, used on the smoothing and polishing machines, which allows the tool to float and to pick up its own axis in the correct plane, thereby producing a true sphere.

After D.P. smoothing the R.P. is left raised above the D.P. This

Fig. 100—General view of machine for smoothing and polishing solid bifocals

is then smoothed down with a small floating tool set exactly to the ridge until it is nearly level with the D.P. again. It is sometimes necessary to adjust the curve of the R.P. This is done by setting a tilt on the top spindle, as illustrated in Fig. 101, which introduces a slight bias by the friction of the driving pins and the ball point, and increases the pressure of the tool on the centre or the edge, according to the direction of the tilt, thus shortening or lengthening the radius slightly.

The principle of the diamond-roughing tool, which was patented by J. A. Moore, has recently been applied to an ingenious tool which enables the D.P. and R.P. to be smoothed at the same time. Two

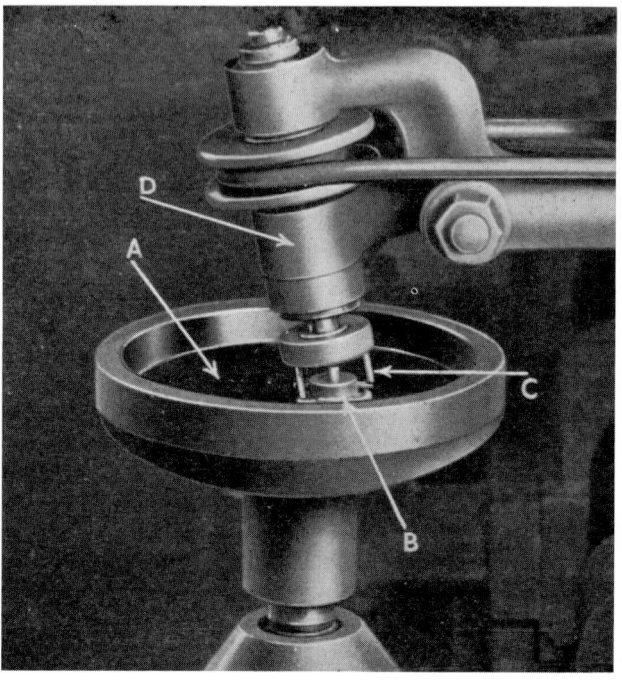

Fig. 101—Close-up view of machine for smoothing and polishing bifocals
The illustration shows " tilt " applied to the R.P polisher

adjustable segments are fitted to the inside of a D.P. smoothing tool (Fig. 102). The R.P. curve produced by this tool is not truly spherical in practice, but only a short run with a floating tool is required to true it.

This tool has two great advantages : (1) it reduces the R.P. smoothing time, (2) it ensures that the lens is entirely free from lopsidedness at the smoothing stage.

Polishing

Polishers used for solid bifocals (Fig. 102) have to be extremely hard to preserve a sharp dividing line between D.P. and R.P. There are numerous recipes for solid bifocal polishers, including hard pitch, hard waxes, shellac, wood, vulcanised fibre, and plastics. Hard waxes mixed with rouge and wood flour are generally the most satisfactory.

A piece of polishing material is stuck on a metal runner which can be fitted to the top spindle. The polisher is turned in a lathe so that it has a sharp edge and its diameter is a little greater than the width of the D.P. It is then set with its edge coinciding with the ridge, and the machine is started and fed with rouge and water until the grey has been removed. By this time, the edge of the polisher has lost its sharp edge and it leaves a narrow wavy band round the

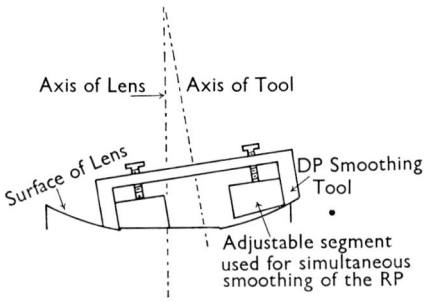

Fig. 102—Principle of a combined D.P. and R.P. smoothing tool

R.P. This is known as aberration or "aber". By re-turning the polisher and running the machine for a few minutes with the smallest trace of rouge, the band of aber is polished away leaving a clear, sharp ridge.

The R.P. is polished in a similar manner with a polisher approximately half the diameter of the R.P. When the lens is received from the polishing, there is still a perceptible ridge and the R.P. has to be polished until it is nearly level with the D.P. During this process as much as 0·001–0·002 in. of glass is polished away and, as can be imagined, it would be quite impossible to convince a solid bifocal maker that glass is not removed by polishing. When the R.P. is nearly level the polisher is returned and run for a few minutes with very low rouge concentration to remove "aber" left by the preliminary polishing.

The bifocal surface is then complete, with the exception of a small pip which is left in the centre of the R P. In the case of 38 mm and 45 mm segments the dimensions of the blanks are such that the pip is not included in the lens when it is fitted to the frame, but it is necessary to remove the pip in 22-mm segment blanks. This is done with a small polisher made of soft pitch moulded to the R.P. curve. It is driven by a small crank fitted to the top spindle of the polishing machine, and it is set so that the polisher does not run over the ridge. This action will cause the curvature to alter quite rapidly, but as it takes only 30–45 seconds to remove the pip, the alteration of the curve is quite negligible.

When the lens has passed the final inspection it is returned to the sticking-on department and removed from its holder and cleaned ready for a further inspection.

At all stages of manufacture the very greatest care is necessary in setting the tools and polishers, otherwise an ugly deformed ridge is produced.

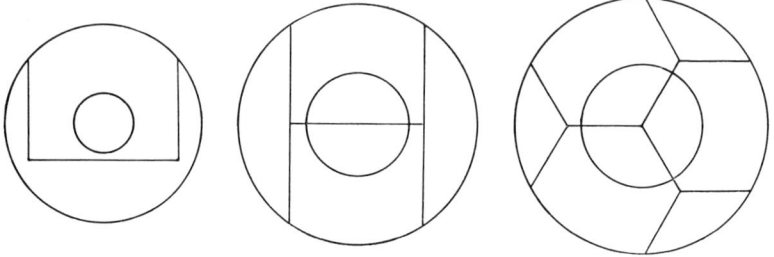

Fig. 103—How solid bifocals are cut from a complete circle
A 22-mm segment yields one blank ; a 38-mm, two blanks ; a 45-mm three blanks

After the final inspection the lenses are cut into suitable shapes for surfacing. The three types of blanks are cut from the complete circle as shown in Fig. 103.

It is claimed, quite rightly, that the manufacture of solid bifocals is one of the most delicate of all lens-working operations, and in addition to the production of the stock type of bifocals just described many other special forms of lenses can be made.

Some of these types are illustrated in Fig. 104. These are made to reduce the weight of very high-powered lenses.

The bilentic is an interesting sample which is a combination of a solid lenticular and a fused Univis " D " segment, used in cataract cases where a bifocal is necessary.

Such are the types of machines and methods of working appropriate for the mass production of stock lenses. There are a number of machines made for working single surfaces which are used for the second side of fused and solid bifocals and for making lenses which are normally outside the stock range of the mass manufacturers. This type of work, however, is not as a rule carried out by firms making ophthalmic lenses in quantity.

Testing of ophthalmic lenses

159 The agreed limit of error in the power of ophthalmic lenses is 0·06D. To keep the lenses within the tolerance it is necessary to work each surface of the lens to 0·03D. This limit has to be reduced still further to allow for slight variations in the index of the glass, and 0·02D can be considered a reasonable maximum.

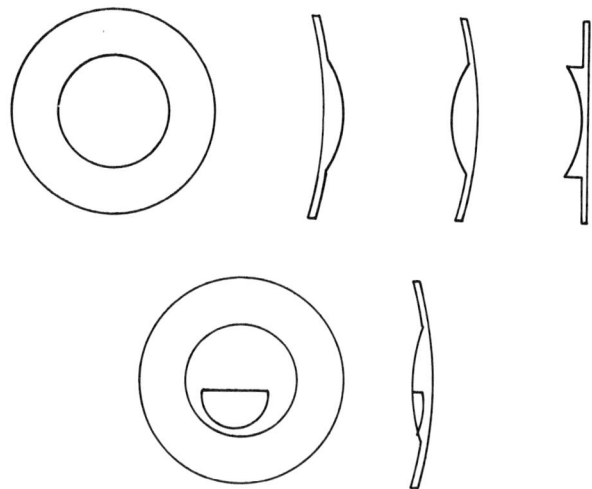

Fig. 104—(*Top*) **Various forms of lenticular lenses**
(*Bottom*) **The " Bilentic "—a bifocal cataract lens**

The smaller blocks used for deeper curves are checked with accurately-made profile gauges. The difficulty of producing this type of gauge with long radii makes the use of a spherometer more convenient for checking shallow curvatures. It is necessary to work to far greater accuracy than suggested above when using large blocks, because the polishing cycle would be greatly extended if a smoothed block with a slight inaccuracy is used with a correct polisher. To check within a suitable tolerance, a spherometer with two fixed legs 150 mm apart is used to compare the curve of the blocks with accurate master surfaces. The ring spherometer described in another section is not used widely in the ophthalmic industry because it is not suitable for checking cylindrical surfaces.

The base curve of a toric block can be easily and accurately checked by measuring the diameter with a slide gauge, and the curvature of the cylinder is measured with a spherometer with two fixed legs 40 mm apart.

An optical device is used for checking the R.P. curvatures of solid bifocals because, owing to the small size of the segment, a spherometer is rather clumsy and its accuracy is considerably reduced.

This instrument is in the form of an optical bench, and the surface of the lens is brought into contact with a fixed point. On the other side, mounted on a sliding carriage is a cube made up of two right-angled prisms cemented together. The hypotenuse of one prism has a semi-reflecting coating. The top of the block is fine-ground, and the rear vertical surface has a graticule on it and is illuminated by a lamp and condenser system. Light passing through the graticule is reflected by the polished surface of the glass under test and then by the semi-reflector on to the upper surface of the block. When the carriage is moved so that the graticule coincides with the centre of curvature of the surface under test an image of the graticule comes into focus on the ground surface. By fixing a scale on the slide the radius of curvature can be read very accurately.

When the lenses leave the workshops they are inspected for power, thickness and position of optical centre and surface. A very high standard of surface is required for ophthalmic lenses.

Planning

160 Properly planned production is vital in the manufacture of ophthalmic lenses owing to the large quantities involved. The stock list of finished single-vision lenses in various forms exceeds 2,000 combinations. In addition to this there are about 500 round segment fused bifocal blanks and a similar number in each of the three Univis shapes together with 140 solid bifocal blank combinations.

A large percentage of these may be duplicated in some three or four of the more popular tinted glasses.

To keep stocks in balance, and to ensure that all requirements are dealt with in rotation, the orders are handled in monthly batches and laid out for the machines so that tool changing and resetting is reduced to the minimum. If demand does not exceed production, an average month's stock is held and orders can be fulfilled immediately and the stock replenished during the following month.

Prescription work

The previous pages of the section describe the general principle of the mass manufacture of spectacle lenses. There are, however, a considerable number of spectacle lenses needed which are not included in the stock range of the mass manufacturers and, generally, these lenses are made singly by prescription houses, or the wholesale branch of the ophthalmic trade. The scope of the prescription houses is therefore very wide. Apart from the spherical single vision lenses they have to make nearly all finished bifocals from partly finished blanks supplied by the mass manufacturers, and in addition many special types of spectacle frames have to be made together with a large percentage of repair work.

The mass manufacturers have well-equipped machine shops to design and build the machines most suitable to their requirements, but in most cases the prescription houses are too small to have these facilities and several British firms are building excellent machinery for this class of work.

These machines include diamond grinding machines for producing spherical and toric surfaces, automatically fed smoothing and polishing machines, automatic lens edgers and a variety of frame making tools and equipment. It will be readily appreciated that the design of such machines must enable rapid resetting to be made in view of the varied nature of the work carried out by the prescription houses.

Below is a list of some of the firms specializing in this type of machinery—

    Autoflow Engineering Ltd, England.
    Impregnated Diamond Products Ltd, England.
    Pearl Manufacturing Co. (London) Ltd, England
    Technaphot Ltd, England.
    M. Wiseman & Co. Ltd, England.
    Shuron Optical Co. Inc., Geneva, N.Y., U.S.A.

CHAPTER 8

MICROSCOPE LENSES

By R. J. BRACEY, F.Inst.P., F.R.M.S., Head of Optical Department of the British Scientific Instrument Research Association, 17 Princes Gate, London, S.W.7.*

161 This type of work falls into a slightly different category and is therefore enlarged upon in this chapter. The outstanding features to be considered are those which are inherent in the small dimensions and, often, the large solid angular value of a surface with reference to its centre of curvature. To obtain a proper perspective of this type of work it may be worth while recalling some commonly accepted industrial tolerances. It is usually considered that, if a steel shaft is to slide in a steel bearing, a clearance of 0·001 in. will be required if the shaft is under 0·5 inches in diameter. It is obvious however, that this tolerance will be related to the length of shaft engaged and could be reduced for shorter lengths. For small lenses this diametral tolerance can be reduced to as small as 0·0001 in. For such accurate work it is essential that the lens mounts are truly cylindrical, and the lens itself must be edged very accurately. Also it must be held square to the axis of the cell when mounting, or have a chamfered edge to act as a lead, otherwise it will be impossible to get it into the cell. The preceding remarks are in the nature of circumstantial evidence, and it is very difficult to be more specific; but there is no doubt whatever that the last decade has seen a great deal of progress in matters of precision.

Another aspect of the matter is that all microscope lenses of high power and large numerical aperture can have their residual aberration balance easily disturbed by an error in thickness of certain of their component lenses. In the following table are given values in wavelengths of the deformation from sphericity of a wave-front due to the introduction of a parallel plate of glass 0·001 in. thick, and of refractive index 1·52, for homocentric beams of light of various numerical apertures incident normally upon it. Thus a table of this sort can be used to determine the effect of incorrect thickness on the wave-front emerging from a lens.

Numerical aperture	0·5	0·6	0·7	0·8	0·9
Deformation in wavelengths	0·15	0·35	0·73	1·51	3·22

* Mr. Bracey wishes to acknowledge the help he has received from his colleague Mr. D. L. Rigby, M.A., and from the following makers of microscope lenses: Messrs Chas. Baker (of Holborn) Ltd, Messrs R. & J. Beck Ltd, Messrs. Hilger & Watts Ltd (Watts Division), Messrs W. Ottway & Co. Ltd, Messrs W. R. Prior & Co. Ltd, Messrs W. Watson & Sons Ltd.

The term " numerical aperture " is defined as follows—

If a point source of light is placed on the axis of a lens system, then a beam of light with a conical boundary is received by the lens and the sine of the semi-angle at the vertex of this cone is known as the numerical aperture of the lens with regard to the point source.

If the angle which edge rays, passing through a lens make with the axis is known, then the sine of this angle when multiplied by the refractive index of the glass of the lens is the numerical aperture of the beam within the lens. If the thickness is varied, then the corresponding change in deformation of the wave-front can be assessed from the preceding table, always providing that none of the other circumstances of refraction are changed as well. This last condition is not usually fulfilled except in the case of the plano-convex front lens of a microscope object glass.

The table then may be used to determine the effect of incorrect thickness on the wave-front emerging from a plano-convex front lens of a microscope object glass, without any reservations.

To measure the thickness of the lens, the workshop usually adopts a unit of 0·001 in. and profitably makes use of dial gauges (" clocks ") reading to 0·0001 in. by estimation, but the commercial article usually requires a little modification before it can be safely used. As one would expect, the use of " tweezers " and a watchmaker's eyeglass or high-power magnifier are necessary to complete the picture.

It is necessary to enlarge a little on the performance required of an object glass. Its wave-front should not exhibit departures from sphericity of as much as one quarter-wavelength (preferably less) under actual working conditions. Reference to the preceding table will show that this means that minute variations of thickness are important with large numerical apertures. It is a fact, however, that lenses which are highly sensitive to thickness variations are also highly sensitive to a change in magnification, which can be varied by altering the position of the adjustable calibrated draw tube with which all standard microscopes are supplied. If, therefore, a lens of large numerical aperture has a deformed wave-front due to residual spherical aberration, this can usually be corrected by a small alteration in the position of the draw tube. For most biological work, the object is mounted under a cover glass, and the complete objective system consists of the object glass, the cover glass, and any layer of material between the cover glass and the object proper. It is clear then that to ensure good definition at all times, the microscopist must have a means of controlling the continuously varying residual spherical aberration. This is achieved by using an adjustable draw tube.

Cover glasses are sold as follows—

No. 0 thickness	... ...	0·004 in.
No. 1 ,,	... ...	0·007 in.
No. 2 ,,	... ...	0·010 in.

This is, however, only a very rough calibration. The area of the cover glass surface which intervenes between the object and the objective is very small and this perhaps accounts for the fact that they are not optically worked. It is well worth while, however, to inspect them before use. The manufacturers adjust object glasses for use with a particular thickness of cover glass at a given tube length, and these values are often engraved on them. The microscope itself is provided with a " fine adjustment " of a calibre sufficient to deal with the exceedingly small movements. If carefully used, this adjustment will remain free from perceptible " back-lash " for a long time.

Fig. 105—Microscope objectives
(*Left*) 4-mm dry objective
(*Right*) 2-mm oil-immersion objective

Construction of a hemispherical or hyperhemispherical front lens

162 Fig. 105 gives the construction of two typical lenses, a 2-mm oil immersion objective and a 4-mm dry objective, and the procedure adopted in manufacturing the *front lenses* of these two objectives will be described. It should be realised, however, that with lenses of this type, one must expect a higher percentage of failures than with more normal sizes, against this may be set the almost negligible cost of the raw material.

Details of the procedure described below vary somewhat from glass-shop to glass-shop and the general description may be taken as an averaged one.

The optical glass is prepared first of all by being cut with a slitting saw into flat plates of convenient size, somewhere about an inch square. One side of the plate is ground and polished flat, and the other side is ground down until the thickness of the plate is somewhat greater than the thickness required in the finished lens ; usually about 0·002 in. in excess. This plate is then cut into squares by means of a light-weight diamond, the size of the squares being somewhat greater than the diameter of the finished lens. The small square plate is mounted with its polished side on a steel peg. The end of the steel peg, being smaller than the finished flat surface will be, is flat or slightly concave and the cement used is usually a shellac mixture. Using the peg as a holder, the plate is then applied to a flat-ended, rotating tool and rotated by hand until it is roughly cylindrical, number 302 emery being applied to the rotating tool. This procedure is carried on until the cylindrical blank is about 0·02 in. oversize. The procedure is then repeated with finer emery (number 303) until the size is about 0·007 in. oversize. A rough curve may also be put on top of the flat surface so as to save time, this operation is usually called " rolling."

The starting procedure may be varied as follows—The plate after being worked to a thickness a little in excess (0·003 in.) of the finished dimension, is left grey on both sides and cut up into squares with a wheel cutter and it is then edged circular but large (again 0·003 in.), and a number of these cylindrical blanks are made up into a block and one side ground and polished flat to within 0·0005 in. of final thickness. The cylindrical blank is then mounted, polished face in contact with a steel peg, and work proceeds as above.

A variation of the " rolling " technique would be to use a " Hooster "; this is a rotating plate with a groove of circular cross section cut into the flat surface near the edge. After this tool has been charged with emery, the peg with the cylindrical blank is applied to the groove, rotating the peg meanwhile and so generating the required rough curve.

A variation of the edging process to be applied to lenses which are slightly under the hemisphere and not quite as small as a hyper-hemispherical lens is as follows—

The square plates are cemented together with beeswax to form a stick of glass, which is cemented with toughened shellac to a spindle. This is placed in a headstock, and the blanks are edged cylindrical by means of a carborundum wheel with a bevelled edge. (This operation is sometimes called " sausaging.")

The lens is now ready for the first roughing tool. This usually consists of a brass rod which has a nearly hemispherical hole in the end. The axial portion of this cavity is removed to give good edge contact, which is an essential feature. The effective radius of the tool is determined from the size of the work it produces (for the first roughing tool this is 0·004 in. greater than the finished lens). Since the wear is considerable, frequent trueing is required.

Care must always be taken in the early stages to see that there is a " spot " or " witness " left on the glass on the axis (*i.e.*, a portion of unworked surface). This is an indication that the substance of the lens is still sufficient. A second and third tool are used, each being about 0·001 in. smaller in radius. The lens, when finally smoothed and the " witness " removed, is about 0·0005 in. to 0·00075 in. large on diameter. Sometimes more than three tools are used with correspondingly smaller differences in their radii of curvature and roughing is sometimes started with 302½ emery instead of 302. For the roughing operations 302, 303, 303½ emeries are used in succession.

These emery numbers are in the British–American Optical Company's notation for " Watford Abrasives " and are given as follows—

Number	Grading
BM 60	Extra coarse
BM 180	Coarse
BM 302	Medium
BM 302½	Finishing
BM 303	Fine
BM 303½	Extra fine
BM 304	Super fine
BM 305	Ultra fine
BM 309	Polishing compound

The lens is now ready for polishing. The polisher may be made by lining a slightly oversize tool with pitch, the layer of pitch being very thin, of the order of 0·003 in. in thickness. The polisher is shaped by using either the smoothed lens or a rejected lens as a former. If a rejected lens is used, one that is a bit " starved " in diameter is chosen, thus ensuring that the polisher is a good fit on the diametrical portion of the lens to be polished. It is *usual* to provide a small hole at the apex of the polisher with the same end in view, *i.e.*, the obtaining of a good diametral fit. Two or three polishers may be used having pitch with various degrees of hardness and, in some cases, shellac may be used as the polisher. The polishing agents are usually E type cerirouge (a trade name for cerium oxide), followed by jeweller's rouge. There is a great deal of individuality allowed or allowable in this work.

The figure obtained by polishing is checked from time to time with a proof glass or similar equipment and the figure is controlled by varying the pressure at different angles and rotating the holder between the fingers and varying conditions from time to time. The whole operation requires a very high order of manipulative skill and intelligence.

Proof plates

163 The examination of the perfection of sphericity of a curve by means of a proof glass causes some little trouble if the curved surface has a large angular value. Proof plates are usually constructed in the manner shown and it will be seen that if the lens and proof plate have a small difference of radius, one will be looking at a uniform spherical film of air. It will be apparent that the different foreshortenings of the various elements of the film will cause the effective thickness in the line of vision to vary considerably, hence a ring system may be seen unless the film is of zero thickness. If there is a first order contact then the surface appears blue-black all over and there is little wrong with it. If the surface is perfect spherically and within the dimensional tolerance, but not quite the size to give a first order contact all over, then rings will be seen. In this case, considerable experience and manipulative skill is needed to apply it properly to the proof glass, and to decide whether or not it is perfect.

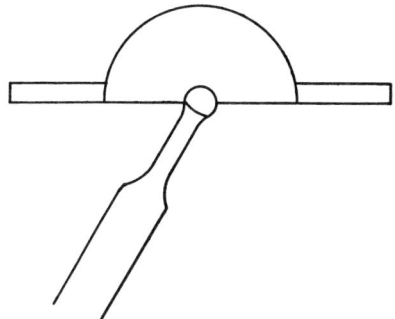

Fig. 106—Testing front lens of microscope objective with a proof sphere

A method of producing a convex curve to test a concave proof plate which is sometimes used, is to turn up a disc of glass to the required diameter, to round and polish the edge of this disc of glass and then to test the proof plate on the rim. When the proof plate is of the correct radius, an interference figure, consisting of a black line, bordered by coloured figures can be seen extending along the line of contact.

It has been found worth while to construct apparatus whereby the surface can be inspected with the proof-sphere very nearly as a whole and the light used for the inspection purposes is made to traverse normally the thin film of air between the lens and proof plate over nearly the complete extent of the film. Precise details of the methods used will vary from maker to maker, but the testing is usually based on the simple principle of the inspection of the Newton ring pattern with the appropriate instrumental modifications which are necessary to render this inspection foolproof.

There is no doubt that nowadays the hyperhemispherical front lens of a microscope objective is a portion of a very perfectly made sphere, in spite of the fact that it is one of the most difficult surfaces to make and that any small error shows up to a marked degree.

Larger microscope lenses

164 For lenses of somewhat larger dimensions the procedure will be different, and is *sometimes* as follows : The glass is cut up into square rods, the cross section being greater than the required lens diameter. The rods are then ground cylindrical in a centreless grinder. The cylindrical rod is next set in a matrix of plaster and slit with a diamond saw, perpendicular to the axis so that blanks of the correct thickness are obtained. With such lenses, which might be components of the middle and back lenses of a 2-mm or 4-mm object glass, a diamond wheel or cup is sometimes used and the spherical surface generated in the usual way. The machine can be set up so that the lens is brought true to curve and thickness at the same instant. As the wear on the diamond wheel is practically imperceptible with such small lenses, little attention is needed over very long periods. It is possible, furthermore, with a diamond cup, to generate a hyperhemispherical surface ; if, then, the blanks are mounted on a suitable " ball " chuck, a larger number can be accommodated by working right over the hemisphere, and this has the advantage that a direct micrometric reading of the radius of curvature of the lenses that are on the block may be made at any stage. Chucks for such hyperhemispherical blocks, however, require careful making and ensuring their adequate ventilation presents no small problem. Very accurate machining of the shoulders, recesses and clearances is necessary. The block, after being diamond ground, is ready for fine smoothing, the same chuck being used for this and the polishing. Co-ordination between the fine smoothing dimensions and polishing dimensions allows for a very close control of results, and, in the case of large work, the time taken is so constant that it is said to be a suitable measure of the finish produced on the lens.

When the lens has been worked and polished on both sides, it is necessary to edge it with a cylindrical rim which is coaxial with the lens. To do this it is mounted on a hollow cylindrical chuck with a shellac cement, and an inspection is made of the reflected images of a target as the spindle is rotated. The cement is softened by heat and the lens adjusted till the images are stationary. It would be hard to over-emphasize the importance of this procedure being adequately carried out. The lens can now be edged by bringing an edging wheel up to the work, or vice versa. This procedure may be varied somewhat, the edging wheel may be a diamond wheel mounted on the same bed as the centring spindle ; or the centring spindle may be transferred to an edging machine, in this case both machines are fitted with split headstocks. The edging process may be combined with a chamfering process, or the chamfer may be put on separately, but the importance of having a chamfer to act as a lead-in assembly (and also to avoid splintering) has been mentioned above. Generally speaking, all lenses are edged and chamfered except the front lenses of 4-mm and 2-mm object glasses. There may, however, be borderline cases.

Cementing, mounting and centring

165 The cementing is usually done with Canada balsam which has been pre-treated by filtering and baking. Two grades are used, " Medium " and " Hard." The medium is used for large lenses, *i.e.* over $\frac{1}{4}$-inch in diameter, and the hard for lenses smaller than this The medium can be indented with the thumbnail, whilst the hard shatters under this treatment. Matters are sometimes arranged so that no further baking is necessary.

The lenses are placed in the cell with the aid of heat, and the cell is mounted on the centring lathe and gently heated. The front surface is run true with a chucking stick, the back surface gradually coming to rest on the shoulder.

Examination of the " faint " images is made at this stage to see if they run true, if they do not and the lenses have been edged to be a good fit to the cell, then the lenses must be rejected. If, however, the fit is not so good, the lenses may be lifted from the shoulder and rotated slightly and the centring repeated, then the " faint " images may be better centred. The perfection of the final centring depends mainly on the accuracy of the centring lathe and this is usually a specially designed or adjusted piece of equipment.

Centring tolerances

166 If a surface has its centre of curvature displaced from the optical axis, it is said to be out of centre. The general refraction through the defective surface is now slightly oblique and accompanied by all the errors due to obliquity. The optics of the system now become a little complicated, but by making certain justifiable simplifying assumptions they can be dealt with and quantitative figures given for the permissible tilt of any surface (derived from a consideration of the physical optics of image formation). In general the most serious effect is due to the introduction of the unsymmetrical aberration known as coma into the centre of the field. This causes the image of an unresolvable point of light to have a slight " side flare," a most annoying defect.

It follows that some sort of compensation is possible by tilting other surfaces different amounts in the opposite direction, but circumstances have to be favourable for this to be a satisfactory cure of all the obliquity errors due to defective centring.

As an illustrative example, consider the tilt permissible on the flat outer surface of a 4-mm object glass of a given numerical aperture. This is given approximately by—

$$\text{(permissible tilt in radians)} = \frac{\text{One wavelength}}{\text{working distance} \times (\text{numerical aperture})^3}$$

Thus, for a lens with a working distance of 1 mm, a numerical aperture of 0·70, and light of wavelength 0·0005 mm—

maximum tilt = 0·00146 radians
i.e. ,, ,, = 0° 5'

The metal work

167 The mounts fall into two classes, and reference may be made to the accompanying sketches which illustrate *diagrammatically* two ways of mounting the same set of components. The first system may be described as the " screw-on " type and the second system as the " plain fitting " type. It will be seen that the " plain fitting " type is likely to be more expensive to manufacture. The operations involved in making these mounts satisfactorily demand rigid adherence to certain engineering principles which will not be enlarged upon here, except to point out that the most perfect set of components may be spoiled by unsatisfactory mounting. When the lens is mounted, a final test is made to see whether its tube length is correct, or whether the total sum of the individual thickness errors calls for correction. If correction is needed, the distance between the front lens and the next component is varied. With the screw-on type of cell, this may be achieved by shortening the cell carrying the front lens to reduce the distance, or shortening the cell containing the middle lens to increase the distance, the second separation being insensitive. It will be appreciated that this procedure calls for very true threads and shoulders. With the plain-fitting type, the countercells merely need unlocking and readjusting. These cells have the further advantage that faults, if any, can be tracked down to the defective unit by selective rotation where feasible.

It will not be out of place to point out that, in view of the extraordinary precautions taken to control their manufacture, the number of lenses calling for individual adjustment is small. For the same reason, it is quite unnecessary to provide means of adjustment for lenses of low-power.

As previously stated, Fig. 107 is only diagrammatic; with some types of 2-mm and 4-mm object glasses, the rigid lens and cell assemblage is protected from stresses and thermal changes due to the fingers. This is achieved by enclosing it in an outer tube with a small clearance. This type of mounting is thought to be useful for petrographic objectives which must be free from strain.

In the original design of the lenses and their mounts, care is taken to see that, for the main run of objectives, the distance from the shoulder of the " Society screw " to the object plane is constant. Hence, one can change from one objective to another, using objective changers or a rotating nose piece and it will merely require a manipulation of the fine adjustment to refocus the object. When this is the case, the objectives are said to be parfocal.

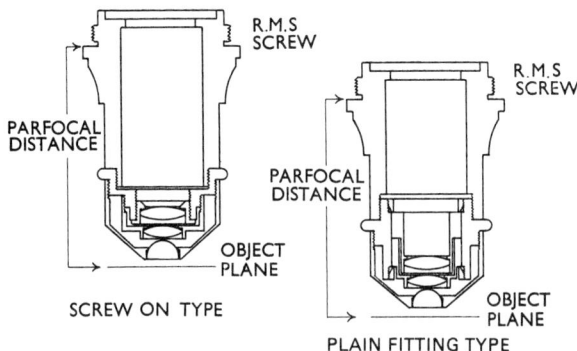

Fig. 107—Microscope objectives and mountings

Care is also taken in the construction of the objective changer to allow means of ensuring that the object remains in the centre of the field of view.

The " Society screw " (Royal Microscopical Society) is a standardized thread having the following dimensions—

	Number of threads	36 per inch
	Form of thread	Whitworth
		Maximum in.	Minimum in.
Male			
	Outside diameter	0·7982	0·7952
	Root diameter	0·7626	0·7596
Female			
	Top	0·8030	0·8000
	Root diameter	0·7674	0·7644

Testing the completed lens

168 The usual (but not universal) test is to examine the object glass on some sort of artificial star. In other words a test object is made in which brilliant pseudo-points of light are available to the observer, but the size of these objects must be below the limit of resolution of the lens. For low powers one may use globules of mercury obtained by splashing from a larger globule. These are illuminated from the side by an intense beam of light, the image of the source being the artificial star. A useful source of clean mercury is a broken clinical thermometer; the mercury should be clean so that its surface tension is unimpaired and good spherical globules obtained. For higher powers, a 3-in. × 1-in. microscope slip may be silvered, and pin holes may be found which when illuminated from below will act as artificial stars. These may be covered with cover glasses of different thicknesses forming a useful test object. The silvering should not be quite opaque as this reduces the chance of pinholes being formed. If there are insufficient pinholes, they may be made artificially by dropping coarse carborundum from a height of a few inches on to the silver. The slide containing the pin-holed silver is often termed a "silver point" slide. Such a slide should be illuminated by focusing on it the image of a bright source and, to get the best results, the condenser used should have a numerical aperture equal to that of the object glass being tested. Testing proceeds as follows : The image of the artificial star is examined below, above and at the true focal plane ; this is easily achieved by the use of the fine adjustment. The appearances are carefully studied for symmetry of form, intensity and colour. A high-power eyepiece is most useful for this purpose. For dry lenses of large numerical aperture the use of different thicknesses of cover glasses also forms an invaluable adjunct. This test can be made a quantitative one with the aid of the proper technique, but the microscope interferometer if available will give better results in a shorter time. The main use of the star test at this stage is to determine whether the lens is properly centred and properly adjusted, *i.e.* whether the sum total of inevitable incidental errors is negligible.

The procedure with a dry lens such as a 4-mm is briefly as follows : examine a " silver point " having a cover glass of the correct thickness, inspect the ring pattern in the neighbourhood of the focal plane. If the ring system is in evidence inside the focal plane, but not outside, then the lens system is undercorrected for spherical aberration. If the ring system is visible outside the focal plane but not inside, then the lens system is overcorrected. Spherical aberration can be reduced by the use of a thicker cover glass if the system is *undercorrected* ; or with most types of lens by increasing the draw tube extension. If the lens is overcorrected the cure is a reduction of cover glass thickness or a shortening of the draw-tube length. By these means it is possible to determine the relation between cover glass thickness and corrected tube length (*i.e.* tube length at which the system is free from spherical aberration) for any particular lens.

Then to a first approximation this relation is of the form—

$$1/T = At/F^2 + B$$

where A and B are constants

T = the corrected tube length graduation
t = the cover glass thickness
F = the focal length

All expressed in millimetres

The constant A does not vary much from one lens type to another with the normal run of object glasses, whilst the constant B is merely the reciprocal of the tube length for an uncovered object.

In other words the reciprocal of the corrected tube length varies directly as the cover glass thickness and if the lens is of relatively long focus, varying the cover glass thickness will have little effect on the corrected tube length. For homogeneous immersion lenses, the question of cover glass thickness does not arise, and the corrected tube length is checked by actual trial. It is worth while pointing out in this connection, however, that the refractive index of the immersion oil changes fairly rapidly with temperature ; this change introduces some spherical aberration which requires a change in tube length for its removal.

The corrected tube length or cover glass thickness having been ascertained, the lens may now be set at this position and the star image in the centre of the field examined for freedom from coma and astigmatism. It is of no use undertaking this examination until the object glass has been corrected for spherical aberration. If the central image has a slight haziness on one side, the centring is defective and the lens must be readjusted. Sometimes there is inevitably a little residual coma in the design, and this may balance the out-of-centre coma so that a coma-free image is obtained, but not exactly in the centre of the field. The central or nearly central coma-free image is next inspected with the aid of a rapid movement of the fine adjustment up and down through the focal plane. This will detect residuals of astigmatism which are not large enough to show up as deformations (ellipticities) of the ring system. There is an evanescent impression of a short bar of light rapidly rotated through a right-angle on focusing through the focal plane, whereas a static examination shows nothing. By this means, astigmatism can be traced down to the equivalent of a quarter-diopter cylindrical spectacle lens as used by the eye, an amount which is imperceptible to many observers—the actual astigmatic deformation of the wave-front emerging from the lens, being of the order of $\frac{1}{8}$th of a wavelength. It is usual to reject any lens which has as much astigmatism as this, although it is not a very obnoxious fault. In testing an apochromatic object glass the corrected tube length functions are ascertained for three colours which may be obtained by isolation with filters, and particular attention is paid to the elimination of all defects outlined above. In addition, a certain chromatic difference of magnification is required to allow for use with compensating eyepieces, but this is more a matter of design than of adjustment.

Other test objects

These are available in considerable variety, but the most popular ones are diatoms ; to use them satisfactorily a fairly wide experience is required. In general the satisfactory resolution of a *suitable* diatom into its appropriate dotted structure, where this exists, is a check on the resolving power of the lens. The coincidence of the focal plane of the dotted structure with the focal plane of the main outline is a test for the residual spherical aberration. The residual secondary spectrum can be judged on more coarsely dotted structures as these are brought in and out of focus. The actual blackness of the dotted structure is again a very severe test—if too near the limit of resolution it will be lacking in contrast. The diatom test is rendered more extensive by mounting the specimens in media whose refractive indices vary from that of the diatoms, thus varying the contrast in the object. It is clear that defects such as coma and astigmatism in the centre of the field of view may prevent a full resolution into dots, and one may get the impression of a fine-lined structure. It will be apparent from the foregoing remarks, that this type of test is qualitative rather than quantitative.

The microscope interferometer

169 This instrument, which is described in §§250 and 251, will give directly quantitative results in terms of the fundamental unit in image formation, *i.e.* the wavelength of light. It is moreover, in the writer's opinion an instrument of highly educative potentialities. Experience, lasting over many years, has served to confirm this view. With a little practice it is easy to set up and use. The instrument is so arranged that the light traverses the object glass precisely as it would in ordinary use, but this passage occurs twice, so that all errors are doubled. It would be out of place here to give the many details of the instrument, but some little discussion may be of interest. **For those who are unfamiliar with object glass testing by interferometric**

methods, the following analogy, although very imperfect, may be a help. Single lenses are tested by means of a proof glass, and an inspection of the ring system gives all the necessary information about the surface. If one could imagine a proof glass which could be applied to a system of several lenses at one and the same time, and give readings of their total errors at once, then the action of such a proof glass would be rather like that of the lens-testing interferometer. In effect it applies a proof glass to the image forming properties of the complete object glass. It will be apparent from the foregoing that a perfect lens will have a very uninteresting interferogram. Most microscope object glasses are very good, but not perfect, and there are usually some points of interest. As is well known, the interferogram may be likened to a contour map, and the elevations in the most important

Fig. 108—Axial residual spherical aberration

Fig. 109—Oblique residual astigmatism and coma

one, that of the axial pencils, do not usually reach much beyond the first contour, if as far. If, however, in a contour map the ground is sloping, then hills and valleys will be indicated by waviness in the contour lines. This principal can be used in the microscope interferometer by tilting the comparison beam. The axial interferogram for test purposes is usually a series of say 5 or 6 more or less straight bands, and the departure from exact linearity tells at once its own story about the residual spherical aberration in the lens. If the contour bands are not symmetrical about a diameter perpendicular to their length, then some of the component elements of the lens are out of centre, or one of the surfaces may not be spherical due to an actual mechanical strain in mounting. The most important monochromatic aberrations of a microscope object glass are spherical aberration and coma. The spherical aberration governs the contrast in the image and the defining power, whilst the coma to a large extent limits the size of the usable field. There is also, except in very special objectives, curvature of field and astigmatism. Coma, astigmatism and field curvature are extra-axial phenomena, and a slight rotation of the object glass will suffice to produce an interferogram containing these errors in addition to the axial spherical aberration. The pattern is now complicated and there is no need to tilt the comparison beam since the pattern contains all the necessary elements for its own analysis. It is worth while pointing out that the coma error has a single axis of symmetry whilst the spherical aberration, astigmatism and field curvature errors have double axes of symmetry in the interferograms. It is customary, at this stage, slightly to readust the focus of the object glass so that the pattern consists as far as is possible of vertical lines. Then in a very simple fashion the curvature of the central vertical band is a measure of the coma in the lens.

Figs. 108 and 109 give photographs of interferograms which show (Fig. 108) *axial* residual spherical aberration with a tilted comparison beam ; and (Fig. 109) *oblique* residual astigmatism and coma.

CHAPTER 9

THE PRODUCTION OF PRISMS IN QUANTITY

170 We shall now consider the manufacture of prisms in quantity, in so far as the methods and machines differ from those already described in the previous chapters.

Included in this chapter will be found some remarks on the making of parallel plates of high accuracy, as these can fairly be regarded as prisms of zero angle.

Raw material

171 Prisms in large numbers are now usually made from mouldings, and what was said in §123.1 about lens mouldings applies equally to prism mouldings, except that the mass of glass and the path length traversed inside a prism are usually greater than is the case with a lens, and the glass makers have therefore to exercise commensurately greater care as regards homogeneity and annealing (but see §180).

Surface flaws and inclusions caused by moulding have been reduced to such an extent that roughing off 2 or 3 mm from each surface is usually enough for their complete removal.

The wastage of optical glass in manufacture is considerable. Records kept over a considerable time, and for a variety of prisms, show that the loss during various processes amounted to the following from all causes—

		Per Cent
(a)	Sawing and roughing	10
(b)	Smoothing and polishing (by damage or otherwise)	10
(c)	Edging (in lenses)	6
(d)	Faulty material	12
	Total	38

If moulded prisms are used some of the loss coming under (a) is avoided.

It is right to add that the record was made during a period when much unskilled labour was employed.

Machining

General considerations

172 In the machining of a lens we were concerned only with the generation of the required curves and the reduction of the blank to a given thickness.

In the machining of a prism account must be taken of the angle, not only between the faces which are subsequently to be polished, but also the angles which those faces make to a common base. Where two surfaces only are to be polished an unpolished surface is frequently relied on as a base for fixing the prisms ; where three or more polished surfaces are concerned lack of truth may produce what is known as " pyramidal error " which becomes apparent when, in the finishing stages the base is ground to the prescribed limits of squareness.

When two faces are square to the base, the third (and fourth if present) must also be so if the prism is to be free from pyramidal error. Another effect of pyramidal error is that the sum of the angles which the prism encloses add up to more than the appropriate number of right-angles : each angle is measured in turn in a plane normal to both surfaces. In a prism with pyramidal error these planes are not parallel. The magnitude of the errors which result from the presence of pyramidal error has been fully described by Guild (Glazebrook's *Dic. App. Phys.* IV). In general it can be said that for work of good quality all surfaces should be square to a common base to less than 2 minutes of arc.

Prism dimensions are usually measured and defined as lengths of face to the intersection of the surfaces. This entails some uncertainty in working as, in the course of manufacture, edges are rarely left sharp for any length of time, and certainly when finished and in the testing department all edges and corners will have been chamfered away, usually by hand. Gauges are a partial solution to this difficulty and in particular more use might be made of the fact that most prisms have an inscribed circle, the diameter of which is not affected by the chamfers.

Machining to angle

173 As far as I am aware, the first arrangement for the accurate angling of prisms by means of cupped grinding wheels was made at the Hilger works by the adaptation of a Brown & Sharp grinding machine. The prism was mounted on a turntable with an accurately divided circle and the faces were ground with the cupped diamond lap one after the other, each being brought into position by reference to the divided circle. I find from my notes that this machine was in regular use in November 1914 and was turning out prisms with an angle of accuracy of about ± 3 minutes.

Diamond laps of the home-made kind were used, but very good results can be obtained with cupped wheels of corundum or carborundum. We have found the most useful to be the carborundum wheels described by the makers (the Carborundum Company) as

Bond E.O., Grade S., Grit 90 and 220—the former for roughing operations, the latter for fine work. The linear cutting speed should be from 2,000 to 4,000 feet per minute, the work being amply lubricated with the ordinary soap emulsion as used for turret lathes.

However, carborundum, corundum and home-made diamond laps are now superseded by those described in §127.

Twyman angling jig

174 A grinding attachment based on these lines has been successfully employed in the manufacture of prisms in numbers.

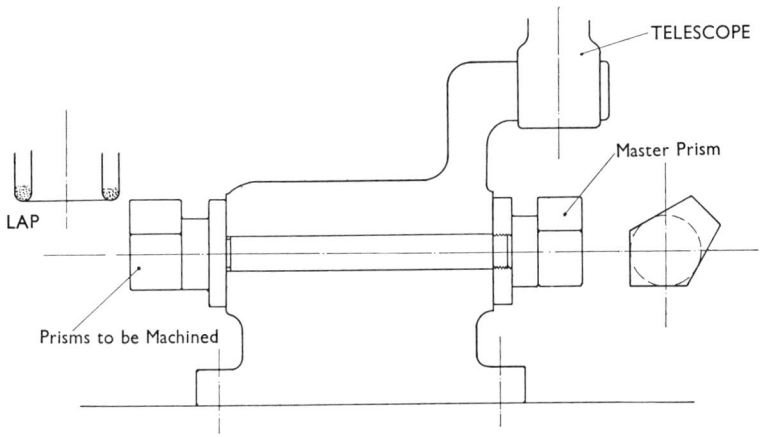

Fig. 110—Jig for angling prisms to match the angles of a master prism

The method depends on a master prism, and we consider the cost of such a prism is justified if the number of prisms whose angles are to be the same is 48 or more. Glass enough for such a number can usually be obtained from the glass-maker all from the same melting and therefore of nearly enough the same refractive index to allow the angles of all the prisms to be the same, even when those angles depend on the refractive index and dispersion of the glass.

The jig is shown in Fig. 110. A rigid shaft is carried in good bearings and arranged to be exactly parallel to the table of a vertical milling machine. It can be clamped at any angle, and carries on one end the master prism and on the other a fitting to take the prisms to be machined which are cemented to metal chucks by their bases. The centre of the circle inscribed within the faces of the prism should coincide with the axis of the shaft.

The Angle Dekkor telescope (see §232), is of the back-reflecting type, and is fitted with its axis normal to the axis of rotation of the shaft. Each face of the master prism is in turn set to zero in the telescope, and the face of the workpiece machined by the diamond lap. The table having once been set to the right height all that is necessary is to traverse the prism below the lap once for each surface. The size of the master prism is unimportant.

Blanks machined in this way will be free from pyramidal error to the accuracy with which the shaft is made parallel to the machine slides, an adjustment which can be checked by machining the opposite faces of a parallel plate. If the surfaces are not parallel

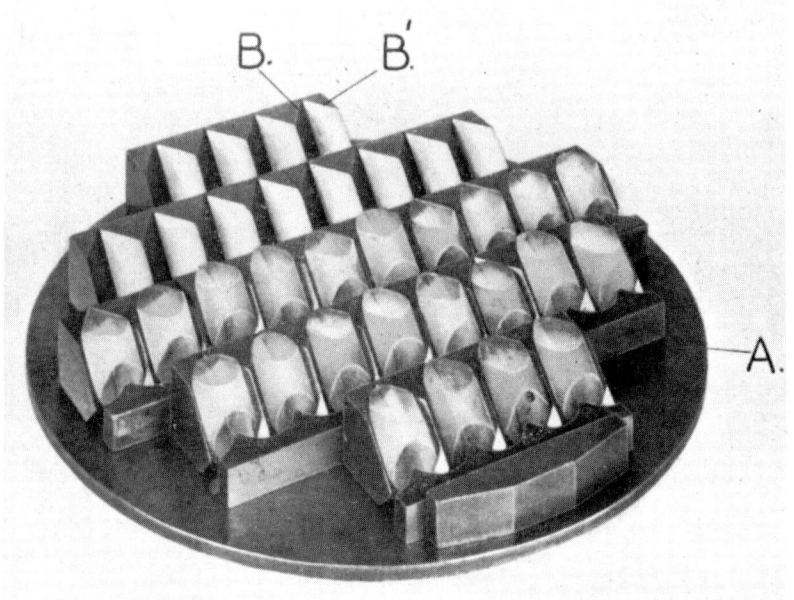

Fig. 111—Accurate metal blocking fixture

along the direction of traverse, the jig should be adjusted with packing.

This system of angle machining is to be recommended where facilities for the construction of accurate metal jigs are not available; it will cope with any prism possessing an inscribed circle. The chuck consists of a flat disc of brass turned parallel, with provision for attaching it to the end of the shaft in approximately the orientation required. The prisms are cemented to one face of the chuck with cement No. 10, §94. The outside diameter of the disc must, of course, be slightly smaller than the inscribed circle of the prism in order to clear the diamond lap. The average angle accuracy attained with this jig is 1¼ minutes, the operator being unskilled.

Accurate grinding fixtures

175 An early example of an accurate steel grinding and polishing fixture is shown in Fig. 111. This was in use in 1916.

The prisms, which are seen in position, are the well-known " roof prisms " used for erecting the image of an object and at the same time turning the path of the rays through a right angle. The fixture to be described was used for polishing the diamond-shaped end faces, the 90° angles having been polished previously.

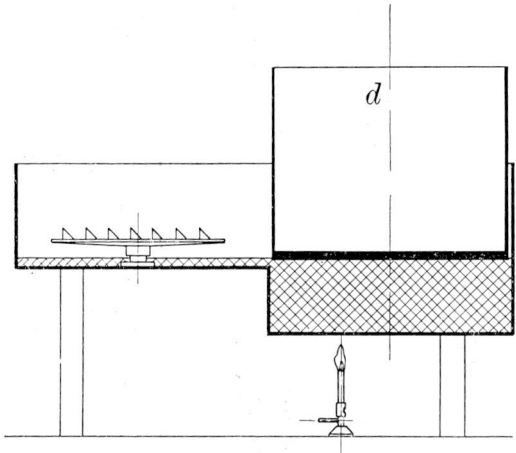

Fig. 112—Paraffin wax trough for blocking prisms

The roof is at 45° to the diamond-shaped end faces, while each of the roof faces makes an angle of 60° with the last-named surfaces. In order that the prism may give a single image the angle between the roof surfaces must be 90°, correct to within one second, but the angle between the diamond-shaped end faces may be several minutes out without affecting the performance. The method of making these prisms now to be described secures a sufficient accuracy in most instances for the latter angles which are rarely more than 5 or 6 minutes out.

The jig has for its base plate a flat optical tool, and a number of strips are milled with interior surfaces B, B, etc., accurately at 90° to each other and at 60° to the surface of the tool. All the component parts of this jig were made accurate to 1 minute of arc, and in assembly were fitted together with like accuracy.

The prisms are machined somewhat oversize, cleaned and laid on the angular recesses with their lower ends bearing on metal strips, whose function it is to set all the prism faces at the correct height, thus ensuring the same length of prism throughout the block. The tool is then placed in the trough shown in Fig. 112.

This consists of a deep cylindrical portion d, the lower part of which is filled with melted paraffin wax, while the upper is filled with a hollow drum adjustable in height on a long screw about $\frac{3}{8}$ in. diameter attached to the bottom of the trough. When the drum is screwed down it causes the melted wax to flow into the side trough in which is the block of prisms. The wax is melted by heat applied from below and this, of course, has the effect also of heating up the tool with its charge of prisms. When a sufficient time has elapsed for the tool and prisms to become thoroughly warm, the drum d is screwed down and this may be done fairly rapidly. This forces the wax to rise until the tool, with its prisms, is completely immersed and, after leaving for, say, 5 minutes, the drum is screwed up again and the tool with its prisms allowed to drain and cool.

If the underside of the base tool has been made parallel to the top during manufacture, the block can be clamped on the table of a large optical grinder and the prisms ground parallel to the surface of the tool. This parallelism is carefully maintained during the trueing, smoothing and polishing, and if these operations are carefully performed throughout, the prisms will not move more than about 2 minutes, and all the polished faces will be correctly to angle with the roof faces to 4 or 5 minutes. This is sufficient accuracy for most purposes for which these prisms are used.

A way of making roof prisms used in one of the optical shops of Hilger & Watts Ltd

176 We have at different times used a number of different ways of working roof prisms. The following has been found appropriate for use by previously untrained women workers for large roof prisms, and is a further illustration of the use of accurate metal jigs.

The prisms are to be of 1 in. aperture. Moulded slabs of boro-silicate crown approximately $3\frac{1}{4}$ in. square and $1\frac{1}{4}$ in. thick are slab milled to a thickness of 1·02 inch.

The slabs are slit across the diagonals to give 4 prisms, as shown by the lines in Fig. 113A.

Each prism is then put in a jig and the cathetus faces are machine milled on the glass milling machine.

The prisms are then blocked in plaster and the cathetus faces polished.

THE PRODUCTION OF PRISMS IN QUANTITY 299

The prisms are now ready for roofing, which is done in an accurate jig, Fig. 113B, in the form of a cube, preferably in aluminium. The prisms are mounted with shellac at each of the four corners of the jig as shown in the figure, which shows one prism in position. With this method of mounting an accuracy between the cathetus and roof surfaces of ± 3 minutes can be assured.

The loaded jigs are now clamped, two at a time, on a holder as shown in Fig. 113C (in which one only of the two jigs is shown). This holder is clamped on the bed of the milling machine and each of the four sides of the cube becomes in turn a base for a cut to be taken from the top side, the jig being rotated through 90° for the successive faces to be milled.

Fig. 113—Jigs for making large roof prisms

The roof faces are then smoothed and polished to the requisite accuracy while still in the jig, measurements of angle being made in the first instance by the Angle Dekkor (see Chapter 11), and then by the Hilger N.26 Interferometer. There are holes, H, H, in the jig, by means of which observation can be made on the Angle Dekkor or Interferometer by internal reflection at the roof of each prism.

Protectors are used for supporting the dividing line of the roof. The surplus of the height noted at the beginning of this section is used to equalise the widths of the roof faces when that is necessary.

The prisms are finished by milling in accordance with whatever method of mounting the prisms is adopted in the instrument.

Making tetragonal prisms

177 Another application of accurate steel jigs is illustrated in the manufacture, during the recent war, of some hundreds of tetragonal or " corner cube " prisms (sometimes called " Triple mirrors "). The finished prism may be regarded as one corner of a glass cube, all the angles forming one corner of the cube being polished to an accuracy of 1 second of arc ; the angles formed by these faces and the fourth face are of much less importance, 10 minutes accuracy sufficing.

Fig. 114—Jig for milling tetragonal prisms

Mild steel cubes were prepared of 2 in. length of face, and ground to within one minute for angle ; they were then chucked up in Woods' metal and one corner turned off so that the resulting face made equal angles to the original faces, which were slightly reduced in length of edge below the original machine size.

The 90° angles were corrected by hand lapping to within one-quarter of a minute, and the centre of the equilateral triangle face recessed so as to leave three triangular facets (see Fig. 114).

The prism mouldings were roughed and trued by hand on the large equilateral face, and finished in 320 Aloxite ready for blocking.

Fig. 115—Adcock and Shipley 2 VS flat and lens-grinding machine
The grinding head and the milling fixture are in position

This trued face was then cemented to the corresponding face on the steel cube with shellac-venetian turpentine cement (No. 9 in §94). Another cement useful for such purposes is 2 parts Montan wax and 3 parts each of beeswax and rosin.*

The machining was carried out on a vertical-spindle surface grinder, holding the steel cube with a magnetic chuck (see Fig. 114).

Each face of the steel cube was laid on the chuck in turn, and the upper face of the prism ground; when the prism was of the desired

Fig. 116—Milling fixture with constant deviation prisms

face length of 2 inches plus grinding allowance, the feed stop of the machine was set, and thereafter the cycle of operations was repeated, using unskilled labour for the most part. Twelve minutes apiece was a fairly comfortable machining rate, using an Abwood surface grinder (by the Abwood Tool & Engineering Co Ltd, Prince's Road, Dartford,

---

* Montan wax is a fairly common commercial article. It is a lignite wax and is obtained from dried lignite by extraction with suitable solvents. Its colour varies from white to black, and the material has a melting point from 70°C to 80°C. It is not a true chemical wax but a mixture.

The commercial use of Montan wax is mainly in the production of candles, polishes and water-proofing materials. It is listed in the catalogue of Harrington Brothers.

THE PRODUCTION OF PRISMS IN QUANTITY   303

England) and a 4 in. diameter diamond cutter of the impregnated type. The subsequent correction of angle was carried out on the Twyman-Underhill machine described in §188, the testing being carried out on the Hilger Interferometer. The subsequent polishing of the equilateral triangle presents no difficulty.

Grinding machine for large surfaces

178   The machine illustrated in Fig. 115 is very versatile, and was developed specially to meet the demand for a flat and curved surface

Fig. 117—Milling fixture with strips for roughing deviating prisms

grinding machine powerful enough to grind large blocks of work quickly, and rigid enough to do so without flexure.*

A widely useful accessory to this machine is the milling fixture illustrated in Figs. 116 and 117 consisting of a cast-iron plate with a machined rim, cast with reinforcing ribs and a boss which enables

---

* A similar glass-grinding machine and prism surfacing fixture are made by the Blanchard Machine Co, 64 State Street, Cambridge, Mass, U.S.A. (No. 11 Blanchard Surface Grinder).

it to be screwed direct to a conventional smoothing and polishing machine.

The tops of the milling-fixtures are themselves machined parallel to the feet on which the fixture stands, and consistent to 0·001 in. for height ; the shoulder enables the fixtures to be clamped to the table with bent dogs.

In use, the glasswork is cemented to the milling fixtures with beeswax and rosin cement, and will be ground parallel to about two or three minutes, or as parallel as the fixtures are made.

In the manufacture of some achromatic deviating prisms, of which we use a number every year, the frequent change of refracting angle due to changes in the index of successive meltings led us to adopt a method of fabricating accurate grinding attachments from glass strips, using one of the milling fixtures as the base tool.

The strips are machined to angle on the tilting magnetic chuck (§180) with due precautions to ensure consistency of thickness parallelism lengthwise ; they are then cemented to the milling fixture (Fig. 117). The mouldings from which the prisms are to be made can be cemented to the strips in the same heating ; if the assemblies are to be used repeatedly it would be better to cement the strips to the fixture with H.T. cement (§146) and thus eliminate the risk of subsequent movement.

The faces of the mouldings are previously polished, and these of course are the sides laid down on the strips.

After milling on the 2VS machine until the thick end of the prism is reduced to the dimension required, the assembly is transferred to the polishing machine and smoothed and polished, an accuracy of three minutes for angle being obtained by this method.

Alteration in angles involves only the re-grinding of the strips.

Clamping type milling fixtures

179 A further type of machining fixture in common use is that in which the prism is clamped in place during machining, the jig being so constructed that the prism is held at the required angle to the machine table.

Fixtures of this type suffer from one main disadvantage, the liability of the prism to move during the cut due to the vibration imparted by the cutter ; if no more serious damage occurs the prism will at least have to be machined again on the affected face.

In some German optical works all prism and lens grinding operations were carried out by the use of mechanical clamps to hold the work, the use of thermo-plastic adhesives being avoided.   (*U.S.N.*, 1945)

Tilting magnetic chuck

180 A very useful item of equipment for the optical machine shop is a magnetic chuck so constructed that the magnetic face can be rotated around an axis parallel to its major dimension. Such a chuck is made by J. H. Humphreys and Sons, Blackriding Electrical Works, Werneth, Oldham, England.

The chuck face swivels 90° on either side, enabling all angles to be machined. Glass being unfortunately non-magnetic, one surface must be ground by hand, albeit quite roughly, and the prism cemented with beeswax and rosin cement to a piece of flat ground steel plate, which is available commercially to an accuracy of 0·002 in., or better,

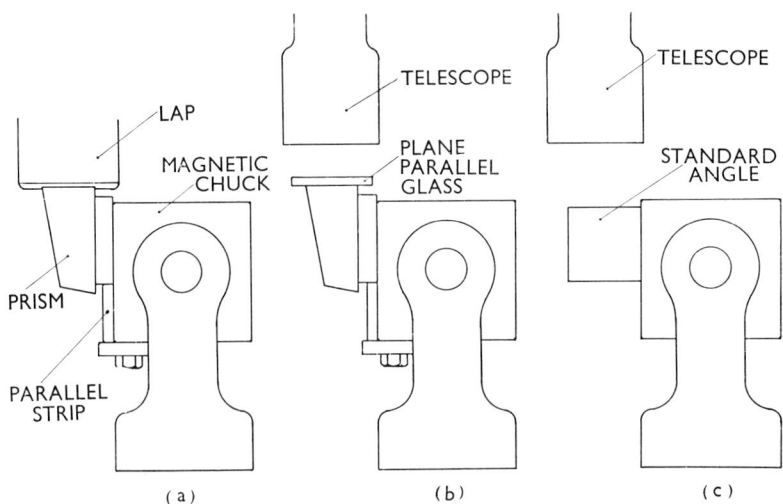

Fig. 118—Tilting magnetic chuck

and in a range of widths and thicknesses. We find the most generally useful plates are one-quarter inch thick, and for average work 8 in. × 4 in. is a convenient size. The description which follows of the machine layout for a constant-deviation dispersing prism (Fig. 118) will illustrate the principles involved in the use of the tilting chuck.

The prism mouldings are laid down on milling fixtures as detailed for plane parallel milling in conjunction with the No. 2 VS machine (§178, Fig. 115), and reduced to within 0·03 in. of the finished height, one of the bases so generated being roughly polished before removing from the milling fixture.

Note that the word "roughly" refers only to the physical appearance of the polish, since in later stages the base is used to correct the angle with a back-reflecting telescope. Thus, although a high degree of polish is not necessary, the bases must be flat to a fringe or so.

The prisms are placed on a hot-plate, cemented in stacks of three with beeswax and rosin cement, and the widest face of the stack ground by hand on a flat roughing machine (Fig. 59) until the moulded surface has been removed. After fining down with 320 Aloxite, the surface is cemented to a strip of flat ground steel stock 6 in. × 2½ in. × ¼ in., one end of which has been milled square to the long edges, these latter being already parallel. While the cement is still warm, care is taken to see that the polished base face is flush with and parallel to the squared end of the steel strip.

One should now refer to Fig. 118, which shows the magnetic chuck with prism.

The tilting magnetic chuck is set to 90° by its scale, which can be effected to about 15 minutes, one stack of prisms is placed on the chuck, and a parallel strip of steel is inserted to enable the support plate to be located with reference to the strip provided on all good magnetic chucks (Fig. 118). A light cut is taken, just enough to generate a clean surface, and the machine is switched off.

A plane parallel piece of glass is now " breathed down " on to the ground surface, which can be regarded as the plane of reference of the machine, whether or not it is parallel to the machine table ; normally it is. A back-reflecting telescope is set up to receive the image from the parallel glass on the zero position of the reference scale (Fig. 118b). The prism stack is now removed from the chuck and an accurate 90° angle gauge substituted in the same spot. The image from the standard angle will probably lie within 15 minutes or so of the zero mark, and the chuck can now be adjusted until not more than, say, a minute of residual error is left (Fig. 118c).

Setting to a greater accuracy is hardly worth while, as errors of a minute may creep in due to lack of parallelism between the surfaces of the cement film, and the steel plates may also be this much out of truth.

The head of the machine is now lowered by the amount it is desired to remove from this face, say 2 mm, and the whole batch of stacks put through the machine at this setting. If the cementing has been done with average care, and the end of the stack well lined up with the end of the support plate, the small face (shown in the diagrams) will be at 90° both to the longest face and to the base, within two or three minutes.

The long reflecting face is next machined in a similar manner, the amount to be removed in this case being such as will reduce the face-length of the previously machined short face to the required dimension.

The fourth face, although not to be polished and of no real accuracy, is machined in the same manner in order to bring the face length of the longest (hand ground) face to blueprint size.

Thus, all three faces have been machined to size, angle and base, and will be free enough from pyramidal error for most purposes if the accessory parts have been prepared with care.

If the steel plates are consistent for dimensions then no dimensional checks need be applied after the setting has been completed, and the process is quite easily carried out with complete success by workers, unskilled but with some *ad hoc* training.

The prisms are left in the stacks of three during smoothing and polishing, as the larger surfaces are of assistance in accurate laying down.

These stacks of prisms will now be followed through the subsequent operations to the finishing.

Trueing to angle

The machining layout already detailed will turn out prism blanks to an accuracy of 3 minutes for angle, and for many purposes this would be good enough for use without improvement. The prisms we have chosen for this illustration, however, require the 75° angle to be correct to 15 seconds, while the 90° angle can be in error by 30 seconds; all polished surfaces must be square to the finished base to within one minute.

The stacks are removed from their steel plates by gently tapping the latter with a copper hammer, as the use of heat would invite the prisms to move relative to one another. During the initial stacking-up, one polished base was aligned with the square end of the steel plate, and this surface is now used as the reference base for the stack. The first face to be polished is the smallest one, and this is ground by hand with 320 Aloxite to reduce the error in the basing to $\frac{1}{2}$ minute or less; no attempt is made at this stage to correct either the 75° or the 90° angles, as if the small face does not make precise contact with the tool on which the block is made, the work would be wasted. When a convenient number of stacks have been trued on the small face, they are passed over to be made into a plaster block for polishing.

The above-mentioned machines and fixtures for milling prisms to shape from mouldings can also be used for sawing prisms to shape and angle from large blocks. The settings are exactly the same except that instead of a milling tool a sintered I.D.P. diamond blade is used.

During the past 40 years my opinion has, owing to changing circumstances, varied as to the advisability of cutting prisms and lenses from blocks or from mouldings. At times the mouldings were satisfactory for several years and then trouble eventuated. Quite recently (since §§123.1 and 171 were written) we on the whole prefer cutting prisms and lenses from blocks or plates. I attribute the troubles which have caused us to adopt this opinion to imperfect control of the heat treatment during annealing. This uncertainty is particularly unfortunate in view of the possibilities of high-accuracy moulding referred to in §123.1.

Plaster blocks

181 The accurate jigs described in §§175 to 177 are too expensive to be used unless the numbers of prisms of a kind run into thousands.

There are various other ways of holding prisms in a block so that a number of them can be polished together. The one I have found most generally useful is the plaster block.

The operator responsible for plaster blocking must be provided with several weeks' supply of plaster of Paris and of hydrated lime ; several optical tools which have been carefully flattened ; several more flat tools of the polisher holder category, *i.e.* not ground at all, for use as backing tools, and several brass rings bent from 1/16th in. brass strip to a diameter slightly less than the true tools in use.

The plaster of Paris must be kept dry by enclosure within a box, say 6 ft × 3 ft × 3 ft (or, if the quantity is large, in a separate room) which is always maintained, by an electric heater or a sufficient number of electric lamps, at a temperature of at least some 5 or 10 degrees higher than the space outside the box or outside the room as the case may be.

One of the flat tools is screwed to the bench screw, and cleaned with petrol, then rubbed with the least possible smear of clean machine oil. (The prism can be made to stick in place by the use of a thin smear of water or soap solution, but as the block is usually left to dry out for 24 hours the tool is then liable to rust). The trued surfaces of the prism stacks are dusted and rubbed down into close contact with the tool surface ; lack of care in this stage is one of the main causes of angle errors in the finished prism. The arrangement of the surfaces should be as symmetrical as possible as, if there is an excess of glass on one side of the block, the other side will grind away more quickly and the angle will alter in the grinding process.

A brass ring large enough to enclose the prisms is then laid on the tool, supported on three slips of glass about 2 mm thick ; this is to prevent the brass ring from touching the smoothing tool and the polisher. Melted paraffin wax is then painted round the gap between the tool and the ring to prevent the escape of the wet plaster. A small quantity of plaster is mixed with much more water than usual, making a very thin mixture, and enough of this poured in to form a layer about 1/10th in. thick.

While this is drying out, plaster and lime are mixed together in the proportions of 2 plaster : 1 hydrated lime in sufficient quantity to fill the brass ring; the mixing of the lime and plaster must be thoroughly carried out before the water is added, a bricklayer's trowel being very suitable for the purpose. If thorough mixing is not effected the block will be liable to sink in places during working.

The powder is then spread on a board in a heap with a depression in the middle, and water poured, a small quantity at a time, into the depression, the outer parts of the heap being scattered on top with the trowel just as in mixing concrete; as the heap becomes moistened throughout, the process of scattering the dry powder should be changed to one of chopping the moistened plaster with the edge of the trowel and turning it over until a lump uniformly moist is produced of about the consistency of mortar as used by a bricklayer.

The operator then puts on a pair of rubber gloves, which are desirable for cleanliness, and taking portions of the mortar in the fingers thrusts them and presses them into the interstices between the prisms in much the same way as one forces the stuffing into a chicken or turkey. The block is filled right to the top of, but not above, the brass ring with the plaster which is made to settle down snugly about the prisms by patting, when, in the same way as wet sand does, it becomes wet and fluid. A block so prepared with the minimum of water to produce a workable mortar will dry out in 12 to 24 hours, when the block can be removed from the tool by heating the latter.

As the plaster sets some heat is generated, which helps the drying process; the block should be left for a couple of hours, after which time a flat tool of blockholder quality is heated to about 100°C and covered with molten pitch. A few reticulations are cut in the upper surface of the plaster, and the pitch pad pressed firmly on top and left for from 12 to 24 hours. It is essential that the block should not be removed from the flat tool until the preparations for smoothing and polishing have been made. Plaster blocks are usually made up in the morning, so that they may be smoothed, polished and another surface laid down the following day.

After the block has stood the above-mentioned time, it is pulled off the flat tool, and screwed to a post screw by the thread in the backing tool. It will be found that the layer of very wet plaster which was poured in first will lift off quite easily, leaving the prisms projecting from the mass of set plaster. The exposed plaster must now be painted with either shellac varnish or hot paraffin wax, as if left unprotected it will absorb the water used in the smoothing and polishing stages. Fig. 119 shows the finished block.

Fine grinding (or smoothing) is carried out on the tool on which the block was made up, and all the prisms should " come up " during the first emery; if parts of certain prisms exhibit the coarser grey of the 320 Aloxite after smoothing for 10 minutes in 400 grade abrasive, then one of two things has happened. Either the prisms concerned were not laid in good contact with the tool, or the surfaces were not flat to start with; surfaces ready for plaster blocking should be slightly concave for preference, never on any account convex. If the block is seen to be " taking " all over in a satisfactory manner, the usual smoothing routine of two or three ten-minute grindings in each of the finer grades of abrasive will follow. It is more important

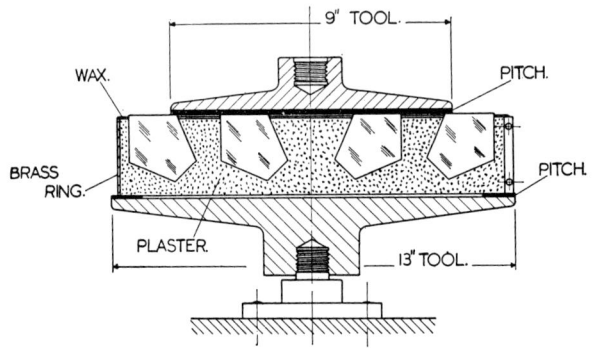

Fig. 119—Plaster blocks
Finished and inverted

than ever that traces of the previous grade should be removed before passing to a finer abrasive, and a powerful spray of water is of great help during the washing between emery grades.

The block can occupy either the upper or the lower position on the smoothing machine, according to whether the operator wishes the tool to assume a more convex figure or vice versa. It is worth repeating here that with discriminating reversal of block and tool, the latter can be maintained within a few rings of true flatness for long periods without grinding the tools together.

The mixture of plaster of Paris and hydrated lime is preferred by us to plaster of Paris alone or any of the other mixtures we have tried. The following remarks are however worth adding.

Plaster of Paris expands on setting, and one of the objects of the addition of the lime is to minimize the effect of this. Portland cement contracts on setting, and a mixture of plaster of Paris one part and

Portland cement six parts gives a very satisfactory result except that its extreme hardness and strength of adhesion makes it difficult to get the prisms out of the block and to remove the cement from the prisms. Soaking in a strong solution of caustic soda is said to help the latter process.

Another old dodge perhaps worth a trial once more, is to dissolve some alum in the water used for making up pure plaster of Paris. The mixture expands at first, but eventually contracts to normal. Some workers prefer Keene's cement to Portland cement. Keene's cement is softer than Portland and does not adhere so strongly to the glass—on the other hand it requires 48 hours to set. A mixture of Keene's cement seven parts to plaster of Paris one part is said to have the required properties. (Both plaster of Paris and Keene's cement are prepared by calcining gypsum, the former at about 204°C, the latter at about 500°C.)

Plaster blocks of prisms are also used in Germany, but, whereas in the United States it is customary to keep the plaster below the surfaces to be worked by pouring hot wax round the prisms (after they are laid down on the flat plate), in Germany sawdust is laid between them, thus avoiding the use of heat with its liability to cause tilting of the prisms. (*U.S.N.* 1945).

Polishing plaster blocks

182 Plaster blocks are usually polished by machine, and the routine is very simple. A loaded polisher of the wood-flour type is warmed from the back to about 60°C, painted with cerium oxide and rubbed by hand on the block until a good, even, pull is felt, free from edge binding. The machine pin is lowered into the pin cup in the back of the polisher and the machine switched on.

After one or two " wets " the block is allowed to dry fairly thoroughly and cleaned for examination. If the block is polishing evenly and the test-plate figure is satisfactory, the polisher can be replaced and polishing continued for half an hour before re-testing takes place.

If the polish is uneven, say bright in the centre but much more grey around the edge, then the stroke should be moved more over the edge; if the central rings in the polisher have closed up they should be opened. If the polisher is still warm it will quickly assume the shape of the block and, once cool, will retain that shape for some hours.

In the third case, where the test-plate shape is not good, some discretion must be exercised; there is not a lot to choose between a well-polished block of the wrong shape and a block of good shape which is grey in places. As a general rule, if the departure from proof plane is not serious, the grey should be removed before any attempt is made to correct the shape as, once a part of the block acquires a full polish, the rate of removal of the remaining grey is very much reduced. If the shape of the prisms after one or two wets is really bad, then the only satisfactory remedy is to re-smooth it on a much flatter tool, which will tend to alter the angles.

The manipulation of the machine to improve the flatness of the block is somewhat easier than in the case of the lens block, since the plaster can be relied upon to hold the prisms without moving for 12 hours or so ; this makes possible such resources as the substitution of a smaller polisher on a block which, although well-polished, is found to be more convex than the blueprint allows.

It will be found that the device of reversing block and polisher will be liable to give trouble, as the pin of the machine is much further from the sliding surfaces when the block is on top, and the amount of friction present is very much greater than in smoothing ; a torque results which tends to make the polisher convex.

The use of a smaller polisher to depress the centre is therefore to be preferred ; the further off-setting of the stroke with a full-sized polisher will always make the block more convex.

In Germany large blocks of prisms up to 30 ins. in diameter are polished on single spindle machines ; such polishing is completed in four to five hours, and each operative takes care of six or seven machines. (*U.S.N.* 1945).

Removing the prism from the plaster block after polishing

183 When the polish and shape are both satisfactory, the brass ring having been removed, the block can be broken up by blows from a hammer on the back of the block, or by a hammer and cold chisel. The breakage of the block will often occur along one of the surfaces of the prisms and those prisms which are so uncovered can often be removed forthwith almost free from plaster. There is no difficulty in removing any small pieces of plaster still adhering to the prisms.

It is convenient, by-the-bye, to stock paraffin wax of different melting points, since this enables one to perform composite operations in which one piece is cemented with a softer wax to a combination of pieces which has previously been cemented with a harder wax. The same applies to the advantages of having available hard, medium and soft Canada balsam.

The surface already polished is protected by painting with a hard varnish, leaving clear in one corner a small area of 2 mm square for use with the telescope when trueing the next face. (This painting may with advantage be done before the block is broken up).

The next longest face is now ground with 320 Aloxite so as to correct the 75° angle and at the same time adjust the basing angle to within ½ minute. The finished prism has to be correct to one-quarter minute on the 75° angle and, using the Angle Dekkor, the surface is trued as closely as the telescope will read (about 6 seconds).

If the previously detailed blocking process is repeated with due care, a large proportion of the prism stacks will pass for angle straight from the block. Residual errors, if small, can be polished right either by hand, or on the machine detailed in §188, which was designed especially for the individual correction of angle and flatness on single elements.

120—Blocking with waxed felt

For most purposes an accuracy of one or two minutes of angle and one fringe for flatness are amply good enough, and prisms of this type would be cleaned and finished without further process.

All the stacks which pass the floor inspection for angles, appearance, and flatness are now carefully warmed apart and cleaned ready for hand finishing.

Polishing blocks of prisms held in the block with waxed felt
184 This alternative method is sometimes advantageous. It is necessary to have a formed tool but it need not be accurately machined. Taking the case of blocking pentagonal prisms, one first of all chamfers all the edges of the prisms which it is presumed have been trued to angle. They are then blocked in the shaped holder (see Fig. 120) as follows : A flat tool is heated and smeared over with paraffin wax,

the tool being of such a temperature that the paraffin wax runs like water. The prisms are then formed up in rows on the tool allowing a suitable separation between each prism to correspond with the formed holder. The prisms should be pressed as close down to the tool as possible and the whole then allowed to cool. The faces that have to bear on the felts should be smeared with paraffin wax to protect them.

While this operation is in progress, heat in a saucepan cement consisting of three parts of rosin and one of beeswax. Dip into this melted cement pieces of felt, and when they have done bubbling place the felts on the cold former, so that when the prisms are in position they will be held on two faces.

The prisms having cooled, the tool holding them is inverted and the prisms and tool lowered on to the felts. The formed holder is then heated until cement drips off it, then allowed to cool. The block is then ready for smoothing and polishing in the ordinary way.

The prisms can be knocked off with a smart tap applied through the intermediary of a piece of wood, but to avoid the prisms being knocked together a piece of card should be placed between the prism which is being detached and the one next to it.

Plane parallel glasses

Polishing blocks of plane parallel glasses

185   Plane parallel glasses are often made from plate glass, in which case nothing is required prior to blocking but to edge the glasses to shape. If they are to be made from blocks of glass, the slitting, roughing and trueing to thickness and mechanical parallelism are first carried out and the discs then blocked for polishing the first side. It is our opinion that it is somewhat more difficult to polish plate glass free from marks than either B.S.C. or H.C.

The method we prefer for this is to block them on pitch. This should be poured on a tool as in making a polisher but the pitch should be $\frac{1}{2}$ in. thick. It should be flattened as in making a polisher. Then the tool is turned upside down while still warm (but not too warm) and allowed to sink on the plates (laid on a flat tool) and cooled. One side having been polished in this way, a flat optical tool which has been polished is taken and the glasses laid thereon, having first been carefully dusted. It is desirable that this tool should be of glass, and the same kind of glass as that to be polished. If it is of a different kind of glass, appreciable distortion may take place owing to difference of expansion and as a result imperfect surfaces will be produced. By saying that the tool should be of glass I mean that a plate of glass about 1¼ in. thick is fixed to the ordinary cast-iron tool by means of a layer of pitch, the glass then being smoothed and polished. The tool must be carefully tested to see that it is thoroughly flat, since even a slight departure from flatness may mean that the edges are tilted by a material amount relative to the middle of the tool. It is customary to make plates of this type from plate glass so that the top surface, having its original polish, permits one to see the Newton's fringes due to interference between the tool and the undersides of the plates. It is best to perform this operation in a dust-free enclosure, at least when the work is being done by other than skilled operatives, since the whole precision of the method depends on the under surfaces of the plates being parallel to or in contact with the tool. One then scrapes paraffin wax into shreds which are placed in between the plates, and on slowly heating the whole tool the wax melts and spreads in a thin film underneath the plates, between them and the tool. The tool is then allowed to cool slowly and the top surface ground and polished parallel to the tool. In order to ensure that the top surfaces are parallel to the tool, an Angle Dekkor (§232 *et seq.*) may be used to obtain reflection from a glass plate put on top and a small space cleansed of paraffin wax in the under tool, or alternatively a depth gauge may be used on the opposite sides of the tool.

186 Plane parallel plates, say 2¼ in. diameter and ¼ in. thick, can be laid down on a tool by carefully working in this way to an accuracy of about 5 seconds. The chief error in parallelism in the resulting windows will usually be determined by the lack of flatness of the tool on which the plates are blocked. This can be corrected by rubbing the three flat tools together in pairs, always grinding together the pair that either bind most strongly or swing most strongly. The closeness of approach to contact may be easily estimated by use of tinfoil, which is available in leaves 0·0005 in. thick, of remarkable consistency. If a pair of tools swing about the middle, and three small pieces of foil placed symmetrically round the periphery of the lower tool are sufficient to annul the tendency to swing about the middle, then the tools are in contact to within this limit (0·0005 in.). If the tools bind at the edge, a small piece of tinfoil is placed in the centre of the lower tool, and if this makes them swing about the centre then again the tools are in contact to this limit.

If all three combinations of the three tools pass these tests with tinfoil 0·00075 in. thick, then we can be certain that all the tools are flat to 0·00075 in. A 9 in. tool with a sagitta of 0·00075 in. has a radius of curvature of 1,125 feet and a slope at one inch from the edge of the block of 54 seconds relative to a plane surface.

An error can arise from the surfaces not being polished quite flat. The sagitta (or sag) of a spherical but nearly flat surface varies as the square of the diameter. Thus, if a 2 in. diameter proof plate applied to the block shows that the surface is $2\frac{1}{2}$ Newton's rings out of flat, the 9 in. diameter block will be 50 rings out of flat ($=0{\cdot}0005$ in.). This indicates a radius of curvature of approximately 1,300 ft and an inclination to the average surface in a position 1 inch from the edge of the block amounting to 40 seconds.

This error must be added to or subtracted from that due to lack of flatness of the blocking tool according to the sign of curvature of each. A test-plate large enough to cover at least half the block is useful. If the wax is cleaned from a central area of the blocking tool, enabling a reflection to be obtained therefrom, the Angle Dekkor may be used to measure the (mean) error of parallelism to an accuracy of about 10 seconds.

Since the errors referred to above are those that can easily occur in ordinary working it will be seen that errors of parallelism may occur up to $54+40+10=104$ seconds $=1'\,44''$ unless care is taken to get the tools flat to a higher accuracy, using a large test-plate ; and in that case a more severe mechanical test can be obtained by using thin plates of mica instead of tinfoil. If the measurement of parallelism be made not by the Angle Dekkor but by an interferoscope, the parallelism can be measured and corrected to within 1 second of arc.

To sum up, in order to produce block work of a high order of parallelism, say 30 seconds, it is necessary not only to observe the most scrupulous care in sticking the plates down, but also to ensure that the blocking tool, smoothing tool or tools, and the polisher forming tool, are flat to a very high order of accuracy, and to provide an interferoscope for checking the parallelism.

Plates of glass can be laid down on a glass tool and fixed by wax, as described above, to an accuracy of about 4 seconds. The Angle Dekkor enables the top surface to be controlled to within about 10 seconds. It is therefore worth while devoting a good deal of care to getting the tools flat and in contact. If they are corrected, and the surface polished flat enough to result in a 35 second error, the procedure would assure a parallelism to within about 50 seconds, or if meticulous care were taken to get the tools in contact and flat, one might get down to 30 seconds accuracy straight from the block. A slight increase in parallelism may be obtained by using an interferoscope (§225) instead of the Angle Dekkor, and a small portable interferoscope (Fig. 166) can easily be made which can be used without moving the blocks from the machine.

Optical contact methods

187 Some increase in the accuracy of parallelism detailed above may be attained by the use of optical contact instead of wax for holding the plates on to the glass tool. In order that the plates may go into optical contact they must be very flat and absolutely free from grease and dust of any sort; when laid on the glass tool under their own weight a symmetrical system of Newton's fringes should appear at once, and unless the plates are thin for their diameter they need to be flat to a fraction of a fringe on the side already worked when they have settled down to a temperature uniform with the piece to which they are to be contacted.

The following is a well-tried routine for contacting—

Use an old soft handkerchief, fresh washed in distilled water and little handled. Clean the surfaces of the plates successively with petrol, methylated spirit and distilled water in that order. (Nitric acid is found helpful sometimes.)

Choose a dustless bench under a mercury lamp.

Finally smear the surfaces of the plates with absolute alcohol and wipe off with as few rubs as possible without pressure, because pressure deposits fluff. When a polished surface is perfectly clean it feels slippery.

Place the two surfaces together at the corners and when the Newton's colours indicate that the surfaces are close, slide the plates into position, the surfaces floating on the thin film of air in between, and contact should then take place with extremely slight force.

"Osman Linette" (a long fibre dust-free fabric), if obtainable, is valuable for cleaning surfaces preparatory to contacting. Personally I use an old linen handkerchief which has been washed in distilled water. This is kept in a tin with a loose fitting lid into which is fitted a camel hair brush. In this way both handkerchief and brush are kept free from dust.

A useful means of getting dust from between two such surfaces has been found by Mr Winterflood of Hilger Division of Hilger & Watts Ltd. A piece of fresh tissue paper whose edge has been cut with sharp scissors is placed between the two surfaces and withdrawn. This nearly always entirely removes every piece, however small, of loose dust. If, as the tissue paper emerges from between the two plates, a little piece of its edge is torn off, it can easily be removed, as it is quite visible, by sliding the one surface over the other until the piece of tissue paper moves over the edge.

The handle of the dusting brush is used to press the plates closer and closer to the glass tool, until the appearance of a uniform greyish tint indicates that the plates are too close for the first order interference colours to show. A strong pressure in the centre of the plate will then cause the surfaces to touch in the molecular sense, and the reflection from the interface will disappear where contact is made,

The rest of the plate should go into contact without much persuasion. When all the plates are in position a thin layer of shellac varnish is painted round the edges of the plates to seal them against the ingress of water which would quickly break the contact.

It has been found that well-cleaned blocks of glass brought together with small jets of distilled water playing on the surfaces can be kept free from particles of dust that would prevent the close contact of the pieces. The water film thus trapped disappears in the course of a few hours, sometimes much more quickly. The glass used was a borosilicate crown and the blocks were held together under moderate pressure in a suitable clamp during the process. The process was not developed for optical contacting, but might be worth trying.

After polishing, the plates are easily removed by inserting a safety razor blade under the chamfer on the lower face.

The Twyman effect

There is a phenomenon which sometimes becomes of importance in working the plates and of great importance if the plates are very thin. This was observed and described by myself (Twyman, 1905) and is as follows—

> If a very thin piece of glass, grey on both sides, be polished so that it can be examined through its thickness and polarised light be applied, stress can distinctly be seen near each surface; if now one side be polished strain disappears from that surface, and the glass blows up (*sic*); on polishing the remaining grey side, the glass becomes parallel again, and strain entirely disappears.

The reader will probably guess that the word " blows " was a misprint for " bows " and was meant to indicate that the plate takes a convex form on the ground side. The mistake was an unfortunate one because I have seen a reference to the phenomenon in which it states that the glass explodes! The phenomenon was described 44 years later by E. May (Kreidl, 1950). The name of " Twyman effect " was given to this phenomenon by F. W. Preston (Preston, 1921).

Correction of angle and flatness

188 Until recent years the correction of angles, parallelism and flatness to the highest perfection was the exclusive province of the skilled hand polisher. The demand which arose during the recent war for large numbers of optical parts to a very high standard of definition caused us to develop a machine for the finishing of single elements to the required accuracy. The final design of this machine is shown in the illustrations (Figs. 121 and 122).

One feature of this type of machine which departs from the conventional design is the ring R introduced by Mr Underhill, Head of the Optical Department at Hilgers, in 1937. This rides on the polisher and serves to retard the local deformation of the polisher

Fig. 121—Detail of Hilger single surface machine for high accuracy

and also to transmit to it pressure from the rocker arm which alters the general shape of the polisher at will. The rocker arm and other distinctive features of the machine were designed by myself.

At the bottom of the ring R is a shoulder about 1 mm thick, on which is laid a brass ring cut to suit the shape of the workpiece. This

Fig. 122—Detail of Hilger single surface machine for high accuracy

is the only restraint to which the work is subjected, and the fact that the forces are applied very close to the surface of the polisher ensures that there is no tendency for the work to tip.

The heavy ring is moved to and fro over the polisher surface by the hinged frame F, which is itself driven from the backshaft; this latter is actuated by two cranks so that the two sine motions are superimposed on it, a slow one carrying the frame right across the polisher, and a smaller and much more rapid one which simulates the action of a skilled hand worker touching up a flat surface on a polisher much larger than the work.

On the top of the work (in Fig. 123 a parallel disc P) is laid a pressure block B, cushioned with a sheet of sponge rubber; alternatively the plate can be attached to the block B by pads of pitch, as shown

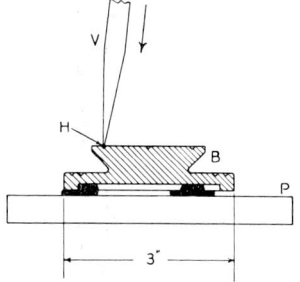

Fig. 123—Mode of attaching parallel plate to pressure blocks

in the figure. In the upper surface of the pressure block are drilled a number of small recesses into which the point of the weight rod V can be placed. If the point is in the centre hole, then the angle will remain fairly constant for long periods, and the parallelism of the plate is corrected to any desired precision by moving the point of the rod V to a position above the thick part of the plate, as shown by the interferoscope (q.v.).

The function of the rocker arm A is to control the shape of the polisher, and so the figure of the glass surface, and operates as follows. We will suppose that after running the disc for an hour or so with rocker arm swung back out of use, the test-plate discloses that the disc is three fringes concave. This means, of course, that the polisher itself is a corresponding amount convex.

The cam G is inserted as shown with the concave side uppermost, and the rocker arm and weights lowered into position.

The rocker arm will now operate so that when the frame carries the ring to the left-hand edge of the polisher, the right-hand weight will be bearing on the ring, and vice versa.

The centre of the polisher will thus be subjected to more weight and friction than the edge, and will slowly collapse the few fringes necessary to become flat ; while this is taking place the disc will be following the shape of the polisher, and will itself become flat at the same time.

Some experience is necessary to determine the amount of weight to be used on the rocker arm ; if the rate of alteration of polisher shape exceeds the rate at which the material can be polished away, a double surface will result, whereas if the weights are too small the process will be too slow.

If a preliminary test shows the polisher to be already too concave, the cam G is reversed, and the weights will then dwell on the outer zones of the polisher only.

Despite the relatively crude construction of this machine, consisting as it does of fairly massive pieces of iron and brass, the accuracy that can be attained is remarkable. During the war many hundreds of 4 in. diameter parallel windows flat to less than a quarter of a fringe and parallel to less than 1 second were produced by women with no previous experience of glass polishing.

By the provision of suitable fittings, prisms can be worked in the same way to very close tolerances for angle, and it has even been found possible to polish shallow concave surfaces to a sufficient accuracy for use as grating blanks for diffraction gratings.

One small observation may serve to illustrate the ideal conditions which may be obtained with care and patience. It is rarely that a surface polished by hand is quite free from " edge turn-down," and many ingenious ways have been devised to render this harmless, such as the edgings of flats after flattening has been completed, leaving the edge perfectly sharp. We have found that if the polisher is in good condition there is every chance that the surface will exhibit an edge slightly turned *up*, and we attribute this to the effect of temperature, in that the edge of the glass is free to radiate heat and is therefore cooler than the main body of the disc. As the disc acquires a uniform temperature the edge expands, thus producing the effect described.

The turned-up edge may be removed by replacing the disc on the polisher for a very short time, say one minute at a time, so that the difference of temperature has not time to build up.

The localizing of prism surfaces for correction of definition, being essentially an individual operation, will be dealt with in the course of the chapter on the Hilger interferometers.

CHAPTER 10

NON-SPHERICAL SURFACES

(A part of this chapter is translated from a paper contributed by the Author to the " Réunions d'Opticiens " held in Paris from 14th to 19th October 1946, printed in *Revue d'Optique*, **26**, 461).

Introduction

189 Although spherical surfaces are the most easy to make, yet they will not in general, without correction, result in the formation of a good image. Fig. 43 shows the way rays are refracted through a bi-convex and a bi-concave lens respectively.

Lens systems can be computed in which, by the use of different kinds of glass, the ordinary purposes of most optical instruments can be achieved by the use of spherical, or nearly spherical surfaces, but certain developments during the past thirty years or so could not have taken place without the use of surfaces departing from sphericity by amounts not to be attained by ordinary retouching.

It is supposed that Descartes was the first to consider what kind of surfaces would result in freedom from aberration. In his discourse *La Dioptrique*, published in Paris in 1638, he gives the geometrical construction for a lens free from spherical aberration, so simple and elegant that I will give it here in translation. He says in the section headed *Des figures que doivent avoir les corps transparents pour détourner les rayons par réfraction en toutes les façons qui servent à la vue* (" Concerning the shapes which transparent bodies should have to refract rays in every way serviceable for vision ").

All the shapes of which I have to speak here will be composed solely of ellipses, hyperbolas, circles or straight lines. The ellipse, or oval, is a curve that mathematicians elucidate to us by cutting across a cone or a cylinder. I have also seen it used sometimes by gardeners in making their flower beds, where they draw it in a way which, though crude and inexact, makes its nature better understood than the section of a cylinder or cone.

They plant in the ground two stakes (Fig. 124), the one at H, the other at I, and having knotted together the two ends of a cord they pass it round them as at BHI ; then, putting the finger in this cord, they follow round about the two stakes, keeping it uniformly stretched, and thus trace on the ground the curved line DBK, which is an ellipse.

If through the point B (Fig. 125) one draws the straight lines LBG and CBE which cut each other at right angles, and of which the one LG divides the angle HBI into two equal halves, the other

CE will touch this ellipse in this point *B* without cutting it, of which I will not give you the proof, " because geometers know it already and others would only be bored to hear it."* If one draws from

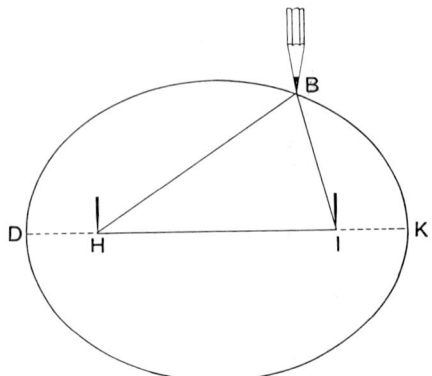

Fig. 124—The gardener's ellipse

this point *B* outside the ellipse, the straight line *BA* parallel to the major axis *DK* and equal in length to *BI* and draws from the points *A* and *I* to *LG*, the two perpendiculars *AL* and *IG*, these will be in

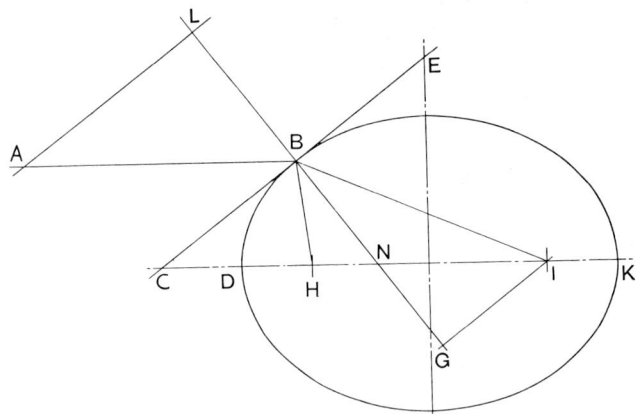

Fig. 125—Aplanatic refraction at an elliptical surface

the same proportion to each other as *DK* and *HI*. Hence if the line *AB* is a ray of light, and the ellipse *DBK* is the section of a solid transparent body through which the rays pass more easily than

* " Pour que les géomètres la savent assez, et que les autres ne feraient que ennuyer de l'entendre."

through air, in the same proportion as the line DK is greater than HI, then this ray AB will be so deflected at B that it will travel towards I. And since the point B can be selected at will, at any point of the ellipse, therefore all the rays parallel to DK which fall on the ellipse will be deflected to pass through the point I.

After giving the proof of this property of the ellipse of focusing parallel rays Descartes continues—

Because every ray which is directed towards the centre of a sphere suffers no refraction, if with centre I (Fig. 126), one draws a circle BQB of any desired radius, then the lines DB and QB turning about the axis DQ will describe the shape which a lens should have to focus, in the air, at the point I, all the rays which shall have been parallel to the axis before falling on the lens.

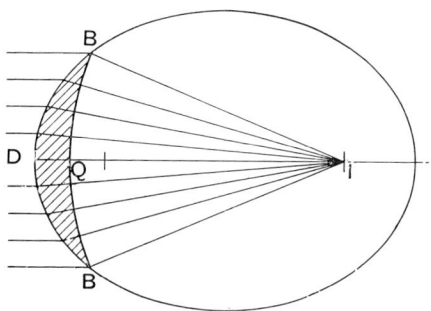

Fig. 126—Aplanatic lens with one elliptical surface

You will no doubt have been struck by the phrase strange to our ears about rays passing " more easily " through the solid transparent body than through air. He does not say " faster " and I have seen it stated that he did not believe light to have a finite speed. Galileo did, and had already tried unsuccessfully to measure it by getting two observers at a distance to signal to each other by flash lamps, but the idea was not generally accepted until Bradley, nearly a hundred years later, explained the astronomical phenomenon of the aberration of light by its finite speed, which he measured by observation of that phenomenon.

So just what Descartes thought is obscure; none-the-less, his geometrical construction is founded on the correct law of refraction, so that we have, at this early date a description of a lens which with one aspherical surface is free from aberration for rays parallel to the axis; the first geometrical design for an aplanatic lens, one, that is which is free from spherical aberration.

326   PRISM AND LENS MAKING

Later he describes the hyperbola and the use that can be made of it in dioptrics. He even describes a machine for producing lenses with hyperbolic surfaces; a machine which I believe could be made to work by adopting modern methods of precision engineering.

It must be realised in considering these passages that there is no hint of any experiments having been made by Descartes nor of his being aware of any experiments made by others on the subject of refraction.

The law of refraction which he assumes in this way, but which is better known to us under the form $\sin i / \sin r = n$ is generally acknowledged to be due to Snell, although Descartes makes no mention of that fact, and it was only published after Snell's death, from manuscripts left by him.

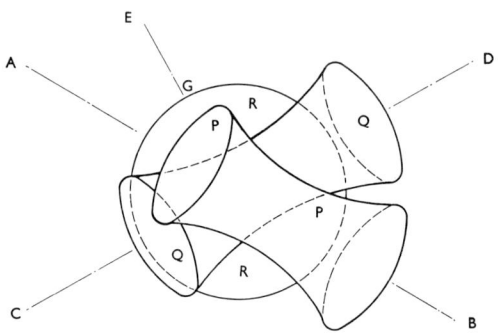

Fig. 127—Wren's suggested engine for hyperbolic surfaces

One must not leave un-noted Dr (After Sir Christopher) Wren's suggestion of an engine for grinding hyperbolical optical surfaces (Wren, 1669). He says (see Fig. 127)—

Let there be three bodies P, Q and R fit for grinding of which let P and Q be equal and of the shape of a pillar and R resembles a lens. Let P have a rotation about the axis AB, Q about CD and R about EG. Let AB and CD be in different planes but so posited that EG being produced may be at right angles both to AB and CD. Lastly let the bodies approach to each other as much as necessary; still however presenting the same situation and position of their axes.

I say that by the rotation and mutual attrition of these bodies, new geometrical figures will arise of which P and Q will be equal hyperbolical cylindroids and R an hyperbolical conoid, given both in species and magnitude.

I should say that quite half the work on non-spherical surfaces from 1920 to 1939 derived from this work of Descartes. W. B. Rayton (1923) refers to the advantages of Cartesian surfaces (*i.e.* elliptical, hyperbolic or parabolic) and gives examples of the application of these to
- (a) The elimination of astigmatism of oblique pencils in spectacle lenses of high positive power.
- (b) Condenser lenses for motion picture projection where the high magnification and long projection distances make it impossible to get sufficient and uniform illumination without very high consumption of electric current.
- (c) Lenses for motion picture photography.

The author says—

With the increasing experience we developed more and more accurate methods and machinery until we felt justified in extending the application of such surfaces to larger lenses and to lenses requiring very accurate surface figure.

The author gives no indication of the methods used for producing them.

Recently a new method of grinding and polishing parabolic surfaces has been described by Dourneau and Demarcq (1949). This is based on the single properties of paraboloids. The surfaces obtained so far are sufficiently accurate for use in eyepieces where they are very useful for correcting astigmatism. They can also be used in objectives wherever precision greater than a few wavelengths is not required.

The position in 1918

190 To come now to modern endeavours to exploit the use of non-spherical surfaces.

A patent by Carl Zeiss (application dated 27th April 1899) deals with lens systems, characterized in that good definition is obtained not only on the axis or for a relatively small field, but also for images well off the axis, and involving the use of non-spherical surfaces in the place of rigidly spherical, or nearly spherical, surfaces. It says—

When in a lens system of any particular design, as for example a photographic objective or a telescope eyepiece, the images of the different parts of the field are formed by passages of the light through various parts of its aperture, then, if the surfaces are accurately spherical the various aberrations for the oblique pencils, such as distortion, curvature, astigmatism, and coma, if not corrected can, at least as regards one of them, and at least as concerns a particular distance from the axis, be corrected by non-spherical deformation of one lens surface ; and in general two, three or more of these errors can be removed at the same time by the correction of two, three or more of the surfaces.

The patent reveals no method of making such surfaces.

Later Zeiss took out a further patent (U.K. 199,013, Convention date 10th June 1922) in which they claim—

Photographic objective consisting of three lenses separated by air distance, of which the two outer ones are collective and the middle one dispersive, characterized by the feature that the last surface of the objective is deformed.

The grant was opposed, with what success I do not know.

[I ought to say here that if, as has been stated of British patents in general, only about one in ten could be maintained in the Courts if attacked, this must certainly be true of patents concerning lenses. By a kind of tacit agreement, however, we respect the claims of others, hoping that they will do the same for us. After all, what is the use of destroying the patent of someone else at a possible expense (if it should be carried to the House of Lords) of £1000 or more, with the risk of retaliation if we are successful?]

Fig. 128—Formation of aspheric lens by subsidence into a mould

In 1908 Zeiss applied for a patent (British Patent 3444 of 1909) for producing aspherical lenses by subsidence. In this process they take a spherically shaped lens, put it on a mould the surface of which has a radius suitable for the purpose of approximating the upper surface to the shape desired (Fig. 128a), heat the whole until the lens subsides into the mould (Fig. 128b), and after cooling, rework the under surface spherical again. A second claim of the same patent is to use a mould which is a suitable non-spherical surface of revolution. It would be interesting to know to what extent this method was used; it would appear to give one a considerable choice of shapes. Some short while after the 1914-1918 war I had occasion to examine an ophthalmoscope made by Zeiss which had non-spherical lenses in it which to my mind might well have been produced by this method. The aberration in axis amounted to about seventeen wavelengths, quite good enough for the purpose, and identical lenses were sold by Zeiss as condensers on a raising and lowering tripod stand for the price of £3, so that the lenses must have been produced very cheaply.

Gleichen (*Theory of Modern Optical Instruments*, translated by Emsley and Swaine, 1918), gives a number of interesting notes on applications of non-spherical surfaces. For example, Schwarzschild stated general conditions for the aplanatism of reflecting systems, following which Siedentopf explained a particularly interesting special case. He showed that for parallel incident rays, aplanatic image-formation can be effected by reflection at a cardioid in conjunction with reflection at a spherical surface (1905).

With regard to the uses of aplanatic systems, Gleichen says—

The employment of aplanatic, semi-aplanatic or Cartesian systems might be adopted more widely than heretofore with the further development of the technique of glass working, and it is within the bounds of possibility that along these lines a revolution in technical optics lies before us. It may be remarked that the fulfilling of achromatism appears to offer no insurmountable difficulties. If, for example, aplanatic lenses made of varieties of glass of equal refractive indices but different dispersion are combined, the aplanatism is not disturbed, whilst the conditions for the fulfilling of achromatism are not restricted, as we have already explained.

He proceeds to give some then recent applications of these principles. For example—

In searchlight systems of two separated convex lenses, a greater spherical correction is produced by the use of deformed surfaces.

and

. . . in complex condensers for projection purposes deformed surfaces are likewise used.

These examples given by Gleichen were embodied in German patents, the numbers of which are wrongly cited in the English translation.

But our purpose here is not to describe *why* we should make non-spherical surfaces, but *how* to make them. Nobody, for more than 200 years, seems to have followed up Descartes' work by actually making lenses of the types he describes, although it became generally appreciated that object glasses of telescopes were subject to errors arising from the spherical figure of their surfaces. It was not until the time of Newton that any other form of aberration was thought to be important, but Newton observed that the chief fault in lenses was the *chromatic* aberration, and since he believed that the dispersion of all refractive substances was proportional to their refractive indices he did not conceive the possibility of making an achromatic object glass. Thus he made no attempt to form surfaces which would correct the spherical aberration of *lenses*, but, to avoid the chromatic aberration, turned his attention to the reflecting telescope. Among those which he made was one of 6 inches focal length which he presented to the Royal Society in 1671. The formation of a parabolic surface was first achieved in 1732 by Short who made reflecting telescopes with parabolic and elliptical figures, presumably by retouching. These were called "Gregorian" telescopes after James Gregory who in 1663 made it clear that spherical surfaces produced aberration but that conic sections could correct this, which, as we have seen, had been shown earlier by Descartes.

According to Sir David Gill (Article "Telescope" *Enc. Brit.*, 11th Ed., 1910) Chester Moor Hall, in 1733, was the first to make an achromatic telescope. Simultaneous correction of chromatic and spherical aberration in lenses with spherical surfaces was due to Dollond (about 1758) after he had independently also made an achromatic telescope, the corrections being obtained by annulling the aberrations caused at some surfaces by opposite ones at other surfaces.

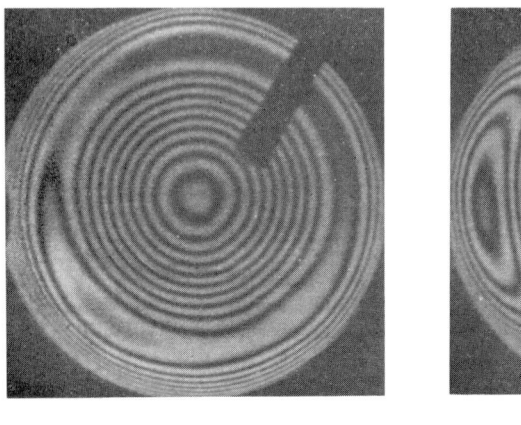 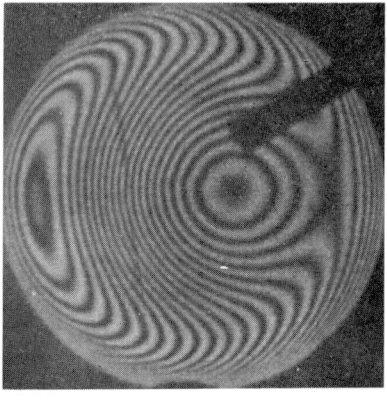

(a) (b)

Fig. 129—Degree of figuring required to render aplanatic a simple quartz lens of f/7

But, with the exception of reflecting astronomical telescopes with parabolic surfaces produced by local polishing (retouching or figuring as it is called), nothing seems to have been done in producing lenses with non-spherical surfaces before the advent of photography. Then, for a long time, attempts to make photographic lenses of very large aperture led to very complex designs which yet failed to achieve the desired perfection.

Figuring was thus used to perfect the images of telescopes, but this is very different from trying to make strongly aspherical surfaces.

To give an idea of the degree to which such surfaces may require to depart from spheres, here is an example (Figs. 129a and 129b). These photographs are showing the figured surface of a lens which was figured

* I have found no support for the claim that John Hadley, inventor of the sextant, parabolized mirrors for his telescopes as early as 1724.

by hand on one surface until it was aplanatic. These Newton rings have been obtained by putting in contact with the figured surface the proof sphere to which the lens was worked before figuring (in Fig. 129b the proof sphere has been slightly tilted). To produce this amount of departure from sphericity is laborious, but quite practicable, provided one has a pattern to work to. But this is a comparatively mild example of figuring compared with what requires to be done today for certain purposes.

For example, the correcting plate of a Schmidt telescope may have to depart as much as one-half millimetre from the true surface and some idea of what is entailed by this may best be shown by attempting oneself to polish in a flat plate a hole a few rings deep. I have found in retouching that, with a polisher one inch in diameter, one hundred circular strokes per square inch of surface remove about one Newton's ring (equivalent to 1/100,000 inch).

A half-millimetre would, then, correspond to two thousand Newton's rings, requiring about 200,000 strokes. The difficulty is not in merely removing that amount of glass, but in doing so to an accuracy of one or two rings at each part of the figured surface. To do this figuring with any certainty, the optician must have means of testing the lens to see where and how much to polish off, and for the final stages of this the interferometer is indispensable.

The way of using the interferometer is described in Chapter 12.

I shall describe later how one approaches this final stage of testing by the interferometer, which should not be attempted until one has got within, say, two dozen rings of the desired shape. When that stage has been reached, the procedure is as follows. First of all a test is made on the Twyman and Green Interferometer, marking the high places and rubbing them down. I say " marking the high places " but what one does is not quite so crude as that ; one makes a copy or a photograph of the appearance on the interferometer and uses it as a chart. It is, of course, a contour map of the amount by which the lens is too thick or too thin in particular places.

It is possible, although the method has not been adopted by lens calculators at all widely, to determine by calculation the departures from sphericity which are desirable in any given type of lens, and Mr. Perry, the Head of the Applied Optics Department at the Hilger Division of Hilger & Watts Ltd, has been doing this for a number of years for certain instruments made by that firm.

The position of the subject in 1918 was described by me in a lecture before the Photographic Society in 1918. I said—

In spite of the great advances in the technique of optical manufacture during the past thirty years, the definition of actual optical systems still leaves ample room for improvement. Even the best of photographic lenses are very defective when judged in relation to the theoretical limit of resolving power. This failure of definition is sometimes condoned by the makers of photographic lenses on the plea that the definition is as good as is justified by the grain of the photographic plate. This is not true, and furthermore, the plate makers might reasonably maintain, and actually do, that the grain of the plates is as fine as is justified by the definition of the image given by the lens.

The imperfection of definition is largely due to the fact that with the choice of dioptric materials available the results desired cannot be achieved by the utilization of spherical surfaces alone. Among the most interesting of new resources is the ability to produce, so far only in a very limited way, non-spherical optical surfaces. A beginning has been made by the firm of Carl Zeiss in the production of lenses with definitely non-spherical surfaces, best known of which are the special spectacle lenses designed by von Rohr, for those whose eyes have been operated on for cataract. This firm has patented a very ingenious method of producing such surfaces, and is thought to be keeping other methods secret (see British Patent 3444 of 1909). I am not aware that any lenses with non-spherical surfaces have been made by them with sufficient perfection for telescopes, camera lenses and so forth, but experience shows that with the interferometer the optician can face, without dismay, the task of making with precision quite considerable departures from the sphericity of his surfaces. It is therefore very likely that valuable results may be obtained by using apparatus of this kind to correct with high accuracy lens systems wherein definitely aspherical surfaces have been generated by machines essentially different from those ordinarily used by opticians.

In abstracting the foregoing passage I have altered the wording in order to compress it a little. The point was that if a machine were designed which would give the surfaces approximately the desired shapes, the interferometer provided a means of finally correcting them by retouching.

It seems to be the general experience that whatever means of mechanical grinding and polishing are used, the final retouching must be done by hand or by a machine simulating hand polishing, if lenses of high quality are required.

Some developments from 1918 to 1948

191 Three lines of work, then, were suggested in my lecture of 1918.
  (*a*) The production by machine or other device of aspherical surfaces

to as high an accuracy as possible ; this work had already been initiated by Zeiss.

(b) The retouching of lenses, either with spherical surfaces or with definitely aspherical surfaces, under the guidance of the interferometer.

(c) The calculation of lenses in which it was assumed that the surfaces might be non-spherical, as a preliminary to producing them in the ways I have just mentioned.

It has not been possible until recently to ascertain what has been done by various firms in these directions but, by the courtesy of Messrs Taylor, Taylor and Hobson and Messrs Ross Ltd, I am able to give some account of what they have done, and some other information on the subject that has also come to light through papers in *J.O.S.A.*, *Proc. Phys. Soc.* and elsewhere. I shall therefore now give a sketch of some of the various methods which are being used, including some in use or in course of development at Hilgers.

The work now going on falls into five divisions—
1. Retouching, in which handwork is replaced by machining, successive lenses of the same kind being therefore repeated with a minimum of trial and error.
2. Machine production of conic surfaces.
3. Machine production of surfaces more complex than conics.
4. Subsidence methods.
5. Smoothing out irregularities arising in any of these methods.

I shall make no endeavour to arrange the descriptions under a rigid classification.

In recent years lenses in the shape of hyperbolæ have become of interest in connection with the improvement of binoculars, the hyperbola enabling a wider angle of vision to be attained.

Messrs Ross Ltd before the 1939–45 war had already made lenses with hyperbolic surfaces for use in high-aperture lenses for television. A hardened-steel cone was sectioned to form a hyperbola, the cone being accurately ground and the sectioned surface also being accurately ground. The edge formed by the intersection of the plane surface with the cone was used as a guide for a pantograph and the movement of a roller around this edge was reduced in the proportion of 5 to 1 to move a diamond-impregnated wheel, rotating at high speed, which generated a reduced replica of the hyperbola on the lens which was to be worked, the lens, of course, rotating about its axis the while. Smoothing was effected by using a thin flexible sheet of brass fed with loose emery which was moved to and fro perpendicular to the axis of the lens. When the smoothing was fine enough polishing was produced by a polisher consisting of pads of pitch backed by a flexible support such as sorbo, so that polishing was evenly distributed over the whole surface.

This machine was dismantled during the war to make room for more urgent work but, I understand, has been reinstated for the production of moulds for making such lenses in plastics.

Messrs Ross Ltd intended also to employ the same mechanism to make the eyepiece lenses of binoculars with hyperbolic surfaces, thus obtaining a wider field of view. As we now know Zeiss, about 1926, had built and was using a machine specially designed for producing non-spherical surfaces for the eye-pieces of binoculars (§208).

Adam Hilger Ltd tried several methods. For a number of years the lenses on many of their instruments have been corrected by hand retouching. Where wide departure from a spherical or plane surface was required the retouching was usually effected by machine, using grinding and polishing pads supported on a flexible backing.

The lens under correction was mounted on a vertical spindle and the polisher had the usual kind of cross motion; the stroke could be varied and set to the desired zone. The polisher consisted of a number of small pads, separated from each other, with a backing of soft rubber, so that each little section of the polisher could follow the contour of the surface being polished. Having by a preliminary series of experiments determined the amount of polishing and the kind of stroke necessary to get the required shape, the process could be repeated with fair accuracy.

When a lot of material had to be removed, as, for example, in a Schmidt plate, it was necessary to resort to grinding, which was done with the usual loose emery on the same machine, and using similar pads for the process. Such laborious methods were excusable for a firm like Hilger's where perhaps only one lens might be needed for an instrument that might cost several hundred pounds. More expeditious, or more certain, means are now employed by Hilger & Watts and by others. Some of the more promising methods will now be described.

Figuring mirrors by the controlled deposition of aluminium

192 Strong and Gaviola (1936) describe a method of figuring mirrors by the controlled deposition of aluminium in vacuo. A number of sources of aluminium are so arranged that, without a screen, a coating of uniform thickness would result; then a screen is designed and cut out of a sheet of metal and placed close to and in front of the surface to be modified. Fig. 130 shows the apparatus and some of the screens used. The screen is rotated by an electro-magnet situated outside the vacuum chamber.

NON-SPHERICAL SURFACES 335

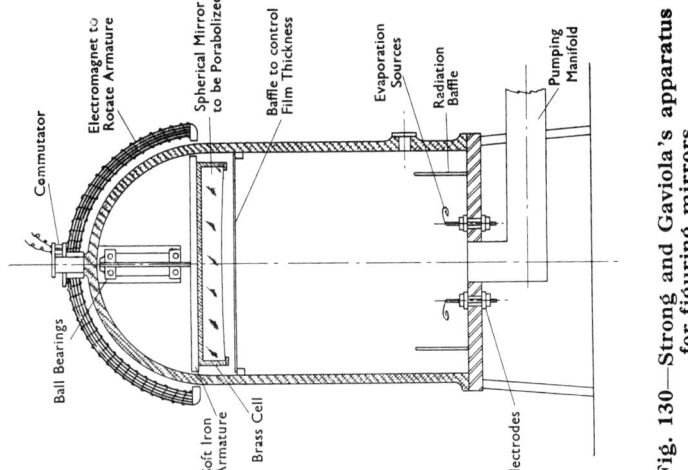

Fig. 130—Strong and Gaviola's apparatus for figuring mirrors
Some of the screens used are shown

The first trial at parabolizing a spherical mirror was made using a 12⅜-in. spherical test mirror of Pyrex glass of 152¼ in. radius of curvature. The surface of this mirror was true to a sphere within one-twentieth of a wavelength of green light ($=5 \times 10^{-5}$ cm). Fig. 131a is a focograph at its centre of curvature taken with a Foucault knife-edge on the testing bench. Fig. 131b shows the appearance of the spherical mirror at the mean focal distance, taken with the help of a testing flat. Both represent the mirror before any correction was attempted.

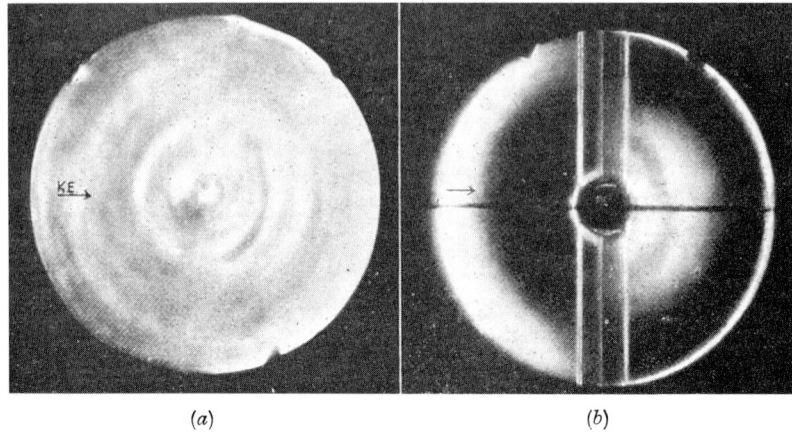

(a)   (b)

Fig. 131—12⅜-in. uncorrected spherical mirror under Foucault's test
  (a) Tested at the centre of curvature. Notice the zones at 0·37− and 0·79− radius from the centre. Their elevation is less than 1/20th of a wavelength, but the use of contrast film and paper has made them conspicuous.
  (b) The same mirror tested at the parabolic focus. The 0·37 zone is visible, the other has disappeared.
  In these, and in the following focograms, the direction of movement of the knife-edge is indicated by an arrow.

The maximum thickness of deposit required for parabolizing was 2·6 wavelengths. This thickness could not be obtained in a single operation. It was, therefore, necessary to reload the tungsten coils several times with aluminium. The deposit thus obtained consisted of films of aluminium separated by layers of aluminium oxide which formed spontaneously when the vacuum was destroyed to reload the coils. Fig. 132a, b, c, reproduces focograms of the parabolized mirror, taken with the aid of a testing flat, for three typical cases: In one the amount of aluminium deposited is just right, in another it is too large and the mirror is "overdone," in the third it is too small and the figuring is "underdone." It is interesting to compare Fig. 131

with Fig. 132 : it can be seen that the more prominent " zones " of the parabolized mirror were already present in the spherical glass disc ; they have not been introduced by the deposition of aluminium. These zones are of heights smaller than one-twentieth of a wavelength. It is to be remembered that the zones will appear deeper in the present case since there are two reflections from the mirror as compared with one at the centre of curvature.

In a second paper (*ibid*, **26**, 163) Gaviola describes a systematic use of the Foucault knife-edge test so that, by refinement of the measurements a survey of such mirrors as described in the previous paper could be obtained with an accuracy better than 1/20-wavelength with visual observation and with photographic measurements carried up to 1/100-wavelength.

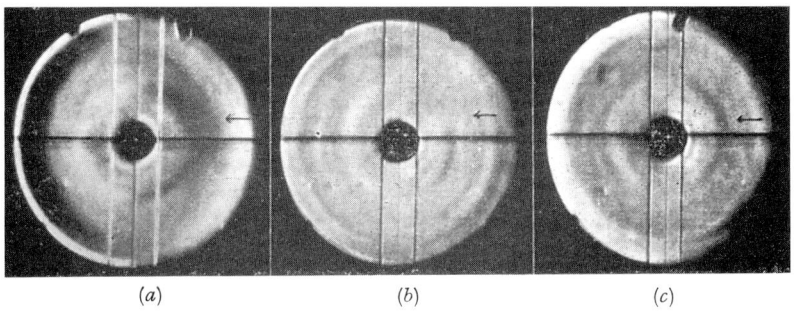

(a)   (b)   (c)

Fig. 132—The mirror of Fig. 131 after parabolizing
(a) Undercorrected. Insufficient aluminium deposited.
(b) In this amount of aluminium deposited is just right and the figure is a good paraboloid. The 0·37 and 0·79 zones both appear.
(c) Overcorrected. Too much aluminium deposited.

The paper describes errors to which the Foucault knife-edge test is liable, especially the one produced by parallax.

Figuring lenses by controlled deposition of transparent substances

193 Adam Hilger Ltd, some years ago, drew up a scheme for treating lenses in the same way by coating with a transparent substance and embodying an interferometer arrangement whereby control of the deposition could be obtained during the process of coating (Fig. 133). The project is described fully in U.K. Patent 569,046/42.

Subsidence or sagging of plates or of lenticular shaped discs

194 This method is closely related to that by which Schmidt made his first corrector plates (see §201).

Plates or lenticular discs are supported on an accurately made ring within an electric furnace and heated to such a temperature that they sag either under their own weight or with the aid of a vacuum. Calculation shows that a great variety of shapes can be obtained in this way, some of which are of utility in the manufacture of camera lenses, in particular of Schmidt plates. The method, the development of which is proceeding at Hilger & Watts Ltd under Twyman, Weinstein and Winterflood, seems to present some advantages over the subsidence method of Zeiss.

A disc of optical material, *e.g.* glass, is ground and polished to appropriate spherical shapes on either side, the calculations being

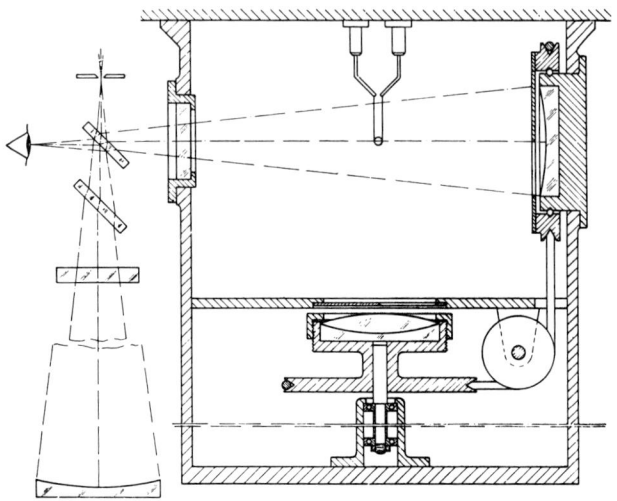

Fig. 133—Deposition of transparent material
The Hilger apparatus for interferometer control

made by one of us (Weinstein) and is then supported near its rim upon an accurate ring of heat-resisting material (Fig. 134).

Both disc and ring are heated until the yield range of the glass is reached and it begins to sag. When the disc has sagged the necessary amount, the heating is discontinued. Cooling is allowed to take place at such a rate as not to introduce strain in the disc. One of the surfaces is then polished flat or spherical as indicated by the calculations.

The method differs from that described in the Zeiss Patent No. 3444/1909 in that the glass sags freely instead of coming into contact with a mould surface, and the amount of sag is controlled by timing the heating.

By a suitable choice of the radii of curvature and thickness of the original disc, the temperature, the time of heating, and the radius of the supporting ring, it has been found possible to produce aplanatic lenses required for one of the Hilger & Watts instruments.

The process is applicable to thin or thick plates, mirrors, or lenses.

Fig. 134 shows a meniscus of glass arranged to produce an aspherical surface on the underside suitable for use in a Schmidt camera system.

Perry and Weinstein universal aspherical grinding and polishing machine
195 This machine can produce *any* kind of aspherical surface which is mathematically continuous. It was designed by J. W. Perry and W. Weinstein, and is giving good results in one of the optical laboratories of Hilger & Watts Ltd (U.K. Patent No. 615,649/1946).

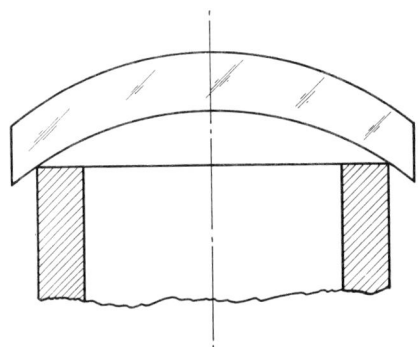

Fig. 134—Formation of non-spherical surfaces by sagging

The lens or mirror is first ground spherical to approximately the curvature required and is mounted on the rotating spindle of the machine. The machine (Fig. 135) produces upon the previously ground surface an aspherical surface as the envelope of the path of the meridian section of a rotating spherical tool, which may be the grinder or the polisher and which rotates about an axis passing through its centre of curvature and is moved in the required manner by the aid of a double pivot linkage. The linkage is controlled by two cams and is based upon a rotating lever MP which has pivots at L and P, the space between the pivots L and P along that lever being a variable factor, so that in the limit, in a simplified form, L and P can coincide so that there is only one pivot. The first member of the linkage pivoted at L, a variable position on the lever, swings about a fixed centre, O, at its farther end. The second pivoted member bears against cam No. 1, which is fixed, the lever being controlled in position at its end remote from the second pivot by cam No. 2. The two cams are so positioned that the lever and the first and second members are all made collinear simultaneously. This takes place on the axis of symmetry of the lens, which is, of course, the axis of the lens spindle.

The axis of rotation of the spherical tool is in rigid connection with the second pivoted member, *PR*, and is inclined thereto and so arranged that the axis of the second member passes through the centre of curvature of the tool. The motion of the linkwork is in its own plane and extends from the position of collinearity of the lever and pivoted

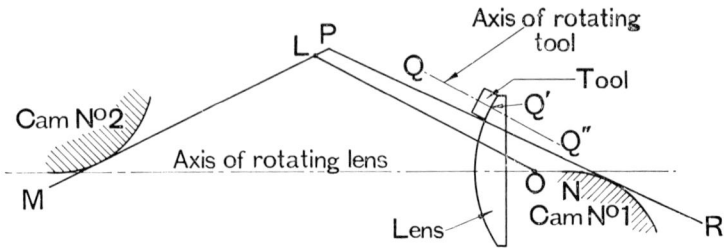

Fig. 135—**Perry and Weinstein universal aspherical machine**
Set-up for convex surfaces

members above referred to, as far as it is requisite on one side of the lens axis.

It may be pointed out that the first cam may be the evolute of the aspherical curve so generated, but this is not in general a desirable condition for operation. The constants of the machine may be

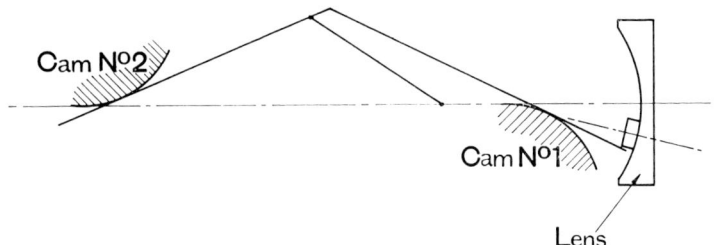

Fig. 136—**Perry and Weinstein universal aspherical machine**
Set-up for concave surfaces

adjusted for any given asphericity so that the required aspheric curve is obtained and the wear on the tool is most suitably distributed.

One arrangement is shown in Fig. 135.

OL and *MP* are linked at *L*, and *O* is the fixed centre; *RP* is linked to *MP* at *P*, and is kept in contact with the cam shown; *QQ"* is the axis of the tool and *MP* is controlled in position by Cam No. 2. Cam No. 1 similarly controls the position of *PR*.

The alternative form of arrangement for producing concave curves is shown in Fig. 136.

In special cases P may coincide with L and cam 2 become inoperative, or the pivots may be arranged in a different order from that here indicated.

Machine for production of non-spherical surfaces by reduction from a cam
196 One may mention, though it was never actually pushed to success, another method that seems fairly promising and was the first actually to be tried at Hilger's, some 30 years ago. This machine was designed by Dowell and the author, and may, I believe still come into the running as one of the approved types.

The lens roughing machine (Fig. 59) made by Adam Hilger Ltd about 1916, was so modified that the grinding lap A was moved in the direction between its axis and B in the following manner. The wheel was mounted on the short end of a lever, and a roller, on the corresponding long end of the lever, pressed against a cam of appropriate shape. Such cams can be made by hand to an accuracy of something like one-thousandth of an inch and the ratio of the short to the long end of the lever was 1 to 50. It seems possible that with care such a machine could be made to give a surface accurate to about one-fifty-thousandth of an inch and that by the now accepted practice of using flexible smoothers and polishers one could attain the accuracy required for high-quality camera lenses.

It must be remembered that the accuracy of surface needed for camera lenses is much less than for, say, telescope or binocular objectives. The difficulties of getting rid of the aberrations for oblique pencils simultaneously with those of the axis results in there being always a considerable number of wavelengths of aberration in the outermost parts of the field.

*Production of elliptical surfaces by Messrs Taylor, Taylor and Hobson**
197 The method depends on the following property of elliptical surfaces. ABA, Fig. 137, is an ellipse, ACA the circumscribing circle. Any line TT is drawn tangential to the ellipse and on to this a line SS is drawn perpendicular to it through focus F, meeting it at the point P. Then the property of the ellipse referred to is that the point P lies on the circumscribing circle. If then we arrange a flat-faced grinding wheel the face of which contains the tangent TT and rotate this wheel about the axis RR perpendicular to the face, of course ; if, further, we fix the distance between O and P by a link, and arrange

* Published by permission.

that the line *FP* is always perpendicular to *TT*, then if we move *O* about in whatever freedom of motion is left to it while rotating the piece of glass we are grinding about the axis *AA*, we shall generate an ellipsoid of revolution. Surfaces made by the application of this principle are used by Taylor, Taylor and Hobson in making condensers for 16 mm cinema projectors.

The Schmidt camera in astronomy

198 The following passages (§198 to 201) are abstracted by permission from the 38th Traill-Taylor Memorial Lecture delivered 2nd

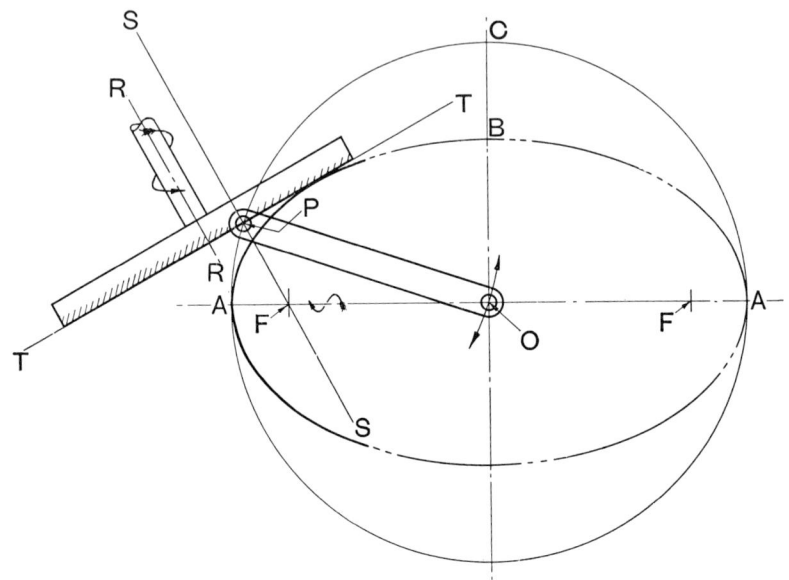

Fig. 137—Taylor, Taylor & Hobson machine for producing elliptical surfaces

December 1947 by E. H. Linfoot, *Photographic Journal*, 1948, 88B, 58.

The lecture gives descriptions of various improved modifications of the Schmidt camera. The following extracts will give the reader a general idea of the nature and importance of this new type of lens system.

It is convenient first to consider one or two basic properties of ordinary spherical concave mirrors and of paraboloids. When a parallel beam of light is reflected from a spherical concave mirror, as shown in Fig. 138, the rays near the axis *CA* are brought to a focus *F* at a distance from the mirror surface equal to half its radius of curvature. However, the rays reflected from the outer parts of the mirror aperture meet the axis CA slightly nearer to the pole A of the mirror surface than those reflected from the parts near the axis. The rays of light arriving from a star may be regarded as parallel when they reach our telescopes; and a spherical mirror held with its axis

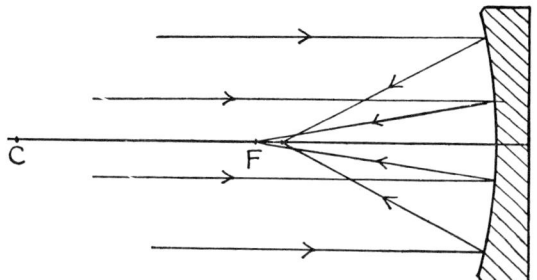

Fig. 138—**Image formation by a spherical mirror**

pointed directly towards a star will form an image which is not quite sharp, but suffers from the defect known as spherical aberration.

Ordinary reflecting telescopes therefore do not use spherical mirrors, but paraboloidal ones, to receive the parallel rays. When a corrected paraboloidal mirror is pointed directly towards a star, the rays are all brought sharply to a single focus (see Fig. 139). The outer parts of

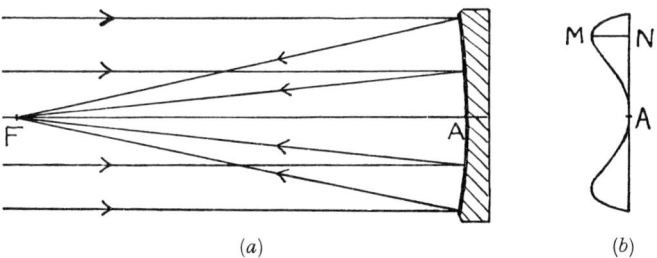

Fig. 139—**The paraboloidal mirror**
(a) Image formation
(b) Difference between paraboloid and sphere.

the paraboloid are in fact slightly bent back or "turned down" relatively to those of the nearest sphere, and this causes a lengthening of focus in the rays reflected from the outer zones of the mirror surface, while the focus of the inner zones is slightly shortened (Figs. 139a and b). The amount of the discrepancy between the paraboloid and the sphere is very small; for a mirror of diameter 12 inches and focal length 60 inches its maximum only amounts to $2 \cdot 3 \times 10^{-5}$ inch.

Nevertheless it makes a big difference to the performance of a 12-inch f/5 Newtonian telescope (Fig. 140), which consists of a paraboloidal primary mirror M_1 together with a diagonal flat mirror D to bring the image into an accessible position F^1 at the side of the telescope tube. The operation of converting a spherical mirror into a paraboloid is known to telescope-makers as " parabolization " or " figuring." The greatest gap between a paraboloid and that of the sphere which touches it at the centre and meets it at the edge is given to a good approximation by the expression

$$\frac{f}{4096F^4}$$

where f is the focal length of the mirror, d its diameter and F the focal ratio f/d.

If we try to photograph a field of stars by placing a photographic plate at the prime focus F of a Newtonian telescope mirror (Fig. 140), the images close to the centre of the field are very good, but those farther out show an unsymmetrical, comet-shaped form ; the optical

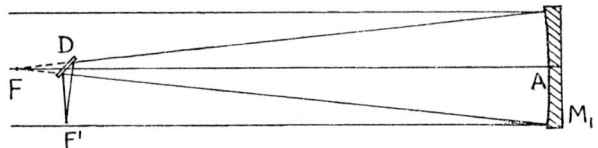

Fig. 140—Newtonian telescope

system suffers from " off-axis coma." This coma not only upsets measurements of position, but also results in many faint star-images being lost altogether, through the intensity of their diffuse image-patches falling below the threshold sensitivity of the photographic emulsion.

The Schmidt camera

199 A new era in astronomical camera design began in 1931, when Bernhardt Schmidt introduced the new type of optical system which bears his name. Schmidt (see Plate 4) was a remarkable personality. Born in 1879 on Nargen Island in Esthonia he attempted as a boy to make a concave mirror by grinding together the flattened lower parts of bottles with sea-sand from the beaches of Nargen. Later, he studied engineering in Götenburg and Mittweida and, about 1900, began to make telescope mirrors for amateur astronomers. His mirrors were soon recognized as outstandingly good ones and in 1905 he created a sensation with an $f/2·26$ paraboloid of 40 cm aperture, made for the Potsdam Astrophysical Observatory, which in the perfection of its figure far surpassed other astronomical mirrors of that time. Schmidt carried out all his work with the simplest means, and did his polishing and figuring by hand—with his left hand, indeed, for he had lost his right arm in an accident in early youth. He was not only a supreme master of the difficult art of mirror making; he was also an enthusiastic astronomical observer. During the last part of his life, he was a voluntary worker at the Bergedorf Observatory near Hamburg, and it was there that he invented, constructed and put into use his new camera.

The novel idea embodied in the Schmidt camera may be explained as follows. As we have seen, the image formed by a spherical mirror of a star situated at a great distance on its axis is imperfect, because the rays from the outer zones of the mirror come to a shorter focus

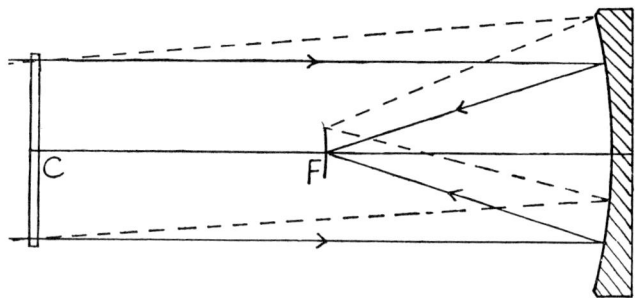

Fig. 141—Schmidt camera

than those from the inner zones; the image suffers from "spherical aberration" (Fig. 138). This aberration can be cured by subjecting the mirror surface to the deformation shown in Fig. 139b. The effect of the deformation is to correct the spherical aberration by slightly lengthening the focus of the rays reflected from the outer zones of the mirror, while it slightly shortens the focus of rays reflected from the zones near the centre (Fig. 139a). Now there are other ways of producing this effect besides figuring the mirror. We could, for example, place a plane-parallel plate of glass (C, Fig. 141) in the incoming beam and "deform" one of its surfaces by optical figuring so that the rays passing through the outer part of the plate were made slightly divergent, while those passing through the inner parts became slightly convergent. To secure this, the surface profile of the plate would have to be made convex near its centre and concave near its edge. Since it operates by refraction instead of by reflexion, the geometrical deformation of the plate surface would need to be about four times the parabolization-deformation of the mirror surface; for an f/5 mirror of 12 inches aperture it would amount to about $1 \cdot 5 \times 10^{-4}$ inch.

So far, it is not apparent that anything has been gained by carrying out the figuring process on the plate instead of on the mirror. In both cases, the effect is to bring sharply to a focus the rays of a parallel beam of light entering the system parallel to its axis. But when we consider, besides this on-axis pencil, the parallel pencils of rays which, entering the system at various angles with its axis, form images in the outer parts of the field, the advantages of the aspheric corrector plate become clear. For when the plate is placed at the centre of curvature C of the mirror (as in Fig. 141) then an off-axis pencil will be focused very nearly as sharply as the on-axis pencil, *i.e.* very nearly perfectly. In fact if we were to turn the plate square on to any particular off-axis pencil, this pencil would be focused perfectly. The defects in the

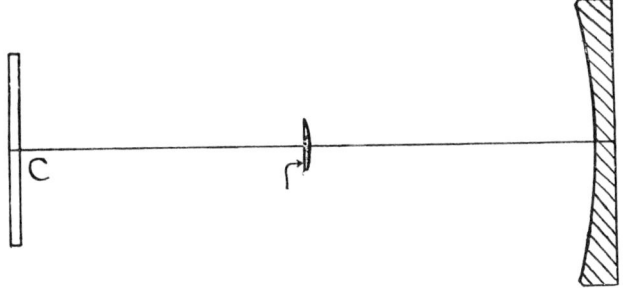

Fig. 142—Field-flattened Schmidt camera

off-axis images are therefore due to the fact that the plate does not lie squarely across the off-axis pencils; and since the plate is very nearly plane-parallel, the effect on the images of this lack of squareness is very small.

Thus the system consisting of a spherical mirror and a figured corrector plate located at its centre of curvature will give good images over a wide field. This system is the Schmidt camera; its off-axis aberrations are small even at very high apertures and are of a symmetrical character.

Modified Schmidt cameras

200 In some of the large Schmidt cameras now in use in the U.S.A. thin glass photographic plates replace the cut film; these can be bent to the required curve without breaking and spring flat again when removed from the camera. Nevertheless the curved field is a serious inconvenience, and it was not long before modified designs were suggested in which a flat image field was obtained at the cost, however, of some loss of image quality. The simplest of these modifications consists in the addition to the system of a plano-convex lens of suitable curvature placed immediately in front of the photographic film, which compensates the curvature of the system (Fig. 142).

Under the stimulus of the results obtainable by the Schmidt camera many modifications have been introduced to solve special problems. Some of these are given in Dr Linfoot's lecture ; the reader may also find further information on the subject in Linfoot (1949).

It is possible to make achromatized corrector plates in the manner shown in Fig. 145.

Solid Schmidts

Still faster cameras for spectrographic work are provided by the solid Schmidt system, shown in Fig. 143 in which the space between corrector-plate and mirror is filled with a solid block of glass. The solid Schmidt, constructed in glass of refractive index n, is n^2 times faster than the ordinary Schmidt of similar dimensions, and effective speeds of f/0·35 are practicable in small sizes. Its astronomical

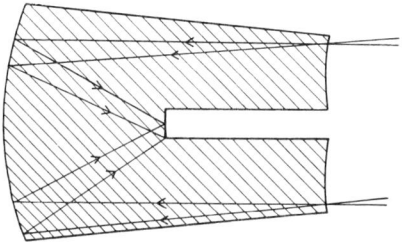

Fig. 143—Solid Schmidt system

usefulness is restricted to spectroscopy, since the initial refraction into the glass is accompanied by dispersion, resulting in a chromatic difference of magnification. The glass is cut away behind the field surface to allow access to the photographic film, which is oiled on to the spherical field surface.

200.1 A paper (mathematical) by Linfoot and Wolf (1949) discusses the problem of designing the aspheric correction plate of a Schmidt camera so as to obtain the best result over the field taken as a whole.

Corrector plate techniques

201 Schmidt made his first corrector plate in a simple and ingenious way (Hodges, Paul C., 1948 ; *Amer. Jour. of Roentgenology*, 59, 122). He cemented a thin plane-parallel glass plate to a metal drum, as shown in Fig. 144, so that its border was supported by the flat rim of the drum, and pumped out some of the air in the drum. This caused the plate to sag by a few hundredths of an inch, and the distorted plate was then ground and polished to a near-flat spherical curve in the ordinary manner; the dotted line in Fig. 144b shows the new plate surface. Finally, air was readmitted to the drum and the plate removed; as the strain in the plate was removed, its upper surface took on the shape shown in Fig. 144c. Schmidt's knowledge of elas-

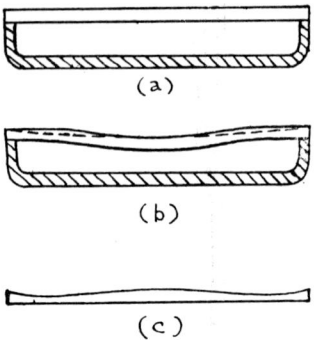

Fig. 144—Schmidt's method for making corrector plates

ticity theory enabled him to foresee that the final form of this surface would be a sufficiently good approximation to that needed for the corrector plate of his camera.

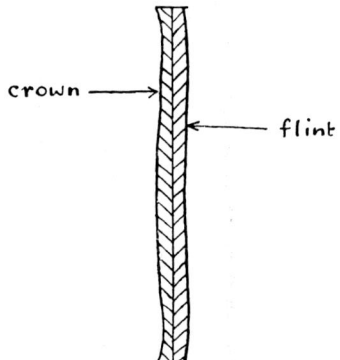

Fig. 145—Achromatized corrector plate

Schmidt's method has rather fallen into disuse, and a more modern technique for making corrector plates of moderate size may be of interest. A plane-parallel glass plate is mounted in a rotating holder and ground with successively finer grades of carborundum by means of a flexible lap, which is traversed backwards and forwards with a rather short stroke as the plate rotates. The working surface of the lap is built up of lead facets, cemented with gold size to a disc of sponge rubber about ⅜-inch thick. The facets do not cover the whole of the rubber disc, but are arranged in petal-shaped areas (see Fig. 146) so that the effective grinding area which works on each zone of the plate is proportional to the depth of glass to be removed from that alone.

Another method of making correction plates, that seems likely to be useful, is described by Brockman (1947). The author says that as it is, he believes, new, and since it is largely the work of

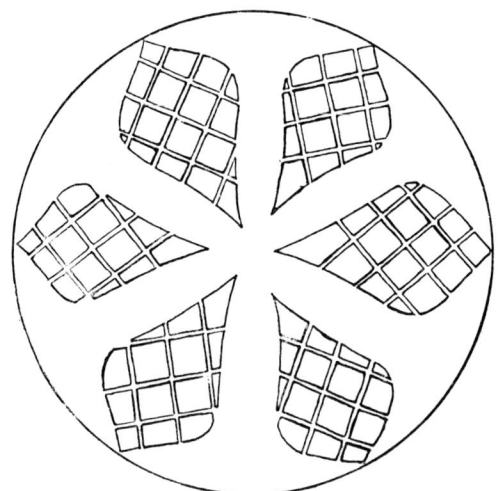

Fig. 146—Flexible lap for grinding corrector plates

Mr. R. J. Beaven and Capt. P. Wojciechowski he refers to it as the Beaven–Wojciechowski method. In this method the approximately plane-parallel disc which is to become the correcting lens (and by this method, the thinner the disc the better) has cemented to it a series of concentric iron rings as shown in Fig. 147. These rings differ slightly in thickness so that the locus of their upper surfaces forms a reverse of the desired Schmidt lens surface. This assembly is inverted and placed glass face uppermost on a magnetic chuck. Regulated vertical pressure is then applied to the plate, until it has been so deformed that the lower faces of the rings all touch the surface of the chuck, which is then switched on. Next the deformed upper surface of the plate is ground and polished flat. Finally the magnetic chuck is switched off (precautions being taken to avoid a sudden release of the stresses in the glass) and the rings removed from the plate. Most grinding has taken place in the zones where the rings were thicker and we have, if the ring thicknesses have been correctly chosen, the aspheric surface of the Schmidt correcting lens. Variation in the figure of the lens may be obtained by the use of suitable packing between the rings and the chuck, or by reducing the thickness of the rings.

Use of Schmidt cameras and telescope systems in astronomy

202 Interest in non-spherical lenses was greatly stimulated by the

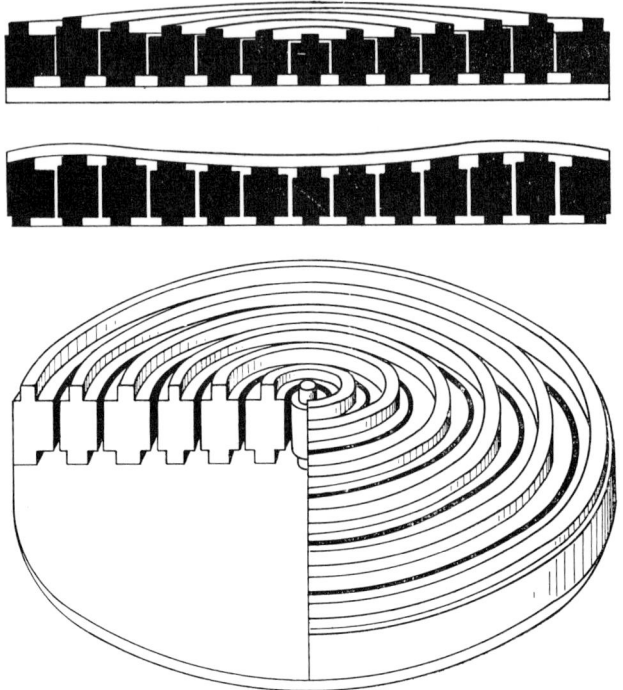

Fig. 147—Beaven–Wojciechowski method for making Schmidt plates

introduction of the Schmidt system, which is coming into widespread use in the big observatories. For example, at Mount Palomar a systematic photographic survey of the whole area of sky visible from the observatory is in progress. The survey is being made with the observatory's 48-in. Schmidt camera, which is designed to record on a single photographic plate an area of sky seven degrees square. This is more than 1,200 times the area covered by a single exposure with the observatory's main 200-in. telescope, which was designed primarily for stellar spectroscopy and only secondarily for direct photography of the sky.

Apart from its own special investigations, the Palomar Schmidt camera will act as a guide to the most effective use of the 200-inch telescope. The usefulness of such a partnership was demonstrated two years before the war by the smaller 18-in. Schmidt camera, also at Mount Palomar, working in conjunction with the 100-in. telescope at Mount Wilson. Up to 1937 only three examples had been observed of supernovae, the type of star—usually in a distant nebula on a chance distribution—which flares up to a brightness of about 40 million times that of the sun and emits a shell of incandescent gas with enormous velocity.

Search with the Schmidt camera led to the discovery of three more within a year, one of which enabled Dr M. L. Humason at Mount Wilson to obtain the first satisfactory spectrum of such a star.

A Schmidt camera of suitable size can thus do valuable work in discovering faint objects for the larger telescope to examine, and also on its own in tracing out the faint edges of nebulae and other photometric work. It can also be used, with an objective prism, or grating, for spectrographic work.

One drawback, not a serious one in practice, is that it is necessary to use photographic plates made to a specified curvature characteristic of the particular Schmidt camera. Another is that the tube-length is twice the focal length of the system.

Schmidt's original camera had a mirror 17 in. in diameter. One of 60 in. diameter is under construction for the observatory of Harvard University.

The Schmidt camera in commercial work

203 Messrs Taylor, Taylor and Hobson are using Schmidt cameras for television sets. They are using corrector plates in two types of system ; one a Schmidt, the other a dioptric system. At the present time they are making these plates by mechanical retouching. A sheet of thin rubber is covered over part of its surface with small diamond-shaped or square-shaped pads of brass to form a kind of mail-clad flexible surface. This is used as a grinder, and is applied by a frame to a grinding machine of the ordinary vertical spindle type. The reciprocating arm moves the frame to and fro over a selected range, the pad being appropriately weighted to bring the pressure on to the point where most material needs to be ground off. The best weight distribution is found by experience. The machine not only saves labour, but makes it possible, having once established the routine, to repeat the desired surface at will. This arrangement is almost identical with that used at Hilger's and by Dr Linfoot in his experimental work on Schmidt-Cassegrain cameras.

The surface is tested during the grinding process by removing it from the machine; placing upon it a flat surface inclined to it by a slight angle, this being made by resting the flat plate on three little blocks of glass of suitable thickness, and allowing steel balls such as those intended for ball bearings (selected to be of exactly the same diameter) to roll down between the two surfaces so that they lie in a curve which indicates the shape of the surface (see Fig. 148).

(This method of test is Dr Linfoot's variant of the one due to Roger Hayward of California.)

The grinding process is continued until the balls assume the desired distribution between the two surfaces. This test can be applied in the grey.

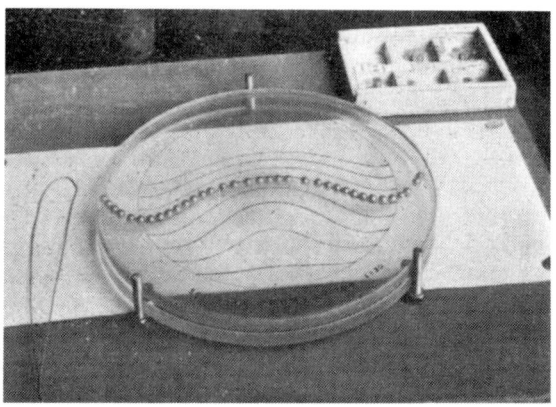

Fig. 148—Steel ball test for control of grinding corrector plates

The same process is carried through with the polishing, except that in place of the rubber a piece of fine canvas is used and in place of the brass pads, pads of pitch (although Linfoot, *loc. cit.*, says that beeswax-coated asphalt pads are better as being more elastic.)

The Schmidt lens is then mounted to form the complete Schmidt telescope system and a test of definition made in a way to be described later. Further correction is carried out if necessary and when the plate is giving good results a master plate is made to fit it exactly. This plate is then used instead of a flat plate to apply the steel ball test, and the balls should lie in a straight line when any subsequent Schmidt plate is made and tested in this way. The advantage is, of course, that it is easier for the eye to detect a departure from straightness than the departure from a specified curve.

Testing Schmidt plates is carried out in three stages. In the first the balls are used as described; in the second a star image is observed, a Schmidt telescope being used to examine a small light reflected in a very small steel ball. The balls for this test have to be very small, about a millimetre in diameter, and I believe these are obtainable at present only in Switzerland.

Schmidt projector systems are required to produce not an image of a distant point but a conjugate image of an object at a finite distance—actually the cathode ray tube of a television outfit. The knife-edge test is therefore applied to the image at the desired working distance. The knife edge is moved side-ways by a fine micrometer screw and the reading on the vernier of the screw is taken as the dark places appear when the observer looks at the object glass, or I should say the field of the telescope system in the ordinary way applied in the knife-edge test.

A refined and quantitative application of the knife-edge test has already been referred to (§192).

Curvature of field in Schmidt systems

204 As has been mentioned elsewhere, the curvature of field in Schmidt systems can be corrected by the addition of an extra lens, but Messrs Taylor, Taylor and Hobson prefer not to use such a lens in the systems they are building for television since it increases the aberrations; moreover the end of the cathode ray tube can readily be given a curvature appropriate for the production of a flat field in the projected image. They test these systems on a curved field in the following way. The image of a light source falls on to a concave surface of the appropriate radius of curvature on which are placed a number of small steel balls stuck to the surface with black paint. This is used with the complete telescope system and with it an image of the small images of the light source formed within the steel balls is projected on to a flat surface. This enables them to see the definition of the outermost parts of the field and the distance between the Schmidt plate and the remaining part of the system is adjusted until the best definition is obtained all over the field.

Schmidt and other aspherical systems as spectrograph cameras (R. Minkowski, 1944)

205 Small chromatic aberrations, large field of good definition and, above all, short focal ratio make Schmidt systems particularly suitable for spectrograph cameras. The outstanding disadvantages—spherical focal surface and loss of light due to the position of the plate in front of the mirror—play different roles in cameras of large and small focal lengths. For focal lengths above 30 inches, the photographic plates on thin glass can be bent to the necessary curvature without great risk of breakage. For cameras of shorter focal length, curvatures necessitating the use of film can be avoided by the introduction of a field-flattening lens. A plano-convex lens placed close to the focus has proved completely satisfactory for cameras of relative apertures f/3 and f/1·5. The use of flat plates without field flattener restricts the field of good definition to an angular extent of 114·6 (d/A) $\frac{1}{2}$-degrees (d permissible diameter of the circle of diffusion, A diameter of the collimator); this is sufficient for many purposes. The loss of light due to occulting by the plate can be avoided in smaller cameras by an off-axis construction. This is completely satisfactory for cameras of large focal length and relatively small field where the plate can be bent to the curvature of the focal surface. Field flattening lenses are usable, but the aberrations introduced by them will be those corresponding to the normal system of larger aperture of which the off-axis system is part. Consequently, the aberrations will be relatively large and the usefulness of field-flattening lenses with the off-axis construction will be restricted to the relative apertures of f/4 and smaller. The use of flat plates in off-axis systems results in a sidewise curvature of the spectrum on the plate, with a radius of curvature $kf2/A$; the factor k depending on details of the construction will usually be between 1 and 1·5. Such a curvature is objectionable. As the use of film generally reduces the accuracy of wavelength measured and is, therefore, equally undesirable, off-axis systems of short focal lengths are of restricted usefulness. For cameras of very short focal lengths, the small scale of the spectrum permits the use of small plates, thus minimizing the loss of light in a normal system, but a range of focal lengths below 30 inches will always remain, for which the full advantages of Schmidt systems are not available without some loss of light.

It should be remembered that the basic idea of Schmidt was to remove the coma due to a spherical mirror by a diaphragm in the centre of curvature of the spherical lens. The author may remind the reader that a plano-convex lens placed close to the focus with the plane side towards the photographic plate (§200) may be used to remove the curvature of the focal surface. This was mentioned by Schmidt but has been widely overlooked (see Ross, 1940). Minkowski concludes: " The definition obtained with a well-made Schmidt system is unsurpassed."

Hilger & Watts Ltd use a Schmidt camera on one of their spectrographs.

A paper by Cojan (1947) describes several Schmidt systems, for the cameras of spectrographs, with very large apertures.

Decentred aspheric plates

206 A paper by Linfoot (1946) deals with the effects of errors of centring on the optical performance of plate–mirror systems. Such errors may occur during the aspheric grinding of the plates and also in the lining up of the system. It is first shown that if a grinding technique is used which builds up the asphericity at a proportionately equal rate all over the surface, the result of a large number of small centring errors at different stages of the grinding is very nearly equivalent to a simple decentring of the asphericity on the surface, combined with a small amount of primary astigmatism. Next, general formulae are obtained for the effects on the Seidel errors of a centred system of decentring and tilting its components, and it is shown that in the case of a plate–mirror system the formulae can be used to estimate the practical tolerances for disturbances of this kind. Lastly, the general formulae are applied to discuss the lining up of a two-sphere one-plate Schmidt-Cassegrain camera.

I make no mention of the other numerous publications on the subject which have appeared since 1923, principally in *J.O.S.A.* because they are very easy to find by means of the excellent indexes in that Journal, and in any case it is very rarely that indications are given by the writers of the methods and apparatus used.

Other Schmidt systems

There are various patents for modified Schmidt systems, among which the following may be mentioned on account of the main claim according to which the corrector plate is correctly spaced from the concave mirror by struts, the whole being composed of thermo-plastic synthetic resin and being joined together by means of cement.

In the space between these struts a cathode ray tube face can be placed. The system has been devised for television projection. *U.K. Patent* 630,653/47 *Greenwood, Liddell and Imperial Chemical Industries.*

Machine used by Carl Zeiss, Jena, to grind aspherical lens surfaces*

General

207 Zeiss constructed five identical sets of machines which were used to grind and polish aspherical surfaces on lenses. A considerable number of lenses were made by Zeiss on these machines and used by them and other German firms. The lenses fall mainly into three

---

* Official report C.I.O.S., Item No. 9 file XXVII–24. This description is included with the permission of the Controller of H.B.M. Stationery Office.

groups; those of the lowest precision are for reticle and scale illumination systems, next are lenses for eyepiece systems, *e.g.* rangefinder and binocular oculars. The last and most precise group consists of objective system components. No objective components were manufactured by these machines for military instruments, but one of quite high performance was made in quantity for a 2½–power ophthalmic instrument.

Schmidt corrector plates up to 150 mm diameter were also made on this machine. Correction by hand was still needed after polishing, but only one-hundredth of the time usually spent on hand correction was required.

Concave surfaces had not been made at the date of this report.

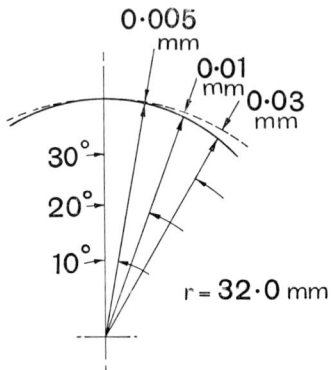

Fig. 149—Showing the amount of glass removed by the Zeiss machine from a typical lens

A rough idea of the amount of glass removed from a typical lens may be gained from Fig. 149. Glass is invariably ground from the outer area rather than the central area of the lens.

Polishing was performed on the common type of machine known as a " Stick machine " in America and as an " American machine " in Germany

Description of the Zeiss aspherizing machine—general layout (*Fig.* 150)

208 The grinding tool is an aluminium wheel rotated about axis B–B. The aspheric surface (mirror image) is cut into the rim of this wheel and maintained by means of a diamond point, 8, activated by a trueing machine. The work face is dressed before the start of the grinding operation and must be redressed (trued) after approximately six lenses have been ground. The area shown as p in Fig. 151 represents a channel 20 to 30 mm deep cut into the face of the machine and filled with wood pitch. Emery is brushed on to the pitch during grinding. Abrasive of grain size 15 millimicrons is used for the first three to five minutes (depending on the size of the lens) and 8 millimicrons for the final three to five-minute grind. The lens receives both grindings in place on the same machine, but is carefully washed with water at the change of abrasive.

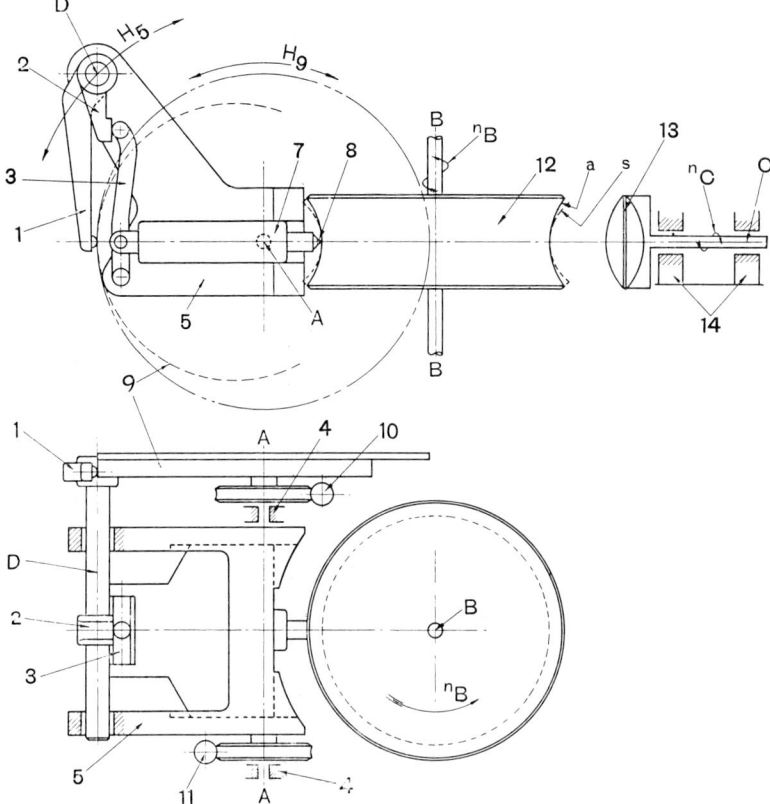

Fig. 150—The Zeiss machine for grinding aspherical surfaces

Mechanism for trueing the lapping wheel (Fig. 150).

The radius of the vertex curvature of the aspheric surface is set from point A. The axis of the trueing mechanism 5 passes through point A. The master aspheric cam 9 is mounted on a continuation of the shaft which supports the trueing mechanism. The master cam represents the aspherical surface desired on an enlarged cam (according to the gear ratio).

The trueing mechanism 5 swings about the axis A–A in the indicated direction $H5$ (may be moved either side) being driven by gear 11 which is indicated in the drawing as a wormwheel drive but, on the machine seems to have been a pulley and weight system.

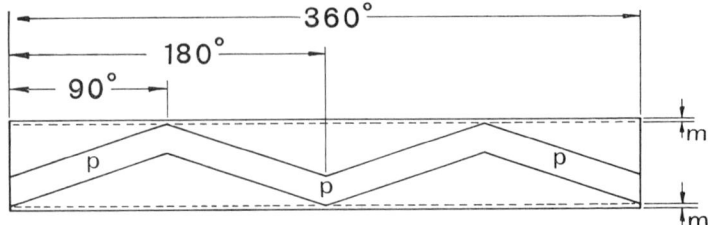

Fig. 151—Lapping wheel for smoothing on the Zeiss machine

Lever 1 slides by means of a ball point over the cam 9. The cam consists of a flexible metal strip (see Fig. 152) which may be set to the desired curvature by means of adjustable screws. Lever 2 is mounted on the same shaft as Lever 1 and moves lever 3 which in

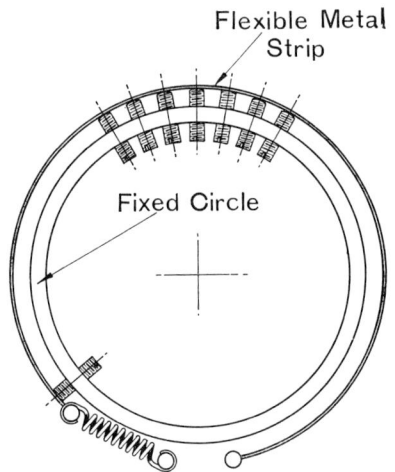

Fig. 152—Cam on the Zeiss machine

turn imparts the slope of the aspheric curve to the diamond tool 8 through the tool holder 7. This movement of the diamond cuts the mirror image of the desired surface into the tool 12. In Fig. 150 s indicates the radius of the spherical lens and a the aspherical curve produced in the tool 12 by the diamond tool activated by the lever mechanism.

Both the tool trueing mechanism and the grinding spindle C are mounted in the same slide. The different operations are set up by movement of the slide.

Polishing

The lens 5 (Fig. 153) remains cemented to the adapter which held it during grinding. The adapter is screwed into spindle A. The polishing lap, mounted in spindle B, consists of a metal body 4 to

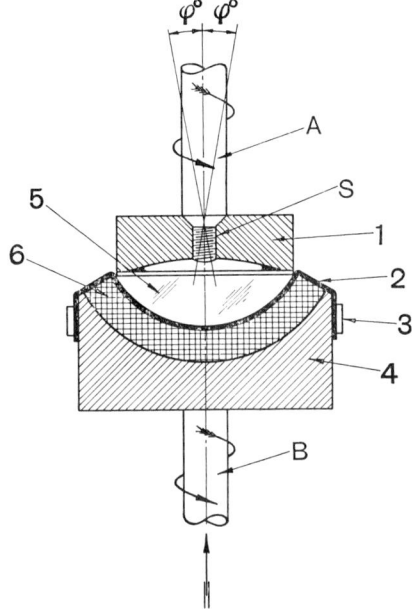

Fig. 153—Lens and polisher of Zeiss machine

which an elastic felt or rubber piece is cemented. For the more precise work, a polishing cloth 2 is clamped over the felt by means of ring 3. On the other lenses, part 6 is the polishing element.

209 These machines were designed and worked out during the years 1924 to 1927, and about that time Messrs Zeiss acquired from Adam Hilger Ltd a complete set of Hilger interferometers.

These interferometers were used by Zeiss in the workshops or factory for testing the lenses polished on the machine just described. The best of these lenses, approximately 2 in. diameter, showed about two rings aberration.

The time for grinding six ordinary lenses simultaneously was a few minutes only; polishing, which was done on ordinary polishing machines, took from one to one and a half hours, rouge being used. Each workman looked after 40 spindles, thus the output was very large.

The polishing was done with sorbo, which appears to hold the rouge quite satisfactorily. For the best lenses considerable care was

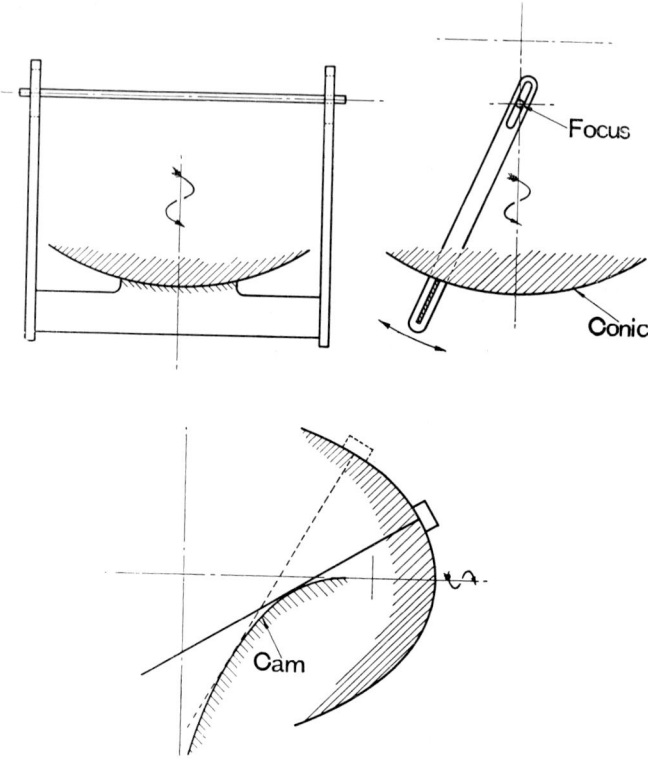

Fig. 154—Burch's angle-control aspherizing machine

taken so to mount the sorbo that it pressed evenly over the whole surface. The sorbo was cemented with shellac to a cup the shape of which was found by experience. A die of the same shape as the lens was used to press the sorbo into the cup during cementing. It was found that sorbo sheet could be selected sufficiently uniform in thickness and hardness to polish uniformly provided it were tested to avoid places of inequality. The pressure on the die was the same as that used in the polishing of the lenses themselves. It was only by taking considerable care in this mounting of the sorbo that the best surfaces were obtained. Most of the lenses made did not attain anything like the same accuracy.

Dr Burch's angle-control aspherizing machine*

210 The angle-control machine (Fig. 154) embodies a departure from the viewpoint that a figuring machine should be simply a mechanical means of carrying out the operations of hand figuring. The work is rotated on its axis and grinding or polishing is done with a small pad which is given a reciprocating stroke in the meridian plane, plus a " creep," *i.e.* a gradual change of the central position of the stroke, between adjustable limits so as to cover the whole surface (with some overhang at the edge) or a chosen zone, as may be desired. This pad

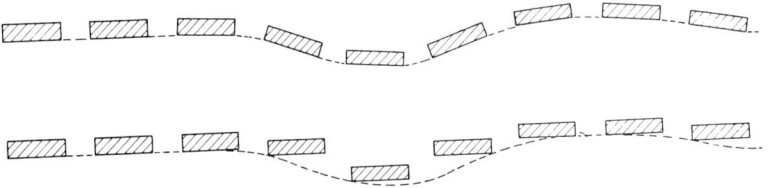

Fig. 155—The contrast between the action of freely-floating and angle-controlled grinding and polishing pad

is driven by a linkage which leaves it positionally free in a direction normal to the work, so that it may rest on the work under gravity or spring pressure. But the linkage exerts orientational constraint— " angle-control ", so that the pad is not free to tilt as it moves up the side of a " hill " on the work, but must lie nominally at, or more nearly at, the angle a freely floating pad would take were the hill absent. The respective actions of a freely floating pad and an angle-controlled pad are contrasted in Fig. 155 in which the meridian section of the work is shown " developed " as a flat with a high and a low zone.

It will be appreciated that the effect of angle-control is to prolong the action of the active zone of the pad on or near the tops of hills, and to reduce it on or near the bottoms of hollows. Thus although angle-control does not automatically generate the desired shape, it enhances the tendency of the machine to make smooth approximations to the desired shape.

* U.K. Patent 593759 of 24th October 1947 (assigned to Metropolitan–Vickers). I am indebted to Dr. C. R. Burch of the H. H. Wills Physics Laboratory of Bristol for this description of his machine.

The linkage should be such that the driving force is applied to the pad substantially tangentially to the surface of the work, so that changes in coefficient of friction do not affect the pressure of the pad on the work. Tangential drive, together with a sufficient approximation to "ideal" angle-control, may conveniently be achieved by mounting the pad on one end of a (sufficiently long) bar pivoted, at the other end, on an axis carried by a linkage in the required non-circular curve. (This is an approximation to, rather than "ideal" angle-control, as the pad orientation is affected by errors of height of the work. But it is not affected by errors of slope as such; and this is the important point.) Means are provided for testing the work by any of the well-known methods (knife-edge, interferometer, etc.)—preferably without removal from the spindle—so that the operator may readjust the stroke, creep rate and creep limits, until a sufficient approximation to the desired figure is achieved.

Many factors govern the performance of machines of this type, and it would be premature to try to define closely their scope. They have proved useful in the laboratory for aspherizing wide-angle (0·7 NA) singlets, used for microscope sub-stage condensers and similar purposes. In this application, the inevitable colour error renders the highest possible accuracy superfluous. The first application of the angle-control principle to work in which high precision was really needed took the form of a hand-operated angle-control device, used to remove highly localized error, not of revolution symmetry, from the concave mirror of the 0·65 NA reflecting microscope objective described in *Proc. Phys. Soc.*, Dec., 1947.

In this "figuring aid" the polisher consisted of a piece of soft pitch on the end of a narrow rectangular brass rod which was held in the hand like a penholder. The upper end of this rod was laid in a slot in a vertical shaft coaxial with the mirror, and the rod was gently pressed with the fingers of the free hand against an evolute cam, fixed in the shaft, which was free to rotate in a bearing well above the operator's hand. One could thus make highly localized strokes with angle control, and with two-dimensional freedom, so that it was practicable to confine the figuring to irregularly shaped areas of width as small as 1 mm or even less. The mirror could be rotated in its mount so as to bring the general area to be worked into a position convenient for the "pen-hand." The cycle of operations was: (1) Paint a pattern of rouge dots on an area the size of a postage stamp on the (8-cm dia.) mirror surface, (2) place the mirror on the interferometer and spread out the fringes, (3) sketch the high areas plus the dots, (4) transfer the mirror to the figuring-aid, and figure the areas shown in the sketch.

The use of the angle-control figuring-aid was found to speed up very considerably the removal of irregular " cobbled pavement " error from this (speculum metal) mirror—the gain in speed being of the order of 10 times.

It may be noted that, so far from making machines to imitate hand operations, the stage had been reached of helping the hand to imitate a machine operation.

Obviously the next step was to conduct the whole aspherizing operation on an angle-control machine ; and this is one of the developments being followed up by a small group in the H. H. Wills Physics Laboratory of Bristol University, under a grant from the Nuffield Foundation.

CHAPTER 11

TESTING OPTICAL WORK

Introduction

Surface marks

211 A lens or prism should be well polished, and free from bubbles and other defects in the glass, scratches and other surface marks. Faults such as these are detected by simple inspection, the lens being held in a strong light, from which the examiner's eyes are screened, and with a dark background.

The surface defects occurring in optical glass work may be due to:

A. Scratches occurring in the roughing or trueing process not removed by the smoothing.

B. Deep grey left from the early grades of emery; this can be distinguished by being coarser than *C*.

C. Uniform fine grey due to insufficient polishing.

D. Sleeks. These are strong at one end and trail away to vanishing point at the other end. They are usually due to the polishing pitch being too hard for the temperature of the shop. Hence, if the shop is very cold in the morning, there is no harm in gently warming the polishers provided they are formed again on the true tool.

E. Polishing marks caused by the polisher not fitting in good contact with the work. This happens especially when fresh rouge is applied, and for this reason, during the last half hour of polishing, no fresh rouge should be applied—only enough water to keep the surface moist.

F. Marks made by using the test plate, which always result from insufficient care in cleaning and dusting the test plate or glass to be tested. Everyone who uses the test plate must be provided with a tin box with loose fitting lid and camel-hair brush inserted in the lid. Inside the box is kept an old linen handkerchief which has been boiled in distilled water. This must always be kept in the box when not in use and never on any account be laid down on the bench, unless the latter has been wiped clean of all dust.

G. Marks occurring in the cleaning owing to gritty material on the cleaning rags. For this reason everyone should keep a special supply of clean rag in an enamel kitchen canister. The rag should never be put on a dirty or dusty bench and must always be replaced in the canister with the lid on when not in use.

H. Marks due to various special means of blocking the glass work; for example, one method of holding in position a number of pieces of glass to be polished in a block is to attach them to an iron holder by means of pieces of felt which have been steeped in hot wax (see §184). If there is grit in such waxed felt it may cause digs in the surface. The avoidance of conditions likely to cause surface marks is dealt with in §63.

Marks are not excusable when work is in the hands of skilled men, but they are the chief cause of low output when the work is being done by unskilled workers.

For this reason, in times of crisis, when output is imperative and the use of unskilled labour unavoidable, it is important that the greatest leniency should be authorised and exercised by the inspectors.

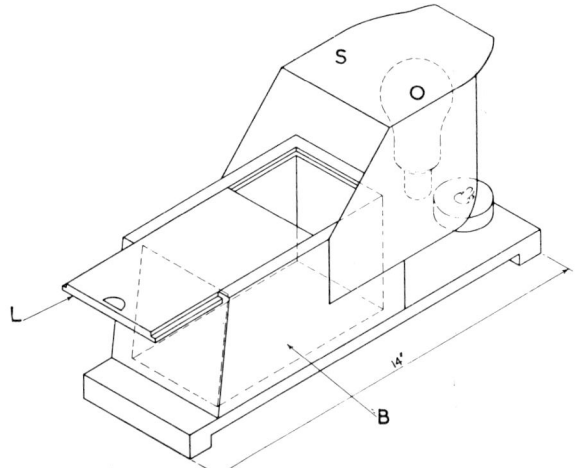

Fig. 156—Box for examining optical work for scratches, etc.

Finished work of the utmost optical perfection is often rejected, by official or unofficial inspectors, for surface marks which would cause no defect whatever in the finished instrument, by reason of rigid insistence on an aesthetic degree of perfection which has been achieved in tranquil times.

It is stated that Fraunhofer, when one of his customers complained that an achromatic objective had surface defects, replied that he made his objectives to look through, not at.

Limits of tolerance for bubbles and other defects of material in optical glass

In a paper by Räntch, 1947, the author considers the effect of such defects on the loss of light in optical instruments. He considers both elements in the neighbourhood of the pupil and those in the neighbourhood of the image plane and gives numerical examples.

The expressions found are elaborate, including the effect both of the size and number of the impurities, but the transference of the expressions into practical rules for examination of the glass as it comes from the makers should not be too difficult and would permit glass with various imperfections to be assigned to appropriate purposes.

A short paper by H. Klarmann, 1948 also discusses the influence of such defects and of scratches and cracks. The article does not concern itself with defects which cause measurable failure of quality, such as veins.

Among the conclusions arrived at by the author are these: that in a lens of 20-mm diameter air bubbles of less than $\frac{1}{4}$-mm diameter do not matter, and that towards the edges of such a lens they may be twice as big in diameter without seriously affecting the quality.

A very useful aid in examining for surface marks or interior flaws or bubbles is provided by the box shown in Fig. 156. A lamp O enclosed within a sheet metal frame S illuminates the object which is thrust with the left hand into the opening at the left hand side of the box and examined at the right hand edge through the aperture which is effected by sliding out the lid L to a sufficient distance.

The inside of the box is blackened, thus the eye while shielded from direct light from the lamp sees the brightly illuminated optical object against a dark background. This simple contrivance has been found extremely convenient and useful.

Specially high quality polishing

The question of high quality polishing has been studied at the Institut D'Optique, Paris. In an account by Fleury (1946 and 1949) it is pointed out that with a special instrument such as the Lyot Coronograph (Lyot, B., 1937) a specially high quality polishing is necessary.

The great difficulty in the Coronograph is that the brightness of the sun is about one million times that of the corona and in such conditions diffusion by the optical surfaces has to be very small to prevent error due to stray light from the mass of the sun. This involves specially high quality polishing. The quality of polishing can be tested by phase contrast; small irregularities of glass surface appear as variations of illumination in the image of the objective and at the same time lack of homogeneity is detected.

By taking two photographs at different orientations it is possible to locate the defects.

Tests concerning definition

212 The defects which affect the definition of the images formed by a lens, prism, mirror, or plane parallel glass, are more important.

It must be said at the outset that among the tests applied during making and on completion, there is one that can rarely with safety be omitted by the optician who wishes to maintain a reputation for good work, and that is to try the work exactly as it is intended to be used.

But in addition to this he should also apply tests more severe than any which the lens will be subjected to in use, and these should include tests which will tell him not only whether the work is faulty or no, but if defects are present will give him at least an approximate measure of its badness and an indication of the origin of its faults.

Tests for the definition of lenses may be divided into two kinds, those which are directed towards the verification of calculations, particularly those of complex systems, or for determining the characteristics of a particular type of lens; and those for testing whether the curves described by the computer have been correctly produced in the workshop, and whether the total result is in accordance with that expected of the type. The former are outside the scope of the optical workshop and are not included here.

Testing lenses

The star test

213 To deal first with telescope objectives, a good test is provided by examination of a star, or of an artificial star produced by a distant ball of black glass, or a bulb of mercury, in which the sun is reflected.

Actually, if the wave theory of light is taken into account, calculation shows that the image formed by a telescope lens is not a point, but a disc of finite diameter surrounded by rings of rapidly decreasing intensity.

Examination of the image with a lens of high enough power confirms the result of calculation—indeed Herschel knew and used this phenomenon in testing his object glasses long before any explanation of it was given (Foucault, 1858, 1859).

Examination of these rings, particularly as they appear a little within and without focus, can yield with experience most valuable information concerning any faults that may be present; and this method of test has been applied with excellent results by Mr H. Dennis Taylor, who gives the best description which I know of the various appearances seen in a telescope when a star is examined in his book *The Adjustment and Testing of Telescope Objectives* (1921, T. Cooke and Sons Ltd, Buckingham Works, York). The first edition was published in 1891 but the following extracts are quoted from the 1921 edition with the permission of Messrs Grubb Parsons, the owners of the copyright.

The least experienced of observers will have noticed that the luminous disc visible when a star is thrown out of focus is not of continuous brightness, but is broken up into a system of interference rings ; the farther out of focus and the larger the luminous disc the greater is the number of rings visible.

Squaring on

The operation called squaring-on means the adjustment of the optical axis of the objective until it passes accurately through the centre of the eyepiece. The symptoms resulting from any such maladjustment of the objective are strikingly evident if the objective is tilted to any very serious extent ; but if only to a very slight extent, then the observer will need very careful and discriminating exercise of his eyesight in order to give the final touches to the adjustment. It must be remembered also that an objective must be carefully squared-on before the observer is in a position to form a just estimate of the quality of the glass unless it is a very bad one ; for there is no type of object glass in existence which will give a perfect image of a point of light or star at a point situated even $\frac{1}{2}$ a degree only from the optic axis. In the first place the oblique image is never a round star disc but a sort of linear formation caused by the astigmatism which marks the oblique performance of refractors and reflectors, but besides this inevitable astigmatism there is also an oblique effect known as coma or eccentric flare whose amount very essentially depends on the form of the object glass. This defect is superimposed upon the inevitable astigmatism and in many cases is violent enough to completely disguise the latter.

The author then describes the different effects observed as a result of faulty squaring-on with different types of telescope object glass and the procedure by which the operation of squaring-on is to be performed. With telescopes of less than three inches aperture it is to be expected that the maker will provide a cell and mounting sufficiently accurate to obviate the necessity for means of adjustment, but with larger objectives adjusting screws should be provided by the maker whereby the cell can be tilted in any direction. The appearances concerned with squaring-on are shown in Figs. 10*a*, *b*, and *c* (Fig. 157, which reproduces Figs 10–24 of Mr Taylor's book).

The first type considered by the author can take several forms, all of which are characterised by good correction for spherical aberration but inward coma, or coma in which the flare lies inwards towards the optic axis.

When such an objective has been finally adjusted (squared-on), it will be found on examining the image of a small star that the luminous disc or ring system will expand itself concentrically with regard to the position of the star when in focus as in Figs. 12a, b, c, d and d_1 (Fig.157), where a little cross marks the position of

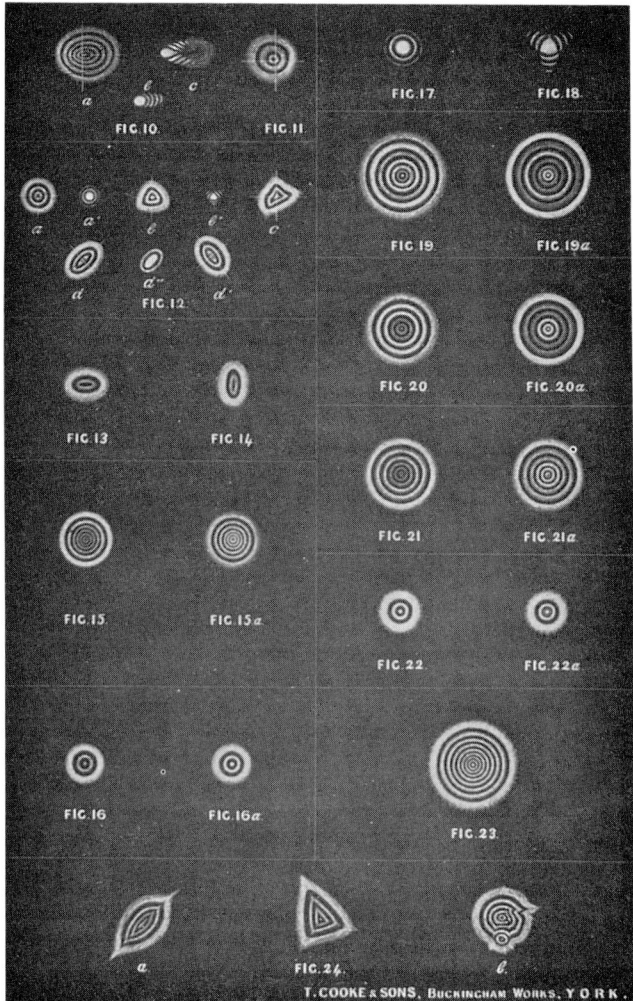

Fig. 157—Appearances in the "star" test for telescopes

the star when in focus. This point is the only one to be regarded at present, for it may be found that although the luminous disc expands itself concentrically with regard to the focused image, nevertheless the disc is not round but oval (d and d_1) or even of some irregular shape, c. Such appearances indicate defects either in the objective or in the eye of the observer or both.

Astigmatism

(The following section is abstracted from pages 30–39 of the book referred to ; the figure numbers are those of the reference, *i.e.* of the subdivisions of Fig. 157 in this book.)

214 The effect of astigmatism is the formation of a short focal line in one plane on one side of the focus and a similar line, but at a direction at right angles to the first, on the other side of the focus. Hence the image of a star, instead of being a minute round disc like $12a'$, may show either as an elongated line in one plane or a similar line at right angles to the former on the other side of focus. Fig. $12d''$ represents such an elongated star image. The cross-section of the rays half-way between these two focal lines will be a circle whose diameter is equal to half the length of either of the two focal lines. This is the circle of least confusion and will in practice be the point actually focused upon. At other positions the image appears an oval, Figs. 13 and 14.

To determine whether the astigmatism is due to the telescope or to the observer's eye the former can be rotated in its mounting. If the focal lines rotate with it then it is the telescope which is at fault.

Spherical and zonal aberration

Any well-marked aberration can best be detected by racking the telescope sufficiently out of focus to cause three or four diffraction rings to be visible. If within focus it is found that the central rings look very feeble and the edge rings, and especially the outermost one, look massive and luminous then the inference is that the edge rays fall short or come to focus nearer to the objective than the focus for the central rays, in other words there is positive aberration. Fig. 15, represents the appearance within focus in this case and Fig $15a$ the complementary appearance outside focus.

The author describes these appearances and other similar ones resulting from zonal aberration in great detail, and the reader who follows out his description with a few actual telescopes will find that a very great deal of information can be obtained as to the performance, defects and causes of any defects in the telescope. The extra-focal method of test described by the author is a very useful one indeed.

Perfect figure

If all the conditions necessary to perfection are fulfilled in an objective it will be found that the appearance in focus is as Fig. 17, while the ring systems which are observed when a bright star is thrown out of focus are perfectly circular in outline, the individual rings growing gradually and regularly stronger and further apart as the outside ring is approached, this outer ring running a little out of proportion in its brightness and breadth. Above all, the appearance and arrangement of the rings should be exactly the same on both sides of focus, if allowance is made for the blue flare which somewhat enhances the brightness of and disguises the more central rings when outside focus.

Artificial star

Mr Dennis Taylor is describing the appearance seen when a telescope is directed under favourable atmospheric conditions at an actual star. The telescopes of small size used in most optical instruments cannot wait for their test until time and climatic conditions permit such a test to be used in production. Nothing is quite so good as an actual star but a sufficiently good test can be provided by placing at the focus of a collimating lens of good quality a piece of tinfoil in which a number of holes have been made in the following way. Thin tinfoil, say 0·0005 in., must be used. It must be laid flat on a backing which is neither glass-hard nor soft—ebonite is about of a favourable consistency—and a needle, previously sharpened with great care to a real point as seen by a magnifier, very lightly jabbed down on to the tinfoil a number of times with different degrees of light force. With luck and perseverance one will be able with such a piece of tinfoil to find a " star " which, while permitting sufficient light to pass through, has dimensions small enough to be considered a point for practical purposes.

The selected hole is illuminated with the image of a Pointolite lamp.

Optical testing bench

215 Such a collimator is embodied in an optical testing bench which has seen continuous service for forty-eight years in the optical testing room at Hilger's. It is shown in Fig. 158. The artificial star or other test object, S, is attached to the end of the collimator C in the focus of a 4 in. achromatic objective O. A telescope T rotates on a substantial axis A, the bottom end of which is supported on a ball-bearing sunk into the cement floor. In this manner the telescope can be swung either into alignment with the collimator or to make any desired angle with it, consequently it can be used for testing prisms of any shape such as a 60° prism, P (shown in the plan) which stands on the levelling table L. The end view shows a triangular bar, B, on which can be mounted a variety of stands for eyepieces, a microscope, or mounts for lenses which require testing.

Filling the aperture

In order to get a proper test of an object glass it is essential that light from the object should be received by every part of the object glass, under an arrangement known as " filling the aperture." In looking at a natural object this occurs as a matter of course, but when

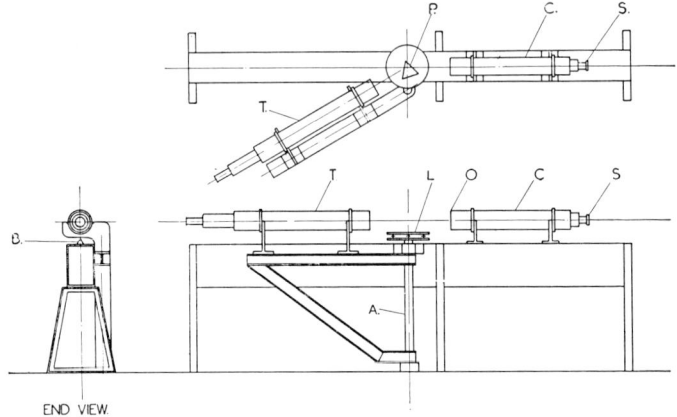

Fig. 158—Optical testing bench

one illuminates a hole such as that described above, or the slit of a spectroscope by means of a local light source the image of which is focused on a hole or slit, it may be that the condensing lens is so defective or the light source so out of line that the aperture of the lens is not filled. This can be easily observed by removing the eyepiece and putting the eye in its place, when it will at once be seen whether the aperture is filled or not.

For those dealing regularly with tests on camera lenses the optical bench designed by Kingslake (1932) should be considered. It takes into consideration the need for dealing quickly and practically with measurements of curvature of field, distortion, the difficulties of testing telephoto lenses, etc.

The Foucault test

216   A test for lenses which is highly informative is that of Foucault (1858, 1859 a and b ; adopted also later for other purposes by Toepler,

1866–67). In this " Schlieren " test, the object glass is examined without any eyepiece. With one eye in the focus of a star (or artificial star) one sees the object glass uniformly illuminated. If a knife edge is passed slowly across the focus from right to left, this uniform illumination disappears over the whole object glass simultaneously if, and only if, the rays all pass accurately through the focus. If they do not, the object glass first darkens over regions from which rays pass to right of the focus. Foucault's test thus affords direct evidence of the course of the rays from different parts of the object glass.*

Foucault's explanation, which is substantially as set forth above, is based on the ray theory. It ignores diffraction, and leaves certain conspicuous appearances unexplained, but has served ever since as the basis of interpretation by optical workers. On the basis of Foucault's simple explanation the method has been brought to great perfection by Strong and Gavioli (1936)—see Chapter 10.

Kingslake (1936) gives knife-edge shadows for primary spherical aberration and for zonal aberration.

In applying the test, the interpretation of the knife-edge shadows is largely intuitive and uses the " grazing light " fiction. The mirror surface is replaced in imagination by a flat disk on which parallel light falls, at nearly grazing incidence, from the side opposite the knife-edge. The Foucault shadows are interpreted as relief-shadows and the bas-relief figure " seen " on the disk is taken to represent the error figure of the mirror. If error slopes are small enough, the error-figure that is inferred agrees in shape with that which would be obtained by photometric analysis of the Foucault shadows on the basis of ray theory.

A valuable contribution to the theory of the Foucault test has been provided by E. H. Linfoot (1948). This carries the investigation of the subject farther than it has been pushed heretofore, the author having applied the diffraction theory to a particular case. The paper is not an easy one, but it seems to explain the difficulty, experienced by practical workers when correcting mirrors by local polishing with the help of the Foucault test, of preventing the polisher going outside the limits of the local error under treatment. This is usually regarded as a difficulty in the manual control of the polisher,

* The test of Foucault (1858) is described by him in the following passage of admirable brevity. He claims, rightly, that the method reveals in one comprehensive view the extent to which the mirror departs from the ideal shape.

The mirror is so placed that it produces an image of a small hole in an opaque screen brilliantly illuminated by artificial light. To make the test this image is almost entirely masked by a second opaque screen with a straight edge. The rays which just pass this edge are received by the eye and the surface of the mirror is then seen in light and shade and one sees, in strong relief, all the reflections which do not participate in the exact focusing of rays. One can thus recognize the parts needing correction.

but the author shows it may be attributed in part at least to the errors inherent in the usual method of interpreting Foucault shadows.

Linfoot's paper is likely to be of much service to those using the Foucault test in correcting mirrors or object glasses. In other papers Linfoot had developed a diffraction theory of the Foucault test (*Proc. Roy. Soc.*, 19, **186**, 72) and had applied it to the appearance arising when the test is carried out on a circular mirror suffering from astigmatism (Monthly Notices of R.A.S., 1945, **105**, 193). His theory is shown to predict in a simple way the salient phenomena observed in practice, and a solution is found for the shadow intensities corresponding to central and non-central settings of the knife-edge. This solution is used by the author to estimate the limit of sensitiveness of the test as a means of detecting astigmatism.

The Hartmann test

217 More direct knowledge of the course of the rays is to be had from the method of Hartmann (1900 and 1904). In this, a diaphragm is pierced with rings of holes at the various zones. The diameter of the holes is between $1/200$ and $1/400$ of the focal length of the objective. This diaphragm is placed close to the object glass, and the image of a star photographed within and without focus. Such photographs consist of dots, each of which corresponds to one of the holes in the diaphragm; and from the distance apart of corresponding dots in the two photographs the course taken by the rays from the corresponding parts of the object glass can be deduced. Highly accurate results have been obtained, but the method is laborious.

Kingslake (1926) points out that the Hartmann test may be complicated when applied to the measurement of the aberrations of oblique pencils and shows how to deal with the determination of coma and astigmatism.

Lehmann (1902 and 1903) describes his experience in carrying out this test on a number of objectives (see also §334). The application of the Hartmann test to large telescope objectives will be dealt with in §334.

Other tests

218 In quite a different category from the above are the methods of examination founded by Twyman and Green (1916, 1918, 1923) on the interferometer of Michelson, to which a separate chapter is given.

Methods which may be of value in special cases have been developed by Waetzmann (Bratke 1924) (founded on the Jamin refractometer), Ronchi (1926), and Lenouvel (1924).

Methods for determining the colour curves of objectives are given by Eberhard (1903).

Testing surfaces

219 Before dealing with the testing of polished surfaces, it is worth recording how a test of some sensitiveness can be applied to grey surfaces. If a piece of greyed glass which is to be tested is placed in contact with a test plate and Stephen's blue ruling condensed ink run between the two, it will be found that there is a pink border surrounding a white patch which denotes a bump on the glass and this is detectable with the unaided eye to an accuracy of 2λ. This ink possesses the advantages that : (1) its colour varies with the thickness in the following sequence—white, pink, blue, black, and these are sub-divided into shades ; (2) it is readily soluble in water. Sensitivity is improved at a certain density by putting in a No. 22 Kodak orange filter and in effect doubling the film thickness by placing a mirror behind. It is possible that this liquid and filter test could be used with advantage in metal work in cases where the surface is bright enough to give as much as 20 per cent reflection, but so scored as to render Newton's ring test out of the question, as for example ground stock, scraped surfaces, slides of lathes and other machines.

A note on an absolute standard of planeness will be found in §239.

Testing polished surfaces

Newton's rings and the test plate

If a shallow convex glass surface is laid on a flat one a system of rings is seen around the point of contact. These are called " Newton's rings." When two flat surfaces are put together, the one being very slightly inclined to the other, the colours are not arranged in rings but in more or less parallel lines or curves, and they are then called " Newton's bands." Although others had observed that transparent substances when made very thin by being blown into bubbles or otherwise exhibit colours according to their thickness, it was Newton (*Opticks*, Book II) who first found a relation between these colours and the corresponding thicknesses.

220 The thickness of the film of air at any point of such a system of rings or fringes can easily be determined by the colour at that point, or, better, by counting the number of bands from the point of contact. Near the point of contact, where the plates are very near together, the colours are brilliant, and up to four or five rings or bands one can assume that the increase of thickness from one band to another is about 1/100,000 in. Beyond this the fringes gradually become fainter unless monochromatic light, such as is provided by a low pressure mercury vapour lamp, is used.

Accurate plates, whether flat or curved, which are used for the purpose of testing surfaces which should fit them, are known as test plates or proof plates. They may be used either in comparing a surface with one of assured flatness, or for ascertaining whether a surface has a desired radius by the use of a proof plate of the same radius but opposite curvature.

These plates have been used for many years; according to von Rohr (1929) such test plates were used by Fraunhofer, although he kept the device rigidly secret. Voit (1887) gave an account of various methods introduced by Fraunhofer, which includes the use of test plates.

Flat test plates are best made of quartz, since they then last much longer without getting scratched. Quartz is not suitable for spherical proof surfaces, because it changes in shape with temperature. For example a quartz test plate of sag 0·2 in. alters in sphericity by 1/5th Newton's band for every degree centigrade. Wherever a number of identical lenses are to be made, proof plates, or proof spheres as they are sometimes called, should be made to correspond exactly to the forming tool; they are not only needed to test the surfaces of the lenses, but also to ensure that in the course of time the true tools do not depart from their nominal curvatures, which are liable to alter gradually owing to wear.

When the test plate is put in contact with the lens or flat surface to be tested, great care should be exercised to see that both the surfaces in contact are perfectly clean.

This should be effected, first by thoroughly cleaning both surfaces free from grease by means of methylated spirit or other similar cleansing solution. Then, before laying the test plate in contact with the lens, a camel-hair brush should be used to flick away any particle of dust which may float on to the surface during this operation.

A good way of getting rid of the dust in applying a proof plane to a flat surface is to place a piece of tissue paper between the two surfaces and, while maintaining a slight pressure on the upper plate, to draw the tissue paper away. It will be found, unless a shred of the paper is broken off and stays between the plate, that they immediately go into good contact.

Care should be taken to avoid prolonged handling of the lens and test plate and thus causing distortion of the surface by heat. Before finally judging the quality of the surface by the appearance of the coloured Newton's rings, time should be allowed to ensure that the test plate and lens have both acquired the same temperature—a lens or plate of 2 in. diameter may take five minutes to settle down. If only one colour is seen over the contact surface, this area of the lens or block of lenses is correctly polished.

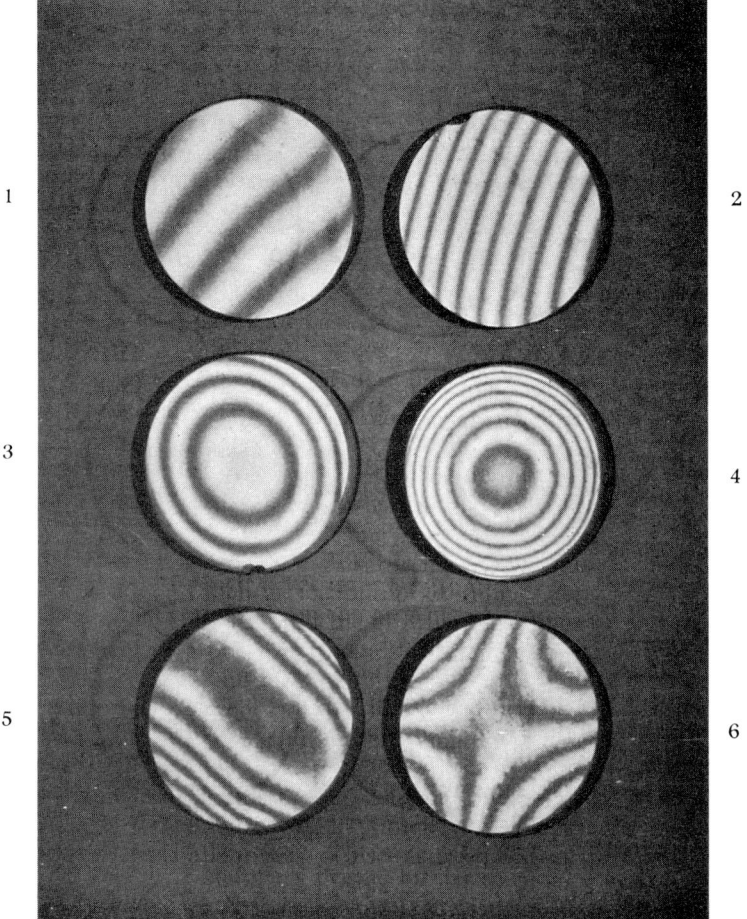

Fig. 159—The use of the proof plate
1. Nearly perfect, the proof plate slightly tilted.
2. The same, but the proof plate tilted about $2\frac{1}{2}$ times as much.
3. Astigmatic (surface slightly ellipsoidal).
4. Surface concave or convex according to the colours.
5. Strongly astigmatic.
6. Astigmatic of the saddle-shaped variety, i.e. concave in one direction, convex in the direction at right angles.

If uniform circular rings or colour bands appear most brilliantly near the periphery of the test plate, contact is towards the outside; whereas, if the brilliant bands are in the central area, contact is in that area.

Although white light is often employed for observing the fringes, yet in the workshop it is desirable to have ample illumination by sources of monochromatic illumination. That most generally used is the low pressure mercury vapour lamp; a neon lamp is also good for focal use and is conveniently portable. One can then see many more fringes and this reduces the temptation, when the two surfaces do not readily come into contact, to press them forcibly together in order that the fringes seen by white light may be distinct and of the

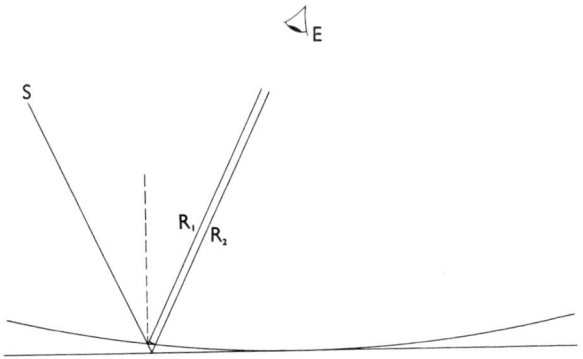

Fig. 160—The production of Newton's rings

brilliant coloration which makes the test a severe one. Such failure of the surfaces to come into contact is always due to dust or dirt of some kind and the optician, in attempting to bring together by pressure surfaces which are not thoroughly dust free and clean, may produce scratches requiring the re-smoothing of the polished surface.

Fig. 159 shows some characteristic appearances which may be seen in the use of a proof plate.

There is a large number of tests which all derive ultimately from the phenomenon of Newton's rings, and we will therefore give a simple explanation of this. Figure 160 shows a very shallow surface in contact with a flat one. Light from an extended source S is reflected from the two surfaces in the beams R_1 and R_2 and reach the eye situated at E. Then if we consider the enlarged diagram (Fig. 161) of the region whence the rays R_1 and R_2 emerge we see that the path difference between them is $2t \cos i$ where t is the distance between the surfaces at the point in question and i the angle of incidence.

Since reflection at the denser medium D introduces a half-wave phase advance in the reflected light then, at the point of contact where $t=0$, there will be destructive interference and the surface will appear black or without colour. In practice it is never quite black unless one presses the surfaces together, owing to imperceptible dust at the point of contact.

At some little distance from the point of contact we get colours, since the optical path differences in the film increase more rapidly for the shorter wavelengths. One usually observes the phenomena of Newton's rings at something like normal incidence so that the formula above mentioned becomes reduced to—

Effective difference of phase $= 2t - \lambda/2$.

The reflected rays therefore cancel each other by destructive interference when $2t$ is zero or any whole number of wavelengths. On the other hand, when $2t$ is a half-wavelength, or a half-wavelength

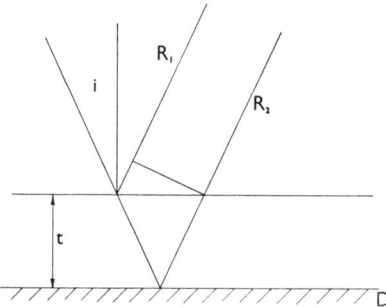

Fig. 161—The production of Newton's fringes

plus any whole number of wavelengths, there is no destructive interference. So that if we take a red wavelength and a blue wavelength, of say 0·6 and of 0·4 micron respectively, the radiation due to the former will be reduced to zero and that due to the latter not reduced at all when $2t$ equals 0·6, because at this thickness the effective difference of path of the red wavelength is $\frac{1}{2}$ wavelength and that of the blue ray a whole wavelength. In effective path difference is included, of course, the advance in phase of $\lambda/2$ due to reflection at the denser medium D.

A more complete account of the phenomenon is given in Martin (1948); what is said above however is sufficient to make it clear that there will be a black spot in the middle surrounded by coloured rings. When the source is an extended one of white light the order of colours shown in these rings follows exactly that of crystal plates in polarized light for which a table has been calculated by Quincke.

This table is given in Martin (1948, p. 210). But without any instruction whatever the optician soon gets to realise that the rings even with white light can be spoken of as first, second, third and fourth, and so forth, although it becomes less easy to assign numbers to the rings as they soon become pale in tint. The phenomenon becomes much simplified if one uses, as is customary in optical workshops to-day, the light of a low-pressure mercury vapour lamp diffused by a ground surface or reflected from a white one. The green radiation of such a lamp is so much more powerful than the others, that the light is virtually monochromatic, and the difference of thickness between the position occupied by one ring and that occupied by the next is nearly 1/100,000 in.

If instead of applying a shallow curved surface to a flat one we apply a truly spherical surface to one of slightly different curvature, the width of separation of the rings indicates how nearly the second surface matches the first; while any departure of the rings from circularity indicates a departure from sphericity of the second surface. Such procedure, therefore, forms a ready test for the accuracy of spherical or flat surfaces.

Haidinger's fringes

221 If, instead of a thin film of air formed between a convex and a flat surface, a thin parallel, or approximately parallel film is used, it was observed by Haidinger (1849) that, on looking through the plate at an extended source, fringes are seen which he called " lamellar " fringes to distinguish them from the rings of Newton. If the plate is very thin and parallel (as sheets of mica can be obtained without much difficulty), there is a uniform coloration. If the plates are a little thicker, one can see rings on looking as nearly perpendicular as possible at the reflection of an extended source of white light when the thickness is up to say one thousandth of an inch. With monochromatic light, such as given by a low pressure mercury vapour lamp, they can be seen up to $\frac{3}{8}$-inch thickness of plate. This forms an extremely sensitive and ready test for the parallelism of a glass plate. With modern methods of polishing plate glass in which the polishing takes place on both sides simultaneously, it is quite easy to find in plates, say, one-quarter of an inch thick, portions in which the rings obtained in this way remain stationary when one moves the plate even as much as one or two inches in its own plane. It is essential that the eye should view the plate normally and be focused for distance, but once this faculty of observing the rings has been acquired it is always very easy to see them.

A convenient way of illuminating the plate is to cut a small hole in a piece of white paper, illuminate the latter with monochromatic light and look through the hole at the plate. By observing the movement of the rings as one moves the plate in its own plane the variations of thickness can readily be measured ; the rings open as one passes from a thinner to a thicker part of the plate.

A very simple variant for examining plane parallel glass, even up to so thick as $\frac{3}{4}$ in., is to lay the glass on white paper, the latter being illuminated by monochromatic light. Then looking perpendicularly at the plate, interference rings are seen with their centre at the reflected image of the eye, these rings opening as the eye moves about to a thicker part of the plate.

Fizeau's fringes

222 Fizeau's fringes follow loci of constant thickness and they can be used to give effectively the same information as Haidinger's fringes. The way in which Fizeau described his system is as follows (Fizeau, 1862):

A convex lens of very long focus was placed on a glass plane so as to give—with white light—large coloured rings as in Newton's well-known experiment ; but here (viz. in *Fizeau's experiment*) the lens was fixed in a metallic mount movable by a micrometer screw perpendicular to the plane of the lens and of the stationary glass plane. By rotating the screw-head one can vary the distance between plane and lens. The observer, looking in a direction parallel to the lens and the system being illuminated by the light from a Brewster lamp (viz. a sodium burner) reflected from the hypotenuse of a small prism placed near the eye, perceives between the two glasses large rings of the greatest beauty ; rings which result, as one knows, from the interference produced between the rays reflected towards the eye by the two neighbouring surfaces.

It is easy to see that to render these appearances visible over the whole surface of the lens at once one should add near it a convex lens of focus equal to the distance from the lens to the eye ; one then sees the phenomenon in its entirety the actual surface of the lens being covered to the margin by rings of the greatest sharpness.

Fizeau, in adding this lens, was only concerned that the observer should see the whole field illuminated. It has another effect, which in other adaptations becomes very valuable. Obviously the eye of the observer situated at the focus of the lens, sees the field illuminated by rays which have passed perpendicularly through the space between the lens and plate. The fringe system now indicates the variation of the separation of the two surfaces without the complexity arising from the obliquity of the rays passing through it, and in this form it was employed by Laurent for testing the flatness of surfaces.

Laurent's modification of Fizeau's apparatus

223 An apparatus, for testing the flatness of surfaces compared with a test surface, which consists of a modified Fizeau apparatus, was described by Laurent (1883, *Compt. Rend.*, **96**, 1035).

Referring to Fig. 162 a base H with three levelling screws carries a raising and lowering pillar which bears the flame (a sodium flame) F and a lens which forms an image of the flame on a piece of white paper placed about 45° to the incident light.

The observer's eye (at O) is placed close to the paper and the illumination from the paper passes through the lens L which collimates it. From there it passes through the plane parallel proof plate T

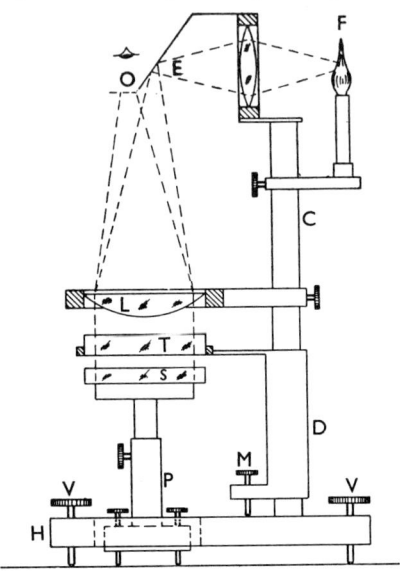

Fig. 162—Laurent's modification of Fizeau's apparatus

the lower surface of which forms the comparison surface and which can be raised and lowered by means of the sleeve D which can slide up and down the pillar C so as to adjust the distance of T from the surface of the plate S which is to be tested. Provided the distance between T and S is not too great one sees the fringes and, by adjustment of the three levelling screws V, the two surfaces can be brought into parallelism when the interference fringes will show how the surface of S differs from that of the proof plate T. The fringes indicate the variation in the separation of the plates from point to point.

The apparatus is cumbrous so that in some of the forms in which it has been used at Hilger's, the plates are automatically brought into parallelism by distance pieces without the necessity of adjusting the parallelism with screws.

There is a distinct advantage in avoiding contact of the plates owing to the fact that when a proof plate is used in contact with the surface to be tested it is extremely difficult to avoid scratches. It will be noticed that there is no optical difference between Fizeau's apparatus and Laurent's except that in the latter observation is being made of the interference between two surfaces of which one is accurately flat and the other is to be tested from departure from flatness.

Although in Fizeau's apparatus, and Laurent's modification thereof, the interference takes place between two surfaces with an air space in between, there seems no justification in withholding the name of these originators from apparatus in which the space between is occupied by a denser medium such as glass. We shall therefore designate all the forms of apparatus to be described in this and §§224 and 225 as " Fizeau's."

A typical Fizeau (or Laurent) apparatus as appropriate for use in an optical workshop (the earliest, I believe, which can be so described) is given by Lummer (1885) who says it was the form used by Abbe. The apparatus consists of a diaphragm in the focus of a telescope of 10 to 20 cm focal length, over the diaphragm aperture is placed a plain mirror at 45 degrees which illuminates it, and the plate to be tested for parallelism is placed in front of the telescope objective and perpendicular to the axis of the telescope. When the eye is placed at the diaphragm and the attention directed towards the plate one sees distinct light and dark stripes, rings and other curves. Along any one of these curves the thickness of the plate is exactly the same, they are indeed curves of equal thickness.

Another apparatus on the same principle was described by Schultz (1912) and was mounted on a stand which could be placed on the tool to which the plate being polished is cemented so that the surface can be tested without removal from the polishing machine.

This apparatus was subsequently made by the firm of C. P. Goerz (Schulz, 1914). The arrangement is such that the object which is to be tested is placed on a table and an upper plate brought into position so that its lower surface which is the test surface is about $\frac{1}{2}$ mm from the surface to be tested. The plates are provided with adjustment so that they can be made parallel to each other or slightly away from parallelism, whichever may be desired.

224 Apparatus for testing plane parallel plates exactly the same in principle but very suitable for workshop use, is shown in Figs. 163, 164, and 166.

Fig. 163—The Hilger interferoscope
A modified Fizeau system

Testing the parallelism of plane parallel plates

225 If a number of plane parallel plates are required all of the same thickness, the purpose can be achieved with a fair degree of approximation by putting the plates down on a flat optical tool with paraffin wax, as described in the first operation of making a plaster block of prisms (§181). If the thickness of the outer glasses on opposite sides of the block is measured by a depth gauge, the accuracy of parallelism will in this way be automatically effected to within 1 or 2 minutes, and for control to this accuracy the Angle Dekkor suffices.

It must be remembered that the scales of the Angle Dekkor are engraved to indicate the angle between surfaces in air. When used for testing plane parallel glass, therefore, the readings must be multiplied by $1/n$, n being the refractive index of the glass, if it is the geometrical non-parallelism that is required. For most practical purposes one can multiply by 2/3.

To test parallelism to an accuracy of 1 or 2 seconds, interference methods must be applied, and the best instrument for this is the interferoscope, a Fizeau apparatus which in the form now made and used by Hilger & Watts Ltd is shown in Fig. 163 and in diagram Fig. 164. Light from a mercury vapour lamp is concentrated on an aperture A by an adjustable mirror and condenser and reflected by the mirror M within the instrument on to a concave mirror, aluminised on the front, at C. This mirror collimates the rays, and reflects them up to the work holder at W. The plane parallel glass under test is put on the work holder and levelled (by means of screws S) to reflect the rays back, when the eye at E sees the surface of the plane parallel glass crossed by fringes due to the interference of the light reflected from the upper

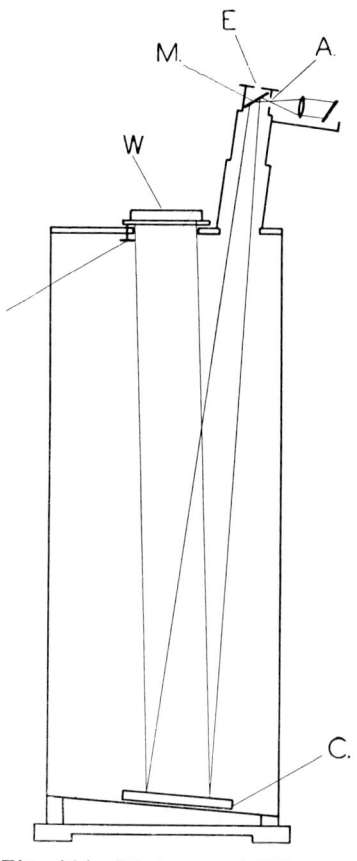

Fig. 164—Diagram of Hilger interferoscope

and lower surfaces. The difference of thickness of two points separated by one fringe is $0{\cdot}00001/n$; n being the refractive index of the glass of which the plane parallel plate is made.

To ascertain which is the thick part of the plate, a small copper rod, mounted in a wooden handle, is used. This rod is kept permanently heated by resting in a metal box within which a 20 watt electric lamp is kept alight. If the warm rod is held for a moment against the underside of the plate the latter is heated locally at the point of contact. The slight expansion which takes place is sufficient to cause a slight displacement of the fringes at that spot, with the result that an

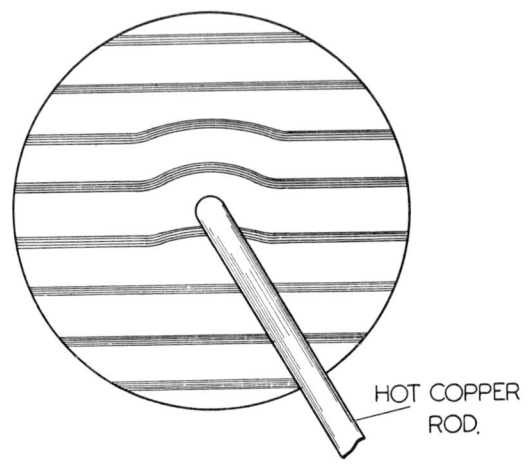

Fig. 165—Use of hot rod with Hilger interferoscope

appearance as shown in Fig. 165 is observed, the fringes being displaced locally towards the thinner part of the plate.

An application of the use of the interferoscope in conjunction with the Hilger prism interferometer to the measurement of variations of refractive index in a plate of glass is given in §287.

Portable Fizeau apparatus

A portable Fizeau apparatus is shown in Fig. 166. Light from a neon lamp L is reflected by a half-silvered mirror (or half aluminized mirror) M through a lens O, which collimates the light approximately, and thence down the tube T to the plate under test. The apparatus stands on the said plate on three rubber feet F which can be adjusted by the screws (that is compressed by the brackets) so that the light is approximately perpendicular to the plate. The light returns through the lens O and passes to the eyepiece of the instrument where the observer sees the interference fringes due to light reflected at the top and bottom surfaces under test. This apparatus finds a convenient application in a case where one has a number of plane parallel plates cemented to a large glass tool. If a test plate is laid on the top surfaces of the windows and interference observed between its lower surface and that of the glass tool a high degree of parallelism can be attained in the polishing.

Michelson's test

226 A form of Fizeau apparatus was used by Michelson (1898) for the purpose of correcting large plane parallel glass plates which were to be cut up for making Michelson echelons. The apparatus, shown in diagram in Fig. 167, consisted of a mercury vacuum tube, S, the light from which was concentrated by the condenser C by reflection from the half-silvered plate P on to the plate under test Q.

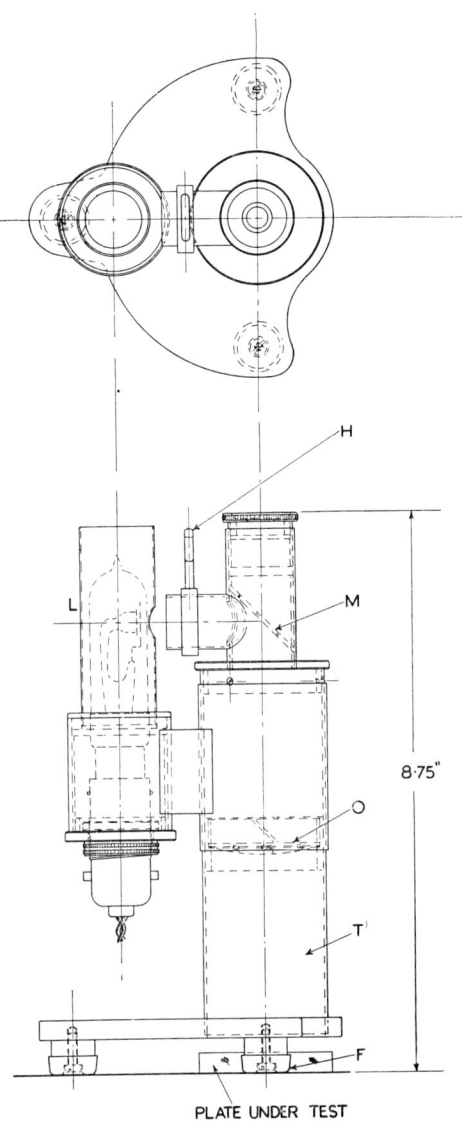

Fig. 166—**Portable Fizeau interferoscope**

The light reflected from the upper and lower surfaces of the plate passed through the plate P with sufficient intensity to form circular interference rings in the focus of the telescope lens T where they were observed by an eyepiece. On passing from a thinner to a thicker part of the plate by moving the latter about in its own plane the rings are seen to open, and the replacement of one ring by another indicates that $2nt$ is increased by one wavelength of the mercury green light, where n is the refractive index and t the thickness of the plate. The relation between $2nt$ and the thickness of the plate can be seen from the table which appears on p. 421. When I was correcting echelon plates from 1898 onwards I did not find Michelson's apparatus very

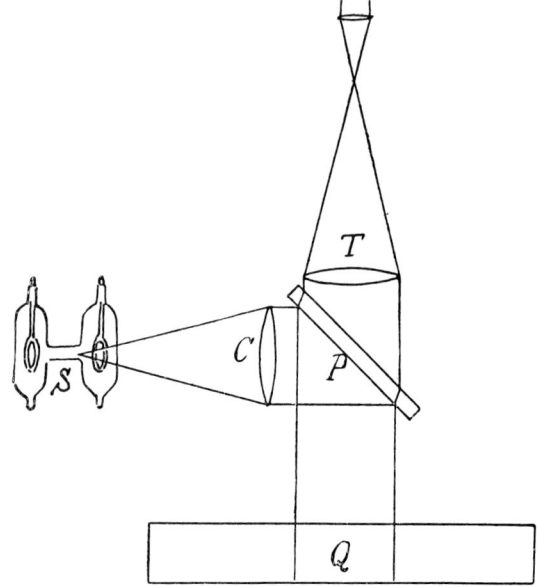

Fig. 167—Michelson's form of Fizeau interferoscope

convenient, and in the apparatus which I built I made provision for marking out easily a chart which would indicate exactly where and by how much the plates should be retouched.

In my apparatus (see Fig. 168) the plate P to be tested was placed in a circular wooden tray T, on the underside of which was a recess R in which a piece of carbon paper was placed with a piece of ordinary white paper below it, these being kept in position by drawing pins round the edge. This tray could be shifted about by hand to bring any desired point of the plate under observation.

Light from a mercury vacuum tube M was focused by the lens A on the plate. The portion of the reflected pencil which passed through the diagonal plate was received by the lens B, which produced an image of the ring system in its focus at S. The ring system was observed by the eyepiece. A graticule at S enabled one by measurement of the successive rings to calculate the differences of thickness corresponding to fractional orders of a ring.

A plunger L immediately below the point of observation could by means of a treadle, actuated by the foot, be forced upwards against the underside of the paper so as to make a black mark on the paper at the point of contact. My procedure was to find the highest or the lowest point on the plate which I was correcting, to move the tray

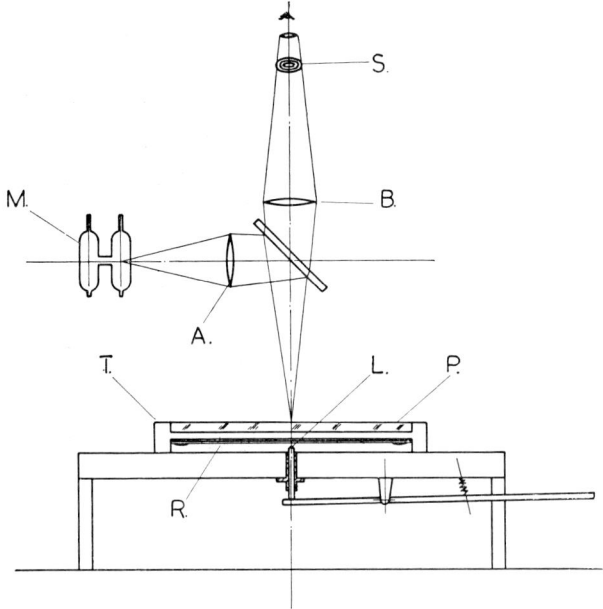

Fig. 168—Twyman's plate charting apparatus

about from that position until the ring system first seen at the lowest or highest place was replaced by the system of the next higher or lower order, and to make a mark by lightly tapping the treadle. I then moved the tray in such a direction that the ring system neither opened nor closed, and by means of the treadle made a number of marks forming a curve which represented a contour line of the imperfections of the plate as measured in wavelengths.

The process being repeated for equal increments of retardation, a chart was eventually obtained such as the one shown in Fig. 169.

I then polished the thick places with a local polisher in the manner usually adopted in re-touching.

Michelson's test results in a plate in which $2nt$ is uniform, t being the thickness and n the refractive index of the glass for the radiation by which the test is made. If n varies the plate will not be correct for transmitted light, the condition for which is that $(n-1)t$ should be uniform.

Increasing the brightness and sensitiveness of interference tests

227 All the interference tests from §219 onwards can be made much more brilliant and more sensitive by silvering the surfaces. Take the case of the interferoscope shown in Fig. 164. If the top surface of the plane parallel plate which is under test is silvered very lightly and the bottom one silvered completely, one can get very brilliant

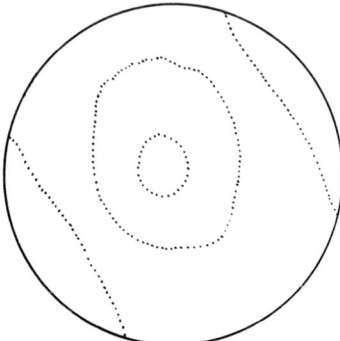

Fig. 169—Diagram obtained with Twyman's plate charting apparatus

effects provided the partial silvering of the upper surface is nicely adjusted.

In this form, however, that is in cases where one is working with reflected light, it is rarely worth while going to the trouble of this special silvering. The case is very different, however, in tests in which one makes the observation by transmitted light, and use was occasionally made of this device by Adam Hilger Ltd.

Using transmitted light the arrangement is that of a Fabry-Perot etalon with the addition of the lens to focus light from a point source on the eye. It was described by Williams (1928) with two flat plates set at an angle of a few seconds to each other, the multiple reflections resulting in the production of very sharp fringes.

The method was independently applied by Rasmussen (1945) in the examination and use of Fabry-Perot etalons and enabled him to detect and correct the plates to the required accuracy of 1/50th λ. Rasmussen's apparatus is shown in Fig. 170. The light from the light source L (an Osram spectral lamp) is concentrated upon the diaphragm B, by means of the condenser C and is then made parallel by the objective O_1. After passing through a half-silvered glass plate G at an angle of 45°, the beam falls perpendicularly upon the two unsilvered interferometer plates $F-P$. The reflected beam will then be reflected at a right angle by the plate G, passing through a second objective O_2, in the focal plane of which is placed the diaphragm B_2 through which the interference fringes can be observed. O_1 and O_2 were achromatic lenses with apertures of 65 mm and focal lengths 700 and 400 mm respectively. The circular diaphragm B_1 could be diminished to 0·5 mm in diameter, giving a sufficient parallelism of

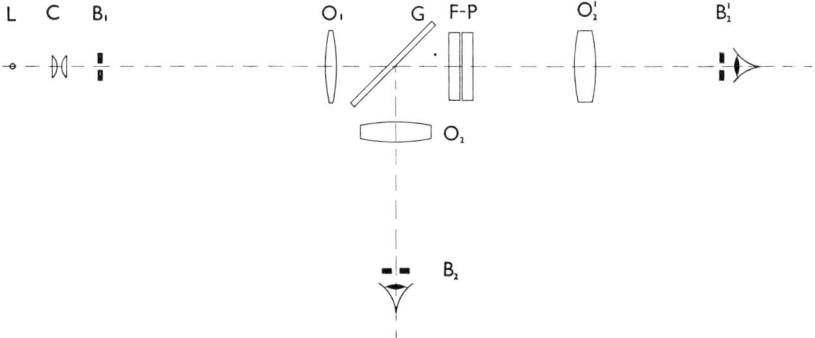

Fig. 170—Rasmussen's apparatus for testing the flatness of plates

the beam even with a thick air interspace. Diaphragm B_2 has a diameter of 2 mm and serves only for fixing the eye position and for removing the light reflected from the front and back of the pair of plates. B_2 can be replaced by a camera in order to photograph the fringes.

The Fabry-Perot etalon itself served as a holder for the interferometer plates a distance ring of 1 mm thickness being used. The use of the Fabry-Perot etalon as a holder for the plates results in a very easy adjustment of the air layer perpendicular to the beam by means of the adjusting screws of the instrument and, moreover, provides a convenient adjustment of the wedge angle of the air layer by means of the three screws which work upon elastic tongues. A rapid setting of the desired fringe distance is easily obtained, and the interference pattern is not sensitive to vibration.

The Fabry-Perot plates should be silvered with the same density of coating as is used when they are employed for a Fabry-Perot etalon, and they are followed by a lens O^1_2 and a diaphragm B^1_2 through which the eye observes the fringes. If the plates are very slightly inclined to each other, very fine fringes are seen as shown for the cadmium radiation 6438A, in Fig. 171. Irregularities corresponding to about 1/50th λ can easily be detected.

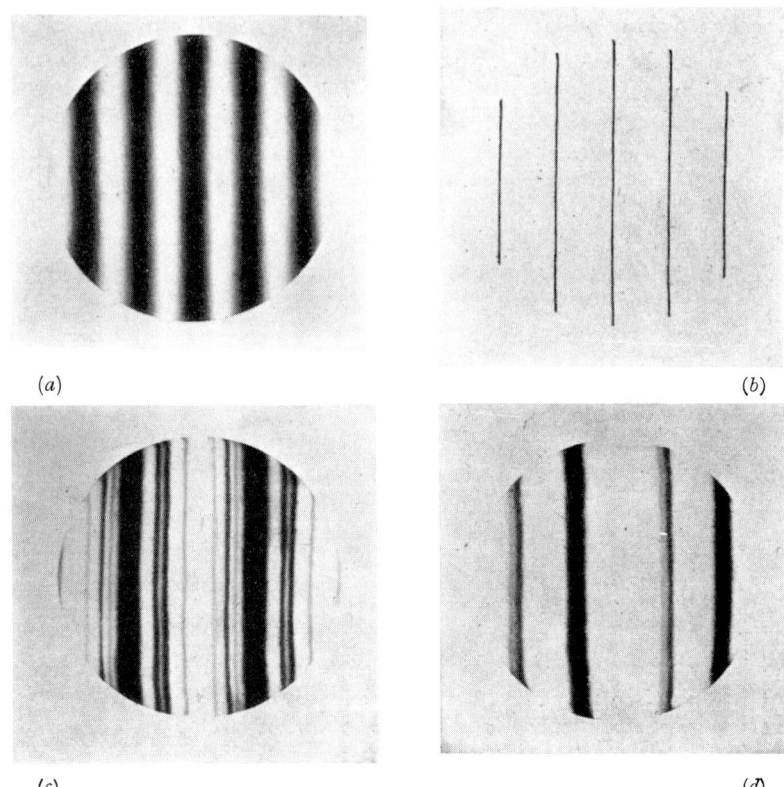

Fig. 171—Fizeau fringes with Rasmussen's apparatus
(a) Fizeau fringes by unsilvered plates. $d=1$ mm, Cd 6438A
(b) Fizeau fringes by silvered plates. $d=1$ mm, Cd 6438A
(c) Hyperfine structure of Hg 5461A, 5-mm wedge etalon
(d) Hyperfine structure of Tl 5351A, 5-mm wedge etalon

As in the Twyman and Green modification of the Michelson interferometer, one advantage of the arrangement lies in the fact that the rays which reach the eye have all passed substantially perpendicular to the plates to be examined. Thus the locations of the fringes depend only on the distance apart of the two operative surfaces and not, as in the ordinary Fabry-Perot ring system, on the difference of paths due to the rays not passing through the plates with the same obliquity.

The adjustment of the light beam to parallelism and of the air layer perpendicular to the beam was carried out by auto-collimation, using the light reflected from the interferometer plates. This could be done very accurately both with unsilvered and silvered plates. In order to facilitate this adjustment the back of the diaphragm B_1 was painted white.

Great use has been made of such silvered surfaces by Tolansky (1947). He also approached the subject from the angle of the Fabry-Perot interferometer and, following up the work of Ritschl (*loc. cit.*, 121), succeeded in getting outstandingly good reflection from silver. He then applied this high reflecting power technique to examining the interference appearance of thin plates by transmission (1945) and was able eventually to perceive differences of thickness of a single molecular layer in sheets of mica. He also applied the principle to studying the topography of crystal surfaces.

Another paper, by Tolansky and Wilcock (1947), draws attention to a point of great interest to the optician. The authors observed a fine structure in the fringes due to defects in the optical flats used for examination of the crystals. The flats used were made of borosilicate crown glass and were specially selected by Hilger as likely to have a smooth surface with high local uniformity. A number of similar flats were available. A pair were silvered and brought into close contact. The combined surface contour was examined by means of fringes of equal chromatic order, and the characteristic fine structure can be clearly seen. This structure represents polish scratch marks on the glass mostly varying from 1/100 to 1/200 mm in width, although some are narrower and are observed only as diffraction patches. No estimate of their depth can be made, for the scratches do not appear to have specular reflection and are simply characterized by complete absence of light within the scratch regions.

Note.—Einsporn (1949) discusses systematic errors which may occur in the determination of flatness of optical surfaces by the Fizeau interference fringes between a test plate and the surface to be tested on the supposition that the former has a slight curvature. The article shows that the approximate fundamental formula customarily employed can introduce erroneous conclusions which, in certain cases, are of importance.

The Zernike test

227.1 Still greater accuracy in testing surfaces was attained by Lyot (1946), and Lyot and Françon (1950), the former paper being concerned with determining the irregularities of polished optical surfaces and the latter with the examination for homogeneity of large discs of glass destined for Dr Lyot's coronograph.

The success of this instrument, which allows the solar corona to be observed without the need of an eclipse, depends on the elimination of every cause of scattered light. The brightness of the corona a few minutes of arc from the edge of the sun is a million times less than that of the sun itself; hence, although in normal practice the diffused light, from all sources, can be considered negligible if it amounts to one thousandth part of the total flux, such a proportion would be quite inadmissible in the coronograph.

The methods of examination described in the papers cited were based on Zernike's phase contrast principle (Zernike, 1934 a, 1934 b).

To realise fully the value of the Zernike test it is convenient to mention another test from which it is derived.

The knife-edge method of testing due to Foucault (1858, 1859 a, 1859 b) suffers in practical use from the difficulties of interpreting the shadow images. Whereas with the Twyman interferometer the fringes form a contour map of the wavefront, the shadows of the Foucault test present a map of the gradient errors of the waveform. Interpretation of these slope-errors requires, therefore, an integration of the gradient contours.

Zernike's phase contrast test does in fact produce a true contour map of the waveform, and furthermore, it is far more sensitive than the knife-edge test. The modification to this latter test, made by Zernike, was simply to remove the knife-edge and to cover the nucleus of the Airy disc (§229) with a minute phase retarding disc. These discs, which are of the order of $\frac{1}{2}\lambda$ thick, may be made from droplets of resin by a method described by Burch (1934).

The test apparatus as used by Lyot was essentially as follows The lens under test formed an image of a fine slit on a glass plate. The plate was provided with a fine phase retarding strip, and was adjusted until the strip lay symmetrically over the image of the slit. Immediately following the glass plate was a lens which was used to image the test lens on to a photographic plate. Lyot used a short slit instead of a pinhole in order to reduce the exposure time. It is known that a slit a few millimetres long does not appreciably alter the test appearance, and yet increases the brightness a hundredfold or more.

The mechanism of the Zernike test may be briefly described using an interference argument. Suppose the test element to be perfect save for a single line's-breadth wave. This defect will act like a single element of a diffraction grating. The centre and left of the wave will produce a single set of interference fringes, and the centre and right of the wave will produce a second set of fringes. The light from the left and right of the wave defect will be out of phase with the light from the centre, and accordingly, the two sets of fringes are displaced towards each other by a quarter of a spacing, with the result that the dark spaces of one set will lie over the light spaces

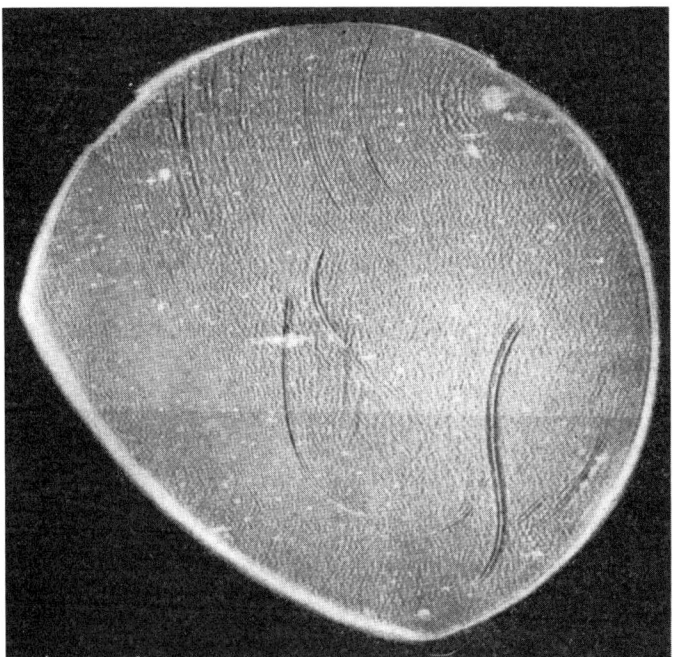

Fig. 171A—The Zernike test

The illustration shows more or less regular veins. One can also see defects, circular in shape, at the top of the illustration, which are the traces of the tool with which the surface was ground. The number of brilliant little points may be either dust, or bubbles in the mass of the glass.

The "orange peel" appearance is due to irregularities in the surface produced in polishing. A pitch polisher was used.

of the second, and the net contrast will approach zero. On inserting the phase retarding strip, the light from the centre of the defect is put into phase with the diffracted light, and the interference fringes will not be displaced; the result being that the sets of fringes will reinforce and produce high contrast. Insertion of the phase retarding strip therefore changes a phase difference into an amplitude difference, making visible the presence of the surface defect.

The reproduced photograph from Lyot's paper shows the orange peel effect due to the polishing, together with other surface and internal defects (Fig. 171A).

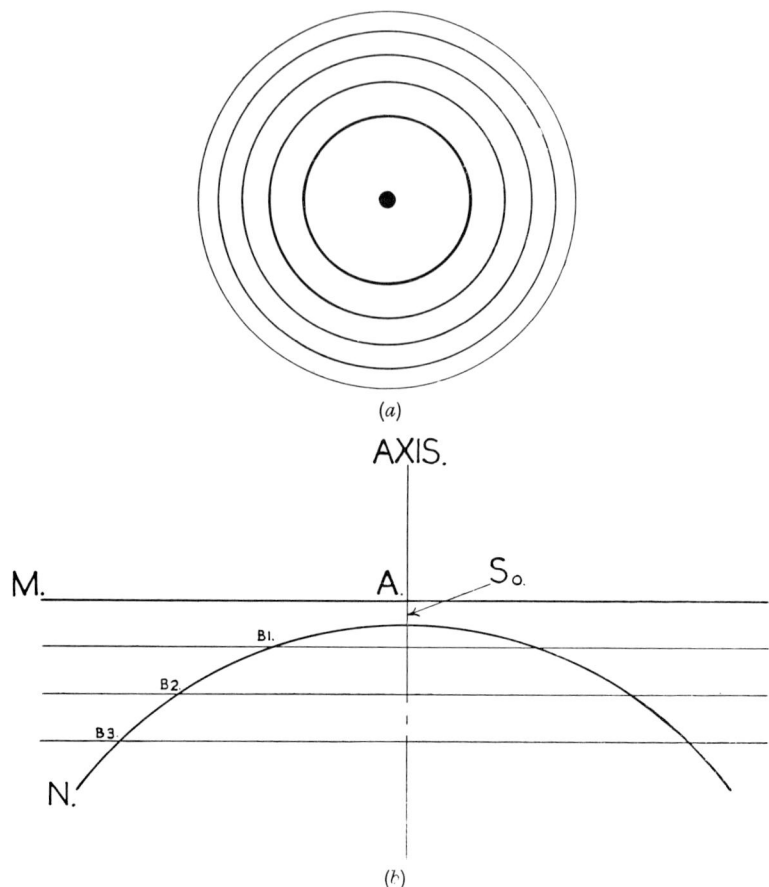

Fig. 172—Determination of power of a shallow lens by interference

Specifications of parallelism

228 In making plane parallel windows and, indeed, in flat work generally, the optician is often faced with the problem of complying to a specification of the following kind.

The window if placed in front of a telescope of 50 in. focal length must not alter the position of the image by more than one-tenth of an inch.

In such case it is convenient to have a simple relation between the appearance of the window on the interferometer and its "power."*

If the window is intended to be plane parallel, but is actually operating as a shallow lens, its appearance on the interferometer is as shown in Fig. 172 a.

In the interferometer we are observing the interference between a plane wave front and a curved one. In Fig. 172 b let M represent the momentary position of the plane wave front and N of the curved one.

The two waves will reinforce each other at B_1, B_2, B_3, etc., points on N separated from each other by a distance along the axis equal to λ, the wavelength of the radiation by which the observation is being made. The radius of the first ring measured from the middle point A is given by the formula
$$R_1^2 = 2fs_0$$
where s_0 is the difference along the axis from A to B_1 and f is the focal length of the curved wave front after double transmission through the interferometer.

Similarly, $\qquad R_2^2 = 2f(s_0 + \lambda)$

Therefore, $\qquad R_2^2 - R_1^2 = 2f\lambda \qquad\qquad (a)$

We have supposed that the accuracy is specified in the form that the focus of a telescope of focal length F should not be altered more than $\triangle F$ by the interposition of the window.

Since the errors are doubled on the interferometer by double transmission through the object under test then we must allow on the interferometer only an error of $\triangle F/2$. Hence f must not be less than $F^2/2\triangle F$. Substituting for f in equation (a) we get

$(R_2^2 - R_1^2)$, must not be less than $\lambda F^2 / \triangle F \qquad\qquad (b)$

To apply this expression to the specific case in question, $F = 50$ in., $\triangle F = 0.1$ in., therefore $F^2/\triangle F = 25{,}000$ in., so that $R_2^2 - R_1^2$ must not be less than 25,000 λ.

Since $\lambda = $ (about) $1/50{,}000$ in., $(R_2^2 - R_1^2)$ must be not less than 0·5.

If then we measure the radii of the first two rings in inches, by holding a scale near the window while looking in at the eyepiece of the interferometer, square them and subtract the squares from one another, the difference must not be less than 0·5. This accuracy is sufficient for most practical requirements.

The same reasoning may be applied to the use of the proof plane. Here it must be remembered that the curvature of the wave front produced by an error of one wavelength in the surface is equal to

---

* The power of the lens is the reciprocal of the focal length. The advantage of the expression lies in the fact that if a number of thin lenses are put close together the power of the combination is equal to the sum of the powers of the separate lenses.

($n-1$) λ, or approximately $\lambda/2$. We may therefore, when using the proof plane, allow $R_2{}^2 - R_1{}^2$ to be equal to $\lambda F^2/\triangle F$.

It is convenient to have a rule for the permissible error of focus caused by a prism. First-class work should fulfil the following conditions. If $\triangle F$ is the alteration of focus of the image of a distant vertical line produced by putting a prism in front of a telescope of focal length F (both measured in metres), then the prism has a power of $\triangle F/F^2$ diopters. A good though somewhat severe rule is that if w is the width of the aperture in *inches*, then $\triangle F/F^2 = 0.0057/w^2$.

Tolerances in the specification of optical components

Tolerances for definition and for surface

229 Lord Rayleigh in his article on Optics in the *Encyclopaedia Britannica*, 9th Edition, 1884, reprinted in his *Scientific Papers*, **2**, 410, etc., considers what tolerance is permissible in the distortion of a wave front without impairing materially the resolving power of an optical instrument. The passage is included in a section on the resolving power of optical instruments from which the following passages are quoted—

According to the principles of common optics, there is no limit to the resolving power of an instrument. If the aberrations of a microscope were perfectly compensated it might reveal to us worlds within a space of a millionth of an inch. In like manner a telescope might resolve double stars to any degree of closeness . . . How is it, then, that the power of the microscope is subject to an absolute limit, and that if we wish to observe minute detail on the overlighted disc of the sun we must employ a telescope of large aperture ? . . .

A calculation based upon the principles of the wave theory shows that, no matter how perfect an object glass may be, the image of a star is represented, not by a mathematical point, but by a disc of finite size surrounded by a system of alternately dark and bright rings. Airy (1834) found that if the angular radius of the central disc (as seen from the centre of the object glass) be θ, $2R$ the aperture, λ the wavelength, then

$$\theta = 1.2197 \frac{\lambda}{2R}$$

showing that the definition, as thus limited by the finiteness of λ, increases with the aperture.

In estimating theoretically the resolving power of a telescope on a double star we have to consider the illumination of the field due to the superposition of the two independent images. If the angular interval between the components of the double star were equal to 2θ, the central discs would be just in contact. Under these conditions there can be no doubt that the star would appear to be fairly resolved, since the brightness of the external ring systems is too small to produce any material confusion unless indeed the components are of very unequal magnitude.

Actually, a double star can just be resolved at an angular separation of one half this. Since the disc is not of uniform brightness the phrase, " the central discs would be just in contact," needs defining more closely. The condition is more correctly stated by Schuster, who says that it is a matter of experience that a close double star may be recognised as such when the images are at such

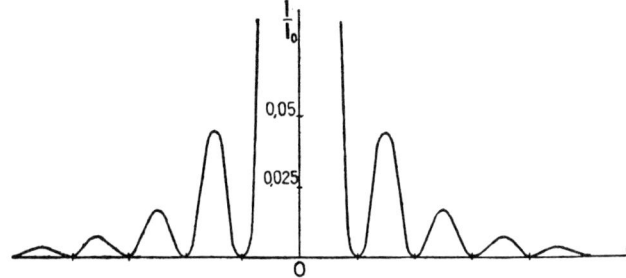

Fig. 173—Distribution of light in a slit image

a distance apart that the centre of the bright disc of one falls on the first dark ring of the other, $i.e$ when $\theta = 1 \cdot 22\lambda/D$, D being the diameter of the object glass.

The theory of image formation is simpler when the aperture is rectangular instead of circular and when the object consists of one or more bright lines parallel to one of the sides of the aperture.

Let us then consider the case of the image of a distant fine bright vertical line, radiating with wavelength λ, produced by a perfect object glass, focal length f and of rectangular aperture, whose horizontal width is a. Then the intensity of illumination I of a point in the focal plane at a distance d from the geometric focus is given by:

$$I = I_0 \frac{\sin^2{(\pi\ ad/f\lambda)}}{\pi\ ad/f\lambda}$$

I_0 being the intensity of illumination at the geometric focus.

If we plot the values of I we get the curve shown in Figs. 173 and 174.

The points of no illumination are at distances from 0 equal to $f\lambda/a$, $2f\lambda/a$, etc.

The abscissae represent the distance from the geometric focus, and the ordinates the values of I/I_0. As it is impossible to represent to scale the central maximum in Fig. 173, the central part of the curve is shown on a smaller scale in Fig 174. (Adapted from G. Bruhat (1935) page 213).

The actual appearance of the diffraction pattern as photographed is shown in Fig. 175 (*top*), which was taken on a spectrograph of 150 cm focal length with an aperture 6 mm wide; the line photographed is the 4358 line of mercury, the width of slit being 0·01 mm. At the bottom is seen the same line with an aperture of 50 mm. Fig.

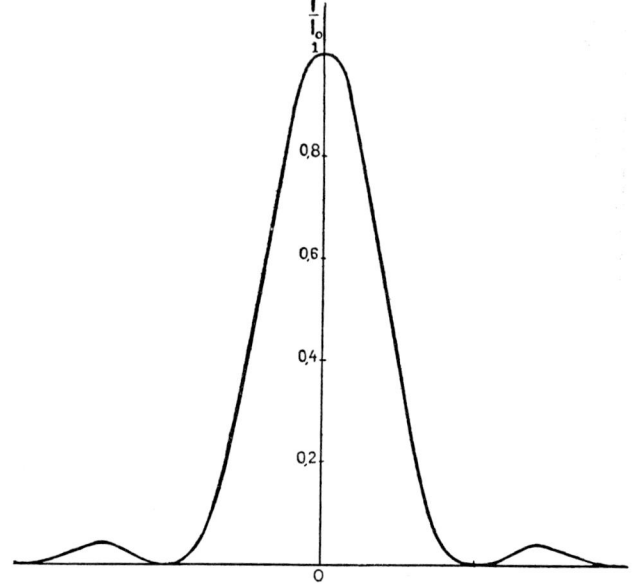

Fig. 174—Central portion of Fig. 173 on a reduced scale

176 shows (full line) the intensity curve produced by a lens of 100 mm focal length of 13 mm aperture free from spherical aberration; the broken line that for a single lens of the same focal length and aperture of the form to give minimum spherical aberration.

In the article referred to above, Lord Rayleigh's conclusion on the subject of tolerance is as follows—

> An obvious inference from the necessary imperfection of optical images is the uselessness of attempting anything like an absolute destruction of aberration. In an instrument free from aberration the waves arrive at the focal point in the same phase. It will suffice for practical purposes if the error of phase nowhere exceeds

$\frac{1}{4}\lambda$. This corresponds to an error of $\frac{1}{8}\lambda$ in a reflecting and $\frac{1}{2}\lambda$ in a (glass) refracting surface, the incidence in both cases being perpendicular.

If the object glass is examined on a Twyman and Green interferometer an error of phase of $\frac{1}{4}\lambda$ will appear as one half band, since the error is doubled by passage of the beam twice through the object.

Let us suppose that there are two lines of different intensities, the distance between whose wavelengths is to be measured. If the error is a symmetrical one, as illustrated in Fig. 176, no error will be caused, although the measurement will be made less precise.

If, however, the error is asymmetrical, the intensity curves are asymmetrical, and although the shapes are similar, the photographed images will in general not be so, for the one may be over-exposed or the other under-exposed. In such case, the measured distances between the apparent middle points of the photographed images may be in error by an appreciable amount. Since the accuracy with which a setting can be made on a line with a micrometer is a very small fraction (some say as little as one-fortieth part) of the resolving power, it can be understood that the presence of a phase error in the beam which will not prevent two close lines from being resolved, may cause an important error if one has to measure their distance apart, as in determining wavelengths.

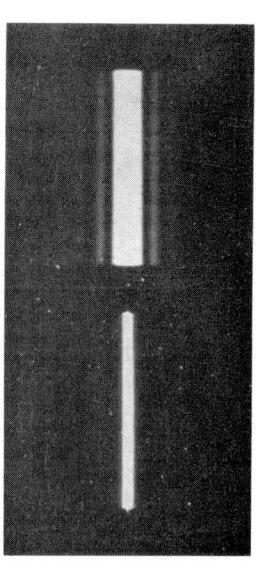

Fig. 175—Diffraction image of slit

Many other illustrations of this point might be given. For most purposes Lord Rayleigh's rule is a sound and safe one ; for example, the phase errors produced by even the best photographic lenses (except in rare instances for the middle of the field) are many times greater than those which can result from ordinary careful working, and an extra $\frac{1}{4}$ wavelength phase error due to faulty glass or workmanship is of no consequence. The rule cannot, however, be relied upon universally ; each case must be considered separately, and the workshop drawing should stipulate an accuracy of surface appropriate to the purpose for which the optical work is to be used.

230 It is sometimes useful to have a test for glass which is a considerable amount out of parallelism, say in a 3 in. plate up to

1 mm. This can be effected if the glass is not too thick by using a sodium flame instead of mercury light. If we consider a plate viewed at 45°, the illumination being by a diffusing screen illuminated with a sodium flame from a Meker burner, one will see fringes at the points corresponding to about 1/20 mm difference of thickness.

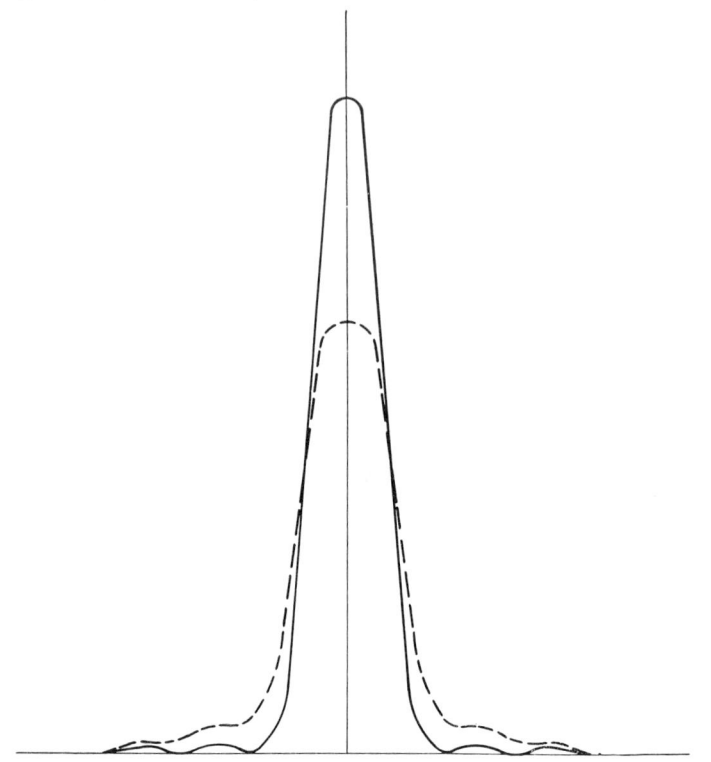

Fig. 176—Intensity curves produced by lenses with and without spherical aberration

Testing angles

Tolerances in angle measurements

231 Errors of angle may result either in defects of colour or of deviation, and both of these must be taken into account in fixing tolerances. For example, it has been found that a wedge of plate glass giving a deviation of $1\frac{1}{2}$ minutes does not cause enough colour to entail the rejection of any optical instrument which utilizes a two-inch aperture at which point the plate glass is interposed. Let us suppose then that in an optical instrument a 2 in. aperture pentagonal prism is used, and that an error of 5 minutes is permitted in the deviation of such a prism, then if

a is the error in the 45° angle
and b the error in the 90° angle
both reckoned positive if too large, then

$3a + b/2$ must not be more than 5 minutes (to keep within the permissible deviation);

$a + b/2$ must not be more than $1\frac{1}{2}$ minutes (to avoid colour), assuming that the crown glass of which the pentagonal prism is made has approximately the same dispersion as plate glass.

Fig. 177—The Guild–Watts precision spectrometer

If then we wish the workshop to have the advantage of the greatest possible tolerance, a set of tables should be worked out giving the permissible range of b for each value of a between $-3 \cdot 25$ and $+3 \cdot 25$—the biggest range of a which will permit the above two equations to be satisfied. One finds on doing this that if a is $-1 \cdot 25$, b can lie anywhere between 5·5 too large and 0·5 too small; while if a is $+1 \cdot 25$, b can be from 5·5 too small to 0·5 too large. This single illustration is given to demonstrate the value of careful consideration in the matter of fixing angle tolerances.

Testing angles and measuring refractive indices

231.1 Every optical factory should be provided with an accurate goniometer for measuring refractive indices. This should be an instrument with which refractive indices can if necessary be measured with certainty to one unit in the fifth decimal place. Such accuracy is essential in, for example, the manufacture of high quality camera lenses.

The Guild-Watts Precision Spectrometer was designed for such requirements twenty years ago by George Watts, the Managing-

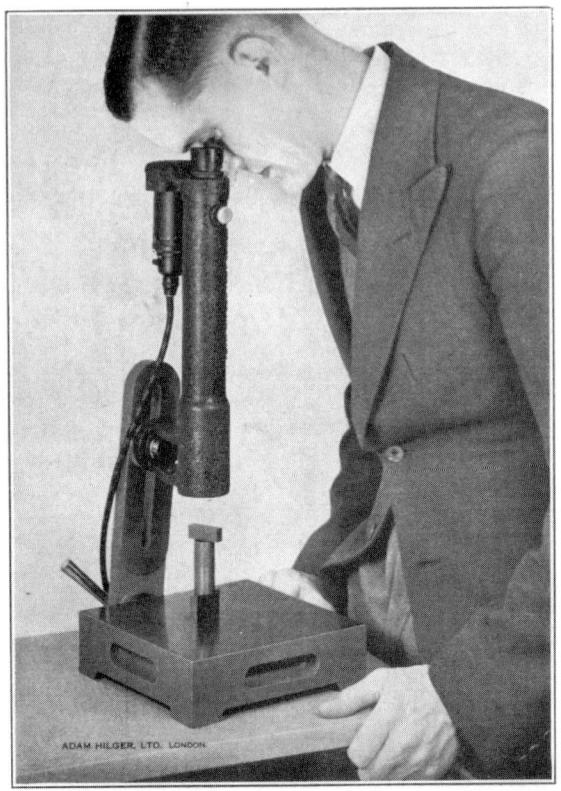

Fig. 178—Checking parallelism of ends of a tube using a Standard Angle Dekkor and Surface Plate Stand

Director of E. R. Watts & Son, and John Guild of the Optics Department of the National Physical Laboratory, Teddington, England.

It is suitable also for research in refractometry and the highest class of goniometric work.

The latest model (Fig. 177) embodies improvements which are the result of continuous development. Providing reasonable precautions are taken, and a correct experimental technique adopted, a refractive index measurement of one unit in the sixth decimal place can be obtained, while certainty in the 5th place is easily assured.

The most useful wavelengths for which to determine refractive indices of glasses for lenses are—

Sodium—D_1, 5895·92A
D_2, 5889·95A
Hydrogen—C, 6562·79A
F, 4861·33A

These are the values given in the M.I.T. tables.

The Angle Dekkor

232 A convenient angle-checking instrument is made by Hilger & Watts Ltd and has been used in our optical workshops for many years. It is known as the Angle Dekkor (see Fig. 178). This instrument consists of a telescope with a brightly illuminated scale at the focus at the eye-piece end, light from which is (after collimation) reflected back into the telescope by the reflecting surfaces under observation, an image of the scale being seen in the eyepiece (see Fig. 179). The position of the reflected scale relative to that of a fixed scale in the eyepiece indicates the relative angular position of the surface under test.

The instrument is employed in the following manner—

The inclination of the telescope to the polished base-plate of the instrument can be easily adjusted and it can be clamped in any desired position. If it is arranged to view the polished surface of the base of the instrument as shown in Fig. 178 the brightly illuminated reflected scale will be seen at right-angles to the fixed scale and with similar dividing. Each division of the scales represents one minute of angle. By tilting the telescope these two scales can be made to cross each other as shown in Fig. 179a. This arrangement of scales is made so that the position of the reflected one can be controlled in two directions. If the reflecting surface (or the telescope) is slightly tilted the position of the reflected scale in the eyepiece is altered.

When *two* surfaces are being compared *two* reflected images are seen. These reflected images are cut across by the horizontal fixed scale, and the difference between the readings on the two reflected scales gives the angle between the surfaces, the fixed scale being used as an index.

Application of the Angle Dekkor

(1) *Testing for parallelism*

233 Parts to be tested for parallelism are mounted on the instrument as illustrated in Fig. 178 in which a test is shown on a tube 0·7 in. diameter, the ends of which are required to be parallel to one minute (a variation in length from side to side of the tube of 0·0002 in which is not easily measured with a micrometer). The telescope is adjusted to obtain a reflected scale image from the surface plate. The tube is placed in position with an optical flat on top of it (a Johannsen gauge is shown in the illustration), when a second scale image is produced and the difference in position between the scale images

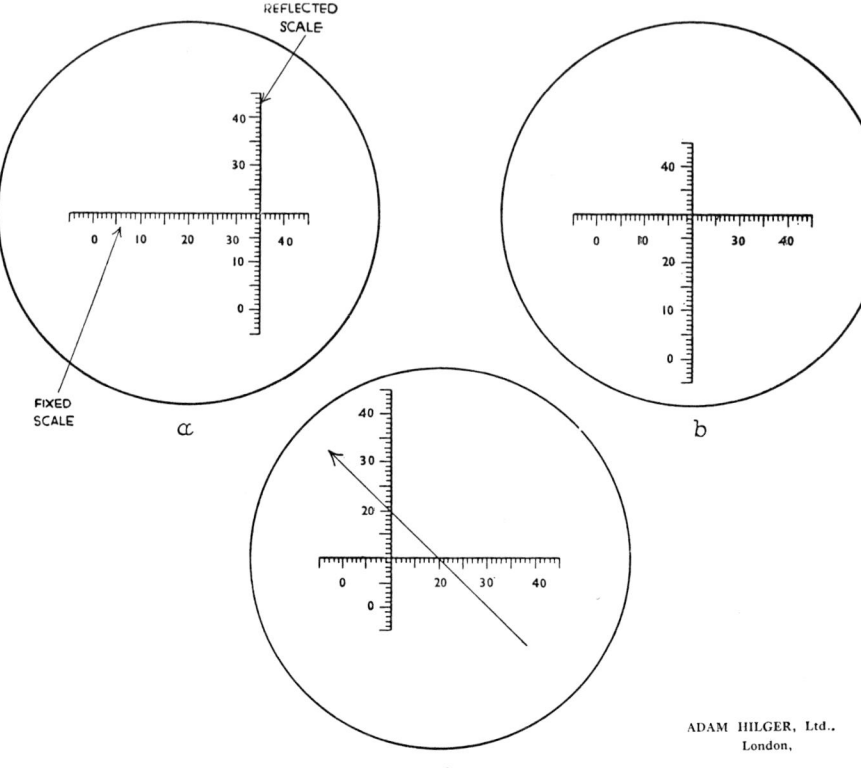

Fig. 179—Scales as seen in the Angle Dekkor
The reflected scale actually has bright lines and figures

(from the surface plate and the reference plate) indicates the inclination between the ends of the tube in minutes of angle. A plane parallel plate is tested in the same way. The sensitiveness of measurement with the Angle Dekkor is about ±6 seconds.

(2) *Testing angles*

234 Angles are measured by using a reference gauge of known angle. The reference gauge and part under test are laid upon the surface plate and the reflections from both viewed in the telescope (see Fig. 180). Then, providing the difference in angle between the standard and the part is not too great (say less than 40'), the angle of the part can be read at once as a difference.

If it is desired to avoid bringing the work into contact with any flat surface for fear of scratching, or to avoid the trouble needed to avoid dust, the reference gauge and part under test are placed one upon another on the surface plate of the Angle Dekkor. The telescope of the Dekkor is then lowered to a horizontal position and the two pieces

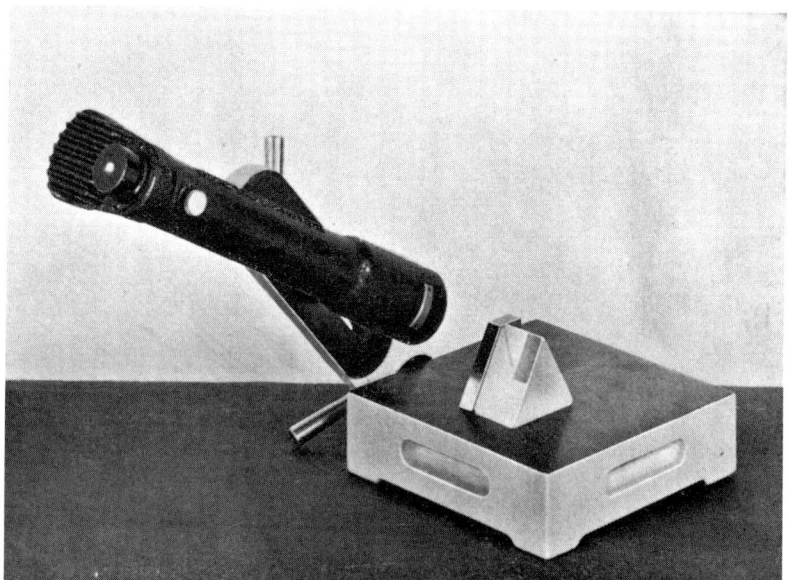

Fig. 180—Direct comparison of angle with a standard angle, using the Angle Dekkor

rotated together so that an image reflected from the lower surface appears in the field of view. The upper part is then rotated by tapping it until it too gives an image in the field of view. This tapping must be continued until these two images lie along each other. When this condition has been obtained, both parts are rotated together until the other sides face the telescope. The difference in angle between the parts will then be apparent from the horizontal separation of the images.

This operation is very much simpler than it sounds, and is much used in our workshops.

It is worth pointing out that the Angle Dekkor can be used on a grey surface provided the latter be so arranged that the light is reflected from the grey surface, at an oblique angle, on to a silvered surface.

The use of sub-multiple angles

235 A method of increasing the accuracy of the Angle Dekkor is afforded by the property possessed by certain angles, which are

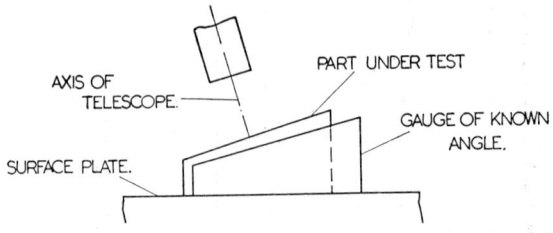

Fig. 181—Comparison of angles by means of the Angle Dekkor

sub-multiples of 180°, of returning a beam of light in its original direction. Sub-multiples of 180° will here be referred to in short as " sub-multiples."

This is illustrated diagramatically in Fig. 182 the angle in this case being 45°.

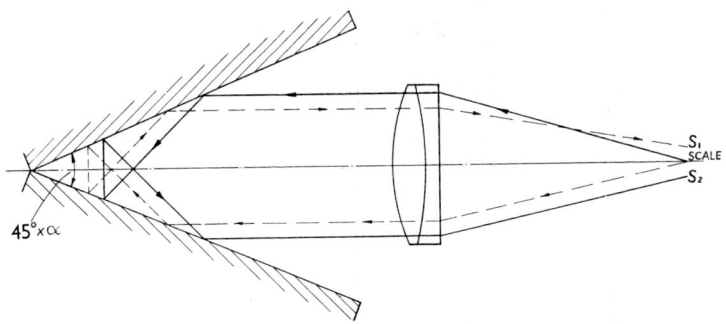

Fig. 182—The use of sub-multiple angles

$S_1 S_2$: Images of scale, only superimposed when the angle is exactly to the stated value

The general principle may be stated thus—

Angles which are sub-multiples will return a beam of light in a direction parallel to the incident beam, provided that the beam is in a plane perpendicular to the line of juncture of the surfaces. If this line bisects the aperture of an object glass used as an auto-collimator, two bright images of the scale will be formed in the focal plane of the objective unless the angle is an *exact* sub-multiple.

In the case of the even sub-multiples, i.e. 180/2, 180/4, etc., the beam at its mid-point is travelling in a direction at right-angles to its original direction, and a ray which leaves the objective from one half re-enters through the other ; this is a constant-deviation system, and the position of the telescope in the plane in which the angle is measured is not important.

For the odd sub-multiples, *i.e.* 180/3, 180/5, etc., the beam at its mid-point is reflected normally from one of the sides, and is consequently reflected back along its own path ; this is not a constant-deviation system, and the telescope axis must lie on the bisector of the angle.

If the angles are not exactly sub-multiples of 180° the error in the angle will give rise to a divergence of the reflected beams equal to twice the " number " of the sub-multiple, that is an error of α in a 45° angle will appear as $(2 \times 180\alpha)/45 = 8\alpha$.

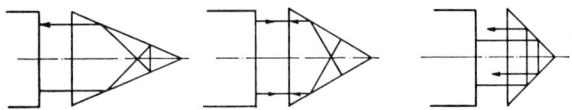

Fig. 183—The use of sub-multiple angles
Internal multiple reflection

As the Angle Dekkor is already calibrated to read correctly under first sub-multiple conditions, the magnification of error obtained may conveniently be remembered as being equal to the sub-multiple number. A table of the magnification values of the first 12 sub-multiple angles is appended—

Angle	180°	90°	60°	45°	36°	30°	25° 42′ 51·4″
Magnification	×1	×2	×3	×4	×5	×6	×7
Angle	22° 30′	20°	18°		16° 21′ 49·1″		15°
Magnification	×8	×9	×10		×11		×12

If the three sides of a prism are polished the accuracy of the sub-multiple angles can be determined by internal reflections as shown in Fig. 183, in which cases the magnification is still further increased by the factor n, the refractive index of the prism.

For the higher magnifications the reflection of the surfaces must be aided by metallization.

The diagrams show some practical applications of the principle which have been used in our workshops during the making and testing of primary angle standards of the very highest accuracy ; the results have been inspected by various National Standards Laboratories and accepted as accurate within their limits of measurement.

The following factors limit the accuracy attainable with sub-multiple principles—
 (a) The flatness of the reflecting surfaces;
 (b) The reflecting power of the surface coatings used;
 (c) The reduction in effective resolving power when the higher orders are employed, due to the decreased aperture available;
 (d) Where the reflection takes place inside a glass prism, the accuracy is affected by the homogeneity of the glass.

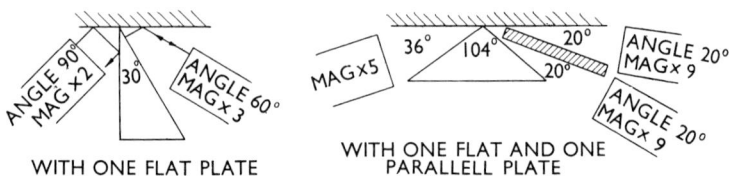

Fig. 184—The use of sub-multiple angles
External multiple reflection

Measurement by multiple reflection of angles which are not sub-multiples

Angles not in themselves sub-multiples can be built up by sums and differences of sub-multiples and set out with the Angle Dekkor. The procedure can best be described by taking an actual example. It requires a sufficient number of plane parallel mirrors silvered or

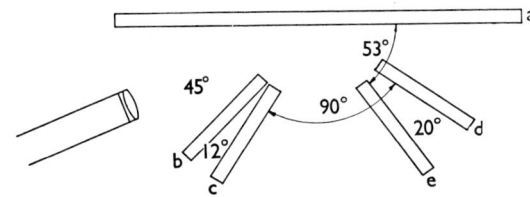

Fig. 185—Use of sums and differences of sub-multiple reflection

aluminized on both sides, each one having one grey edge perpendicular to the mirror faces. These stand on a flat plate (Fig. 185).

$$53° = 180° - (45° + 12° + 90° - 20°)$$
sub-multiples   1   4   15   2   9

The mirrors a, b, c and d are first arranged as shown in the sketch, b, c and d successively being set with the Angle Dekkor so that the corresponding coincidences of the reflected images are obtained. Mirror e is then interposed to set out the angle of 20°, after which mirror d is removed. The angle between a and e is then, obviously, the required 53°.

To avoid movement of the plates when transferring the telescope from one angle to another, we have constructed a jig enabling two or three separate telescopes to be directed on to the same set-up and each is read in turn without moving anything at all.

It must be remembered that, when the higher orders are in use, several reflections take place inside a small space at the inner end of the strips, and this has the effect of magnifying any lack of flatness which is extant in the plates as a focusing error, in addition to the usual error in angle due to the sag of a curved plate.

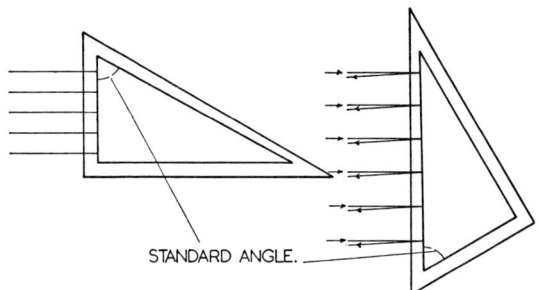

Fig. 186—Comparison of angles by means of the Hilger interferometer

Comparing angles with the Hilger interferometer

236 For a test of equality of angles to a higher precision the Hilger Interferometer is used (Fig. 186). This is described in the next chapter, which should therefore be read at this stage. Let us suppose that two 60° angles are to be compared, one being the standard. If the bases of the two prisms are perpendicular to the polished sides, a method for comparing their angles on the Interferometer is to get one face of each prism in the same vertical plane simply by standing one prism on the other and tapping the top prism round (Figs. 186–7). The parts are placed on a table of convenient height so that one face of each replaces the side mirror F (Fig. 192) in its position normal to the direction of incidence of the light from the collimating lens D. The observer should place his eye close to the lens E. One spot of light, at least, should be seen. This spot is the image of the pinhole as seen by reflection from the back mirror G. By suitably tilting and turning the prism and the standard angle two further spots of light can be brought into view and careful adjustment will superimpose them upon one another and upon the first spot, a condition which is best observed when the eye is placed well away from the eyehole P. On bringing the eye close to P the two faces will be clearly seen, covered with a system of bands whose number is reduced to a minimum by further slight adjustment which can be supplemented to a very slight extent by tilting the diagonal mirror. A typical appearance as seen in the instrument is shown in Fig. 187 and it is necessary to get both sets of fringes nearly horizontal. If no accurate levelling means are available for obtaining this condition the fine adjustment on the diagonal mirror may be utilized—provided the adjustment required is very slight.

Both prisms are now rotated so that their other faces are presented to the beam as shown in Fig. 186*b*. Needless to say this rotation must

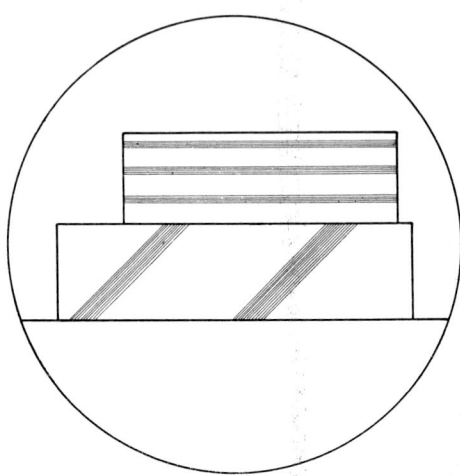

Fig. 187—Comparison of angles by means of the interferometer
Appearance of field of view

be very carefully done to avoid shifting the top prism relative to the bottom one. A slight rotation must be given to both prisms (or the diagonal plane used) in order that the fringes on the standard may be horizontal. Then, if there is any angular difference between the prisms, the fringes seen on the one prism will be tilted, the number of fringes cutting the horizontal edge of the work being a measure of this angular difference. If the interference bands are formed by the green light of a mercury vapour lamp each band represents 1/100,000 in.

Testing the 90° angle of a roof prism on the Hilger interferometer
237 The side mirror F of the instrument (Fig. 192) is swung round to the right-angle position and the roof prism is placed on a table of suitable height to reflect the light on to the mirror. Care should

be taken to see that the hypotenuse face of the roof prisms is centrally disposed with reference to the centre of the interferometer table. To set up the prism, proceed as in the above case of the comparison of angles except that there will only be two spots of light instead of three to superimpose and adjustment is made by tilting mirror F.

A typical appearance of a roof prism is shown in Fig. 188. Two horizontal bands appear, indicating that the angle of the roof is not quite 90°. If the angle is smaller than 90°, then by pulling the telescope rod gently in a downward direction (which increases the path length of the light reflected in mirror G) the fringes will move inwards towards the dividing edge.

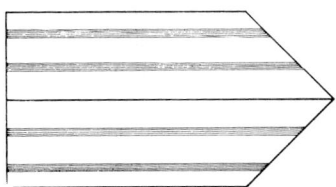

Fig. 188—Diagram of fringes obtained in testing a roof prism on the Hilger interferometer

238 *A standard set of angles* as supplied regularly by Hilger & Watts Ltd, consists of a block of steel whose faces produce the following angles—

30°, 60°, 90°,

together with four angle gauges—

1°, 3°, 9°, 14°.

A patent has been applied for in respect of this set of angle gauges.

By contacting not more than three of these together so that the angles are added or subtracted (i.e. placing them with apex to apex or base to apex), any angle can be set up between 0° and 90° at intervals of 1°.

They are put together in the same way as Johanssen gauges, but without the comparatively violent process of wringing that is employed by engineers in the use of these gauges.

Many applications of the Angle Dekkor to mechanical engineering will be found in a paper by Harrold (1948).

An absolute standard of planeness

The liquid surface interferometer

239 All production of flat surfaces requires a comparison surface of unimpeachable planeness. The production of such surfaces by rubbing together three optical tools has been described in §186, but it is worth while having an independent and absolute standard. Such a standard can be provided by a liquid surface. The following description is abstracted from an article by Barrell and Marriner 1948).*

Although Lord Rayleigh described, in 1893, the use of a water surface as a naturally reproducible basis of reference for verifying flat surfaces by interferometry, surprisingly little practical application has been made of the high quality of optical flatness which, with certain reservations, is possessed by a liquid surface. The allied property that liquid surfaces have of indicating level, is of course, well known and very frequently employed in physical and engineering metrology.

The main reservations concerning the quality of flatness achieved by a surface liquid are based upon the effects of—

(i) The earth's radius of curvature;
(ii) Capillary curvature at the boundary of the surface;
(iii) Vibrational and other external disturbances; and
(iv) Superficial contamination.

For all practical purposes the effect of (i) is negligible, the earth's curvature causing the centre of a liquid surface of 10 inches (25 cm) diameter to project by only 12A (1/500 of a wavelength of yellow light) above the true plane intersecting its perimeter, if the effect of capillary curvature at the boundary region is omitted from consideration. With many liquids the effect of capillary curvature vanishes approximately 1 inch (2·5 cm) from the boundary, and causes no difficulty if the diameter of the liquid reference surface is about 3 inches (7·5 cm) larger than the surface to be tested. Recent experiments in the Metrology Division of the National Physical Laboratory have shown also that the effects of external vibrations and other disturbances are eliminated by employing as the reference medium a suitably viscous liquid with a clean surface shielded from air draughts.

Fig. 189 is a diagram of the optical arrangement of the Fizeau type of interferometer used at the National Physical Laboratory for liquid surface interferometry.

An important item of the equipment is a single plano-convex lens of 15 inches (38 cm) diameter (F/5) which was acquired from war-surplus stocks of optical components. This is arranged horizontally as the collimating lens of the interferometer, and directs a beam of parallel monochromatic light vertically downwards on to the test

* Crown copyright. This abstract and the illustration, Fig. 189, are published with the permission of the Director of the National Physical Laboratory, Teddington, England.

piece—a lapped cast-iron surface plate of 12 inches (30 cm) diameter supported on three levelling feet.

The liquid layer (medicinal liquid paraffin has been found best), about $\frac{1}{8}$ inch (3 mm) thick, rests on the clean metal surface, and is retained either by paper strip gummed round the periphery of the

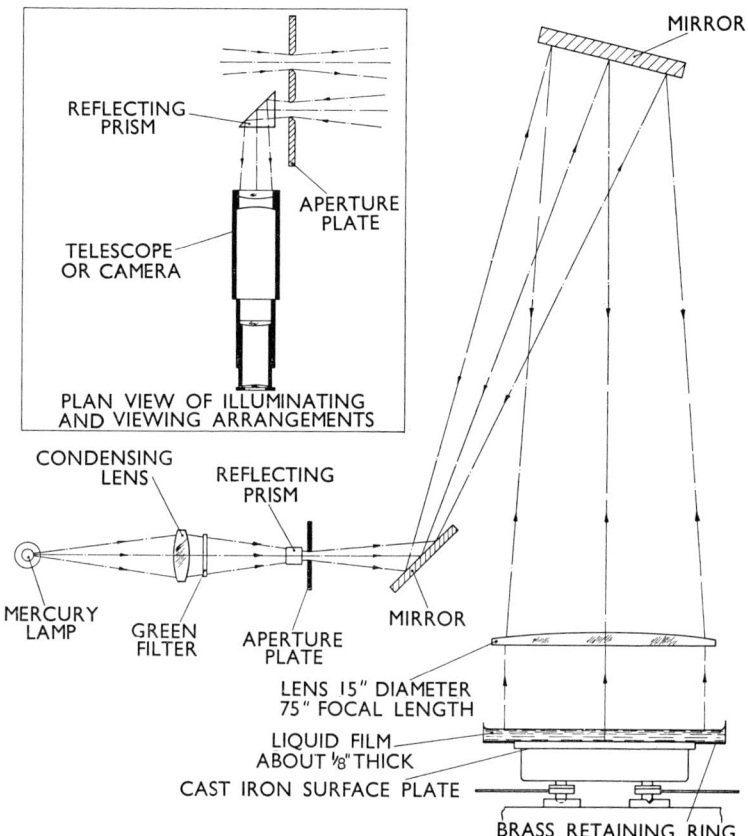

Fig. 189—The optical arrangement of Fizeau interferometer for using a liquid plane surface

plate, or, better still, by means of a projecting brass annulus fitting round the rim of the test piece so that the surfaces of the annulus and test piece are approximately co-planar. The outer edge of the annulus is itself suitably formed to retain the liquid layer. It is convenient if the brass annulus projects about 2 inches (5 cm) outwards from the rim of the test piece so that the whole surface is unaffected by the capillary curvature existing near the boundary of the retaining wall. If necessary, the joint between the annulus and rim may be made liquid-tight by means of " Plasticine " used externally as a filler.

The source of monochromatic light is an Osram mercury discharge lamp of the hot-cathode type operated from the 230-volt a.c. mains. Light from this source is condensed on one of the two chamfered apertures, each of $\frac{1}{8}$-inch (3 mm) diameter, cut in a plate placed in the principal focal plane of the large collimating lens mentioned above. A Wratten filter No. 77 isolates the green line at 5461A for illuminating the interferometer. For convenience in the layout of the interferometer, the monochromatic light transmitted through the entrance aperture is first reflected upwards by a small adjustable mirror towards a larger adjustable mirror which, in its turn, reflects the light vertically downwards to the collimating lens. Both mirrors are of the front reflecting type, and are of good optical quality.

The optical system described is one of the standard arrangements of the Fizeau interferometer in which fringes of constant thickness (in this case consisting of dark bands on a bright field) are produced by the interaction of two beams, one partially reflected from the liquid surface and the other totally reflected from the metal surface. Control of the disposition of the fringes is obtained by means of the levelling feet under the surface plate. The liquid surface fulfils the role of the optical proof plane in the normal type of flatness-testing interferometer, and any departures from straightness and parallelism of the fringes, or of the uniformity of fringe spacing, indicate errors of flatness of the test piece.

Measurement of dimensions

239 Where large numbers of parts have to be gauged all of the same kind, the air operated gauges as made by Solex Gauges Ltd, 223 Marylebone Road, London, N.W.1, should be very suitable. Among their advantages is the fact that, whether for diameter or for thickness gauges, there is a definite clearance between the gauge and the work to be measured, air passing in the space between. They are stated to be reliable to within 1/10,000 inch having once been set on a master gauge ; and another advantage would appear to be that, unless the difference of diameter of the measured object is considerable, no plus or minus gauges are necessary.

Developments of these gauges for optical purposes are dealt with by the French Solex Company, 190 Avenue de Neuilly, Neuilly-sur-Seine, France ; in particular they make an elegant and accurate interferometer. An improved form is described in National Physical

Laboratory report S.S. 188 of Sept. 1939 with which an accuracy of comparison of 10^{-6} in. was attainable.

Light sources used in optical manufacture and testing

For general lighting, the use of test plates, and interferometry

240 The most useful illuminant in the optical workshop and in the testing room is the long tube, low pressure, direct current mercury vapour lamp such as is supplied by the Hewittic Electric Co. This lamp gives a mercury spectrum of well-defined lines with little or no background, and test-plate fringes seen by its light are strong and distinct.

Where conditions permit, as in the testing room, it facilitates test-plate tests if a diffusing screen of fair size is fitted up at an angle of about 60° over a board covered with dark velvet cloth or green baize. The board can be large enough to accommodate several test plates and the sloping diffuser above it can be conveniently made from tracing cloth stretched on a light wooden framework.

The same illuminant is also very suitable for the use of the Hilger Prism and Lens Interferometers. When monochromatic light is required the mercury green line (5461A) can be sufficiently isolated by filters.

"*Osram*" *thallium laboratory lamps*—Thallium vapour laboratory lamps are made by the Research Laboratories, The General Electric Company Ltd, Wembley, England. The thallium lamp was initially designed as a replacement source for the Zeiss Micro-interferometer and has been used satisfactorily with this instrument. It provides a bright source of almost pure monochromatic radiation, 99·5 per cent of the light being confined to the 5350A line. This line is free from the satellites which limit the use of the mercury vapour lamp in interferometry; interference bands are sufficiently visible for the purposes of practical testing, and without filters, over a path difference of about 30 mm.

For local monochromatic illumination and refractometry

Gas discharge lamps (*G.E.C.-Hilger*)—These, particularly those filled with neon, can be run on ordinary electric light current at 200 volts D.C. or more, and give a line spectrum which can be used with the interferometer when no mercury lamp is available. Their brilliance is not high and the luminous area is restricted. The form known as the Osglim lamp is a possible substitute for the mercury lamp for test-plate observation since its luminous area is reasonably large but its intensity is low and the spectrum contains more lines. They can be used on A.C. with shortened life.

Osira metallic vapour discharge lamps—These give relatively simple spectra of high intensity and are suitable for refractometry, though their pressure is rather high for interferometry. Forms are available giving the spectra of sodium, mercury, cadmium, mercury and cadmium, and zinc. They have a high intensity and are steady. Zinc lamps run on A.C. only. All others are supplied in separate types for A.C. or D.C.

Gas sodium flame—For emergency working and for occasional refractometry a sodium flame can be used. A simple way of obtaining such a flame of moderate intensity is to place a fair-sized bead of sodium borax on the grid of a Meker burner or one of the upright Bray burners. This needs little or no attention and burns quietly without crepitation or fumes. A chip of rock-salt may be used similarly, but is liable to " spit " until it has fused.

Vacuum tubes—For refractometry, since custom has established the hydrogen lines for refractive index (although the lithium and mercury lines are more easily obtained and better spaced), a hydrogen vacuum tube should be used. This is preferably of the Guild type, with a large attached bulb. The best means of excitation is a small transformer, such as is used for neon signs, giving about 15 milli-amperes at 2000 volts. (Such transformers can be obtained from Claude General Neon Lights Ltd, Transformer Type A.2.)

Lamps for interferometry

Mercury-198 isotope discharge lamps—The use of radiation from a single, even atomic number, mercury isotope as a monochromatic light source for interferometry, was first suggested by Dr. W. E. Williams. He urged the production of Hg_{198} by exposure of gold to slow neutrons for this purpose, pointing out, for example, that the green line would be free from satelites. Owing to the facilities now available in England for the production of this isotope the General Electric Company's Reasearch Laboratories, North Wembley, Middlesex, are now able to market several lamps in which the radiation is from this isotope. The spectrum lines approach the ideal of strict monochromatism ; they are

Yellow	5790·662 A.
	5769·598
Green	5460·753
Violet	4358·337
	4046·571

Three types are available ; air-cooled and water-cooled internal electrode, and electrodeless lamps.

Lamps for illumination by ultra-violet light

The Hanovia mercury lamps supplied by the Thermal Syndicate, Old Pye Street, Westminster, give ultra-violet radiation suitable for the production of fluorescence. Where great intensity of the line 2537 is required the Houtermann type of mercury lamp may be recommended —it also is made by the Thermal Syndicate.

Mercury lamp for the Raman effect

In the Hilger Raman Source Unit, marketed by Hilger & Watts Ltd, the liquid under test slides vertically into the innermost of three co-axial glass tubes. These provide a water jacket to protect the liquid from the heat of the lamps and also when desired, a liquid sheath to act as a colour filter.

Grouped around these tubes symmetrically are four low pressure mercury discharge lamps completely enclosed in a hollow cylindrical metal jacket, whitened internally to act as a diffuse reflector for all four lamps. This arrangement shows an improvement in efficiency of about 4 times over the usual reflecting mirrors owing to the high reflectivity of the coating used at the wavelengths employed and the utilisation of all the radiation from the lamps. After circulating round the water jacket between the lamps and the filter cell, the cooling water passes round the outer, hollow, metal jacket of the source unit.

Mercury lamps to serve as very powerful source of illumination

Very intense mercury lamps are available, which although not sufficiently monochromatic for interference work can be made sufficiently nearly so for some purposes by the use of filters.

Such lamps are made by the General Electric Company, England, and the British Thompson Houston Company. They are bright enough for projection under great enlargement and are particularly appropriate for rapid photography owing to the intensity of the ultra-violet lines to which all photographic emulsions are very sensitive. They are made in a number of different forms.

" Point Source " hydrogen arcs

" Point Source " Hydrogen Arcs made by the Manufacturers Supply Company, 19 Glebe Way, West Wickham, Kent, are designed to provide high intensity " point " sources of ultra-violet radiation for use in spectrophotometric and absorption measurements. They provide a continuous spectrum extending from 2000° to 3500° A while at longer wavelengths the spectrum consists of lines superimposed on a continuous background.

The lamps are made in two main classes. The first is a water-cooled glass type with a waxed-on fused silica window for the U-V radiation and is denoted by the letter W in the type number. The second class is constructed entirely of fused silica, thereby eliminating the need for water-cooling. In both cases the use of fused silica extends the U-V range as compared with lamps made in special transmission glasses.

The lamps are of the low voltage type where the discharge is confined to a small volume, thereby giving a high intrinsic intensity as compared with the older capillary hydrogen discharge tubes. They provide the additional advantage that a homogeneous, wide angle cone of radiation is obtained, the variation of intensity for an angle of 15° on either side of the axis being less than 1 per cent.

The luminous source is presented in the form of a rectangular slit (recommended for the illumination of optical slits) or of a circular aperture. The sizes of these openings vary from 1 mm dia. to 4 × 1·5 mm.

Radiations for refractometry

Although custom has established the use of hydrogen lines as standards for refractometry, the following wavelengths are easy to excite in the laboratory, and should prove useful as standards for that and other purposes.

Element	Wavelength in angstroms	Method of excitation
Sr	7070·1	Iron Arc (D.C.)
Cd	6438·4696	" Osira " Discharge Tube
Li	6103·642	Meker burner
Na	5895·923	Meker burner
	5889·953	
Hg	5460·740	Cooper Hewitt Discharge Tube
Hg	4916·036	Cooper Hewitt Discharge Tube
Hg	4358·35	Cooper Hewitt Discharge Tube
Hg	4046·561	Cooper Hewitt Discharge Tube
Th	5351·13	" Osram " thallium laboratory lamp

These values are from the Massachusetts Institute of Technology *Wavelength Tables*, measured and compiled under the direction of George R. Harrison.

Notes on parallelism of glass plates

241   1 second   = 0·000,0048 inches in 1 inch
             1 minute   = 0·000,29   ,,   1 ,,
             1 degree   = 0·0175   ,,   1 ,,

TESTING OPTICAL WORK

Testing by interference (reflection) from two nearly parallel surfaces separated by air or by glass—

1 band by mercury green light (air between the surfaces) indicates 0·000,0107 inches out of parallel.

1 band by mercury green light (glass, refractive index 1·52, between the surfaces) indicates 0·000,0071 inches out of parallel.

Angle Between Surfaces (A)	No. of Dark Bands per inch by Reflection (Air Between Surfaces) $=2t/\lambda$	No. of Dark Bands per inch by Reflection (Glass Between Surfaces) $=2nt/\lambda$	Deviation (produced by Back-Surface Reflection) $=2nA$	Deviation (Single Transmission) $=(n-1)A$
1 sec.	0·45	0·7	3·0 secs.	0·5 sec.
5 secs.	2·2	3·4	15·2 ,,	2·6 secs.
10 ,,	4·5	6·8	30·4 ,,	5·2 ,,
15 ,,	6·7	10·2	45·6 ,,	7·8 ,,
20 ,,	8·9	13·6	60·8 ,,	10·4 ,,
25 ,,	11·2	17·0	76·0 ,,	13·0 ,,
30 ,,	13·4	20·4	—	15·6 ,,
35 ,,	15·6	23·8	—	18·2 ,,
40 ,,	17·9	27·2	—	20·8 ,,
45 ,,	20·1	30·5	—	23·4 ,,
50 ,,	22·3	33·9	—	26·0 ,,
55 ,,	24·6	37·3	—	28·6 ,,
60 ,,	26·8	40·7	—	31·2 ,,

Refractive Index, $n = 1·52$.

CHAPTER 12

THE HILGER INTERFEROMETERS FOR TESTING AND CORRECTING PRISMS AND LENSES, AND OTHER INTERFEROMETERS COGNATE THEREWITH

Introduction

242 My justification for the rather full account I propose to give of Hilger interferometers is that their introduction at the Works of Adam Hilger Ltd established an epoch in the development of the firm. Not only so, but their importance in the control of optical production was early recognized by some of the foremost optical firms in the world.

These interferometers received recognition in three unusual ways. The author received on their account, in 1927, the Duddell Medal, awarded by the Physical Society for "meritorious work on scientific instruments and materials," in 1926 The John Price Wetherill Medal awarded by the Franklin Institute of Washington for "the great scientific value of his interferometer for the testing of optical parts," and was, in 1931, with the co-inventor, Alfred Green, granted a ten-year extension of the two principal British patents. Once before only had so long an extension been granted for British letters patent.

W. Taylor, Governing Director of Taylor, Taylor & Hobson of Leicester, said of the camera lens interferometer (§248) in "The Times" of 24 April 1929—

> We have had one of these interferometers in use in our works upwards of four years and it has been one of a number of things which have enabled us to make advances in the design and construction of our optical systems.

This opinion supplemented that of A. Warmisham, Optical Director of the same firm, who said in a letter to me dated 9 May, 1925—

> We see in the Hilger photographic lens interferometer a new and most powerful means of revolutionizing the method of optical design. The present state of the art of photographic lens design is determined by the absence of specific means of dealing with the spherical aberrations of oblique pencils. In spite of all the mathematical work that has been done on this subject there exists no means at present of forming a definite judgment, based on algebraic processes, of the quality of definition given by a lens throughout a semi-field of 26°. We suppose that a fairly complete idea of the

quality of definition could be obtained by a sufficient amount of trigonometrical computation, but an enormous amount of labour would be involved, especially to trace the effect of small departures from the spherical in any surface, for it would be necessary to compute a large number of rays skew to the axis. The lens interferometer will provide quick and certain means of determining what departures from sphericity are required, and in what surfaces they are best applied, in order to reduce to a minimum, at any selected angle, the outstanding oblique spherical aberrations.

Ross Ltd had a like experience and informed me (May 1925) that they found that with its aid they could definitely determine the cause of bad definition in optical instruments and, in any doubtful case, examination on the interferometer showed at once the location and the extent of the fault, and gave them the means of correcting these faults by local re-touching of one or two surfaces.

Wilfred Taylor, of Cooke Troughton & Simms, Buckingham Works, York wrote (1 Oct. 1930)—

We have now used one of your interferometers for a number of years, and regard it as a most remarkable weapon in the hands of the optician, particularly, as regards our own work, in connection with the figuring of large prisms. If it were possible to obtain consistently large pieces of perfect optical glass, the correct angling and working of the surfaces could be ensured by various tests, but, as is well known, imperfections in the material itself may mar the performance of a large and expensive prism, even though the surfaces and angles are free from error.

For example, Dr G. Hansen of the Zeiss-Opton Optischewerke, Oberkochen, Württenburg, sends me the following particulars of the use made of interferometers supplied by Adam Hilger Ltd. (Translated from Dr Hansen's letter of 27 April 1950).

The interferometers were used in Jena to an ever increasing degree for the testing of optical parts. In particular, since the beginning of 1930, such parts for all glass and quartz spectographs in my department were so tested. Further, particular lens systems were frequently examined interferometrically when it was desired to obtain quickly an exact determination of the degree of correction. Later the instruments supplied by your firm no longer sufficed and we then ourselves made similar interferometers for our testing. The large interferometer for photographic objectives was used in the correction of the prisms of the tower spectograph of the Einstein Tower in Potsdam.

The firm of Carl Zeiss, Jena, purchased in 1929 an entire series of these interferometers (prism and lens, camera lens, and microscope objective interferometers). Many years afterwards evidence was afforded that use was made of the instrument to test the performance of *non-spherical* lens systems (see Chapter 10).

These interferometers express the aberrations of optical systems in terms of departure from sphericity or planeness of wavefront arising from passage through the systems. None-the-less they are capable of being interpreted so as to give the aberration in any of the customary forms, namely longitudinal, lateral or rectangular (Perry, 1923). They can also be applied to many other physical problems, and some of these are mentioned in §252.

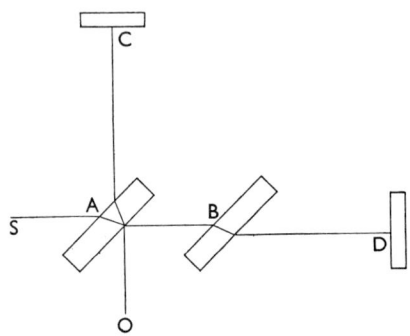

Fig. 190—Arrangement of mirrors in the Michelson interferometer

All these interferometers derive from that of Michelson (1907), illustrated in Fig. 190, as will be described later. This by-the-bye. was not the form originally described by him (Michelson, 1881) which had the disadvantage that the two comparison beams were side by side.

If the reader will turn to §223 he will see that Laurent used a form of Fizeau apparatus in which a lens focused the rays from the source on the eye. By this simple modification it was ensured that those rays which reached the eye of the observer had passed parallel through the plates.

This was, in effect, the modification to the Michelson interferometer that transformed it into the Hilger prism interferometer now to be described, although in the development of the principle many other new devices were employed.

The first types were described in British Patent 103,832 (Twyman & Green, Jan. 1916), of which one of the forms (Fig. 191) was appropriate for testing concave mirrors and telescope objectives only.

The Hilger prism and lens interferometer

243 The most generally useful of the instruments built on the Twyman and Green principle is the Hilger prism and lens interferometer, made by Hilger & Watts Ltd for testing prisms and lenses. This instrument is suitable for workshop use in the operation of retouching.

Numerous methods of testing telescope or camera objectives have been devised with a view to the control of retouching.

The opinions of Schroeder, Grubb, Czapski and Alvan Clark are cited in a résumé by H. Fassbender (1913) of the then known methods of testing object glasses. It omits, however, an ingenious method of Dr Chalmers (1912).

The more recent methods of Waetzmann (Bratke, 1924) (founded on the Jamin refractometer), Ronchi* (1926) and Lenouvel (1924) (these two derived from Foucault's test), are of interest and capable —in experienced hands—of yielding useful results. Of none of them, however, can it be said—as it can of the Twyman and Green forms of

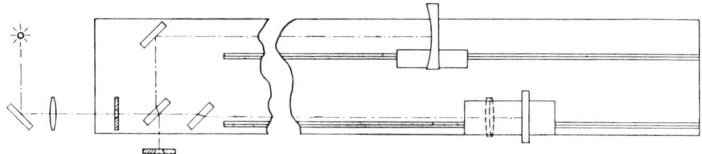

Fig. 191—The first Hilger interferometer

interferometer—that unskilled boys or girls can in a week or so be taught not only to test prisms and lenses and state precisely the nature of their defects, but to correct the optical performance by retouching the surfaces.

It must be added that the phase contrast test developed by Burch (see §252.3) from that introduced by Zernicke (§§227.1 and 336) is also capable of yielding very direct information concerning distortions of wave fronts caused by concave mirrors or lenses.

The Mach-Zehnder interferometer

The Mach-Zehnder interferometer may be mentioned for the sake of historical completeness. In application to the examination of plane parallel plates this interferometer presents appearances almost identical with those of the Hilger Interferometer; and, indeed, but for the fact that it does not lend itself very well to the examination either of prisms or of lenses, there would be little to choose between it and the latter instrument. Doubtless it could be modified to overcome these limitations, but it would then form an extremely cumbrous piece of apparatus.

* King (1934) describes a modification of Ronchi's test which makes it quantitative. The article is clear and well illustrated.

It was described in 1891 by Zehnder and shortly afterwards modified by Mach, and has since received many modifications.

A very complete description of the instrument in its various forms is given in an article by Kinder (Kinder, W., 1946, *Optik*, 1, 413). The paper has a very complete bibliography and a plate of beautiful coloured photographs of the interference fringes, which are like those obtained on the Hilger interferometer.

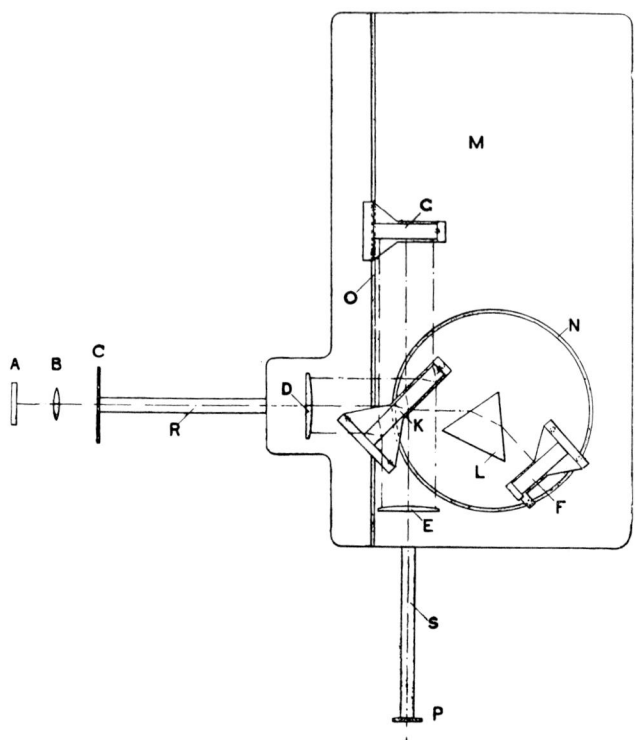

Fig. 192—Diagram of Hilger prism interferometer

The Hilger Interferometers here described produce a series of interference rings which may be regarded as a " contour map " of the imperfections. This contour map can for practical purposes be considered as located at any of the optical surfaces involved and, in the case of the control of retouching, the observer may, if he likes, draw this map upon the surface under treatment. He is then in a position, without further preliminaries, to remove the superfluous material from the prominences by polishing with pads of suitable size and shape, the " contour map " giving all that is necessary for him to know both as to the location and magnitude of the sources of the imperfections.

General construction principles

244 This instrument in its simplest form resembles the well-known Michelson interferometer, the main essential optical differences being that the light is collimated and the two interfering beams of light are brought to a focus at the eye of the observer.

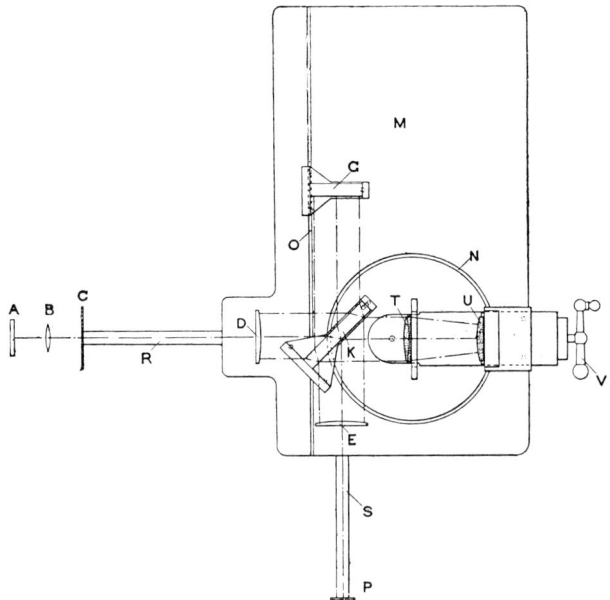

Fig. 193—Diagram of lens interferometer

Optical elements or combinations suitable for examination on it may almost all be classed in two categories. Into the one category fall those combinations which are required to receive a beam of light which has a plane wave front and deliver it again after transmission with a plane wave front, and into the other fall those the object of which is to impart spherical wave fronts to beams which are incident on them with plane wave fronts. The two corresponding arrangements will be referred to as the prism interferometer and the lens interferometer respectively.

The prism interferometer

The prism interferometer is shown in the diagram (Fig 192) as arranged for the correction of a 60° prism, such as is used for spectroscopy.

The light used must consist of very homogeneous radiation. Such light may be obtained from a low pressure mercury vapour lamp with glass tube, such as the " Hewittic."

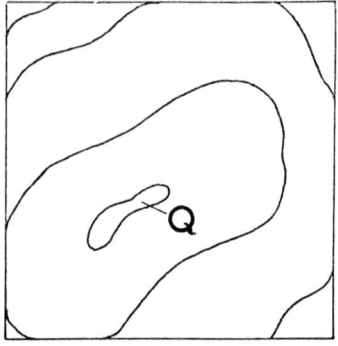

Fig. 194—Diagram of interference pattern

The light from the source is reflected by the adjustable mirror A through the condensing lens B, by means of which it is condensed on the aperture of the diaphragm C.

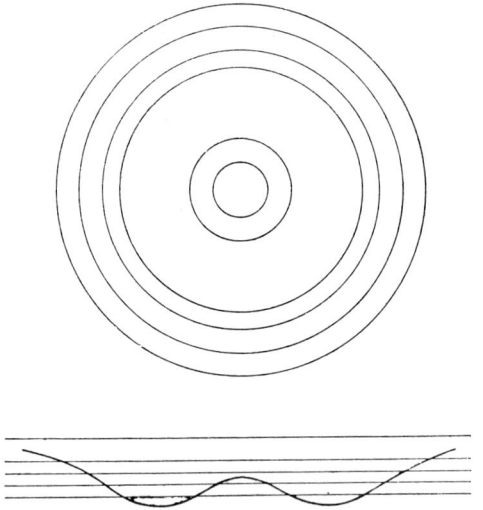

Fig. 195—Diagram of interference pattern for a lens

The diverging beam of light is collimated by a lens D, and falls as a parallel beam on a plane parallel plate K, the second surface of which is silvered (or aluminized) lightly so that a part of the light is transmitted and part reflected. The major part should be reflected. One part passes through the prism L in the same way as in actual use, and, being reflected by the mirror F, passes back through the prism to the plate K. The other part of the light is reflected to the mirror G and back again to the plate K. Here the separated beams recombine, and passing through the lens E each forms on the eye, placed somewhat beyond the aperture in the diaphragm P, an image of the hole in the diaphragm C.

When the mirrors are adjusted, interference bands are seen which form a contour map of the glass requiring to be removed from the prism face in order to make its performance perfect.

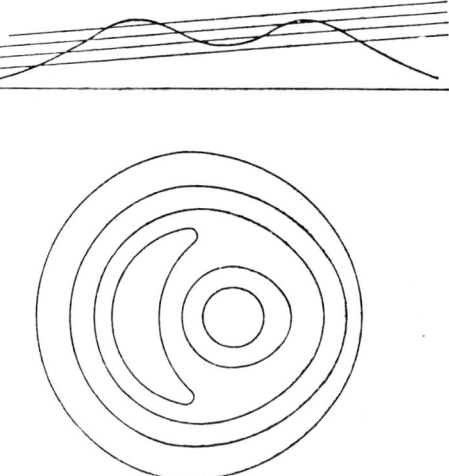

Fig. 196—Diagram of interference pattern of a lens, tilted

Fig. 194 represents in diagram a typical map, where Q represents the highest point of a " hill." The procedure in such a case would be to mark out the contour lines on the surface of the prism with a paint brush and then to polish first on the region Q, subsequently extending the area of polishing, at first partly, then wholly, to the next contour line ; and so on. The marking out of the prism surface can be done while one is observing.

It should be noted that variations in the contour lines are obtained by a tilt of the plane of reference. Thus a slight adjustment of mirror F (Fig. 192) might change a contour map from that shown in Fig. 195 to that shown in Fig. 196. The form of surface is in each case the same (see the sectional diagrams at the top of the figures), but correction can be carried out according to whichever plane of reference is the most favourable from the point of view of the operator. In order to find whether Q is a hill or a valley, the cast-iron table M can be bent with the fingers so as to tilt the mirror F in such a way as to lengthen the ray path. If the contour line at Q expands to enclose a larger area, a hill is indicated, and *vice versa*. Although the words " hill " and " valley " are convenient to use, it must not be supposed that the imperfections necessarily result from want of flatness either of one or of both surfaces of the prism. The contour map gives the total effect on the wave front produced by double passage through the prism, and shows in wavelengths the departure from planeness of the resulting wave surface.

Increased illumination

One can get a greater contrast and considerably more light in the fringes by the use of zinc sulphide coating for the semi-reflecting surface instead of silver. In spite of the better effect obtained, however, we have not found it worth while to use it since zinc sulphide coating is considerably more complicated than coating with silver or (as we now prefer) aluminium ; and, although the zinc sulphide gives a better result than silver and the silver than the aluminium, the last named is on the whole less troublesome because of its durability.

It is quite easy to get photographs of the interference patterns by placing a camera at P in Fig. 192 ; using a high pressure mercury vapour lamp good photographs have been obtained with 1/25th second exposure and still shorter exposures could be used with suitable selection of the optimum conditions.

A very complete set of interferometer patterns due to the primary aberrations are illustrated by Kingslake (1925) ; interferograms obtained by calculation and by photography on the Hilger lens interferometer are shown side by side.

Compensated Twyman and Green interferometers ; white light fringes

245 For some purposes, *e.g.* that described in §288, the interferometer must be " compensated."

Referring to Fig. 192 and supposing the prism to be removed and the mirror F adjusted perpendicular to the incident rays, a mirror identical with K is placed between K and F and parallel to the former.

By adjustment of the mirror G the paths of the two interfering beams can then be made exactly equal when white light interference fringes are seen with a black central band which indicates the positions of exact equality of path.

Thermal uniformity necessary

246 In the final stages of polishing large prisms it is essential that before testing the prisms the temperature should be allowed to settle down. It used to be our practice to stand the prism for this purpose on three projections of non-conducting material, such as ebonite, to allow free access of the air all round the prism.

Although by this means an approximate equalization of temperature throughout the prism is acquired fairly rapidly, yet until the prism has acquired the temperature of the air the equalization is not good enough for the purpose of a critical test. The method that has been adopted, therefore, for a number of years is to stand the prism on a metal plate (which should not be too thick, so that it can rapidly accommodate itself to the temperature of the room, and should be nicely flat, so that it can rapidly convey that temperature to the prism) and to place over the prism a metallic cover nicely fitting the metal plate at the bottom and rough blacked on the outside so that it, also, rapidly acquires the temperature of the room.

In these circumstances, half an hour is sufficient for a 60° glass prism 2 in. high and $2\frac{1}{2}$ in. length of face to settle down appropriately for the most critical test. With increasing length of prism the length of time required increases rapidly; for example, a prism 2 in. high by 3 in. length of face would require three-quarters of an hour.

It is scarcely necessary to add that for very large work a constant temperature room, in which the temperature can be held within $1/10°C$, must be used.

The interferometer is very useful, also, for testing angles to a high precision. This application is described in detail in §§236 and 237. An application of this instrument to measurements of refractive index variations in a plate of glass is given in §286–288.

Additions for the testing and correction of lenses

247 In the lens interferometer all parts are left as in the prism arrangement except that the mirror F is removed and replaced by the lens attachment shown in Fig. 193. T represents the lens under test, U a convex mirror in such a position that it reflects back along their own paths the rays received from T. The mirror U can be moved by a screw motion actuated by the handle V, so that its distance from T can be varied at will. It will be seen that when the adjustment of this part of the apparatus is correct, the whole lens addition will, if the lens T be perfect, receive the beam of plane wave front and deliver it back again with a plane wave front. If it does not do so the departures from planeness of the wave front so delivered will give rise to a contour map of the corrections which have to be applied to the lens in order to make its performance, when in actual use, perfect.

The camera lens interferometer (*Twyman*, 1921)

248 An apparatus which will test for axial pencils only is of course of limited use for testing camera lenses. The modifications essential for the latter purpose are:

1. Means of rotating the lens about a line at right-angles to the axis and passing through the second principal point.

2. Mechanism whereby, simultaneously with the above rotation of the lens, the convex back-reflecting mirror is automatically moved away from the lens in such a way that its centre of curvature always falls on the plane, perpendicular to the axis of the lens, on which the lens is desired to form its image.

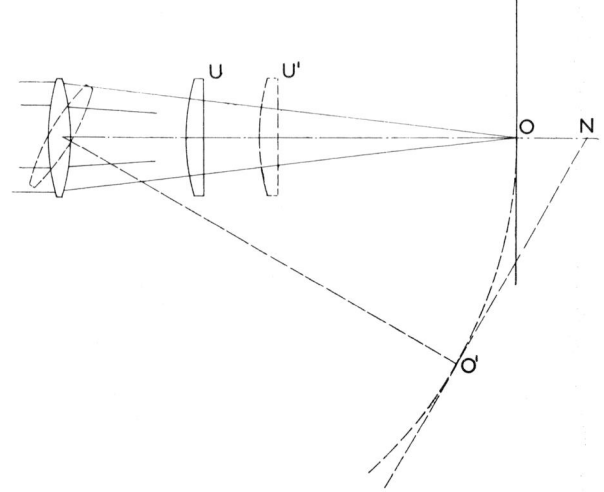

Fig. 197—Principle of construction of the camera lens interferometer

Thus, in Fig. 197 the full lines represent the lens and mirror of the ordinary lens interferometer for axial pencils. The modifications necessary in the relative positions of lens and mirror for pencils oblique to the lens are shown by the broken lines. The mechanism must obviously be such that $UU'=ON$. Such, briefly, is the principle on which the camera lens interferometer is constructed. Full details are given in the reference, in which is also to be found reproduction of interferograms corresponding to fairly pure examples of the five principal aberrations of a centred optical system, viz., a spherical aberration in axis, coma, astigmatism, curvature of image, distortion.

The Universal lens interferometer

249 When the present author first described the original camera lens interferometer (Twyman, 1918) one of which had been installed at the National Physical Laboratory in 1917, T. Smith, Head of the Optics Department there, pointed out that such an instrument did not meet all the requirements of the Laboratory, where photographic lenses of a great variety of types require to be tested. For example, the original instrument was only arranged for testing lenses with an object at an infinite distance. Further, telephoto lenses could only be tested over a very narrow range of oblique angles, because

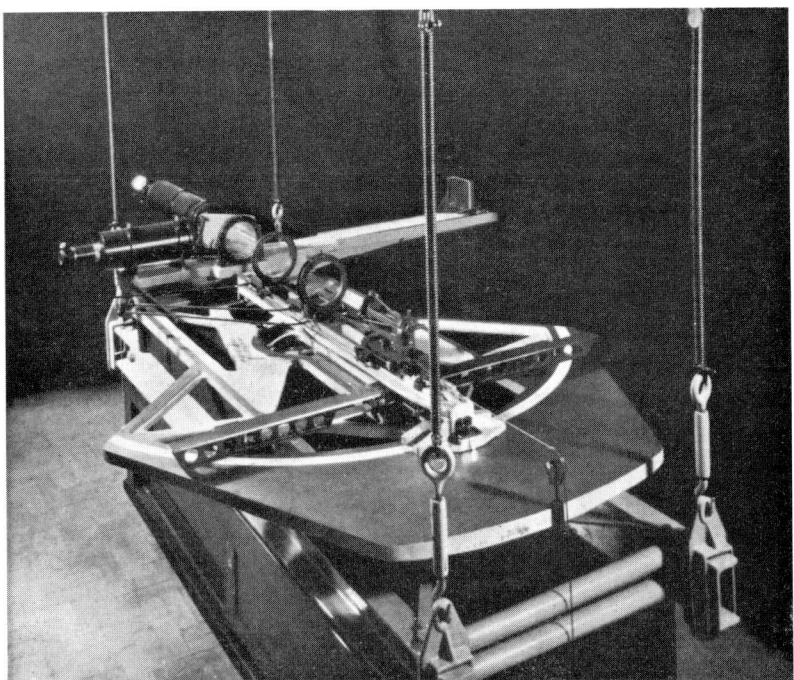

Fig. 198—The Universal lens interferometer

the principal planes of such lenses are a long way in front of the lens, so that on rotation the aperture of the lens will move out of the field unless the aperture of the collimating lens of the interferometer is made very large.

Mr Smith in due course set forth his requirements and arranged them into a scheme of great completeness, designing linkages and optical systems by which the geometrical conditions could be fulfilled.

Sir Herbert Jackson, as Director at the time of the British Scientific Instrument Research Association, supported a proposal in consequence of which H.B.M. Department of Scientific and Industrial Research

434   PRISM AND LENS MAKING

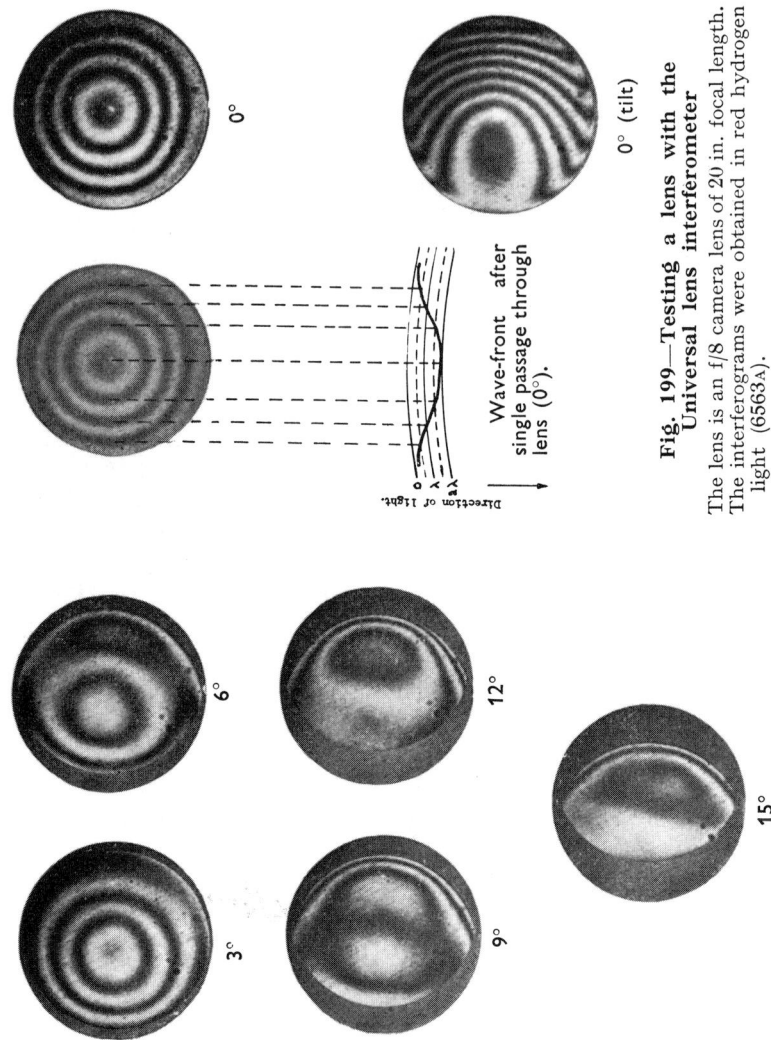

Fig. 199—Testing a lens with the Universal lens interferometer

The lens is an f/8 camera lens of 20 in. focal length. The interferograms were obtained in red hydrogen light (6563A).

THE HILGER INTERFEROMETERS 435

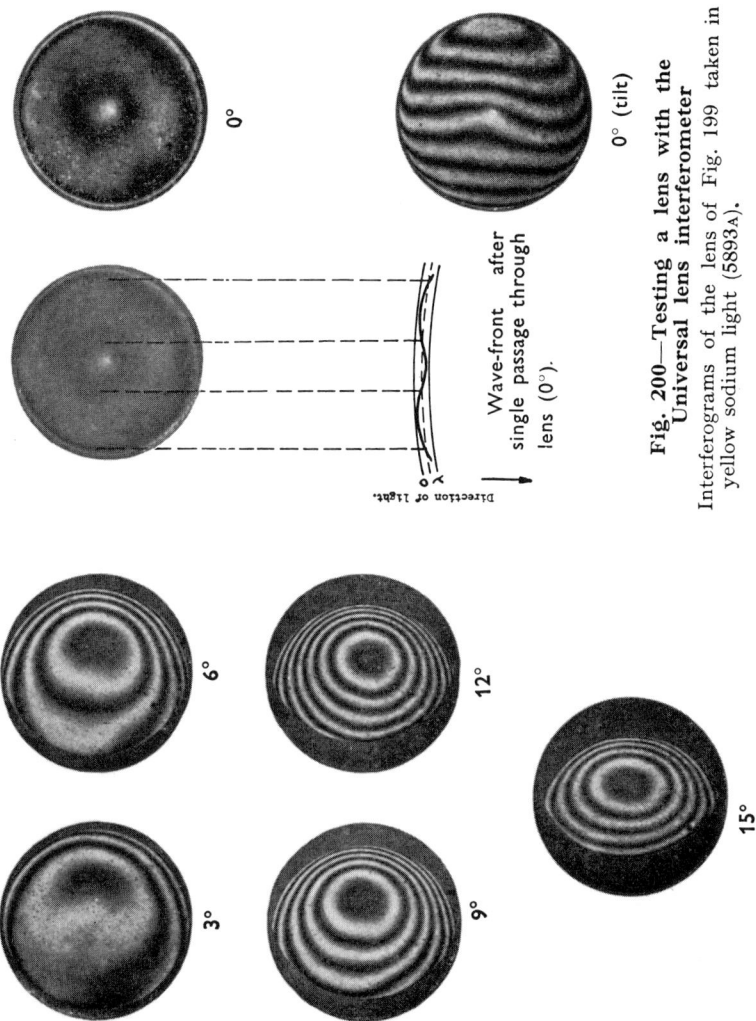

Fig. 200—Testing a lens with the Universal lens interferometer

Interferograms of the lens of Fig. 199 taken in yellow sodium light (5893A).

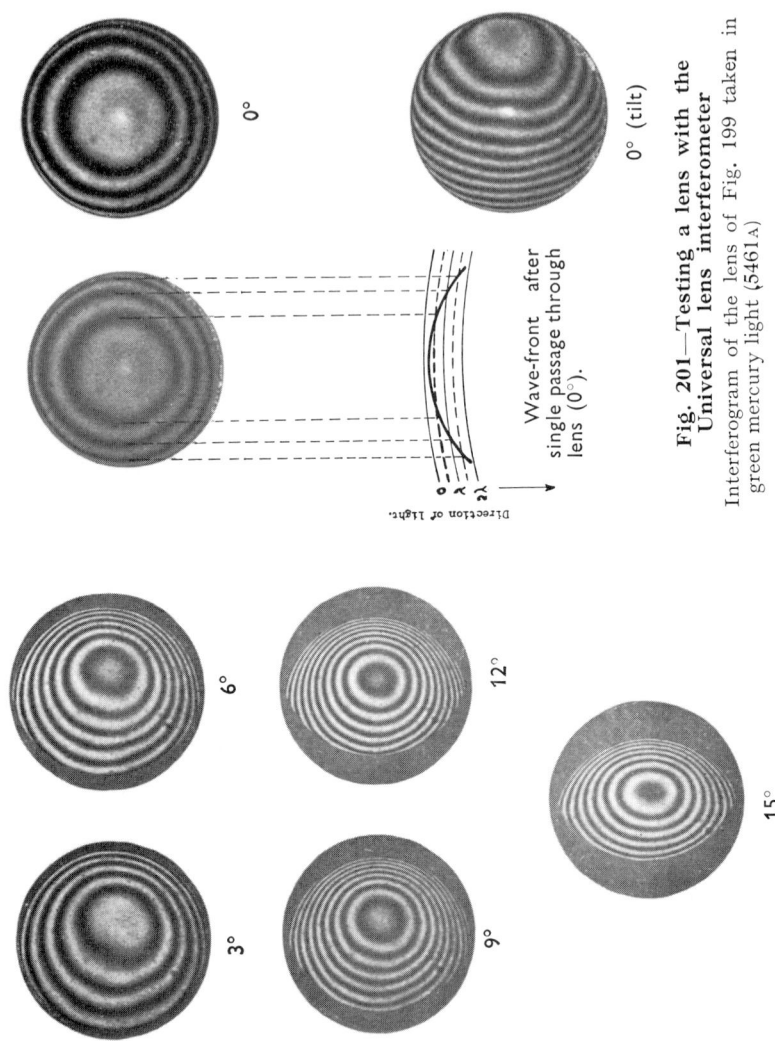

Fig. 201—Testing a lens with the Universal lens interferometer

Interferogram of the lens of Fig. 199 taken in green mercury light (5461A)

made a special grant for the construction of one of these instruments which was installed at the National Physical Laboratory, Teddington, England, in August 1926.

The design of the instrument to realize Mr. T. Smith's conditions was carried out by J. H. Dowell, Chief Designer of Adam Hilger Ltd (Fig. 198).

(The above is abstracted from *Proc. Optical Convention*, 1926, Part II, 1,032, where a very complete description of the mechanism and optical arrangements will be found).

Much work has been done at the National Physical Laboratory with this instrument and Figs. 199–201 illustrate results obtained on this instrument in 1944 with a Taylor, Taylor & Hobson lens for the Ministry of Aircraft Production. The lens aperture was $f/8$ ($f = 20$ in., diameter 6·3 cms). These illustrations and descriptions are published with the permission of the Director of National Physical Laboratory and Messrs Taylor, Taylor & Hobson. The lens was examined on the interferometer. Photographs have been taken with hydrogen red light (Fig. 199), sodium yellow light (Fig. 200), and mercury green light (Fig. 201) at intervals across the field from the axis to a semi-angle of 15°. They relate to the best focus obtained at the centre of the field with sodium yellow light. In each case the pattern obtained on the axis is shown in its symmetrical form and also with a small amount of tilt, in order to indicate the shape of the wave front. In addition the wave front after single passage through the lens is drawn for each axial pattern. A " hill " indicates that the wave front is relatively retarded and therefore corresponds to a " high " region of the lens. A " valley " indicates that the wave front is relatively accelerated and therefore corresponds to a " low " region of the lens.

Other modifications of the camera lens interferometer have been produced to fulfil special requirements.

The Hilger microscope interferometer (Twyman, 1923)

Principles of construction

250 The microscope interferometer is a development of the lens interferometer (Fig. 193).

If such an instrument be provided with sufficiently delicate adjustments for focusing the objective under test (namely, moving it to and from the mirror U), and for moving the mirror U laterally, it becomes immediately suitable for exhibiting the aberrations possessed by a microscope objective when the latter is focused to produce its real image at infinity. Various arrangements are shown in Fig. 202. The objective is represented by T in Figs. 202a and b. If, as is

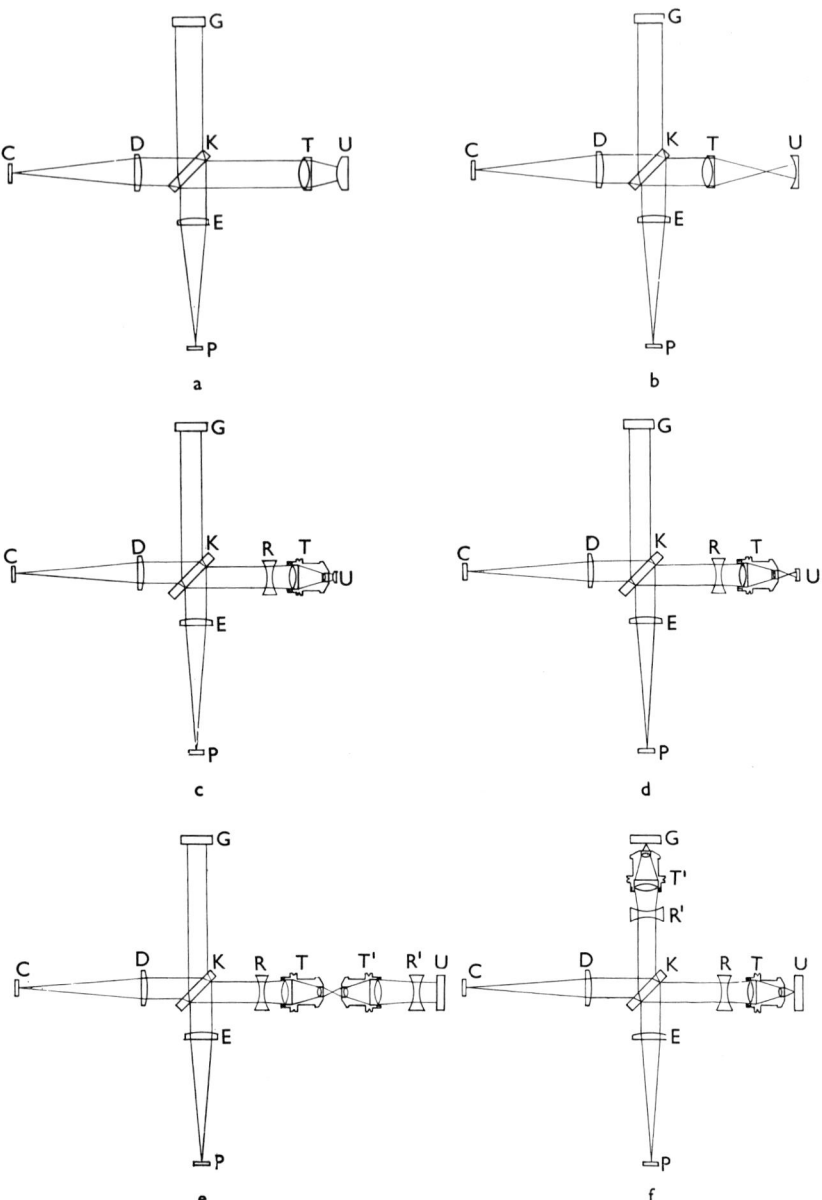

Fig. 202—Various optical arrangements for the Hilger microscope interferometer

customary, the objective is intended to produce a real image at a finite distance, such a condition is simulated by the introduction of a negative lens R (see c and d), corrected for spherical aberration and called, conformably to the nomenclature of the microscopist, the tube length lens. This lens must give to a parallel beam a divergence exactly corresponding to the convergence of the beam which obtains in the intended use of the objective.

The polishing of a concave mirror (Figs. b and d) with sufficient precision not to introduce objectionable aberrations of its own becomes very difficult in the case of objectives of high numerical aperture, while if a convex mirror be attempted (Figs. a and c) it needs to be of very small size, and is on that account very difficult to produce accurately by ordinary procedure. A small drop of mercury may be used as a convex mirror.* In this case also the observer has no direct way of assuring himself that the surface of the drop is sufficiently spherical, and other modifications have therefore been devised, whereby the use of spherical mirrors can be obviated.

Figs e and f show arrangements which may be used where the observer is satisfied if he can compare the objective to be tested with a second one (not necessarily identical) which he regards as a standard. In Fig. e the standard objective T' is in line with that under test, together with a second tube length lens R', while in Fig. f the standard objective and its tube length lens are put in the comparison beam. In the arrangement of Fig. e the interference pattern represents the sum of the aberrations of the two objectives, while in Fig. f it represents their difference. In either case flat mirrors suffice in both the test and comparison beams. It will be seen that in Fig. f each ray returns, not along its first course, but along another course axially symmetrical with the first. This has the disadvantage that only in the case of an objective whose aberrations are axially symmetrical does the interference pattern represent the aberration truly, the aberration of wave front being shown at two axially symmetrical points, a point a in one direction and a symmetrically opposite one on the return path.

Full details are given in the paper cited at the head of this section.

Ultra-violet microscope

This microscope interferometer provided a mounting for an ultra-violet microscope, the mechanical part of which I designed for Barnard (1936, *J. Roy. Microscopical Soc.*, **56**, 365) and which was used by him for some of his observations on viruses. In the previous use of his Kohler and von Rohr ultra-violet microscope (loc. cit.) Barnard had found various difficulties, one of which was that there was no

* See the reference for method of producing.

systematic way of focusing the image for the desired ultra-violet light ; it had to be effected by trial and error.

The microscope interferometer provided the necessary accuracy of measurement by which, after focusing the object by visual light, one could move by exactly the right distance for it to be focused by ultra-violet light of a selected wavelength.

The second defect of the earlier instrument, not mentioned in the paper cited, was the means of ultra-violet illumination. Owing to the long path in air traversed by the radiation and the use of two large quartz prisms for separating the beams of various wavelength, it was not possible to make use of the strong mercury radiation at 1850A, since both quartz and air absorb strongly in that region.

Consequently, at a later date I provided him with a small fluorite prism mounted immediately below the sub-stage condenser which latter formed a microscopic image of the spectrum visible, by fluorescence, with a low-power eyepiece in the microscope. By turning the fluorspar prism any desired radiation could be brought in to the middle of the field. By this means it was possible to gain the increased resolving power resulting from the use of the 1850A radiation.

The Sira-Hilger microscope interferometer

251 The instrument described in §250 is adequate for controlling the manufacture of microscope objectives, yielding as it does full information on the axial aberrations whether resulting from faulty manufacture and mounting or imperfect design. This does not, however, provide for testing off axis, which is particularly desirable when there is a question of perfecting a new design of objective. The requirements for an instrument to meet the purposes of the British Scientific Instrument Research Association (20 Queen Anne Street, London, W.1) were drawn up by R. J. Bracey, Chief Lens Computer of that Association, and realized by *Adam Hilger Ltd* from drawings which like those of the Universal camera lens interferometer, were prepared by J. H. Dowell.

No description of this instrument has yet been published and a somewhat detailed description is therefore included here.

This instrument was constructed for the British Scientific Instrument Research Association and provision was made for the detailed examination of all the optical factors which are involved in image formation by microscope object glasses.

The most important additional features are means of determining the extra-axial aberrations, viz—coma, astigmatism, field curvature, distortion.

Since the field of a microscope object glass, although small, is finite, all these errors play a part in modifying the ideal point-to-point correspondence between the object and the image. The Hilger-Sira interferometer has, then, potentialities similar to those of a photographic lens interferometer and could be used to test suitable small photographic objectives.

The measurement of the listed aberrations can be accomplished in a direct manner if the objective is rotated about its optical centre— the optical centre being a point which divides the distance between the nodal points in the ratio of the magnification at which the objective

Fig. 203—The Sira-Hilger microscope interferometer

is being used. If, when the objective is rotated, the tube-length lens and the test mirror which is used with the objective are held stationary, it will be appreciated that this corresponds to the case when the object and image fields are spheres which are concentric with the optical centre. If, on the other hand, by means of cross bars the focal point of the tube-length lens and the centre of the test mirror are constrained to lie in planes which are perpendicular to the optical axis, this corresponds to the case when object and image fields are planes. In the actual use of a microscope objective, the object is flat and the image field is usually strongly curved. In this last case the most convenient set-up is that in which the test mirror moves in a plane perpendicular to the optical axis, while the tube-length lens is undisturbed and, hence, the whole set-up represents a spherical image-field derived from a flat object-field.

The interferometer is fitted with an eyepiece (A in Fig. 203) and, when properly adjusted, the focal plane of the eyepiece is conjugate to the small aperture stop E in the focal plane of the collimator, and hence a small spot of light is seen in the field of view. This spot of light is formed in two ways; firstly by back reflection from the comparison mirror F, and secondly by back reflection from the system comprised of the tube-length lens L, the object glass G, and the spherical mirror at M (see Fig. 203). It is of interest to trace the path of the

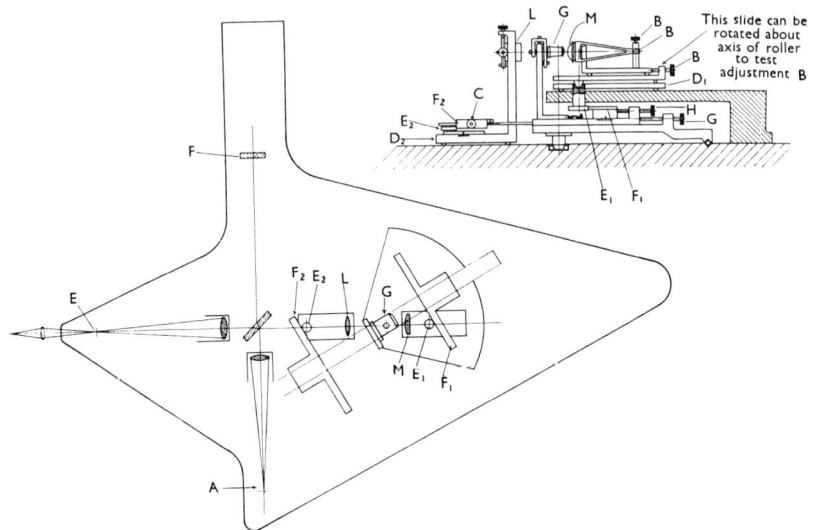

Fig. 204—The Sira-Hilger microscope interferometer

light along the object glass arm of the interferometer. After leaving the tube-length lens it diverges from a virtual focal point at a distance from the microscope object glass corresponding to the normal image distance. After passing through the object glass it comes to a focus, which focus is made to coincide with the centre of curvature of the spherical mirror; and hence the light, returning on its own path, eventually leaves the tube-length lens as a parallel beam in the reverse direction, and is brought to focus in the eyepiece. It is clear that, if the system is incorrectly set up, the two spots of light obtained by these different means may not coincide or even lie in the same focal plane.

The accompanying diagram Fig. 204 may be referred to for the various mechanical controls which are necessary to obtain sharp focus and stability in the image of the second spot on rotation, assuming that the object glass has ideal properties. In order that the chosen object and image fields, which are determined by the distance between the cross bars, may be conjugate to each other when the objective is tested, it is necessary to adjust the position of the objective between these two reference planes, *i.e.* the objective must be mounted on an independent focusing slide. This corresponds to the fine adjustment of an ordinary microscope, and the actuating screw A is fitted with a divided head enabling readings in steps of 0·0001 in. to be made. It is also necessary to ensure that the centre of curvature of the spherical mirror lies accurately on the object field and on the optical axis of the objective before this is rotated. The fulfilment of this need implies a three dimensional control, and might be considered to be analogous to the centring and focusing movement of the microscope substage condenser. The three actuating screws, one for each dimension, are marked B. It is necessary to adjust the tube-length lens so that its virtual focal point lies on the image cross bar and this requires an independent slide, marked C. The spherical mirror slide D_1 and the tube-length lens slide D_2 are fitted with rollers and constraints which keep the rollers E in contact with the cross bars F when they are rotated. These slides run lightly on ball bearings and their rollers are pressed against the optically worked cross bars by means of gravity controls. The system is kinematically designed to be nearly free from mechanical hysteresis.

When the adjustments outlined above are correctly made, and the object glass, tube-length lens and spherical mirror are accurately centered, the two spots of light seen in the eyepiece are both in sharp focus and coincident. If, however, the optical centre of the microscope objective does not coincide with the axis of the spindle of the turntable, then on rotation the spot of light due to the object glass arm of the interferometer is displaced. In order to bring the optical centre into coincidence with the spindle axis, a further independent slide is required which bodily moves the cross bars, the objective, the spherical mirror and their associated slides as a whole until coincidence is obtained. This slide is known as the nodal slide : its actuating screw G is fitted with a divided head enabling readings in steps of 0·0001 in. to be made. This final adjustment ensures that on rotation the two spots in the eyepiece are coincident and in sharp focus, always provided that the object glass is free from any errors whatever. On removal of the eyepiece, an inspection of the field will show the appropriate interferogram. If, however, the object glass is being examined for an oblique transmission, the resulting curvature of the field may be sufficient to give too small an interferogram to interpret, and the fine adjustment may be used to refocus the object glass to allow for the curvature of the field. The amount of refocusing required is a direct measure of the field curvature. This refocusing of the object glass will, however, move the optical centre of the objective out of coincidence with the spindle axis, and the nodal point slide must be used to restore the position. An alternative method of examining the field is to free the tube-length lens from its control and, after any rotation, to refocus the system by a movement of the tube-length lens, such movement being a direct measure of the departure of the field curve on the image side from a spherical surface concentric with the optical centre of the object glass. This method also does not disturb the adjustment of coincidence between the optical centre and the spindle axis. If the object glass has distortion, it will be impossible to find a position, except for an infinitesimal rotation, where the optical centre and spindle axis remain coincident. In fact, an adjustment of the nodal slide is required for every obliquity and is a measure of the distortion. An alternative means of measuring distortion is to introduce a pair of deflector prisms and to use these to compensate for the distortion in the object glass. By this means a direct reading of distortion may be obtained from the collar of the deflector prism mount.

The means described above are adequate to measure the tangential and sagittal field curvature; also the distortion of an object glass. In all cases, however, the spherical aberration and coma will be very small, as every attempt will have been made to reduce these errors to zero and their quantitative disturbance of the fringe system is relatively negligible. The pattern itself as viewed subtends an angle which is very small, and the interferometer is provided with a photographic attachment whereby photographs can be taken for subsequent enlargement and calibration, thus rendering possible the measurement of residuals of coma and spherical aberration.

The interferometer is fitted with a control rod which enables the comparison beam mirror to be adjusted in position from the observer's side of the instrument, and the position of this comparison beam mirror can be read against a scale. The comparison beam arm is longer than usual since provision has been made for a special device which causes the light to traverse the object glass four times instead of the usual twice. The ordinary field of a microscope object glass does not extend much above $8°$, but the turntable is capable of a rotation of $\pm 30°$ so that, if necessary, eyepieces or short focus photographic lenses could be tested. The tube-length lenses supplied with the instrument cover a range of tube lengths from 250 mm down to 60 mm.

When testing a microscope object glass, the thickness of the cover-glass which is used plays an important part in achieving the residual balance of correction. Also the working distance of the object glass may be very short, so that the most suitable spherical mirror for testing lenses of this sort is concave and also has mounted in front of it a piece of glass of the same thickness as the ordinary microscope cover-glass. It is for these reasons that the type of spherical mirror adopted consists of an accurately worked glass hemisphere which is silvered on the spherical side and has, cemented to its front surface, a piece of glass equivalent to the cover-glass for which the object glass is corrected.

Testing may also be carried out with a lens–mirror system or a tetragonal mirror in the comparison beam arm and, in this case, the spherical mirror which is used with the object glass must be replaced by a flat mirror. It is well known that when a system of this sort is used, in which the light passes first through one side of the objective and then returns through the other side, asymmetrical aberrations are removed, *i.e.* no measurements of coma or distortion are possible. It is, however, a very convenient set-up since the flat mirror has merely to be placed in the focal plane of the objective—a much simpler matter than the three dimensional adjustment which is necessary with the spherical mirror.

The setting up of the interferometer is usually achieved in the following manner—

The appropriate tube-length lens is selected. The objective and the spherical mirror are mounted in their approximate positions: These have been determined by previous experiments in the original calibration of the instrument. The eyepiece may be removed and an inspection made of the back lens of the objective, when a number of images of the collimator stop will be seen. A slight movement of the fine adjustment one way or the other will cause one of these spots of light to expand and at the same time to move either to the left or right. The centring screws of the objective mount may now be used to bring this expanded spot of light to the centre of the back lens, and the fine adjustment may then be used to make this expanded spot of light fill the whole of the back lens. The eyepiece may be replaced and in the field of view will be seen a sharply focused image of the collimator stop; also a diffuse image which is obtained from the object glass arm. A further use of the centring screws and fine adjustment will render these two spots of light coincident and sharply focused. The eyepiece may now be removed and the distance of the comparison beam mirror varied until a fringe pattern is obtained of as strong a contrast as is possible. Owing to the fact that different object glasses have their nodal points in different positions with reference to the shoulder of the R.M.S. thread, different positions of the tube-length lens and spherical mirror are required for every variation of focal length of object glass, and this in turn requires that the comparison beam mirror should be adjusted with every different objective.

If the objective is of very short focal length it will sometimes save time to focus it first of all with the flat mirror and then to change over to the spherical mirror; one has then merely to centre the objective without the necessity for the prior identification of the appropriate image of the collimator stop. In an instrument of this sort, adjusted in the manner described, there will inevitably be some small residual errors of adjustment, particularly in relation to the placing of the optical centre of the objective in coincidence with the spindle axis, and the placing of the centre of the spherical mirror in coincidence with the roller axis. These errors do not affect the axial readings but they do affect the oblique readings and would be troublesome to remove. It can, however, be shown that the mean of readings for positive and negative obliquities cancel out the adjustment errors completely.

Other applications of the Hilger interferometers

Testing interior sub-multiple angles

252 In Chapter 11 reference was made to the principle of multiple reflection as a method for increasing the accuracy obtainable with a back-reflecting telescope. The principle is applicable also to interference testing, and will be found of great use in the preparation of primary angle standards by reflection from the interior surfaces.

In the case of the odd sub-multiples, the light is at its mid-point reflected normally, and returns along its own path. In these conditions interference fringes will be formed as with any other prism system, provided that an adequate increase in the path length of the back mirror is allowed for.

The even sub-multiples are slightly more difficult to use, and it is only possible to employ them in a straightforward way if the light passes normally through the entrance face. This entrance face is squared-on to the incident light to within a couple of fringes, and the path length increased by the required amount to compensate for the glass path, when some rather weak fringes should be seen; these are due to light reflected normally from the *inside* of the entrance face.

This was one of the methods employed for the final testing of the corner cubes instanced in the section dealing with accurate metal jigs for machining glass. The function of these prisms is to return each incident ray in a direction parallel to the incident direction, but emerging from a point diametrically opposite the point of incidence, and under these conditions interference fringes are not formed by the directly reflected light. The corner cube is one of the possible

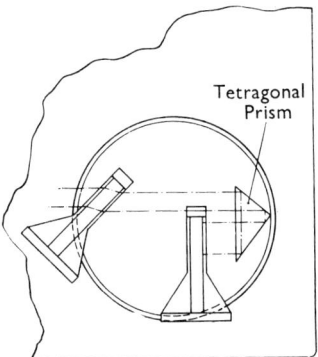

Fig. 205—Prism interferometer setting for testing the angles of a tetragonal prism

methods of " reversing " the wave front from the back mirror of the interferometer when using a plane mirror instead of a mercury globule in the testing of microscope objectives (see §250).

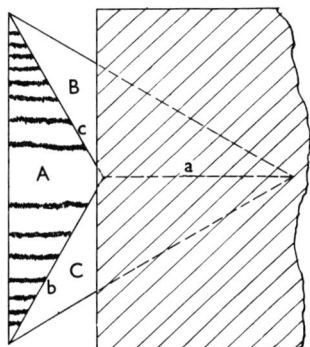

Fig. 206—Interference pattern exhibited by tetragonal prism set for testing the right-angles

Hence another arrangement is used. Fig. 205 shows a portion of the interferometer as so re-arranged, the rest remaining as shown in Fig. 192. The tetragonal prism is mounted with its equilateral face, and one edge of the same, vertical; the right-angled edge opposite to the latter is horizontal.

Fig. 206 shows an elevation of the prism viewed from the observation aspect, the shaded portion representing the back of mirror F.

Light enters surface A, and after reflection at surfaces B and C returns along its own path. In detail the procedure of testing and correcting the prism is as follows—

1. Observing in A as shown in Fig. 206 it will be found that when the lower half of the surface of A is made one uniform colour, the bands in the upper half are horizontal. Get rid of these bands by working surface B or C. After this neither B nor C must be touched again.
2. Observing in B, and working on surface A get angle b right in the same manner.
3. Observing in C and working on surface A get angle c right.

The prism will then be correct.

Bibliography

The following is a condensed basic bibliography on the Hilger interferometers and methods of testing.

Prism and lens interferometer
Twyman, F. (1918). *Phil. Mag.*, **35**, 49.
Twyman, F., and Green, A. Brit. Pat. 103832/1917.
Smith, T., 1927, *Trans. Opt. Soc.*, **28**, 105.

Camera lens interferometers
Twyman, F. (1920–21). *Trans. Opt. Soc.*, **22**, 174.
Twyman, F. Brit. Pat. 130224/1919.
Smith, T., and Dowell, J. H. Brit. Pat. 236634/1925.
Dowell, J. H. (1926). *Proc. Opt. Conv.* II, 1032.
Perry, J. W. (1923–24). *Trans. Opt. Soc.*, **25**, 97.
Hay, O. G. (1929–30). *Trans. Opt. Soc.*, **31**, 91.

Microscope interferometer
Twyman, F. (1916). Brit. Pat. 103832/1916.
Twyman, F. (1922–23). *Trans. Opt. Soc.*, **24**, 189.

General
Guild, J. (1920). *Proc. Phys. Soc.*, **32**, 341.

Other interferometers for testing optical work derived from Hilger's

Shearing and rotating interferometer

I am indebted to Mr W. J. Bates for the following brief description of his interferometers.

252.1 One would have anticipated that after the introduction of the Twyman and Green interferometer interferometric examination of optical wavefronts would have extended to the testing of large lenses and mirrors. Hitherto, however, one has been limited as to interferometer components some of which have to be of the same kind of size as the optical system under test. The Shearing interferometer (Bates, 1947) demonstrates the possibility of testing an optical wavefront or surface against an identical sight of itself, the technique being to obtain two identical views of the same wavefront or surface and by superposition to produce interference between them. If the two wavefronts α and β, Fig. 207 overlap exactly then a single bright

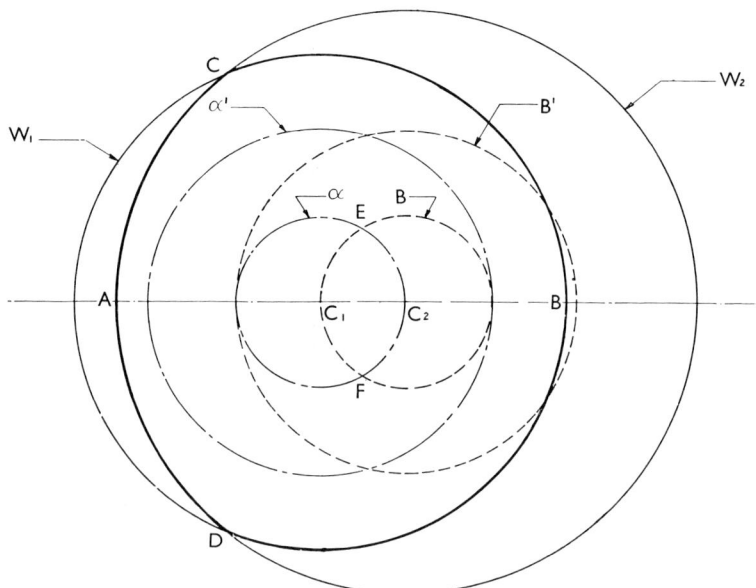

Fig. 207—Wavefront in the Bates shearing interferometer

fringe covers the whole field of view. If, however, they are displaced laterally with respect to one another the errors on the one no longer coincide in space with errors on the other and the interference which one obtains in the overlap region EC_2FC_1 is indicative of the errors involved. The general theorem demonstrating the existence of solutions to the wave-front's asphericity is not difficult.

There are a whole set of interferometers of two types; one in which lateral shear only is used and the other in which rotatory and/or lateral shear is used. One example is shown. The wavefront shearing interferometer (Fig. 208) obtains the two sights of the same wavefront by splitting it at a half silvered film and recombining with lateral shear by another half silvered film which can be rotated to produce the shear.

In the second type—the inverting interferometer—the one sight of a surface is turned upside down with respect to the other by viewing

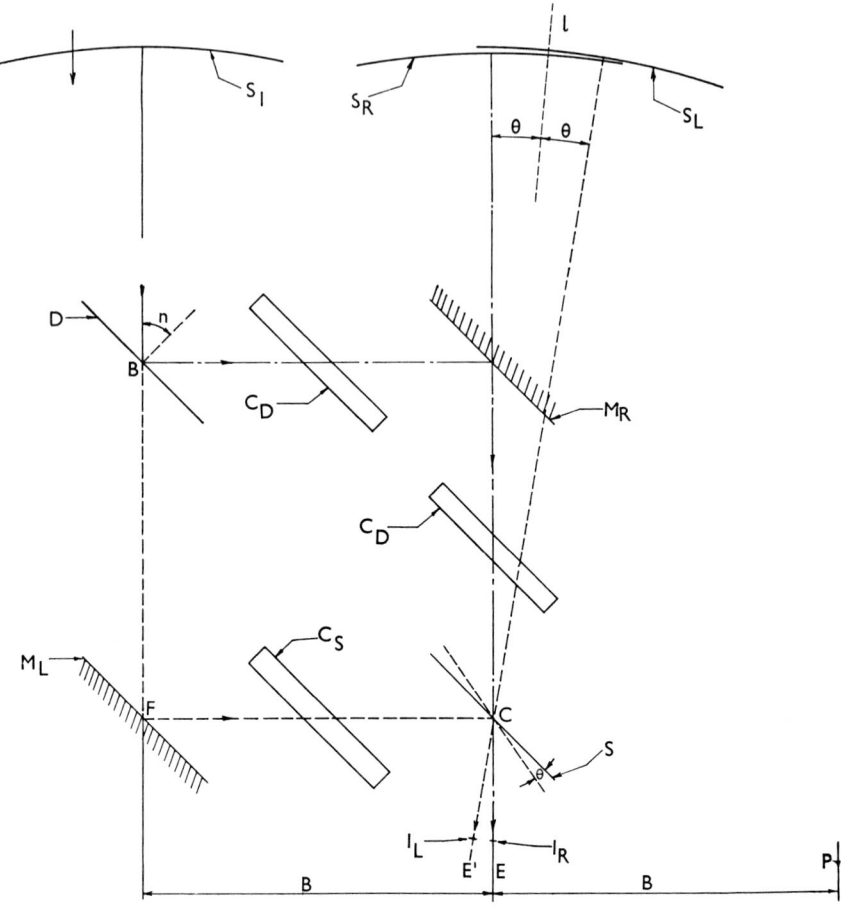

Fig. 208—The Bates shearing interferometer

it through three mirrors which form the corner of a cube. The advantage is that an *extended white light* source can be used.

The nice points about these devices are (1) that aperture for aperture it is as easy to test interferometrically a 200 in. as a 2 in. surface—which should prove useful for astronomical telescopes. Mr. Bates

THE HILGER INTERFEROMETERS 451

has already tested a 72 in. mirror with the first interferometer. (2) Extended source white light fringes are in some cases obtainable, which facilitates projection and observation. (3) One may de-sensitize the apparatus to vibration and air heterogeneities by reducing the shear. (4) There would appear to be some applications in the field of microscopy interference and phase microscopy.

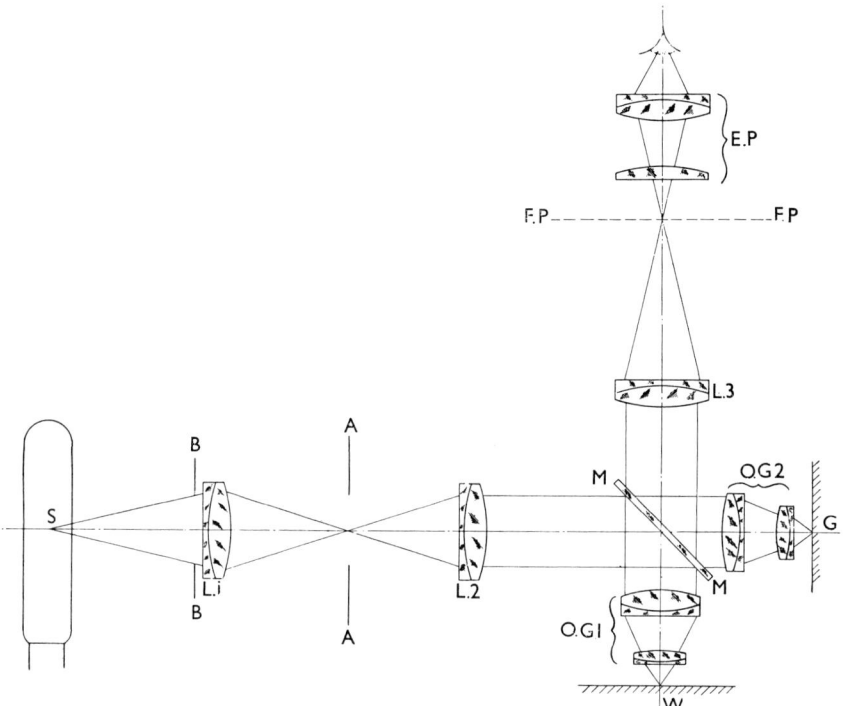

Fig. 209—The Linnik micro-interferometer

The Linnik micro-interferometer (Zeiss)

252.2 An interesting modification of the microscope interferometer has as its purpose, not the determination of the errors of the microscope object glass, but the examination of the irregularities of surfaces observed with the microscope, such as those of gauges. It is known as the Linnik micro-interferometer (Habell and Cox, 1948) and is shown in Fig. 209.

S is the source of monochromatic light which is imaged by the lens $L1$ in the plane of the iris diaphragm A. B is also an iris diaphragm controlling the area of $L1$ that is used. A is at the front focal point

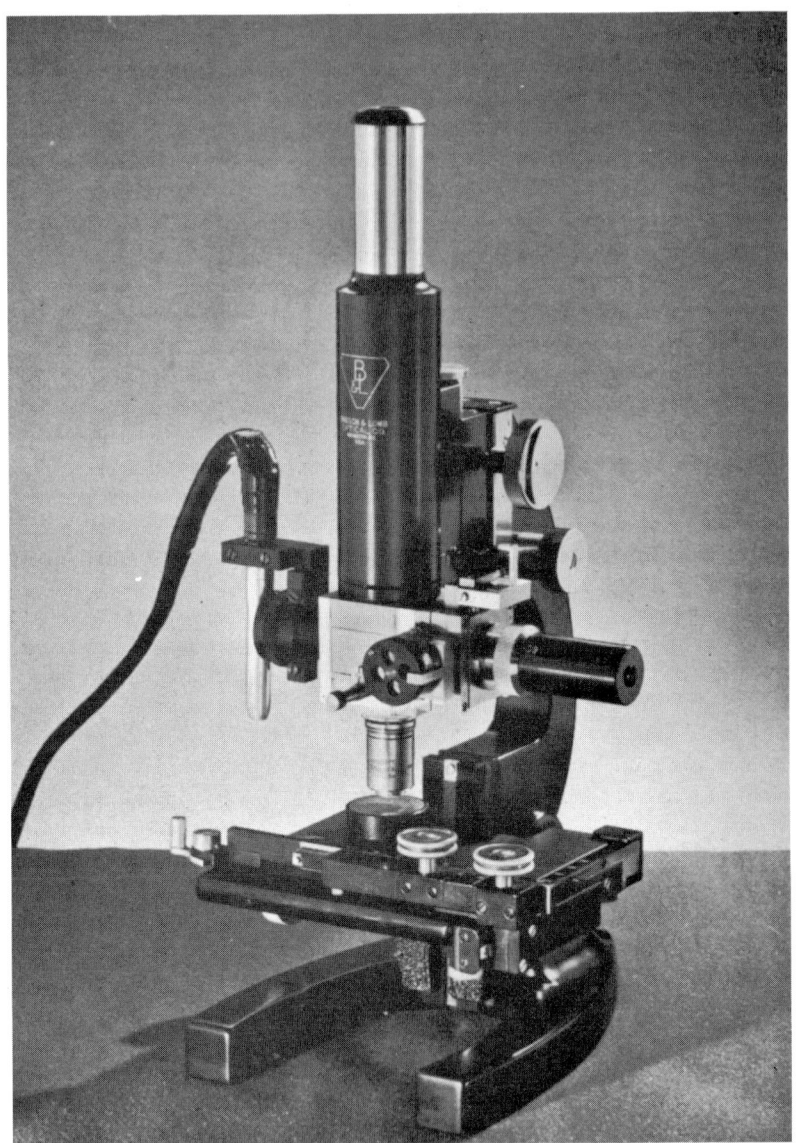

Fig. 209a—Bausch and Lomb micro-interferometer

(or principal focus) of the lens $L2$, so that there emerge from $L2$ parallel beams of light corresponding to points in the aperture A. Half of the light in these parallel beams is reflected at the semi-reflector M and half transmitted and the reflected beams are brought to a focus on the gauge or work surface W by the microscope objective $O.G.1$. Light reflected from W is picked up by the objective $O.G.1$ from which it emerges as a number of parallel beams. Half of this light is reflected and half transmitted by M, the transmitted portion being picked

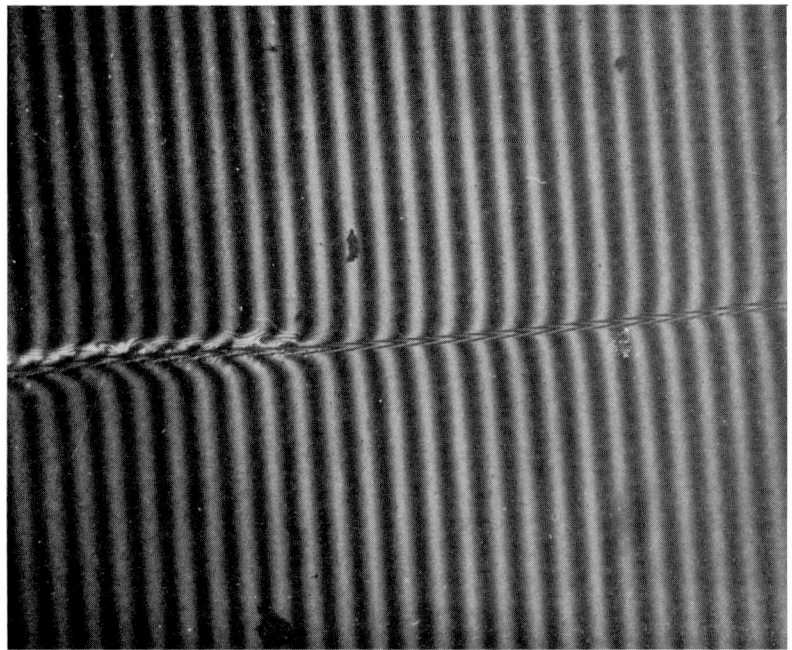

Fig. 209b—Diamond scratch on glass plate
As seen with the Bausch & Lomb micro-interferometer
Magnification ×200 (slightly reduced)

up by the lens $L3$ and brought to a focus in the plane $F.P.$, which is the focal plane (or not far from the focal plane) of the microscope eyepiece $E.P.$ The light originally transmitted by M is picked up by $O G.2$ which is a microscope objective identical in construction and focal length with $O.G.1$. Light reflected from the perfectly smooth surface of the gauge G after traversing $O.G.2$ for a second time encounters the semi-reflecting surface of M where part of the light is directed into $L3$ which brings it to a focus in the plane $F.P.$ The conditions are met that must be satisfied before interference can take place, namely light which has started out from the same point of a light source arrives at a point by two different paths, and interference fringes are observed in the focal plane $F.P.$ when the instrument is in proper adjustment. The iris diaphragm A serves to control the area of the work or reference surface that is illuminated, whilst B controls the brightness of the fringes and to a certain extent their contrast. An overall magnification of the microscope of about $\times 200$ is suitable. One point worth noting is that the microscope

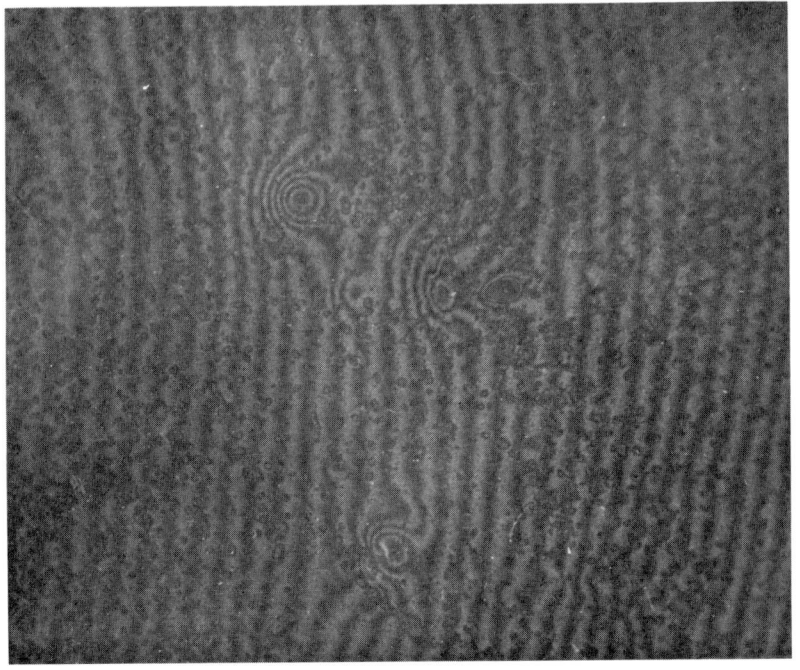

Fig 209c—Surface of rubber sheet
As seen with the Bausch & Lomb micro-interferometer
magnification $\times 200$ (slightly reduced)

objectives $O.G.1$ and 2 are not standard objectives corrected for a tube length of 160 mm, but are specially designed to work in this instrument and are corrected for an infinite tube length.

The fringes seen through $E.P.$ are regular in outline when the surfaces of both gauge and work are perfectly smooth, and have serrated edges when either of these surfaces is not perfectly smooth. The instrument may be used to examine the surface smoothness of either flat or round surfaces.

A form of the instrument is made by the Bausch and Lomb Optical Company, by whose permission the instrument and results obtained with it are shown in Figs. 209A, B, and C.

Phase contrast method

252.3 The phase contrast arrangement originated by Zernicke has been described in Chapter 11. Although quite different from the Hilger lens interferometer (§244) the appearances of wavefront aberrations are much the same. The method has been used by Burch (1934).

In the method as described by Burch an enlarged star image from the mirror to be tested is examined under a magnification sufficient to allow the Airy disc to be isolated. The Airy disc, the reader may be reminded is the bright patch in the middle in Fig. 157, sub-figure 17. For a mirror of $f/8$ it is 0·01 mm in diameter approximately. If then a small area of a transparent plate is just sufficient in diameter to cover the Airy disc (exact coincidence in size is not essential) and of a thickness suitable for causing a retardation of a quarter of a phase, any defects in the object glass will be shown up as a brilliant pattern which is actually a contour map of the imperfections, the appearance being much the same as in the Hilger lens interferometers.

The reader is referred to the papers cited for details of this very beautiful test. So far as my very limited experience goes, it is not so simple in application as in principle and when large sizes are concerned it is not easy to secure the accuracy of adjustment and the rigidity of the mounting.

Dr. Burch's method of producing the retarding spot is to precipitate tiny spheres from a 5 per cent solution of resin in acetone poured into 20 times its volume of water. This yields spheres of all sizes from 0·05 mm down to 0·001 mm. These are then pressed flat between glass plates.

The method is at its best in determining errors which are small; if the errors do not exceed one quarter to one half fringe the interpretation of the test presents no difficulty. For large errors the matter becomes much more complicated.

CHAPTER 13

SURFACE TREATMENTS

The preparation of reflecting surfaces and anti-reflection films
253 Metallic reflecting surfaces from which light is to be reflected directly (*i.e.* front surface reflectors) may be prepared by chemical or physical methods. The deposits may be opaque or translucent.

Chemical silvering

A method of chemical silvering which was extensively employed in the Hilger workshops is a modification of Brashear's method and is described in the following (extracted by permission from *Discussion on the Making of Reflecting Surfaces*, The Physical Society of London and the Optical Society, 1920, pages 18–20, C. R. Davidson).

The method employed is essentially the Brashear process, with slight modifications, a description of which was published by Mr Brashear in the *English Mechanic* in 1893. Its merit is that it gives a hard enduring film which will stand a considerable amount of polishing—

The formula used is as follows—

    A. 10 per cent silver nitrate solution
    B. 25 per cent ammonia (0·880)
    C. 10 per cent caustic potash solution
    D. Reducing solution:

Distilled water	2,000 c.c.
Sugar	180 gm
Nitric acid	8 c.c.
Alcohol	350 c.c.

A, B and C may be made as required. The reducing solution should be made up several months before it is required, as when freshly made it is not very active. It may be improved by boiling, the alcohol being added after it has cooled.

The silvering bath is made up in the following proportions—

A. (silver nitrate)	20 c.c.
B. (ammonia)	10 c.c. (more or less)
C. (caustic potash)	10 c.c.
D. (sugar)	5 c.c.
Distilled water	100 c.c.

145

To prepare the bath:
Of A (silver) take, say, 100 c.c. and to this add B (ammonia) gradually. The solution at once turns brown. Continue adding ammonia, in

quite small quantities, until the solution clears or nearly clears. Now of C (potash) add 50 c.c. The mixture will again thicken, turning dark brown. Again slowly add ammonia as before, keeping the solution agitated till it again clears. The solution will now be a pale brown colour but transparent. This part of the operation is a critical one, as it is important to avoid an excess of ammonia. In fact, it is absolutely necessary to have a slight excess of silver in the solution, and this is secured by now adding silver drop by drop until the solution will take up no more, and a little brown matter is left in suspension.

At this point a word of warning must be given. If at any stage of the above-mentioned procedure the precipitate or solution resulting from the addition of the ammonia becomes dry, the silver nitride (Ag_3N) (silver fulminate) formed by the addition of ammonia to silver nitrate becomes liable to explode with great violence, even on the touch of a feather. An explosion of this kind has happened twice in my experience.

If the ammonia is added to the liquid in an ordinary chemical flask there is a tendency for the substance to dry in the neck of the flask and it should be carefully washed down *immediately*.

To 500 c.c. of distilled water add 25 c.c. of D (sugar).

When the silver-potash solution is added to this the bath is completed, but this must not be done until the mirror is ready for silvering.

A mirror may be silvered either face upward or face down as circumstances decide. Small work is preferably silvered face down, but large mirrors are more easily handled face up. The dish for the bath should be of glass or porcelain, but large baths may be of wood or sheet metal thickly coated with paraffin wax, and for economy should be of nearly the same size as the mirror to be silvered. In the case of very large mirrors it is most economical and convenient to make the mirror itself form the bottom of the bath.

Cleaning

The cleaning is one of the most important operations. Unless the work is absolutely clean failure must result. All dust is removed and the old silver cleaned off with strong nitric acid, using a swab of cotton wool. Considerable pressure should be applied and the swabbing should be very thorough. Wash with water and with nitric acid swab again. Rinse off the nitric acid using plenty of ordinary water followed by distilled water. Finally, leave the mirror standing completely covered with distilled water. It is now ready for silvering. In the cleaning operation it has been recommended that the nitric acid be followed with a swabbing with caustic potash. Our experience is against this, the nitric acid being more easily removed than the potash.

In addition to the varying activity of the sugar solution the temperature has a very large influence on the result. Working under somewhat unfavourable conditions, it is not practicable to exercise much control over this factor.

A temperature of 65°–70° is recommended as giving the best results, but with the 30 in. mirror we have generally to be content with a temperature not much above 55°, and the proportion of reducing solution has to be increased to suit that condition. It may be taken, however, that if the temperature is too high reduction will be too rapid, and the resulting film soft, whilst if too low action is very slow and the film too thin.

The amount of sugar solution required must be found by experiment at the time of silvering by making three small test baths, using—

(1) normal reducer,
(2) 25 per cent more,
(3) 25 per cent less,

and judging by the result.

We left the mirror covered with water. This is now thrown off and the water and sugar poured on; then the prepared silver potash solution is added. At the same time a conical swing is given to the suspended mirror so that a continuous wave passes round the bath. This must not cease until the exhausted solution is thrown off. The drawback to the Brashear process is the formation of sediment which must be prevented from settling on the mirror surface by keeping the solution constantly in motion. This may be further assisted in the following manner—

Immediately the prepared silver is added to the water and sugar it begins to darken and in two or three minutes there will be a visible coating of silver. As soon as there is an appreciable deposit it will be found tough enough to stand light swabbing with cotton wool. Using rubber gloves, the operator takes a handful of cotton wool and draws it lightly over the surface, exerting no pressure beyond the weight of the swab itself. This will disturb the heavy sediment which, as the bath gets thicker, the motion of the solution is unable to prevent falling. As the cotton becomes dirty it is thrown away and a fresh handful taken.

It is a difficult point to decide when to throw off the solution. If too soon, the film will be bright but thin. If too late, the deposit will be thicker but clouded and will require much polishing. The former alternative is preferable. One must be guided by the preliminary experiments and experience.

When the silvering is judged completed, throw off the spent solution as quickly as possible and wash thoroughly with distilled water. If lightly swabbed during the washing much of the cloudy bloom on the surface will be removed, and when dry it will be found to require very little polishing. Stand the mirror in a tilted position to dry and in an hour it will be ready for polishing.

The polishers are made of best chamois leather stretched and tied over a ball of cotton wool. Two are necessary. First with a plain rubber go over the entire surface with light circular strokes, dusting constantly. Then rub a little rouge into the other and repeat. If the film is a good one it will take a high polish with very little rubbing and with very little scratching. The rubber must be scraped from time to time or any particles that may be polished off will cause scratching.

Chemical half-silvering

For such objects as the diagonal planes of Michelson or Hilger interferometers whose silvered surfaces must be semi-transparent a convenient chemical method is available.

Solutions

Prepare the following stock solutions—

A

10 per cent Silver Nitrate

B

40 per cent Formalin

C

Granulated sugar	400 gm
Alcohol	200 c.c.
Nitric Acid	10 c.c.

Make up with distilled water to 2,000 c.c. and allow to stand two weeks before using.

D

Chromic Acid	250 gm
Sulphuric Acid	1,500 c.c.

Procedure

The glass plate is placed in a glass dish, and cleaned first by strong nitric acid swabbed over the surface by small wads of cotton wool twisted round the end of a stick or glass rod, and then by allowing the glass to stand in some of D solution for $\frac{1}{2}$ minute. Pour off solution, rinse the plate thoroughly by a stream of running water, lifting the plates with a glass rod, to allow acid to escape from beneath the glass plate. Take 20 c.c. of A solution, add ammonia until the precipitate is *just* redissolved, add silver nitrate solution (any strength) until the liquid is a faint straw colour. Make up to 100 c.c. with distilled water.

Reducing Solution

5 c.c. of B
5 c.c. of C } mixed.

The mirror can be suspended with the face either upwards or downwards in the ammoniacal solution of silver. Then add the reducing solution. Keep the solution in constant motion until it becomes reddish in colour. Pour off the solution and put in a second quantity of the ammoniacal solution of silver *without* any reducing solution. Allow the mirror to remain until the required density of deposit is obtained— this will take a few minutes only. The plates can then be rinsed with distilled water and dried. No polishing is required, the deposit being bright and uniform.

For the interferometer it is better to have the deposit of silver rather dense.

For the fully silvered mirrors any of the usual silvering processes may be used.

Rapid high efficiency silvering of glass and plastics by the chemical method

Most methods of producing silver mirrors by the chemical reduction of an aqueous silver solution suffer from the defect that only a small proportion of the reduced silver is deposited as a mirror on the required surface. The remainder is deposited either on the walls of the containing vessel or is precipitated in the solution and eventually forms a loose black sludge ; this sludge is additionally objectionable since, besides wasting silver, it may also spoil the mirror.

Upton and Herrington (1950) show how this can be avoided by pre-treating the surface to be silvered ; the method is applicable to producing heavy coatings on plastics. The following recipe is given by the authors :

The surface to be silvered is pre-treated by being wetted with a solution of 10 gm stannous chloride in 20 ml hydrochloric acid (A.R.) and 80 ml water. The surface is then rinsed with a 5 per cent silver nitrate solution and well washed with distilled water and kept under water until required.

Silvering solutions

Solution 1 : 10 ml of pyridine ; 0–50 ml of I.C.I. Fixanol C_2 (cetyl pyridinium bromide) solution (0·005 per cent); 16 ml of 1·25 per cent hydrazine sulphate solution

Solution 2 : 70 c.c. of 1 per cent solution of amoniacal silver nitrate prepared by adding 0·880 ammonia to 1 per cent silver nitrate solution until the precipitate just redissolves

Solutions 1 and 2 are mixed just before use. The amount of Fixanol C solution is adjusted to give satisfactory results. No exact quantity can be recommended as the amount required depends on the concentration of surface active agent already present in each batch of pyridine, and this may change with time.

A discussion on the making of reflecting surfaces took place on 26th November, 1920, under the joint auspices of the Physical Society of London and the Optical Society. A full account was published (London, Fleetway Press, price 5s. 0d.) which includes a very comprehensive bibliography and twelve papers on various aspects of the subject.

" Physical " processes

254 There are two physical methods by which metal reflecting films are deposited on glass. In both methods the metal is volatilized and condenses on the polished glass surface.

The volatilization may be induced by bombarding the metal with gaseous ions in a moderate vacuum ; in this process, called " sputtering," the metal source remains comparatively cool.

The other method is to heat the metal in a very high vacuum to a temperature sufficient to make it evaporate freely.

Sputtering

The general method and apparatus have been fully described in Strong's *Modern Physical Laboratory Practice* (Blackie, 1940), which also contains data on the rate of sputtering of metals in various gases. Special precautions for the sputtering of good silver films are described by Gwynne-Jones and Foster (1936).

The gas discharge is generated in a glass vessel large enough to allow plenty of space between the mirror and the walls of the vessel. The gas pressure is reduced to within the range 1 mm—10^{-2} mm of mercury by means of a mechanical vacuum pump. The chosen gas is allowed to flow slowly through the gas chamber or the system is flushed with the gas several times before sputtering is begun.

A dissipation of about 50 watts at a voltage of from 500v. to 20,000v., preferably D.C., is required to produce the gas discharge.

The metal to be sputtered is made the cathode, and the anode is a ring or plate of aluminium or of the same metal as the cathode (Gwynne-Jones and Foster, 1936). The cathode should be of about the same size and shape as the surface to be coated, and parallel to it.

In the range of pressures used for sputtering, the discharge dark space surrounding the cathode extends to a distance of not more than a few centimetres from the cathode and the mirror surface is placed so that it is tangential to the boundary of the dark space.

A drawback of this method is that the surface area over which reasonably uniform coatings can be obtained, is rather limited. It is a useful method in cases where no satisfactory filament material can be found, e.g. for the production of platinum and nickel films, or where the heat radiated from the filament would damage the objects to be coated.

The rate of deposition is slow, so that even the easily sputtered metals such as silver, gold and platinum may take an hour to form an opaque film, but this has the advantage of allowing accurate control of the density when semi-transparent films are required.

By far the most usual way now is, however, depositing the films by evaporation *in vacuo*.

Deposition of thin films on optical surfaces by evaporation in vacuo
by W. ZEHDEN, Ph.D., Head of the infra-red and vacuum sections of the Research Department, Hilger & Watts Ltd.

Introduction

255 The production of thin surface films has become a very important technique in optical industry. Depending on the material, thickness, and number of these films, various effects can be produced, the most important of which are—
1. High reflectivity (surface mirrors);
2. Semi-reflection-transmission (beam splitters);
3. Reduced transmission (neutral filters);
4. Selected transmission and reflection (colour filters);
5. Reduction of undesirable surface reflection and increase of transmission ("Blooming").

For the production of certain films of this kind, individual chemical methods are available which depend on suitable precipitation from solution or etching. The method most in use, however, is a physical one—evaporation in vacuo. It has the advantage of being, with minor modifications, applicable to any material which develops a suitable vapour pressure and remains stable when heated in vacuo; its origin is set forth on pp. 482-3. Process 4 was invented by Dr Walter Geffcken of Jenaer Glaswerk Schott & Gen. (Geffcken's patent, D.R.P. 716, 153; 14th Jan., 1942). The following is a free translation of the claims of the patent in question.

1. Light filter consisting of a number of transparent or semi-transparent layers in which the filter effect is produced by interference phenomena. It is characterized by the fact that it consists of at least two semi-transparent layers giving metallic reflection with a distance between them not exceeding ten times the wavelength of

the light for which maximum transmission is obtained, whereby this distance is filled with at least one intermediate layer giving non-metallic reflection.

2. Filter according to claim 1, characterized by the fact that the outermost metallic-reflecting layers are coated with one or several transparent layers.

3. Filter according to claim 1 characterized by the fact that some of the metallic-reflecting layers consist of materials for which, in the neighbourhood of the wavelength to be passed, the refraction and the absorption vary greatly with the wavelength.

4. Filter according to claim 1 characterized by the fact that the non-metallic-reflecting layers do not have the same refractive index throughout.

5. Filter in which two or more filters according to claim 1 are combined in such a way that the distance between two filters is large compared with the distance between the metallic-reflecting layers within the single filters, characterized by the fact that each two of these filters include an acute angle between themselves.

6. Filter according to claims 1 and 2 characterized by the fact that, for increased filter effect, at least one of the non-metallic-reflecting layers is coloured, and several may be.

7. Filter according to claim 1 characterized by the fact that, for increased filter effect, the carrier plate is made from an absorbing material.

The plant and details of vacuum deposition are dealt with in §257. The basic technique is as follows—

The material to be deposited is put on or attached to a tungsten or molybdenum filament of suitable shape mounted on the base of a vacuum chamber, and the optical components are arranged at a certain distance around it so that the surfaces to be coated face the filament. The chamber is then pumped out until a vacuum of approximately 10^{-5} mm Hg is reached. The filament is then heated by an electric current until the material evaporates. The vapour molecules spread from the source in all directions in straight lines, as there are practically no gas molecules in the way with which they could collide, and eventually condense on any " cold " surface which happens to be in the way. There they form a film the thickness of which depends on the duration and rate of evaporation, the geometry of the source and the distance between source and coated surface. In order to produce good adhesion between the film and the surface on which the film is to be deposited, it is necessary to free the surface from molecular layers of gas, water, or grease before the film is deposited. This is usually done by exposing the surfaces to the ionic bombardment of a high tension glow discharge run in the vacuum chamber prior to the evaporation.

The reflectivity of a freshly deposited silver film exceeds 90 per cent throughout the visible spectrum but on the short wavelength side of 4,500A it falls at first slowly and then rapidly to a minimum value at about 3,200A; a silver film is practically transparent to light of this wavelength. An aluminium film, on the other hand, has a reflectivity which is nearly as high as that of freshly deposited silver in the visual region and does not show the severe drop in the ultraviolet. Between 3,250A and the limit of atmospheric transmission at about 2,900A the aluminium film has therefore very marked advantages over silver for astronomical investigations. The film does not tarnish, adheres strongly to the glass surface, and is hard, so that it can be cleaned. Under favourable conditions it will last for several years with very slight decrease in reflectivity.

Examples

In the following a few typical examples of surface coating are discussed briefly.

1. *Aluminium.*—Aluminium has a very good reflectivity for infrared, visible, and ultra-violet radiation, and is, therefore, widely used for optical mirrors. The filament is formed by a horizontal tungsten wire which is bent to form a number of vertical loops hanging downwards. In each loop, a short piece of aluminium wire is suspended. When the filament is heated in vacuo the aluminium pieces melt and form beads which cling to the tungsten filament at the bottom of each loop. The surfaces to be coated can be mounted vertically on either side of the filament, or horizontally *above* the filament. It is dangerous to place optical components *below* the filament as the latter may break and drop hot metal on the glass.

2. *Rhodium.*—The reflectivity of rhodium is inferior to that of aluminium, but still good enough for many purposes where radiation is being handled which is not too feeble. It is superior to aluminium in resistance to abrasion and against the influence of sea atmosphere and tropical conditions. In a thin layer as a semi-transparent filter, its optical density changes only slowly from visible to ultra-violet, so that for not too wide spectral intervals it can be used as a neutral filter.

A semi-transparent film of 38 per cent transmission gives approximately the same percentage reflection, the remainder being lost in absorption. Films of this type are used for many beam-splitting devices.

As filament a tungsten spiral is used on which rhodium is deposited by electrolysis from a commercially available rhodium solution. The filament is mounted vertically, and the surfaces to be coated are placed vertically around it at a distance of approximately 6 in. Rhodium evaporates from the solid state, if one over-heats the filament so that the rhodium melts, the filament usually fuses.

3. *Silver.*—Silver films are used where the highest reflectivity in the visible range is required. As a surface reflector silver requires a protective coating, otherwise it tarnishes quickly when left exposed to the air. High reflectivity is of particular importance in cases where multiple reflections are involved. One interesting recent development in this line is the interference colour filter originated by Geffcken (1944). This filter is produced by depositing on a glass surface first a semi-reflecting silver film, then a transparent non-metallic film (see example 4), and finally another semi-reflecting silver film. The action of this three layer system can be understood if one looks at it as at a low order Fabry-Perot interferometer. It produces a series of transmission bands the wavelengths and orders of which depend on the optical thickness of the transparent spacer layer. The higher the number of multiple reflections between the silver layers the narrower the spectral width of the transmission bands will be. There is one interesting point worth mentioning. While for a Fabry-Perot interferometer the exact flatness of the plates is of greatest importance, an interference filter can be produced on good quality window glass because all irregularities in flatness are reproduced in all three layers so that the uniformity in optical thickness of the spacer layer is practically not affected.

Silver is usually evaporated from a molybdenum strip filament mounted horizontally and bent to form a cavity in the middle into which pieces of silver are placed. It can also be evaporated from a tungsten wire filament in the same way as described above for aluminium. The only difference is that the tungsten wire must be coated with a thin film of rhodium before the silver wire is attached, because molten silver will not adhere to a tungsten surface.

4. *Magnesium fluoride.*—Considerable improvement in the transmission of complex optical systems is achieved by "blooming," *i.e.* by a treatment which reduces the surface reflection and increases the transmission of the components. If a glass of refractive index n is coated with a transparent film of refractive index $\sqrt{n}$ and of optical thickness $\lambda/4$, the wavelength λ will be missing, and wavelengths in the neighbourhood of λ will be reduced in intensity, in the light reflected from the coated surface. The intensities suppressed in reflection will be added to the transmission. Magnesium fluoride is widely used as material for these anti-reflection films because of its mechanical strength. Its refractive index is actually somewhat too

high. The optical efficiency of these films, therefore, increases with increasing refractive index of the glass. Among other applications, magnesium fluoride can be used as material for the spacer layer in the interference colour filters mentioned in example 3.

Evaporation of magnesium fluoride can be carried out from a molybdenum strip bent in a similar shape as described above for silver. The magnesium fluoride powder is placed in the cavity, and the surfaces are arranged above the filament approximately tangential to a sphere with its centre on the filament. Another way of evaporating magnesium fluoride, is to mix the powder to a paste with a small quantity of distilled water, and to fill this paste into a tungsten spiral, dry it and pre-heat it in vacuo. After pre-heating, the filament is ready for use. It is mounted, and the surfaces to be coated are arranged vertically in the same way as described for the case of rhodiumizing. If larger quantities of magnesium fluoride have to be evaporated, as in the case of interference filters (see above), a crucible can be filled with the powder and a tungsten flat spiral filament can be mounted just above the powder surface. The magnesium fluoride will be heated by radiation, and evaporation can be carried out for a long period. A number of other substances (*e.g.* silica, cryolite) have been used for making non-reflecting films, but in practice magnesium fluoride has been found the best.

Interference colour filters

256 An interference colour filter is composed of two partially reflecting metal films separated by a film of transparent material (Geffcken 1942, 1948, 1949). A mathematical and experimental analysis of their properties is given in two papers by Hadley and Dennison (1947, 1948). All three films are deposited on a glass base. The narrowness of the transmission characteristic of these filters demands a high degree of accuracy in the regulation of the thickness of the films if maximum transmission is to occur at a specified wavelength.

The principle and mode of making these films are fully described by Greenland and Billington (1950) who describe a method of observing the optical thickness of transparent films during deposition with an accuracy enabling the centre of a fourth order transmission band to be placed within 20A of any wavelength in the visible spectrum. This precision has been successfully achieved in the high vacuum laboratory of Hilger & Watts Ltd, from whom lists of the wavelength peaks available can be obtained.

5. *Zinc Sulphide.*—While the reflectivity of a glass surface can be reduced by a film of lower refractive index, it can be *increased* by

coating it with a transparent substance of *higher* refractive index. For this the film thickness should produce a phase difference of 1 wavelength between the beams reflected from the film surface and from the film glass surface. A practically 100 per cent efficient semi-reflecting surface can be thus obtained, all the incident light being either reflected or transmitted. Zinc sulphide has a very high refractive index (approximately 2·3). A zinc sulphide film deposited on glass, therefore, increases the surface reflection up to 30 per cent. By applying alternatively several layers of zinc sulphide and magnesium fluoride, even higher reflectivities can be obtained so that this method can be successfully applied to the manufacture of beam-splitting devices. The efficiency of these films is superior to that of beam-splitting metal films (see above) owing to practically complete absence of absorption losses. However, owing to interference effects, they are generally tinted, transmitted and reflected light showing complementary tints.* Zinc sulphide can also be used as spacer layer in interference filters. For zinc sulphide, the same evaporation methods can be used as described above for magnesium fluoride.

Titanium oxide and antimony sulphide have also been used (Arnulph, 1946).

Employment of multiple films for the plates of Fabry-Perot etalons

Dufour (1950) has applied to Fabry-Perot etalons triple films of Ag-Cryolite-SZn and other films still more complex, including quintuple films of SZn-Cryolite-SZn-Cryolite-SZn, and has succeeded in improving both the luminosity and the contrast.

Reflection reducing films without vacuum plant

Films for reducing reflection can also be produced without vacuum plant. The films can be of cryolite, sodium silicate or potassium silicate. In the production of the sodium silicate coating the lens is held by a chuck on the spindle of a centrifuge. The spindle is rotated at from 1,000 to 2,000 r.p.m. and a small quantity of the solution is poured upon the rotating lens surface, the optimum thickness being determined by the colour reflected by the film. This type of coating readily deteriorates in contact with greases, so to avoid the possibility of this happening in the finally assembled instrument, the film is removed from around the edge of a lens, as it is rotating, with a small wooden tool. As soon as the desired colour of film is attained the lens surface is flushed with distilled water, heated to 70°C, and then heated by an electric hair drier for a few minutes

* A useful application of this colour effect is discussed in an article by Dimmick (1942). *A new dichroic reflector and its applications to photocell monitoring systems. Jour. Soc. Motion Picture Engineers*, Jan. 1942.

before it is removed from the chuck. The lens is then placed in an electric oven and baked for half an hour at 70°C.

The sodium silicate solution is prepared in the following way.

(a) Sodium silicate (Baumé 28° to 30°), diluted to one part of sodium silicate to seven parts of distilled water.
(b) Hydrochloric acid, specific gravity 1·19, diluted to one part of acid to five parts of distilled water.
(c) " Nekal " sodium-isopropylnapthalenesulphonate saturated solution in distilled water.

The coating solution is then made up as follows—

1·5 per cent by volume of diluted hydrochloric acid.
1·5 ,, ,, ,, " Nekal " solution.
97 ,, ,, ,, diluted sodium silicate.

The solution is held at a temperature of 34°C for a period of 15 hours before using. Inasmuch as the solution is not stable for more than 48 hours, only sufficient quantity is made up for each day's need.

Potassium silicate solution for coating is prepared as follows—

A potassium silicate solution having a titration factor of 1·74 and a specific gravity of 1·225 (26·5° Baumé) is well mixed with distilled water on the basis of 7 parts of water to one part of the solution. As this solution is being stirred there is added 1·5 per cent by volume of hydrochloric acid (concentration 5-normal). This solution is then immediately heated to—

22°C for 148 hours, or
38°C for 23 hours, or
70°C for 19 minutes,

and then cooled quickly to room temperature.

Shortly before the above solution is to be used for coating, the following materials are added—

2·5 per cent by volume of " Nekal " saturated solution in distilled water (Sodium isopropylnapthalenesulphonate).
2·0 per cent by volume of concentrated Ureic solution (51·9 grams in 100 c.c. distilled water).

The solution should be well filtered before using.

The process of coating is the same as for the sodium silicate except that after drying on the centrifuge no further heat treatment is required. This method of coating has, therefore, the advantage that cemented elements can be coated. (*U.S.N.*, 1945.)

Plants for producing surface coatings by the evaporation process

257 The first essential of course is a good pump.* These are made in U.S.A. by Distillation Products, Inc., and in England by W. Edwards & Co. and by the Metropolitan-Vickers Co.

General information on the subject of surface coatings will be found in Strong (1939). The plants used for the purpose may have vacuum

---

* It is said that in Germany phosphorus pentoxide is used in the plants for vacuum coating as a " getter " to facilitate the attainment of the requisite vacuum. (*U.S.N.*, 1945.)

Fig. 210—Vacuum metallizing plant

chambers either of glass or metal. Fig. 210 is a photograph of the metal plant most in use in the coating shop of Hilger & Watts, Ltd.

Since glass bell jars give an almost unrestricted view of what is happening inside they are better for coating anti-reflection and other interference films, for such work it being very desirable to be able to observe the interior conditions continuously.

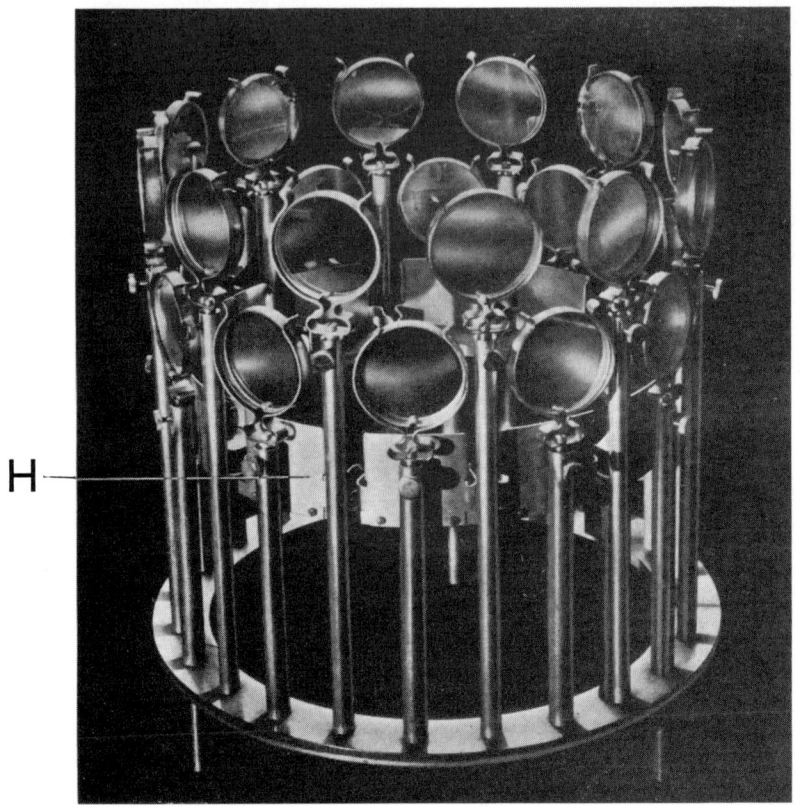

Fig. 211—Frame for windows to be coated

The apparatus will be described in its application to the coating of rhodium mirrors which were required to transmit with a density of between 4·0 and 4·9, and to be supplied in pairs differing by no more than 0·1 in density.* This example is chosen because it illustrates special difficulties and shows how they were surmounted.

* Density=$\log_{10}$ (I_0/I) where I_0 is the intensity of the incident and I that of the transmitted radiation.

Three of the figures have one dimension indicated so that the dimensions of the other parts can be deduced.

The vacuum chamber consists of a cast-iron cylinder, C (Fig. 210), which stands on a circular steel plate, S. The windows to be coated are arranged in the frame H (Fig. 211) which can be placed in position

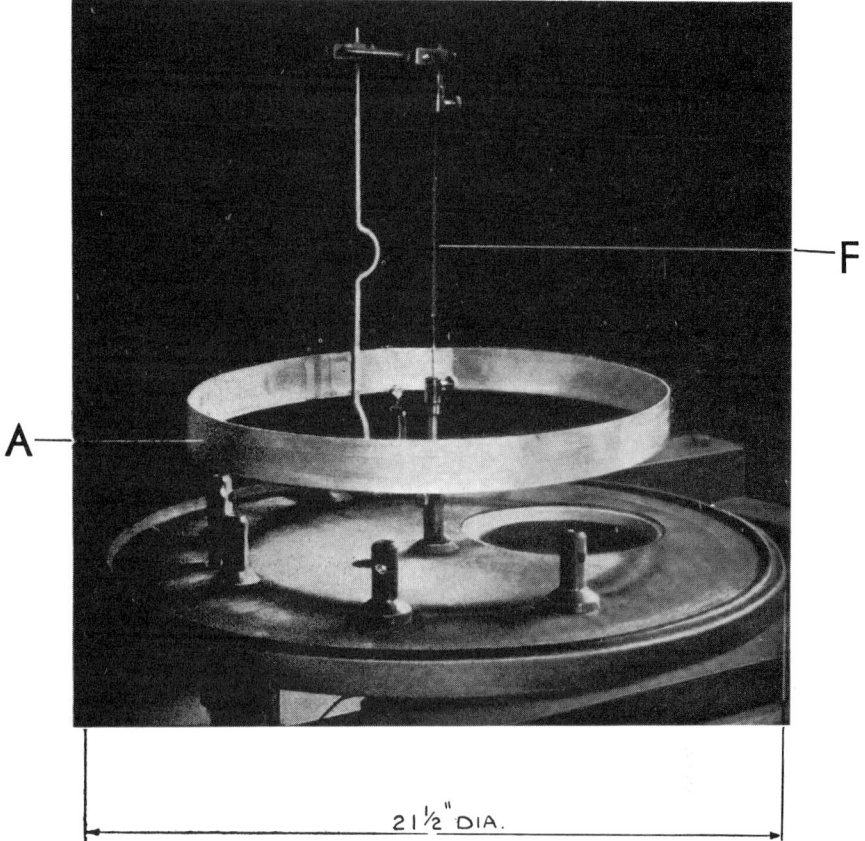

Fig. 212—The aluminium ring and the tungsten filament

within the aluminium ring A (Fig. 212) with the tungsten filament F in the middle.

The vacuum chamber has plane parallel windows, W_1 and W_2, at opposite sides, through which light from the tungsten filament F passes to the system of right-angle prisms P,P,P,P (Fig. 213) and thence by reflection at a fifth right-angle prism to the eye. The eye sees one image of the tungsten filament through the window G on which rhodium is being deposited.

A mask M (Fig. 213), can be introduced into the other beam by an electro-magnetic arrangement actuated from the outside to prevent the window W_1 becoming coated with metal, and this mask is only

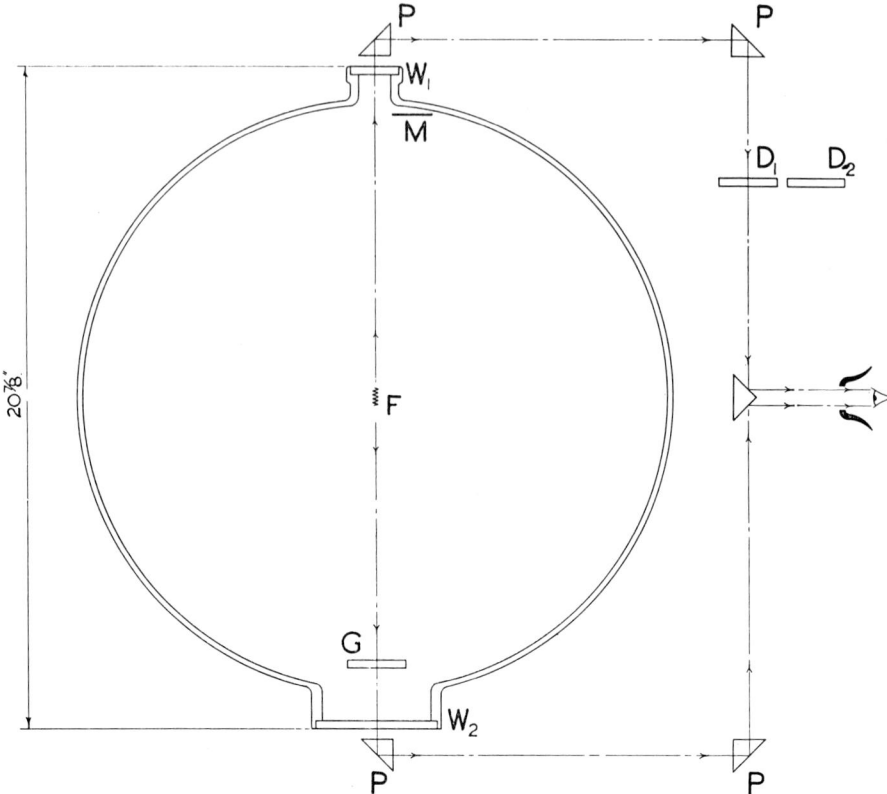

Fig. 213—Lay-out for observing the optical density of a coating

removed when it is desired to make a test of the density of the rhodiumized window G.

D_1 and D_2 are windows which have been coated, the one to the minimum and the other to the maximum density. At various epochs during the coating the tungsten filament is reduced in temperature so that, while still red hot, no appreciable evaporation is taking place, the mask M moved out of position and visual comparison made between the density of the two standards and that of the window.

Visual comparison of this kind has been found sufficiently accurate without the elaboration of a complete photometric set-up.

The results of our investigations on rhodium coating with this apparatus are given in the following pages.

The production of high density rhodium films on glass by the evaporation method

258 In recent years, rhodiumizing has gained great importance as a method of making surface reflectors and semi-transparent filters. This is due to four properties of rhodium films which render this metal particularly suitable for these purposes—

(1) high reflectivity,
(2) neutral colour in transmission,
(3) hardness,
(4) resistance to chemical attacks.

The technique generally used for making these films is the evaporation method. The glass surfaces are first exposed to a glow discharge in a vacuum chamber at a pressure of some 10^{-2} mm Hg, then the pressure is reduced below 10^{-4} mm Hg, and rhodium is evaporated from a hot filament (rhodium wire or electrolytically rhodium-plated tungsten wire) and condenses on the glass surfaces facing the filament.

Similar methods are being widely used for the deposition of many other materials, and if it sufficed for rhodium mirrors to be treated with the same care as, say, aluminium or silver mirrors, no special difficulties would be found in rhodiumizing. But the hardness of the films obtainable with rhodium under the best conditions has resulted in demands for commensurate resistance to drastic cleaning and rough handling. The hardness of the metal alone is not sufficient to satisfy these demands, particularly with thin, semi-transparent films; the hardness has to be combined with strong adhesion to the surface on which the rhodium film has been deposited.

It is generally recognized that this adhesion is brought about by the cleaning effect of the high-tension glow discharge on the glass surfaces, but unless special precautions are taken, this cleaning effect will either be insufficient, or the discharge will have a very adverse effect due to cathodic sputtering. This dilemma has been the starting point for much discrepancy among the opinions of different workers, and it is the main object of this report to show how an efficient discharge can be run without much sputtering.

(a) The effect of mercury vapour

259 For rhodium coating where the highest degree of adhesion to the glass is important we prefer mercury pumps with liquid air traps.

For blooming we use either type of pump. For all other depositions, in particular for routine aluminizing and gilding, we use oil pumps. Although with gold films better adhesion can be got with the mercury pump, yet for routine the gold-mercury combination is too troublesome to deal with.

If the discharge is run in the presence of mercury vapour, the sputtering effect is enormous. It will be found that after a discharge of this kind without any rhodiumizing having followed, the glass surfaces are coated with a brownish metallic deposit. Spectrochemical analysis of this film has shown that it consists mainly of copper and mercury. The copper comes from the brass clips and jigs which hold the glass plates. This is clearly indicated by the extremely clean, almost polished appearance shown by these clips after such a discharge. If the plant has already been used for rhodiumizing, the deposit will contain rhodium as well.

The quality of a rhodium film which is plated on top of such a deposit, will be poor, for the following reasons—

(i) Adhesion is weak since sputtering started the very moment the discharge was switched on, and the glass surface, therefore, had no time to be exposed to the cleaning effect of the discharge,

(ii) Resistance to chemical attacks is low, particularly for thin rhodium films, since the sputtered layer is not a noble metal.

(iii) In all cases where neutral transmission or bright back reflectivity is required, rhodium films obtained in this way are of poor quality.

Thickly coated surface reflectors are useable because a thick rhodium film is self-supporting and depends to a lesser extent on strong adhesion in every single spot of the surface, the chemically weak sputtered layer is well protected against influence from outside, and the back reflection is of no importance ; they cannot, however, be considered fully satisfactory.

It is, therefore, of fundamental importance to exclude the mercury vapour from the vacuum chamber during the discharge, and it should be stressed that this cannot be achieved by simply using a liquid-air trap between the pump and the chamber. This trap, while it is filled with liquid air, will stop mercury vapour from penetrating into the chamber but it will never remove the mercury absorbed on the metal surfaces in the chamber. This mercury will be liberated by the discharge and will maintain a noticeable vapour pressure for a long time, and by the time it has been driven out and found its way to the liquid-air trap by diffusion, damage will already have taken place.

A simple and efficient way to get rid of the mercury, is to allow air to enter the chamber through a valve while the pumps are running and the liquid-air trap is filled. Then the discharge is started with a series of short bursts at reduced current and increased pressure (exact conditions will be discussed later) while from time to time the bursts are being watched with a pocket spectroscope through a window in the vacuum chamber. It is very easy even for an unskilled person who knows nothing about spectra, to learn to recognize the green mercury line and to decide if it is still present or has disappeared. Usually after 10-30 bursts in intervals of 5-15 sec, the mercury line will disappear and as practice has shown, will not return during the discharge. When this state has been reached, the discharge proper at full current is started, and the air leak is reduced until the pressure is obtained at which it has been found best to run the discharge. After the proper time has passed, the discharge is switched off, and then the valve is shut and the vacuum chamber pumped to the high vacuum limit. Then the evaporation of rhodium is carried out.

(b) Operating the vacuum

The procedure of closing down and opening the vacuum chamber should be guided by two important rules—
  (i) Never allow air to enter the chamber through the pumps, but always through the inlet valve directly into the chamber.
  (ii) Never leave the plant at a pressure of less than 0·5 mm Hg without liquid air.

If rule (i) is neglected, much mercury vapour will be blown into the chamber and contaminate it, particularly when the pumps are still hot. It should be noted that a leak in the backing vacuum part of the plant may also act against this rule. Rule (ii) cuts down contamination by diffusion. It should be always observed when the plant is left with vacuum over night.

(c) The self-rectification effect

260 Some sputtering, however, will take place even in a pure air discharge, after the mercury vapour has been driven out, or in plants operated with oil diffusion pumps, and the problem is to reduce this effect to the lowest possible rate.

In oil pump plants this is very simple. Since the sputtering effect of aluminium, in absence of mercury, is practically nil, all one has to do, is to give the inside of the vacuum chamber as well as all metal holders and jigs to be used, a coat of aluminium before starting the rhodiumizing process. To do this, the jigs and holders are put into the vacuum chamber, and then some aluminium is evaporated in vacuo from a tungsten filament to which a few short pieces of aluminium wire have been attached. If this procedure is repeated after every second or third rhodiumizing process, no trouble from sputtering will be experienced. Unfortunately, this technique cannot be applied to mercury pump plants because aluminium eagerly absorbs mercury vapour. With all internal surfaces aluminized, it would take a very long time to get rid of the mercury, and sputtering, therefore, would be very strong as, in the presence of mercury vapour, aluminium offers no protection against sputtering.

One way to prevent the glass surfaces from being coated by sputtering, is to use a D.C. discharge with the vacuum chamber and to have all metal parts close to the glass surfaces anodic while the cathode is kept well away from the glass surfaces. The disadvantage of this method, apart from having to use more complicated electrical equipment, is that with the chamber being continuously used as anode, the greater part of the metal surfaces are never exposed to the cleaning bombardment by the gas ions, and this has a bad effect on the vacuum conditions. Practice has shown that satisfactory results are obtained with an A.C. discharge, provided the half-phase for which the vacuum chamber is cathode, is kept below a certain limit. To simplify the following discussion, the half-phase for which the vacuum chamber and the metal jigs are cathode, will from now on be referred to as the " wrong " half-phase, the other one will be called the " right " half-phase.

As high tension electrode we use a horizontal ring of aluminium A (Fig 212) which has a diameter about 30 mm smaller than the inner diameter of the bell jar. The other electrode is formed by the earthed base plate and walls of the vacuum chamber and all metal parts in it. Owing to the different geometry of these two electrodes, the discharge obtained from an ordinary A.C. high-tension transformer is partly rectified, the current being higher in the direction of the " wrong " half-phase, *i.e.* just in the direction which produces the sputtered films on the mirror surfaces. This effect varies with the gas pressure in the vacuum chamber, and with the buffer resistance in the discharge circuit (internal resistance of the transformer and additional external resistance). The gas pressure has to be chosen to give the most efficient cleaning effect of the discharge. This matter will be discussed later. For the present considerations it is, however, of importance that this pressure approximately coincides with the maximum rectification. If, therefore, in addition the buffer resistance is low, the " wrong " half-phase may be many times stronger than the " right " half-phase, and the sputtering effect will be too high. In this case, one has to increase the buffer resistance (and if necessary, the voltage to maintain sufficient current) until for a discharge sufficient to clean the glass surfaces, the total amount of sputtering remains negligible.

It has been found that satisfactory results are obtained if the ratio of the " wrong " half-phase to the " right " half-phase is kept below 2 : 1. This can easily be checked by putting an A.C. milliammeter and a D.C. milliammeter in series into the discharge circuit (the D.C. instrument must be well damped, otherwise the needle will vibrate). When connecting the D.C. meter, one should remember that the vacuum chamber is preferably cathode. If the currents of the " wrong " and of the " right " half-phase be called I_w and I_r, the reading of the A.C. meter will give, approximately $(I_w + I_r)/2$ while the D.C. meter will read, approximately $(I_w - I_r)/2$. If for instance, the readings are 85 ma on the A.C. meter and 25 ma on the D.C. meter, the approximate currents in the two half-phases are $I_w = 110$ ma and $I_r = 60$ ma, and the ratio $I_w : I_r$ is, therefore, below 2 : 1. But if the readings are for instance 90 ma on the A.C. meter and 60 ma on the D.C. meter, the approximate currents are $I_w = 150$ ma and $I_r = 30$ ma, and the ratio $I_w : I_r$ is above 2 : 1.

(d) *Choice of discharge conditions*

261 It is a good test for the efficiency of the discharge if a glass surface after having been exposed to the discharge, feels rough and dragging if one rubs it gently with a Selvyt cloth. Burnishing makes it smooth again. At the same time, there should be no noticeable deposit on the surface. Before doing any rhodiumizing, it is therefore, useful to test the discharge conditions by exposing a glass plate, part of which is covered by another glass plate, to the discharge at the same current and pressure and for the same length of time as it is proposed to apply the rhodiumizing process. After the discharge, without any evaporation having been carried out, the plant should be opened and the glass plate inspected. When rubbing with a Selvyt cloth, the exposed part of the surface should drag while the covered part of the surface feels smooth, and this difference should be a striking one. At the same time, if this glass plate is laid on white paper, no noticeable difference in colour between the exposed and covered parts should be found. It is well worth while to do this test ; it may produce surprising results. If no, or only weak, dragging is found, the discharge was too short, or the current too low, or the pressure unsuitable, or the discharge was run in a " dirty " atmosphere. If a difference in colour or, maybe, even a metallic deposit on the exposed part is found, sputtering was too strong, and steps have to be taken to reduce it in accordance with the advice given above. For a vacuum chamber of about 45 cm diameter and 45 cm height, the discharge conditions adopted in our laboratory are the following : current 80–100 ma, time 40–50 min, pressure about 0·02 mm Hg, voltage actually measured on the discharge chamber at this pressure 750–900v.

Air is flowing through the chamber during the whole discharge. In mercury pump plants the liquid-air trap is filled.

With regard to the pressure, some further explanations are necessary. A technique widely applied is to pump the chamber to the limit obtainable with the rotary oil backing pump alone and then to start the discharge. The pressure thus obtained, even without using the air leak, is hardly lower than 0·05 mm Hg while the discharge is running, and according to our experience, the cleaning effect is poor. With the air inlet valve sufficiently opened, the pressure would be higher than 0·1 mm Hg. Therefore, it has become common routine in our laboratory to use the rotary oil pump and a diffusion pump for running the discharge. In plants equipped with two diffusion pumps in series, it is sufficient to use the smaller backing diffusion pump. Otherwise, the main diffusion pump has to be run. To start the discharge, the pressure in the chamber is set about 0·05 mm Hg by adjusting the air inlet valve. Then, after the discharge has become " clean," the air leak is reduced until the pressure in the chamber drops to about 0·02 mm Hg. The pressure in the part between rotary backing pump and diffusion pump under these conditions will be about 0·2–0·1 mm Hg. According to our experience, this small air leak does no harm to the performance of oil diffusion pumps, and it may be of interest to add that, with the discharge conditions and precautions given in this report and with the additional precaution of shielding all rubber seals from the discharge, first class rhodium films with excellent back colour have been obtained at production scale in the American Distillation Products (Inc) oil pump plant without using any liquid air or solid CO_2 during the whole rhodiumizing process.

(e) *General Remarks*

Little remains to be said about the actual rhodiumizing. If after a discharge as described above, evaporation takes place in a vacuum of the order 10^{-4} mm Hg or better, adherence and colour of the rhodium film will be excellent. Final surface hardness is obtained by burnishing with a Selvyt cloth or chamois leather. When being burnished, the rhodiumized surface will feel rough and dragging in the beginning, but will become smooth after a few strokes. It should be quite impossible to damage a rhodium film by burnishing. The finished rhodium films should stand up to any reasonable treatment and even to a fair amount of unreasonable treatment, and this applies not only to thick films, but also to thin films. It should hardly be possible to scratch a transparent rhodium film without damaging the glass underneath.

Method of preparing and shape of the filament and the routine of carrying out the evaporation will be of importance mainly for the uniformity of the films, density control, life time of the filament, and rhodium economy, and will have to be chosen to suit the requirements of the individual problems to be dealt with. The quality of the deposited rhodium will not depend on these factors.

(*f*) *Pinholes*

High density filters often become useless if they contain pinholes since the amount of light transmitted by a clear pinhole becomes comparable with, or even stronger than, the total amount of light transmitted by the dense filter.

There are two types of pinholes. The first is produced by the screening effect of loose dust particles left on the surface during evaporation. They appear either at once or after burnishing. It is difficult to remove all dust particles from a glass surface owing to electrostatic attraction. The best way to deal with this trouble when making dense filters, is to build up the density in three separate rhodiumizing runs with burnishing after every run. If a dust particle has produced a pinhole in the first coat, it is unlikely that exactly the same spot will be covered again by dust when depositing the following coats, and, in addition, one will find it much easier to remove dust from a metallized surface than from the clear glass surface. If, therefore, in order to produce a filter of density 4·5, one deposits a first coat of a density 1 to 1·5, a dust pinhole in the first coat will produce a " pinhole " of a density 3·5 to 3 in the final filter, which is hardly noticeable.

The other type of pinhole occurs as a result of bad adhesion of the rhodium to the glass surface. Some of them will appear after burnishing, but many will break through later, sometimes after several weeks. If this bad adhesion is due to poor discharge conditions or to bad vacuum during evaporation, one will find it fairly easy to damage the first coat by hard rubbing with cotton wool or scratching with a needle, and one can do nothing but correct the discharge conditions in accordance with the recommendations made above, and improve the vacuum conditions of the plant. Often, however, bad adhesion is confined to separate tiny spots of the surface. This is due to local contamination by dirt (grease, pitch from the polisher, saliva marks resulting from talking near the cleaned surfaces). Particularly dangerous in this respect are pits and scratches in the glass surface because they always retain some traces of dirt.

Every surface to be rhodiumized, therefore, should be thoroughly inspected, and every doubtful spot should be touched and scraped with a needle. However, it is very difficult to detect microscopic dirt spots on a clear glass surface ; it becomes much easier after the first rhodium coat has been deposited. A good method to deal with these spots is, after the ordinary burnishing, to give the first rhodium coat a hard rubbing with cotton wool and then again to scrape any doubtful spots with a needle. In case there are many of them, breathing on the rhodium film followed by hard rubbing with cotton wool, may help. It should be stressed that spots with bad adhesion *must* be cleaned off. It would be useless to leave them even if they are temporarily covered with rhodium, for they will break off in due course. If, however, they are cleaned off before the following coats are deposited, they will stand a good chance of being covered like the dust pinholes mentioned before. In this connection, the reader may be reminded of an earlier statement that there is no risk of damaging the sound parts of a good first rhodium coat by this treatment.

Some of the suggestions made in this account may, at first sight, seem somewhat academic and unsuitable for industrial conditions. Against this it is emphasized that all the precautions mentioned can be and have been incorporated in our rhodiumizing routine under industrial conditions. The practical elimination of failure in the production of all types of rhodium films has shown that it is well worth while to spend some time and equipment on getting working conditions under proper control.

Large scale industrial applications of vacuum coating

262 An article by Godley (1948) illustrates some of the large plants now in use in Industry for metal coating by vacuum deposition. The National Research Corporation has now adapted the process to the *continuous* application of metallic coating to rolls of cellophane paper and other flexible material. Pilot plant was in operation by April 1948 which could coat continuously 6,000 ft rolls of such material.

Vacuum coating with refractory materials

263 Campbell, Powell, Nowicki and Gonser (1949) describe the conditions requisite for vacuum coating with refractory materials such as refractory metals, carbides, nitrides, borides, silicides and oxides on both metallic and non-metallic surfaces.

Silica coating for protecting metallized mirrors

This process is useful for various metals and plastics as well as on glass. Coating with silica is not a success as the coating always has a " crazed " appearance. The secret is to evaporate silicon monoxide and later convert this to silica *in situ*. The material to be evaporated is sold under the name of Silcote, by the American National Research Corporation, their process being copied from the German original. It can also be obtained under the name Barcote from Daniel Verney, in Glasgow. However, there is no need to buy either, because it is just as good to put a mixture of silicon and silica in molecular proportions (60 gm of silica and 28 gm of silicon) straight into a molybdenum boat for evaporation, and SiO is evaporated automatically. It is helpful to wind some fine molybdenum wire round the boat after the mixture is in, for this prevents the material from hopping out of the boat during the evaporation process.

Having evaporated the protecting layer of SiO on to the mirror, it may be converted to SiO_2 in two different ways, both found satisfactory.

(a) The mirror is baked at 300° for about 3 hours. This is recommended if one is in a hurry.

(b) Do nothing for two months approximately and the SiO will have converted itself to SiO_2. This is a better way for routine production, but, in this case, the mirror must be kept in a dust-free container.

Reflecting pellicles

264 The need sometimes arises for optically flat and very thin films. These can be produced in the following way.

The first thing necessary is a frame on which the film would be supported and we usually make this in the form of a brass ring with a hole of the appropriate size and one surface (the one which is to carry the pellicle) made flat by optical grinding.

A solution is then made of cellulose acetate in methyl-alcohol. Acetone can be used as solvent, but we prefer the methyl-alcohol which does not evaporate quickly. The varnish sold under the trade name of Halac is a very suitable material. To obtain extremely thin films, such as show no colour, the solution should be weak. If the pellicle is required to show colour the solution is made stronger.

Spread the solution over a clean glass plate, either by rocking the plate or if a thinner film is needed by centrifuging it. Soon after the coating is dry cut on it the size required with a safety razor blade. Slide the glass plate, film uppermost, obliquely into a bowl of water which will float off the film. Apply the same solution with a small brush to the mount avoiding the inner edge.

When the solution on the mount is nearly dry, place it on the floating film and pull it out again in an oblique direction with the film attached to it. If the pellicle is buckled or flabby it will straighten out afterwards.

Alternative method for very thin films: Pour a drop of solution on the water in a vessel not much bigger than the frame and when dry mount in the same way as above.

The origins and development of processes for depositing films on optical surfaces by evaporation in vacuo and their use in reducing reflection

265 The reduction of reflection on glass surfaces which had weathered so as to become covered with a purple tarnish was patented as a method of reducing reflection by Taylor (1904). In this he gave the theoretical and practical basis of his method in the following words.

The formulæ for giving the fraction of light reflected back from a freshly polished glass surface where the incidence is perpendicular, or nearly so, and the intensity of the incident light is unity, is simply $\left(\dfrac{\mu-1}{\mu+1}\right)^2$ wherein μ is the refractive index of the glass in question

If $\mu = 1\cdot 5$ then above gives $\left(\dfrac{0\cdot 5}{2\cdot 5}\right)^2 = \dfrac{0\cdot 25}{6\cdot 25} = 0\cdot 04$. That is ordinary crown glass will reflect back about 4 per cent of the light falling upon it and transmit 96 per cent.

The essence of my invention is to bring about this very desirable result by the artificial and intentional tarnishing of the polished surfaces of the lenses. I find that if the freshly polished surfaces of dense barium crown glass is tarnished in an aqueous solution of sulphuretted hydrogen until by reflection it assumes a slatey brown colour, then I estimate that the amount of light reflected is reduced to about 40 per cent, or perhaps to only ⅓rd. That is in place of the 5·5 per cent reflected off a newly polished surface, I then get only about 2 to 2½ per cent, while the transparency of the surface is increased from 96·5 per cent to 97·5 per cent.

With glasses such as dense barium crowns and dense flints the film can be formed naturally by the weathering action of the atmosphere; Taylor (1896) produced this tarnish in 1892 and chemical methods of producing "blooming" in this way are now well established (Jones and Homer, 1941).

In most of these chemical methods as in that of Nicoll (1942) (placing the glass in an atmosphere of hydrofluoric acid vapour) the film is formed by dissolving away all the constituents of the glass except the silica which remains undisturbed, the polish of the surface being preserved. Silica can also be deposited from solution however; for example the glass may be dipped in a solution of silicon in carbon tetrachloride. The carbon tetrachloride is then evaporated off, the surface rubbed, and the process repeated until the desired degree of iridescence is attained.

For high class work, however, the film is produced by deposition in vacuo in exactly the same way as the deposition of metals originated by Strong (1936).

266 Later Cartwright & Turner patented this process (U.S.A. patent 2,207,656 issued 9/7/40 application 27/12/38) but at a later date the claims relating to evaporating non-reflecting films were disallowed and the process is entirely free of patent rights except for a few claims for multiple coating of various salts. I understand that it was claimed by Zeiss that they were using the process as early as 1935 and they undoubtedly applied for patent protection (D.R.P. No. 685767,1935) which, however, seems to have no validity owing to prior publication.

Hiesinger (1948) discusses the relative advantage of the two methods of annulling reflection on glass surfaces—namely (a) by a change of refractive index from the medium traversed by the incident light to that traversed by the transmitted light (b) by interference.

He concludes that the deposition of homogeneous layers is preferable to that of inhomogeneous layers.

Extensive reviews, theoretical and practical, with many references are given by Smakula (1947), and by Bateson (1947), J. *Soc. Glass Technology*, **31**, p.170 (June–August, 1947).

Glass graticules in optical instruments
by CONSTANCE E. ARREGGER, M.Sc., of the Research & Development Department, Watts Division, Hilger & Watts Ltd.

267 In optical instruments used for measuring it is necessary to have, in the image plane of the instrument, a fiducial mark or marks which, for setting purposes, may be brought into coincidence with the image of the object being viewed. These " marks " usually take the form of lines either parallel or crossed. They must be fine and straight and can be made from spider webs, or fine metallic or plastic threads stretched across a frame which is located in the optical system of the instrument. Such fine webs are easily broken and are difficult to replace by the user, so it is convenient in many cases to use, instead, a glass disc engraved with fine lines. These glass graticules or reticules (both of these words being derived from Latin words meaning grids or nets) are more robust and durable and have the further advantage that the patterns engraved can take practically any form and need not necessarily consist of straight lines only. Instead, it is possible, if required, to engrave on the glass circular or linear figured scales. These we shall term " measuring graticules " to distinguish them from the former type which are simply " setting graticules," and which are used in conjunction with a measuring scale on some other part of the instrument. Further, in some instruments designed in recent years, it is found convenient, instead of viewing the image of the object through an eyepiece, to project it on to a screen which carries setting lines or a scale. These screens may be termed " projection graticules " Another type of graticule, termed a " mask," is used when part of the image plane is required to be blacked out and only clear windows left in appropriate places for viewing.

Because of their position in the image plane of the instrument, where they are usually viewed under considerable magnification, the graticules must be small in size, the lines must be fine and accurately placed and the figures, if any, microscopically small. The graticule

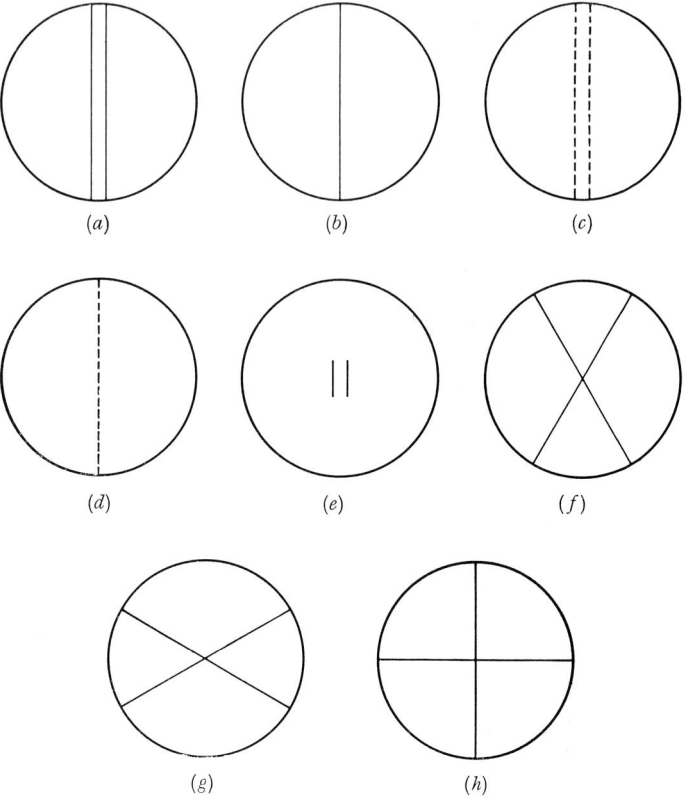

Fig. 214—Types of parallel and cross line graticules

markings must not be such that they disturb the clarity of the object. They should, as a rule, be symmetrical. The eye is able to resolve objects separated by one or two minutes of arc but, where variations from symmetry are observed, the sensitivity is greater and discrepancies as small as twenty times the resolving power are distinguishable to a trained observer. No blobs or irregularities should be present to confuse the main image and to distract the observer's attention. It must be remembered that the ease of observation and elimination of fatigue are of paramount importance in making accurate observations.

The manufacture of the graticules as described above calls for skilled and patient work.

First we shall enumerate some samples of the various types of graticules mentioned, and then proceed to describe the various methods of manufacture.

Types of graticules

268 *Setting graticules*—For setting on a straight line object, such as a scale line or spectrometer slit, there are several types of straight line graticules taking the form either of pairs of parallel lines, or of crosses or dotted lines, singly or in pairs (Fig. 214).

The accuracy of setting depends largely on the magnification of the instrument, the brightness of the field, and the width and cleanness of the edges of the lines to be viewed, but the type of graticule also has

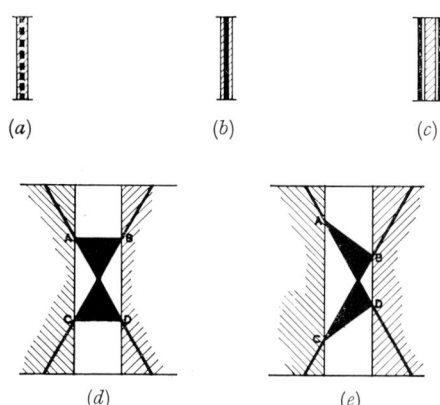

Fig. 215—Graticules
(a, b, c) Parallel line graticules set on scale lines.
(d, e) Symmetrical 60° cross set on slit.

considerable influence. For setting on fine clean lines the double line graticule is to be preferred. The graticule lines are set symmetrically about the object line and the eye balances the two bright areas contained between the edge of the viewed line and the two graticule lines. Tests have been made to find the most favourable graticule line space for accurate settings, and the least variation from the mean in repeated observations is found when this clear space is as small as possible without actually disappearing. This extreme, however, leads to fatigue, and the writer has found that a space which is about one quarter of the line-thickness to be viewed is the most convenient and the least tiring when a large number of observations have to be made. Since the eye concentrates on the parts of the image in the centre of the field of view there is no advantage in having long lines, as in (a) ; short lines, which are approximately three times as long as the space between them, as in (e), are sufficient. Where object lines of several widths are likely to be met with, it is advisable to have, on the same graticules, three pairs of lines of varying spacing and, to concentrate attention, if the pairs are long, another pair of parallel lines at right angles to the others is recommended. The width of the graticule lines should be somewhat finer than those of the image lines (Fig. 215a). Continuous single lines should not be used as setting lines because difficulty is experienced in trying to set the dark graticule line centrally on a dark image. This is mitigated, to some extent, by having a broken or dotted single line so that setting is effected by a " continuity " method, but this again is considered likely to cause fatigue. Pairs of parallel dotted lines are quite satisfactory.

Crossed lines are very effective for setting on a straight line object, particularly on a thick line or a bright slit as in a spectrometer. The cross should be symmetrical about a vertical line and experience has proved the superiority of the vertical 60° cross over the other angles, particularly when setting on slits. The advantage may be due to the fact that the intersection of the 60° cross on the lines gives a pair of equilateral triangles when set symmetrically (Fig. 215d). The eye concentrates on the points A B and C D which form a pair of parallel lines. When the setting is slightly out of symmetry as in Fig. 215e the points A B C D form two lines which are easily recognized as not parallel. When the 60° cross is used these lines are separated by nearly twice the width of the object line and are easily distinguishable, but near enough together to be easily viewed. Other special types of graticules are sometimes used for setting on targets of a particular shape such as V's, angles, thread forms, etc., *e.g.* T. T. & H. Microscope.

269 *Measuring graticules*—Some graticules are used directly for measuring. This may be achieved, as with the conventional micrometer microscope, by causing the cross lines to travel in relation to a comb or scale in the field of view and by breaking down still further on a micrometer drum or diagonal vernier scale or spiral, or by having a scale engraved on the graticule as in the Fixed Focus Microscope supplied by Hilger & Watts Ltd for breaking down divisions on glass scales. The advantage of this is that the eye, while focused for setting a position, can also make the observations without a fresh adjustment, thereby avoiding much fatigue.

The value of a division of a linear graticule scale depends upon the primary magnification of the instrument and the accuracy of measurement will be influenced by the aberrations of the lens system; it is in no case advisable to extend the scale beyond the central portion of the field of view. Ruled rectangular grids or gratings are used as measuring-graticules though, if the screen is too fine, this tends to obscure the object to be measured. When an image is to be projected

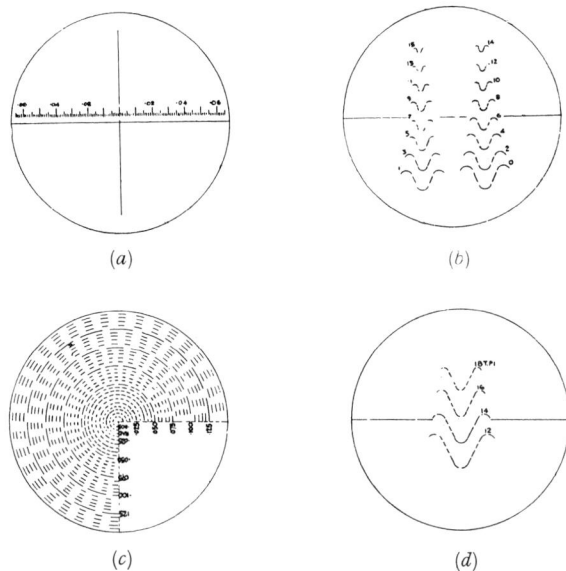

Fig. 216—Taylor, Taylor & Hobson graticules
Scale, radii, and thread forms.

(a) For linear measurement using ×5 objective. One division represents 0·001 in.
(b) For comparison of small thread forms
(c) For measurement of radii
(d) For comparison of normal thread forms

or photographed, a graticule, or a grid located in the image plane which appears also in the projection or photograph, can be used for direct measurement since the spherical or other distortions of the lens system are given to both the image and the grid so that measurements can be estimated directly in terms of grid spacings.

270 *Projection graticules*—When the measuring instrument is designed to project the image of the observed object, the graticule can in effect be the projection screen, and can carry a setting mark or a fine scale for subdividing, say, an object scale. This device is used in various manometers of recent design, in a new improved Jig Borer just introduced by The Coventry Gauge & Tool Co, and in projectors of the type manufactured by Hilger & Watts Ltd. The great advantage of this method is that the observer need not keep his eye at the eyepiece of an instrument but may make observations in greater

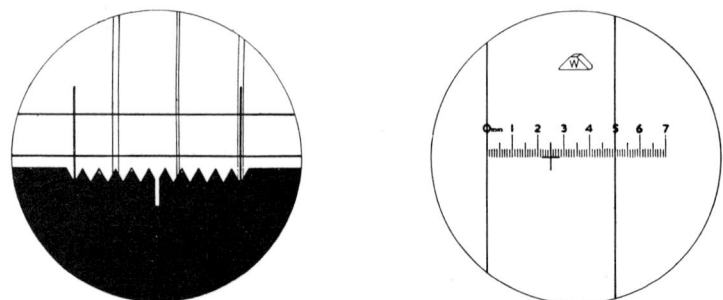

Fig. 217—Fields of view of micrometer microscope and fixed focus microscope

comfort and, if necessary, be able to give his attention to setting his machine at the same time.

271 *Masks*—It is at times desirable to concentrate the observer's attention on a special part or parts of the field of view and to mask out other parts which might cause confusion. This is particularly

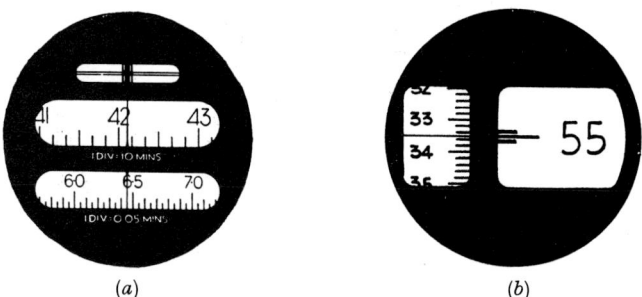

Fig. 218—Eyepiece graticules with masks

the case when an instrument where two optical systems are used to form two images in the object plane and where the junction of the two systems such as images of edges of prisms etc, would be confusing, *e.g.* Watts No. 2 Microptic Theodolite with a two-side reading system. These masks form a special type of graticule which give the effect of clear windows on a black background.

Graticule making

272 In describing the techniques of making graticules we shall consider them under headings which describe the different types of processing ; but, first of all, it should be emphasized that, for all types of glass graticule, the glass itself must be of the highest quality, without bubbles or pits or flaws, which will show up by catching black specks when the graticule lines are filled with pigment. The glass should be optically polished and the faces flat and parallel to each

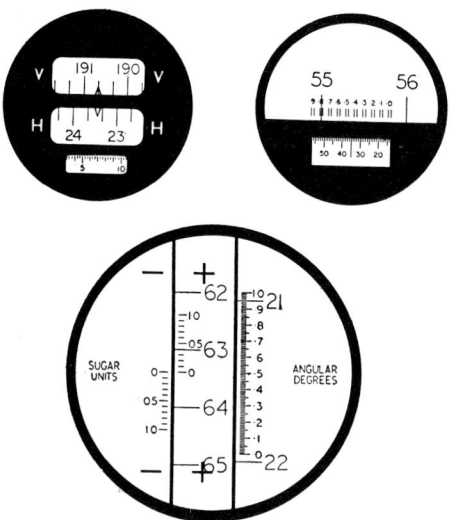

Fig. 219—Projection screen

other because, when the image is viewed through the graticule, any wedge shape or curvature would cause distortion of the image.

It has been found that the most suitable glass for diamond dividing is the best quality plate glass commonly known as green plate. For very fine *etching* work Barium Light Flint (B.L.F.) is recommended, although it has the disadvantage of being soft and easily scratched so that it should be protected with a cover glass when used in a position where it is liable to get dirty and to need frequent cleaning.

The nomenclature of the Barium glasses as they are usually called, may be a little confusing. In France the crowns are called barium crowns, for example " baryum crown léger," but to describe the flints the word " baryte " is used, thus " flint léger baryte." In the Schott catalogues " barit " is used for both the crowns and the flints, *e.g.* " barit–leicht–kron " ; " barit–leicht–flint." In Chance's catalogue " barium " is the word used for both flints and crowns.

It is unfortunate that all Barium glasses are liable to bubbles, and the requirements of makers of graticules are in this respect necessarily exacting.

The glass should be thin but not so very thin that it is difficult to handle during the processing.

Direct diamond dividing

273 Lines on glass are, in this case, ruled directly by means of a diamond cutting tool set up in a ruling machine or pantograph. Some workers prefer to use a natural corner of a diamond or the edge of a chip formed by fracturing a diamond with a blow from a hammer. Such a diamond chip is mounted by means of shellac on the end of a short metal rod of dimensions suitable for inserting into the machine. It requires time and patience to find exactly the correct cutting edge and to orient it in a suitable manner, but good clean lines can be made in this way. Some cutting edges of chips last for a considerable time but others wear away quickly. Owing to recent improvements in the technique of working and polishing diamonds as cutting edges for machine tools it is now possible to obtain scribing tools with an edge as fine as 0·0001 to 0·0002 inch and with almost any shape desired. For scribing fine deep lines with both edges clean the tool must be very sharp and symmetrical, and the cutting edge must be accurately perpendicular to the direction of cutting, and horizontal when the tool shank is mounted vertically. The writer has found that tetrahedral diamond is the best for cutting clean lines in glass. These diamonds are supplied by L. M. van Moppes Ltd, North Circular Road, London. They have the advantage that they are oriented correctly by the lapidary so that they do not require the time and patience needed for setting up a chip. They also have a much longer useful life provided that they are carefully treated and that the cutting edge is not allowed to drop suddenly on to the glass. Diamond engraved lines on glass should have clean edges without conchoidal fractures. It is usually found that the glass comes away from the line in a curled thread when the line is clean, and as a powder when fracturing occurs. It is not advisable to try to make crossed lines with a diamond since the crossing of a line already ruled is liable to spoil the edge of the tool. Any vibration or shaking of the machine during diamond dividing should be avoided since fracture is liable to occur.

F. T. Gilmore & Co. 285 Columbus Avenue, Boston. Mass. U.S.A. specialize in diamonds ground to specified shapes suitable for engraving purposes.

It is sometimes important to ascertain the shapes of the grooves made by diamond cuts. This can be done by the simple device of putting a small piece of microscope cover glass or, better, a +6·0 diopter plano-cylindrical lens on to the ruled area and observing the interference fringes caused by the light reflected from the two contiguous surfaces with a microscope fitted with diagonal plate which reflects downwards on the specimen the light from a bright mercury vapour lamp. The same effect can be obtained, but with greatly enhanced brilliancy by the Linnik micro-interferometer described in §252.2.

Fig. 220 shows photographs illustrating this point taken on the Linnik micro-interferometer, by Dr L. A. Sayce, Head of the Optics Department, of the National Physical Laboratory, Teddington, England.

Recently a demand has arisen for scale graticules with clear lines on a dark background. After a considerable time spent unsuccessfully

 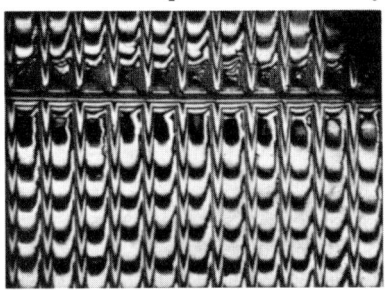

Fig. 220—A ruling on aluminium-bronze Magnification
× 700, *Ruling* 2,000 lines per inch.
(*Left*) The grating photographed with normal vertical illumination.
(*Right*) A Linnik micro-interferogram.
The grooves are intersected by transverse razor cuts to show their shape.
The two photographs show similar but not identical portions of the ruling.

in attempting to rule fine lines on painted glass it was found that excellent results could be obtained by ruling with a diamond or sapphire tool on glass which has been coated with a thin layer of metal —say aluminium, platinum, or rhodium deposited by electrical evaporation or sputtering or with silver deposited chemically. A special chisel-shaped diamond tool was designed for this work which scraped away clean lines of metal leaving the glass clean and unscratched. Aluminium deposited directly on glass produced the best results. Figures may also be engraved directly in some cases and a process has been proved for silvered glass which involves an etching technique.

The photographic method described later under the heading "Masks" has also been used successfully for dark background scales, see Fig. 221.

Etching

274 Lines and figures may be engraved by the method of etching which consists of coating the glass, while spinning at a high speed, with a thin protective coating of " resist " and then of dividing the required pattern through the resist leaving the clean glass showing.

The graticule is then exposed to hydrofluoric acid fumes for a few seconds, or for coarser work, the acid itself is mopped on to the work. The graticule is then washed in water and the resist removed with

Fig. 221—Dark-background scale for use in a projector

benzine or naphtha. The acid will have attacked the parts of the glass which were not coated with resist, leaving indented lines and figures in the glass. These are then filled with a black pigment. Great care needs to be taken to ensure a successful result, and air temperature and humidity play a large part in the success both of the resisting and the etching. Various resists are used and the most popular consists of a mixture of bitumen and waxes in fixed proportions in a solvent such as Naphtha, as prepared by B.S.I.R.A. This mixture is somewhat difficult to manage since it has to be kept at 20°C and also must be heated to 40°C frequently to dissolve out the wax which tends to solidify into small particles. The heating must be effected without driving off the solvent, or the proportions are disturbed.

A resist made of a mixture of 99 parts of English beeswax and one part of Carnauba wax similar to that used formerly by Zeiss in Germany, gives more consistent results. This wax solidifies at normal room temperatures so it should be kept ready for use in a warm oven and the glass and wax should be at about 75°C when the wax is spun on to the glass. With this a fine even coating can be obtained which cuts readily but which is extremely resistant to hydrofluoric acid vapour, and which is not liable to allow spread of the lines to take place during etching due to insufficient adhesion at the edges of the lines. This wax mixture shows some advantage over the bitumen resist when crosses have to be made, permitting the intersections to be cut clean and sharp.

Whatever resist is used a trial piece should be etched first to determine the optimum etching time which may vary from day to day due to variations in temperature and humidity. For deep lines with clean edges a high concentration acid, say 70 per cent, is recommended for vapour etching.

The type of resist determines the type of cutter employed for the engraving. A steel tool sharpened to present an edge which has the same size as the width of the line required may be used on bitumen resist or wax, but a diamond or sapphire cutting tool is not usually successful for use with bitumen resist, while gems are best for cutting wax and metallized surfaces, where a steel tool tends to drag. Some craftsmen find that cutting is facilitated by the use of a lubricant such as oleic acid.

Conical sapphire tools for engraving figures through the resist can be ground to great precision on the Mikron pivot polishing machine. For example, conical points can be ground with a well-defined flat end of about 0·0002 inch diameter.

For graticules of the highest quality the firm of Müller in Hamburg use a double process. The graticule is first silvered and then covered with wax. The lines are scribed with a diamond tool through the wax leaving the silver uncovered. The silver is then etched away by immersion in an alcoholic solution of ferric nitrate followed by washing in alcohol. The clear glass is then exposed. This is etched with hydrofluoric acid vapour, and then both wax and silver removed with nitric acid.

The black pigment used for filling the etched lines may be a paint, ink, or wax. A good filling can be effected by first rubbing across the graduations with a black wax pencil (carefully examined to be sure that no grit is present) then by rubbing gently with fine tissue paper coated with fine graphite—always rubbing across the lines. When the graticule is edge illuminated it is possible to have the lines show up light against a dark background by filling with yellow, while the same graticule will appear to have black lines against a bright background if illuminated directly.

Careful inspection is always necessary after finishing, to make sure that all lines and figures are perfect, and that no accidental " pinholes " or scratches in the resist have etched, and caused blemishes. If the resist is too thin or faulty a general " blooming " will be seen over the glass.

The utmost cleanliness is necessary throughout this process. The wax, in particular, should be filtered several times to remove dust, and after spinning should be kept covered until set because it becomes charged electrostatically during spinning and attracts small particles of dust.

Great care should be taken to protect the hands from contact with hydrofluoric acid, which is extremely corrosive. Rubber gloves should be worn and if the acid should come into contact with the skin it should be washed immediately and for a long period (fifteen minutes) in a saturated solution of sodium bicarbonate. If it is not realized that the hands have been in contact with the acid until irritation is set up, which may be some time later, the burn should be soaked in a 1 : 6 strength solution of ammonia in water for 15 minutes or longer until the burn becomes painful. The wound should then be covered with ointment made from magnesia 1 part, and liquid paraffin 2 parts, and bandaged.

Photographic reproduction

275 Where numbers of similar graticules have to be made it is advantageous to reproduce them photographically. For this purpose it is necessary to make a " master "—either positive or negative, depending on the process used. This master can be a photographic reduction from a drawing in cases where a slight loss of accuracy due to possible distortion during reduction is not important. Alternatively a finished graticule of the correct size can be made by diamond dividing or etching as described above, and this can be used as a " master of masters " *i.e.* other subsidiary masters can be contacted from this which in their turn can be contacted to make the finished graticules. In this way several hundred graticules can be made from one master.

Photo-etching

Where the requirement is for a photo-etched graticule which does not differ in any respect from the etched graticule already described, the process is as follows—

The selected glass blank is carefully cleaned by swabbing with fine cotton wool, using cerium oxide in methylated spirit, followed by washing in nitric acid or chromic acid, and then is chemically silvered giving two coats to avoid " pin-holing." The silvered blank is subsequently coated with glue treated with ammonium bichromate to render it sensitive to ultra-violet light. The glue used is a fix glue used in the printing trade for engraving. After dissolving in water the ammonium bichromate is added, and the whole is left standing for two weeks and then carefully vacuum filtered through cerium oxide. It is centrifuged before use and dropped by means of a glass tube on to the graticule when spinning at 2,000 r.p.m. The graticule is spun until dry and is then ready for use. The next process is to contact to a positive master in a vacuum frame and expose to ultra-violet light from a high-power mercury vapour lamp with a quartz envelope. The light reaches the glue through the transparent regions on the master and here the glue is hardened by actinic action while the rest of the glue remains unaffected. After exposing, the graticule is dipped in water and the unexposed glue is dissolved away leaving the silver clean in the positions which had been in contact with the dark lines on the master. This silver is now etched away with a solution of ferric nitrate in alcohol. This clears away the silver leaving the glass clean which should then be well washed and dried. Here the graticule should be inspected under a microscope and retouched, if necessary, with melanoid resist and then exposed to 70 per cent hydrofluoric acid vapour to etch the lines into the glass. After well washing, the glue and silver are cleaned off with nitric acid, and the etched lines filled with a black pigment.

Dyed glue process

276 For a graticule which is to be fitted into an instrument in such a way that it can be cover-glassed, there is a process of reproduction which leaves dyed glue lines and figures adhering to the glass. First the selected glass blank is carefully cleaned with chromic acid following cerium oxide and meths, and then covered with bichromated glue, without any silver undercoating. This is the same type of glue as described above for photo-etching. The glass blank to be coated is spun rapidly at about 2,000 r.p.m. while the liquid glue is applied, in order to ensure an even coating, and the spinning is continued until the glue is dry. The graticule is then contacted with a master negative and exposed to ultra-violet light, *i.e.* the lines which are to be black on the final job are exposed. The glue is rendered insoluble in water by exposure and now the rest of the glue which was not exposed can be washed away by immersing in water for a few seconds. The glass is thus left with a glue pattern on a clear background. After washing, the graticule is immersed in a dye bath which can consist either of black powder dye dissolved in methylated spirit, or of methyl varnish. After a further washing and spinning until quite dry, the graticule should be inspected and, if perfect, should be baked at 80°C for 15–30 minutes. After further cleaning the graticule with a preparation of glycerine in acetone a cover glass should be cemented on. The cement recommended is H.T. cement as prepared by B.S.I.R.A. After cementing, the graticule is baked at 60–70°F, for 24 hours or until hard. The cover-glass should be kept in contact by means of a clamp during baking.

Certain precautions should be observed when working with bichromated glue. The temperature of the glue should be maintained at approximately 20°C and the humidity should remain at 55 per cent R.H. during the processing. All the work should, as far as possible, be conducted in a dust-free atmosphere, because even minute specks can ruin the work completely. The bichromated glue is sensitive to visible blue light if exposed for a considerable period, so, while the work may be done in ordinary room illumination, very bright lights should be avoided. Of course, one of the recognized advantages of this method is that the inconvenience of working in a completely dark room is avoided.

Composite graticules

277 Lately, in the firm of Hilger & Watts, there have been calls for graticules of extremely complicated design. One shown in the diagram required two fine scales at right angles to each other, intersecting at the centre, and a small circular scale round the edge, the whole to be covered with a fine mesh rectangular grid. The accuracy required would not allow of a reduction from a drawing with the consequent risk of distortion. The graticule could not be ruled and etched because of the numerous intersections required, which would have taken away the edge of the tool.

The graticule was finally made by a composite process using both ruling and photography. The scales were divided on one blank, and the grid on another, and these were both photographed on to one master blank, great care being needed in the lining up. (Fig. 222.)

Masks

278 The mask design is first carefully drawn up on paper and then reduced photographically to the size required. The glass blanks are coated with a mixture of glue and colloidal silver, and after contacting with a negative and exposing to ultra-violet light, they are washed alternately in solutions of mercuric chloride and silver nitrate, until a sufficient density of blackness has been produced. They are then developed in the usual manner in Kodak 163 and baked at 260° for 10 minutes. They are then inspected, and if necessary, retouched.

Another method of obtaining a light pattern on a dark background is that of coating the glass blank with a metal layer by a high vacuum process of sputtering. If the graticule is situated at a sufficient distance from the source, the streams of molecules reaching the target are practically rectilinear. Thus, if a mask is exposed in the beam,

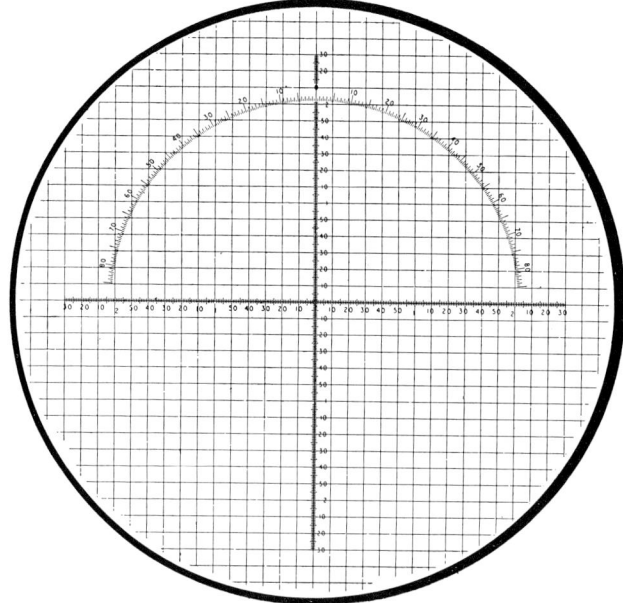

Fig. 222—Admiralty kine-theodolite evaluator graticule

there will be a clear " shadow " left against an opaque background. If a matt finish is required, the metal can be blackened by an accelerated oxidization process. This method has the advantage that by repetition many exactly similar graticules can be made using the same mask.

Other types

279 There are other processes for producing graticules which may not give quite the accuracy required in high-precision measuring instruments. For instance, it is possible to transfer patterns on to glass and then to fire them as in the production of ceramics. Where highly accurate grids or scales are concerned this exposure to extremely high temperature is likely to cause changes in the glass, so that the accuracy of line spacing is liable to be impaired.

Conclusion

280 Graticule types were divided into—
  1 *Eyepiece graticules*
      A. Setting lines or crosses where measurements are made on a scale included in the instrument.
      B. Scales for direct measurement.
  2 *Projection graticules* which divide into A and B as above.
  3 *Masks*
      The processes of production have been outlined as follows—
        1 Direct diamond engraving;
        2 Etching;
        3 Photo-etching;
        4 Dyed glue process;
        5 Dark background scales;
        6 Metallic deposition around a mask.

CHAPTER 14

TESTING OPTICAL GLASS
With some notes on annealing and normalizing

281 It is well that the optician should know how to test optical glass for its various faults, although the optical glass made in this country is now so good that it is not necessary in the ordinary course of manufracture to test it in its raw state, except large pieces for individual working. The broken pieces straight from the pot are examined at the glass works, those being rejected in which veins, dead metal or excessive bubbles can be seen. If the glass-maker is concerned to determine the origin of dead metal or " stones ", the nature of these can be determined by X-ray diffraction (see Rooksby, 1952)

The glass is broken with a hammer into pieces of about the size required ; any large pieces which would form specially large prisms or blocks are often set aside and reserved for such purposes. The pieces are then moulded into lens shapes, plates or prisms and annealed.

The plates or blocks are usually supplied polished on two opposite edges so as to facilitate a thorough examination. The defects which can then be detected are as follows—

Veins

282 Veins are threads or streaks of glass within the mass having a refractive index so different from the bulk of the glass as to be distinguishable to the eye. They consist chiefly of glass which has dissolved at the sides of the pot and has been carried by convection or by stirring into the body of the glass.

A vein may be surrounded by quite good glass but there is often what one may call a hank of veins presenting—as is natural, having regard to their cause—the appearance of water in which sugar is dissolving.

A further source of veins, though of far less importance, is the layer of glass at the top of the pot which acquires a different refractive index owing to the evaporation of the more volatile constituents (Twyman and Dalladay, 1921*a*).

Bad veins are easily detected by what our opticians call " flaring ". The glass (which needs to be polished on two opposite surfaces) is held between the observer's eye and a small and distant bright light. Using a magnifier of say 4 inches focal length at about the same distance from his eye so that the source of light is focused on the latter, the observer then holds the piece of glass at such a distance as to view the surfaces and interior through the magnifier. Even faint veins then become plainly visible as dark lines or bright lines accordingly as the light is focused exactly on the eye or a little to one side of it. This

500   PRISM AND LENS MAKING

simple procedure, which amounts to a rough application of the Schlieren method, alone almost suffices to reveal all injurious veins; but if the piece of work is an important one, more delicate tests are applied. We ourselves examine it by the Schlieren method (*see* §216) and by the Hilger interferometer (§243 *et seq.*).

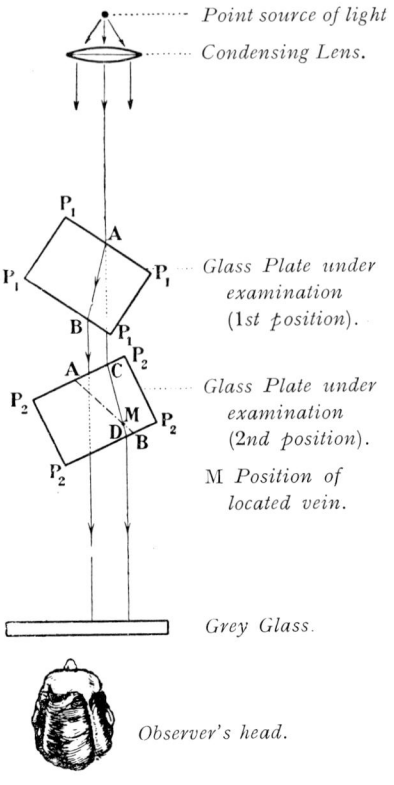

Fig. 223—Location of faults in a plate of glass
(From Dévé)

Another method due to M. Albert Arnulf, practical instructor at the Institut d'Optique, is described by Dévé (1936, pages 27–31). (The following is a free translation of that description.)

Light from a small brilliant source of light such as a Pointolite Lamp is collimated by a condensing lens. Several yards from the lamp is placed the plate of glass or optical piece which is to be examined. Between the latter and the observer is a grey glass which receives the shadow of the piece of glass. In the shadow appear black and sharply-defined shadows even of the smallest defects (Fig. 223).

The phenomenon is the same as that observed in the evening when a thin stream of water runs from a tap in a room lit by a lamp. The water is almost invisible but its shadow on the wall is as black as that of a pencil. The explanation is that the light which falls on the stream of water almost passes through it and, since the remainder is reflected in all directions, only a very small amount reaches the eye. On the other hand, the light which enters the thread of water is refracted in all directions, and the part which arrives in a direct line at the wall is very small in amount so that the shadow is very intense. In the glass, veins act like a stream of water.

The method even reveals imperfect polish on a plate of glass.

Other and more sensitive methods are described in Chapter 11 (§§227).

Double refraction

283 If we apply force to a piece of glass, as for example by trying to bend it, we cause stresses and strains within it. The stress at a point consists of the forces acting on the small portion of glass around that point, and the strain is the amount of distortion of that portion caused by the stress.

Such a condition separates the molecules by a greater distance in one direction than in the directions at right-angles to it and this causes the electro-magnetic waves of which luminous radiation consists to separate into two waves travelling through the glass at two different speeds, resulting in double refraction similar to that caused by Iceland spar. The degree of double refraction varies with the strain from point to point of the glass and may be detected by a simple form of polariscope designed for the purpose, to which I gave the name of the " Strain Viewer." The form of strain viewer mostly used in this country embodies an improvement known as a tint plate or half-wave plate, first described by Brewster in 1830 and applied to the strain viewer by Mr F. E. Lamplough, of Messrs Chance Bros. & Co., about 1914. In this form the instrument shows regions of stress in vivid colour contrast. Well-annealed specimens have no effect on the colour of the magenta* background, while regions of stress become a light blue or red according to the direction of stress.

A popular form of strain viewer is shown in Fig. 224.

The apparatus is so arranged that any article can be held by the hand in a beam of polarized light at a convenient distance from the eye, when the presence of strain is revealed by distinct changes of hue in the portions of the object which are in tension or compression. The field of

---

* Magenta may be described as a pinkish purple, approximating to Purple 8/2 in the Munsell Colour System.

view, including all parts of the glass which are not strained, remains a magenta tint, and the condition of strain is readily judged from the hue of the strained parts.

Before testing for strain the glass should be left for some time resting on the bench to acquire a uniform temperature and, when it is held in the hand, should be covered with a glove or piece of cloth to avoid local heating of the glass which would strain it temporarily and thus falsify the test.

A strain viewer designed for rapid industrial examination of small parts for strain and for surface defects is the Hilger Projection Strain Viewer (Fig. 225). It is particularly suitable where large quantities are to be examined. An enlarged image of the object under test is

Fig. 224—Hilger strain viewer

projected on a ground-glass screen arranged at a convenient inclination and height to suit an operator sitting at a bench or table of normal size. The aperture in which the object is held is also arranged at a convenient distance above the bench and to the right of the screen so that the object can be held with the operator's hand resting on the bench. The eyestrain associated with strain viewers of ordinary form when used over long periods of time is eliminated and, because of the natural sitting and handling position, fatigue of the operator is reduced to a minimum. Strained areas are instantly identified as distinctly coloured patches in the image.

In addition to the rapid detection of strain, defects in or on the surfaces of the glassware are perceived in the projected image of the object more readily than by direct examination. The magnification in the normal instrument is approximately $\times 3\frac{1}{2}$. The projection screen is 5 in. diameter, and the illumination bright enough for the strain viewer to be used in a normally lighted room.

284 In §283 we considered the temporary effect produced by applying an outside force and we referred to that due to local heating, but permanent stress can exist independently of these two causes.

This results from certain parts cooling first and thus becoming rigid while other parts which cool more slowly have their contraction resisted by the already cooled portions. The effect is that, broadly speaking, those parts which cool first are in compression, those which cool later are in tension. These internal stresses are revealed by the strain viewer just

Fig. 225—Hilger projection strain viewer

as those are which are caused by the application of external force or local heating.

Glass which has not been sufficiently well-annealed after the chilling which takes place in moulding shows double refraction, but the effect can be quite marked on the strain viewer without in itself causing optical parts made from the glass to be defective. For example, double refraction amounting to one-tenth of a wavelength gives very brilliant colour effects on the strain viewer but is without practical detriment to the definition. It is, however, evidence that the heat treatment has been such as to leave in the glass heterogeneity of the kind referred to below, and on that account no glass showing this amount of strain should be used.

285 Prior to the 1914–18 war it was customary to spend several weeks in the annealing of optical glass, and it was obvious to me therefore in

1914 that annealing would constitute one among the many difficulties in the way of producing in this country the greatly augmented supplies that were needed. I therefore offered to carry out for Messrs Chance experiments to determine a rapid way of finding the best temperatures at which to anneal the numerous kinds of glass made by them, some of them not previously made in the country. Messrs Chance were sufficiently interested to pay for the necessary apparatus and in 1915 I proceeded with the work on lines I had been considering for some time previously.

Working with specimens provided by Messrs Chance, my assistant, Mr A. J. Dalladay, and I were able to establish the now generally accepted law of variation of viscosity of glass with temperature in the range of temperatures in which annealing can take place. From this I was able to deduce the best temperatures to which to raise the various glasses and the cooling schedules which should be followed in order to get sufficiently complete removal of strain in a comparatively short time. My methods and apparatus were adopted by Messrs Chance with the result that in a few months prisms which had previously taken fourteen days to anneal were being brought to the same state of annealing in $4\frac{1}{2}$ days.

At the present date this time has been still further reduced, and I see no reason why optical glass should not be dealt with in the same way as glass bottles, travelling through a lehr with carefully controlled temperatures, the whole process being completed in six or seven hours.* The limiting factors appear to be, first, the length of time the glass must be held at the annealing temperature in order that the temperature may become uniform throughout and for the glass to become normalized to that temperature, and secondly, the differences of temperature that must necessarily exist in an object which is cooling. I believe, however, that, adopting the exponential cooling schedules originated by me (see §301), the speed I have indicated could be reached. Later on I applied my methods to the annealing of glassware and an account of part of the work was published (Twyman, 1917). A further account of our experiments and the results obtained with them will be found in §§291 *et seq.*

On one point, however, the procedure I recommended gave no guidance—the length of time necessary to " soak " the glass at the annealing temperature in order to remove the effect of chilling introduced in moulding, a process which we shall now proceed to consider.

Heterogeneity of refractive index

286 Veins are, of course, a form of heterogeneity, but under this

---

* For moulded pieces of, say, twice the linear dimensions of the prisms used in binoculars.

heading I refer to changes of refractive index much more gradual in character and arising from a quite different cause. I have never seen heterogeneity in glass which gave the impression of being due to failure to mix the sand and the powders from which the glass is made; the glass in its broken lumps ready for moulding is usually very homogeneous, with the exception of any veins that may be present.

During the 1914–18 war when very large numbers of moulded lenses were used, I had been greatly impressed with the fact that certain glasses when moulded were very defective even after annealing. The effect was specially marked with dense barium crown. Dr Simeon and I therefore investigated the matter (Twyman and Simeon, 1923). It had long been known that fine annealed glass has a refractive index higher than the same glass not fine annealed. In the experiments referred to we studied the effect of chilling, that is to say, rapid cooling from a high temperature as opposed to controlled cooling from the annealing temperature. We found that a dense barium crown allowed to cool in air from a temperature at which it began to flow on a sheet of iron upon which it was heated had a refractive index of 1·5955 whereas the same glass when carefully cooled from the annealing temperature (591°C) had a refractive index of 1·5986.

The same kind of effect, though in a less degree, was caused by chilling other kinds of glass.

The lowering of refractive index caused by chilling was, we found, removable by soaking at the annealing temperature for two hours, whereas more than 90 hours at a temperature of 520°C failed to do so.

It is obvious therefore that two purposes are served by annealing optical glass; as indeed they are in the annealing of metals. One is to remove internal strains, the other is to normalize the glass so that it should be of uniform refractive index throughout the mass. To the optician the first of these problems used to be considered of prime importance. It is now known, however, that this view was mistaken. To quote from a paper by Hampton (1942) (read at the Inaugural Science meeting of the Optical Group of the Physical Society on 6th March, 1942)—

> Sufficiently slow cooling reduces these birefringence effects to a negligible amount and the old test of an annealing schedule was that it should reduce the double refraction to something of the order of 0·01 wavelengths per cm (= 5 millimicrons per cm). It was found, however, on occasion that glass which appeared satisfactory from the point of view of double refraction and which was known to be free from striations still did not give a satisfactory optical image and it was certainly possible to find two pieces of glass which, while

appearing equally satisfactory in polarized light, gave substantially different pictures when examined on an interferometer.

How, one may ask, can this be reconciled with the conclusion of Simeon and myself that two hours at the annealing temperature sufficed to normalize the glass to the refractive index appropriate to that temperature? The explanation and the remedy were provided by work carried out at the British Scientific Instrument Research Association in collaboration with the laboratories of Messrs Chance. To quote once more from Dr Hampton's paper—

If a piece of glass is uniform in temperature while being held prior to the annealing operation, the only index changes introduced into it will be the plus or minus variations due to the temperature gradient during cooling and these will probably be small; probably not more than a few units in the fifth decimal place . . . Without the most elaborate precautions it is very difficult to get any sort of annealing kiln where the temperature differences are so small as not to give a risk of departures across a slab of optical glass of the order of 10°C. Much experimental work therefore has been directed towards the provision of annealing equipment where the temperature differences are neglibly small, and kilns have been designed and are in use where the temperature difference over a length of two feet or more does not exceed 3° or 4° C. These kilns are capable of being cooled at a much higher rate than the old type while still yielding glass which is of the highest standard when viewed on the interferometer although the strain viewer pattern may not be appreciably different from that given by less uniform kilns. Experience has shown that the glass which is satisfactory on the interferometer, due to the absolute uniformity of index, is capable of being used for the very highest type of optical instrument and it is becoming generally accepted in the optical industry that the interferometer is the instrument to be used and not the strain viewer. It is now a routine operation for samples from all kilns of optical glass to be examined on the interferometer and for the strain viewer to be used to a considerably less extent.

There was for many years an objection which was based on painful experience that the highest quality prisms could only be obtained by sawing from large blocks and not by moulding. Since moulding is much more economical in the use of optical glass than cutting up large pieces, recent developments in annealing which have proved beyond doubt that moulded prisms can give results equal in quality to those produced by any other method have been of the utmost importance in amplifying the supply under the present emergency conditions.

TESTING OPTICAL GLASS 507

The interferometer referred to by the author is the Hilger (Twyman and Green) interferometer (see §243 *et seq.*).

For observation on the interferometer two opposite surfaces are polished and to avoid the necessity of polishing them very accurately flat, plates of plane parallel glass are placed over these surfaces with a liquid in between the block under test and the plates of glass, of refractive index equal to that of the former.

The block of glass is then observed on the interferometer. Unless the surfaces are polished very accurately parallel, such a test will

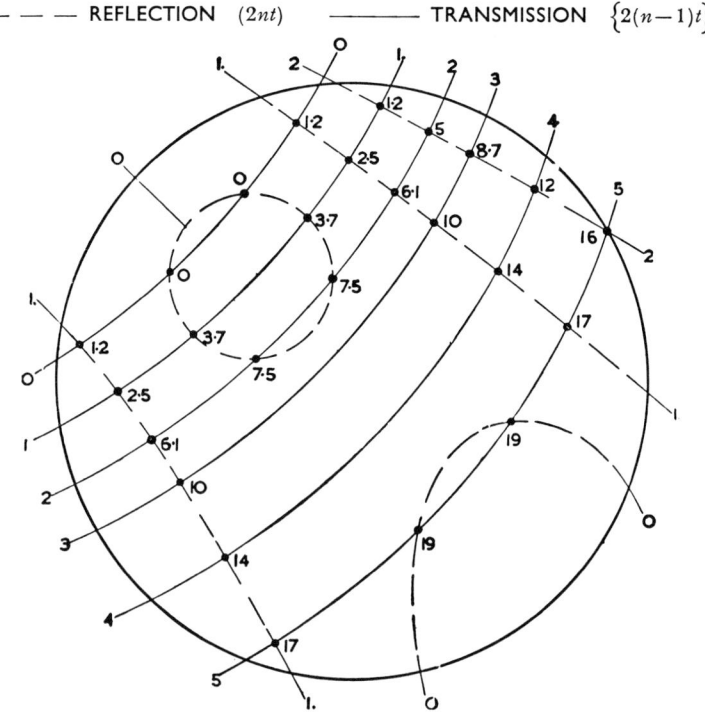

Fig. 226—Measuring small differences of refractive index in a plate of glass

not, of course, reveal a linear change of refractive index in a direction at right-angles to the light path, but for practical purposes it may be assumed that a block which shows no " error " is satisfactory to use even for work of the highest quality.

It will sometimes be found that the system of interference fringes seen on the interferometer is indistinct in places. This is because of double refraction which is due to faulty annealing. This indicates that the degree of double refraction is sufficient to impair the optical performance.

Measuring small differences of refractive index

287 There are occasions when it is desirable to determine the variations of refractive index throughout a block or plate of glass. In this case there is no escape from the necessity of making the surfaces fairly flat and fairly parallel, but extreme accuracy is not necessary. To give an example, it was suspected that a 4 in. diameter disc 0·4 in. thick was heterogeneous in refractive index. The disc was polished fairly parallel and tested on the interferoscope and the N.26 interferometer (see §244.). In Fig. 226 the dotted lines show the dark bands as seen by reflection, the full lines those seen by transmission. The band in each system indicating the least equivalent thickness is marked 0. Then if n is the mean refractive index of the glass in a direction perpendicular to the surfaces at any selected place, t the thickness at that place m_r and m_t the orders of the bands of the reflected and transmitted reflections respectively,

$$\triangle n = \{n(m_t - m_r) + m_r\}\lambda/2t \quad \text{(Twyman and Perry 1922)}.$$

For the points of intersection of the two systems of bands the evaluation of $\triangle n$ is very simple. Full details of the determinations are given in the table below. The refractive indices in units of the fifth decimal place relative to that of the points marked 0 are marked in the figure at the points of intersections.

m_r	m_t	$m_t - m_r$	A $= 1\cdot5(m_t - m_r)$	B $= m_r + A$	$n = B \times \lambda/2t$ $= B \times 0\cdot000025$	$t \times 10^{-5}$ (inches)
0	0	0	0	0	0	0
	1	1	1·5	1·5	0·000037	−1·0
	2	2	3	3	0·000075	−2·0
	5	5	7·5	7·5	0·00019	−5·0
1	0	−1	−1·5	−0·5	0·000012	+1·0
	1	0	0	1	0·000025	0·0
	2	1	1·5	2·5	0·000061	−1·0
	3	2	3	4	0·00010	−2·0
	4	3	4·5	5·5	0·00014	−3·0
	5	4	6	7	0·00017	−4·0
2	1	−1	−1·5	+0·5	0·000012	+1·0
	2	0	0	2	0·000050	0
	3	1	1·5	3·5	0·000087	−1·0
	4	2	3·0	5·0	0·00012	−2·0
	5	3	4·5	6·5	0·00016	−3·0

The variations in thickness at the same points can be found with equal ease from the expression: $\triangle t = (\lambda/2)\,(m_r - m_t)$.

The method is easily accurate to about 5 units in the sixth place for the refractive index, and to about one-millionth of an inch for the thickness. The accuracy for refractive index could, if necessary, be pushed higher.

It will be noticed that there is no sudden jump in refractive index. Nothing, that is, that would indicate that the variations are due to veins or to faulty mixing. The natural inference is that the differences must be due to the variations of temperature during the annealing process referred to in §286.

The glass was therefore annealed again in an experimental furnace with controlled temperature, using the annealing temperatures determined by the apparatus described in §297 *et seq*. The plate was then re-worked and it was found that there were no variations of refractive index exceeding two in the fifth place of decimals. That is, the disc was nine times as good as before re-annealing. The re-annealing was kindly carried out by the British Scientific Instrument Research Association.

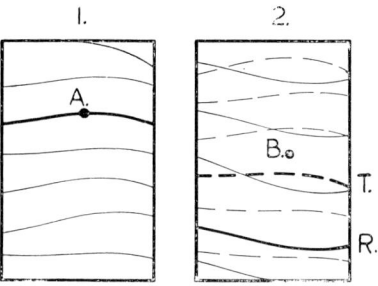

Fig. 227—**Measuring small differences of refractive index in separate pieces of glass**

Measuring small differences of refractive index in separate pieces of glass

288 The above method can be extended to deal with separate pieces of glass which differ only slightly in refractive index and thickness.

The two pieces to be compared are worked nearly to the same thickness but slightly wedge-shaped so as to show, say, 10 to 20 bands on the interferometer by transmitted light. They are then mounted side by side, each on a separate levelling table, in one of the beams of the interferometer. They are levelled until the reflected spots coincide as seen at the eyepiece of the interferometer.

The fully reflecting mirrors of the interferometer are covered and one then sees at the eyepiece the bands produced by interference of light reflected from the front and back surfaces of the two pieces.

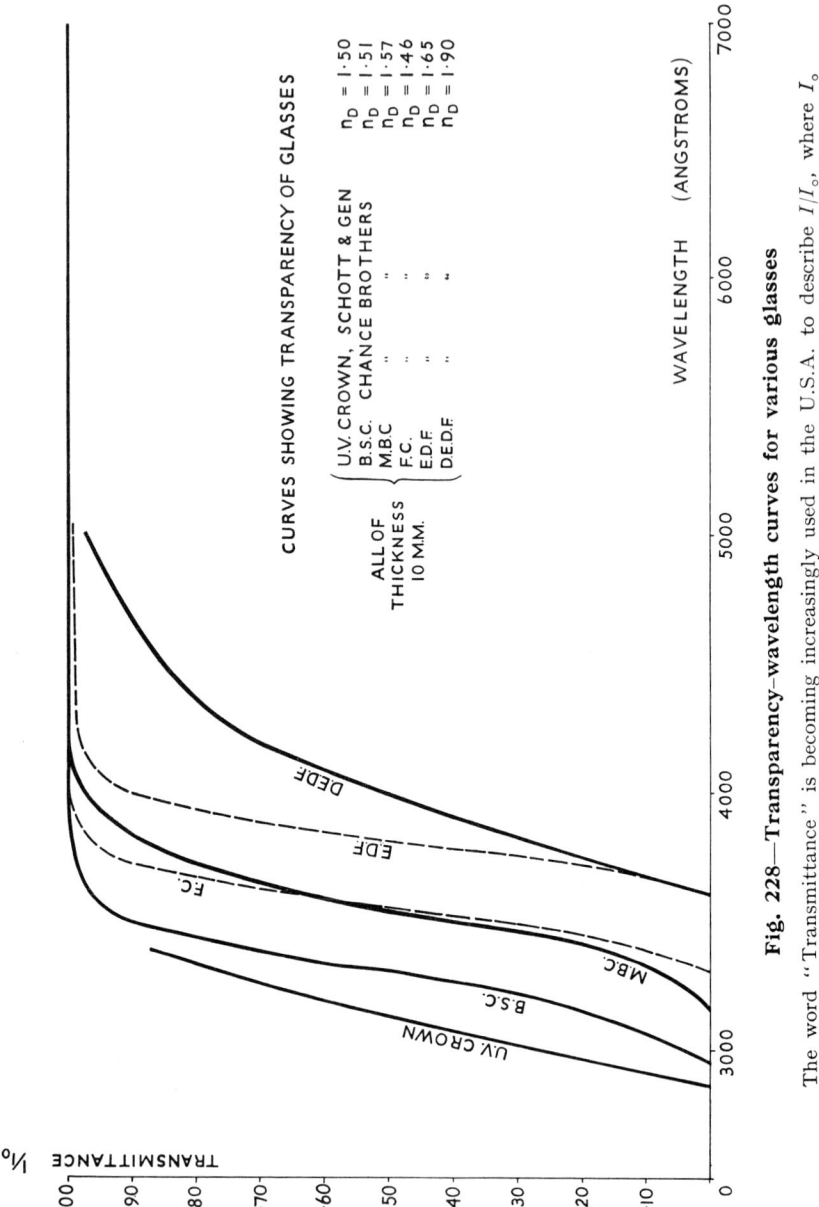

Fig. 228—Transparency–wavelength curves for various glasses

The word "Transmittance" is becoming increasingly used in the U.S.A. to describe I/I_o, where I_o is the intensity of the incident light which enters a medium, and I that remaining after its subsequent passage through the medium (*National Bureau of Standards Circular* No. 484, 1949). The word recommended by the Society of Public Analysts is "transmission" (*Analyst* 67, 164, 1942).

The colour sequence of the bands obtained with mercury light are not such as to enable one to identify in the two plates bands of the same order, but one can do so if a cadmium lamp is used (an Osira Cadmium Laboratory lamp made by the General Electric Co., England, is suitable).

Let n be the refractive index for a point A in one piece (Fig. 227) and $n + \triangle n$ that for a point B in the second piece; t and $t + \triangle t$ the thicknesses for the same two points. The point A is so chosen that an identifiable interference band runs through it. That band having been identified in the second piece at R, the number of bands from it to point B is recorded. Call this m_r, counting m_r as positive if the band at B is of a higher order than that at A.

Fig. 229—The Guild–Watts precision goniometer

The fully reflecting mirrors of the interferometer are then uncovered so that the interference bands of transmission are seen.

The instrument must have a compensating system (§245) and a piece of glass similar to the pieces under examination in the comparison beam. Under these circumstances one can obtain a black band, easily identifiable, and one adjusts the instrument so that it runs through A. As before, having identified this black band in the second piece at T, one records the number of bands from it to point B. Call this m_t, counting m_t as positive if the band at B is of a higher order than that at A.

One can then find $\triangle n$ and $\triangle t$ from the expressions given above.

Transmission and refractive indices of glasses

289 The two qualities which determine the choice of glass for a given purpose are its transparency and refractive index. The transparency-wavelength curve for a number of glasses of interest to the optician are given in Fig. 228.

Measuring refractive index

Where large numbers of highly corrected lenses, such as camera lenses are to be made, it is highly desirable that the refractive indices should be known better than to one unit in the fifth place, and for such purpose a very high precision goniometer is required. To fulfil this need the Guild-Watts Precision Goniometer was designed over twenty years ago by John Guild, of the National Physical Laboratory, and George Watts, of E. R. Watts & Son, Ltd, in collaboration.

Fig. 230—The Watts "Research" model goniometer

The basic design of that original instrument is still retained and, providing all reasonable precautions are taken and a correct experimental technique adopted, *a refractive index measurement of one unit in the sixth decimal place* can be obtained.

The latest form of the instrument is shown in Fig. 229.

Of the two principal functions of the spectrometer, namely the measurement of the telescope rotation and the measurement of rotation of the prism table, the latter all too frequently receives very secondary consideration, although it has been shown that measurement of prism angles are best made by the rotating table method.* For this reason the Guild-Watts Precision Spectrometer utilizes a single accurately divided scale for both measurements. By providing separate bearings for the telescope and table motions on a single axis, the main divided

* Goniometry (1) *Glazebrook's Dictionary of Applied Physics*

circle and reading microscopes can be used to measure each rotation with equal precision.

A smaller instrument (Fig. 230) measures refractive indices to one unit in the fifth decimal place.

The Hilger-Chance Precision Refractometer for transparent solids

290 Measurements of refractive index by goniometer, as with the instruments just described, are not appropriate for the workshop testing room. Further, in their use the specimens of glass require surfaces and angles of great accuracy. This entails considerable expense and, what is worse, a hold-up of the material until it is passed for despatch by the glassmaker or for acceptance and stocking by the optician.

The Hilger-Chance Refractometer, developed in the laboratories of Messrs Chance Brothers Ltd, of Birmingham (Hughes, 1941), and made by Hilger & Watts Ltd, avoids these inconveniences. Readings are obtainable rapidly and with certainty to an accuracy of one in the fifth place, while the specimen only requires to be worked approximately to angle and flatness of surface.

It is desirable to have a member of the staff whose duty it is to examine all materials as they come in. If such controller of materials is provided with simple slitting and polishing equipment, there is no reason why he cannot himself cut and polish a prism and test it on the Hilger-Chance refractometer for refractive index and dispersion within an hour of receiving the specimens. It can usually now be assumed of optical glass made in this country that all from the same melting is of the same refractive index with sufficient accuracy for practical purposes.

Optical system and form of specimen

The instrument is, in effect, a goniometer arranged in a vertical plane. The optical parts are indicated in the diagrams Fig. 231, and the instrument with its adjuncts in Fig. 232.

Material to be tested is inserted in the 90° recess in the top of a vee-block, with which a system similar to an Amici prism is formed, refractive indices being measured in terms of the angle through which the emergent rays have been refracted. Solids are prepared as right-angle prisms which are placed in position with a suitable contact fluid. A large range of fluids are available, covering refractive indices from 1·33 to 1·78, which, with the corresponding dispersions, are listed in the makers' catalogue. Neither the angle nor the surfaces of the prism need therefore be of high quality, nor, indeed do they need to be polished.

An important feature of the system is that, as the critical-angle principle is not employed, it is not necessary to use a prism of higher refractive index than that of the specimen to be measured. Thus the

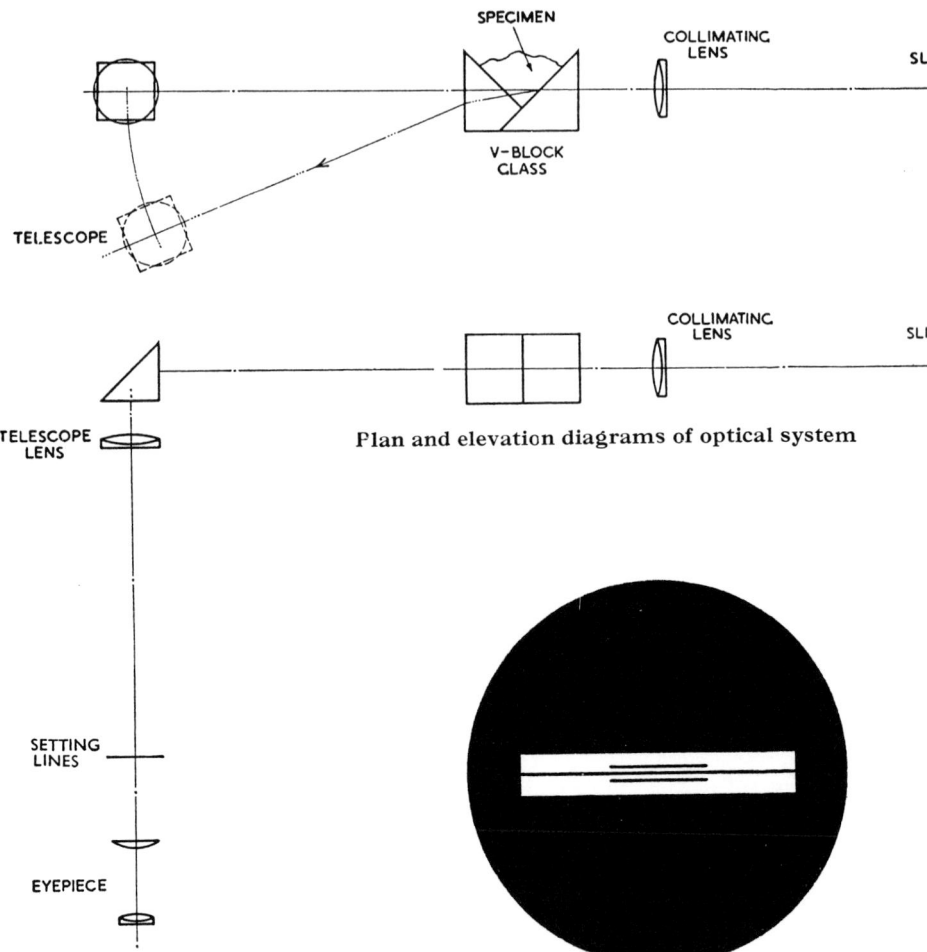

Fig. 231—Optical parts of the Hilger–Chance refractometer

range of measurement is large and a relatively hard glass, resistant to damage by abrasion, can be employed for the vee-block—a matter of some importance owing to the recent production of new glasses with exceptionally high refractive indices.

The refractive index at the surface of a material may differ to an important extent from that of the body of the material. This surface phenomenon plays an important part in critical angle refractometry, but is without effect on readings made with the Hilger-Chance Refractometer, which measures the refraction of the mass of the specimen.

Fig. 232—The Hilger–Chance refractometer

The accuracy of an instrument is materially affected by the means of setting employed. It has long been known that the asymmetrical setting of the typical critical-angle refractometer can be inaccurate. In the Hilger-Chance instrument settings are made upon a fine line which bisects the length of a fairly wide slit. Thus the setting of the parallel reference lines on the slit-line is strictly symmetrical and can be made with a high accuracy. Zero setting need be verified only at comparatively long intervals thanks to the precise means of location of the prism mount ; the method of setting is precisely similar to that in an ordinary reading and consists of observing the angular position of the direct image through the lower part of the glass vee-block.

A transparent glass circle is used and is read by a system (British Patent No. 574295) designed and made by Messrs E. R. Watts for their accurate surveying instruments.

First scientific control of annealing and normalizing glass

291 Before the 1914–18 war there was no precise method of determining annealing temperatures of glasses. Adams and Williamson (1920) say—

> Concerning the annealing of optical glass or any other kind of glass the lack of definite information was especially striking. Schott (1891) had determined for several glasses the lowest temperature at which an undoubted diminution of stress took place within twenty-four hours, and more recently Zschimmer and Schulz (1913) had investigated the amount of stress produced in glass by cooling it suddenly from various temperatures, but altogether the published records up to the year 1917 contain practically nothing on which to base a rational schedule for the annealing of glass.

The improvements made in this direction during and subsequent to that war have been summarized by Chance and Hampton (1926). The following passage from their paper is included by permission of the authors—

> The most important property of a glass in relation to its annealing is the variation of its viscosity with change of temperature. This question has been dealt with at some length by Mr. Hampton in the " Transactions of the Optical Society " but a brief outline of the progress made during recent years will not be out of place here. Mr F. Twyman, F.R.S., of Adam Hilger Ltd, in 1915 carried out investigations for our firm which for the first time pointed to the importance of a knowledge of the rate of release of stress at various temperatures and the variation of this rate with changes of temperature. The result of these experiments was a decided quickening up of the annealing process with an important consequent saving of time at a period when output was an imperative consideration. He found that the rate of release of stress at constant temperature was proportional to the stress present at any time. In 1919, Adams and Williamson of the Geophysical Laboratory of America showed that this is not strictly true ; it is close enough,

however, to give an insight into the physics of the process. Mr Twyman also showed that the rate of release of stress doubles itself for a constant temperature interval, which averages about 8°C. This is known as Twyman's Law, but is again only true in so far as the change of viscosity with change of temperature follows an exponential law. Fig. 233 shows the logarithmic viscosity curves for several glasses plotted against temperature on a linear scale. These curves would be straight lines were Twyman's Law exact. It will be seen that the law is reasonably correct over a short range in the neighbourhood of what is somewhat arbitrarily known

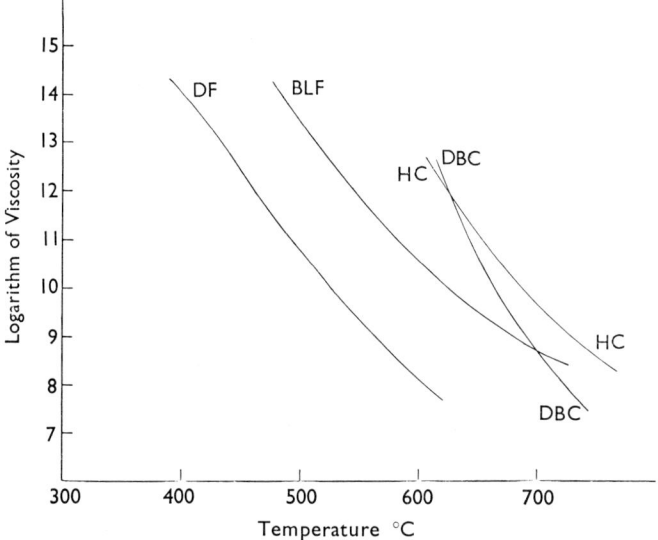

Fig. 233—Variation of viscosity with temperature for various glasses

as the annealing temperature; in any case no alternative to this relation has yet been suggested. The scheme put forward was essentially that at present in use, namely, that the glass should be held at a suitable temperature—high enough for the glass to be distorted—and then cooled at a predetermined rate which would safely be increased as the temperature fell. At about 100°C below the fixed temperature, the rate of cooling is governed only by the risk of fracture of the glass under the strain imposed by the temperature gradient, none of this temporary strain being present in the cold glass. This was an enormous advance in practice, and subsequent investigations have done no more than produce closer approximations to the still unknown laws governing the flow of so-called viscous bodies under stress, and discussions of the mathematical complications of the formula which Twyman deduced.

Reference to Fig. 233 at once indicates that the dense barium crown glasses change in viscosity more rapidly than other types of glass for a given temperature change. This necessitates much greater accuracy in temperature control with consequently increased difficulty in the annealing process. A further factor in the annealing of these glasses is the fact that, although for a given stress the birefringence or double refraction shown by these glasses is not greater than that shown by a dense flint, the accompanying absolute change of refractive index, which is more serious from the lens maker's point of view than the double refraction itself, is at least twenty times as great. Thus not only is dense barium crown more difficult to anneal to a given standard than other types, but the standard itself must be higher.

The success attained in attempting to solve this difficult problem is indicated by the fact that lenses containing dense barium crown are produced by several makers which pass the most exacting tests as regards definition, and the firm has even produced a disc which has been worked to a lens of $15\frac{1}{2}$ inches diameter for an astronomical objective from this type of glass. We believe that a lens of this size in a glass of such extreme properties is unique.

To the above remarks of Chance and Hampton I cannot refrain from adding the following details about the investigations which are referred to above.

At the beginning of 1915 it became obvious to me that there was danger of a great shortage of the optical glass required for national purposes. From some work I had done before the war I felt sure that it would be possible to determine precisely and with certainty appropriate temperatures at which to anneal any given type of glass and I had formed an opinion as to the way in which this information might be arrived at. As a consequence I suggested to Chance Bros & Co., shortly before June 30th 1915, that I should undertake an investigation on the subject and it was agreed that I should do so. By January 1916 I was able to communicate to them annealing temperatures for ten of their glasses, and the use of these temperatures was found by them to result in good annealing. Further, by February 1916, I had derived a general principle of annealing by the application of which to a particular case the time of annealing was reduced from three weeks to three days without loss of quality. The following is an extract from an affidavit sworn in 1924 by Mr A. J. Stobart, then Managing Director of Chance Bros & Co., in support of an appeal made by me for a reward for the inventions involved—

On the outbreak of War, the demand for optical glass from Messrs Chance Brothers became very great and the anxiety of those acquainted with the position was intensified by the following points—
1–5. [Points are not material in connection with the question of annealing.]
6. Annealing plant had to be remodelled and control system devised.
7. Annealing temperatures had to be determined.
8. Cooling rates for best annealing fully investigated.

With all the difficulties and pressure for increased output,* I, as Manager of Messrs Chance's Optical Department, did not hesitate to accept the assistance of Mr Twyman in regard to annealing matters, coming as it did at a most serious and critical time.

By the use of his instrument he determined the correct annealing temperature for the various glasses produced and this information greatly facilitated the output as every moment saved was of vital importance to the country.

Owing to the shortage of chemicals, even the standard glasses which had been produced in this country for years had to be changed, and by such a procedure the annealing range was altered ; so that it may be taken that practically every type of glass produced for war purposes had its annealing temperature determined by Mr Twyman and on the instrument devised by him.

On looking back and reviewing the position from 1914 to 1920 I tremble to think what might have happened had the work undertaken by Mr Twyman been thrown on the shoulders of the already overburdened glass manufacturer.

Annealing bottles and other glass containers

292 A like spectacular success was achieved in applying my method to the annealing of glass containers.

In discussing the matter of annealing glass bottles with the British Hartford Fairmont Syndicate Ltd, who had approached me for advice on the subject, I learned for the first time of a glass bottle annealing lehr—the Hartford lehr—in which it was possible to control the rate of cooling by pyrometers placed along the length of the lehr. I suggested that this should be used in order that my cooling schedule should be adopted.

My suggestion was followed, with the result described in the following letter received by me from them in 1927—

We have operated the Hartford Lehr in such a manner as to comply with the annealing temperature determined with the Twyman

---

* Optical glass production had to be increased 20 times I understand.—F.T.

apparatus for determining the annealing temperatures of glass, made by Adam Hilger Ltd, and so as to cause the cooling to approximate as closely as possible to that indicated by the formula supplied with the apparatus.

It has been shewn conclusively that as we approximate to the curves given by the formula, the annealing improves in quality and the time for annealing is shortened.

In working along this line we have been able, when making pint bottles, to pass 22 tons of glass through the Hartford Lehr in 24 hours, and to reduce the time of passage through the channel to 47 minutes. This was continued for a period of 10 days during which time we used no fuel whatever but relied solely upon the heat in the bottles to accomplish the annealing.

We do not pretend to believe that the Hartford Lehr is likely to be extensively operated without fuel, but the advantages of being able thus to prescribe the ideal curve for any particular kind of glass are obvious. This will be particularly valuable since the method is applicable to glass of every variety of composition.

In our opinion the apparatus is going to enable us to state at once and without any previous experimental work the best annealing temperature for any sample of glass and for any size of bottle, and, basing ourselves on the formula referred to, the exact temperature which should be registered on the various thermo-couples along the lehr.

My methods were adopted by a number of makers of glass containers.

A classification for quality of annealing of glass containers is given in A.S.T.M. Standards on Glass and Glass Products, published by the American Society for Testing Materials, 1916 Race Street, Philadelphia, 3. Pa. This method is dealt with on pages 35 to 37 under the title " Standard methods of polariscopic examination of glass containers," A.S.T.M. designation : C 148-43. The publication was issued in November, 1946.

The method is based on the use of certain standard discs which were originally made and issued by the Glass Container Association of America. The discs are strained radially, and the optical path difference in a single disc is 22·8 mμ at points $\frac{1}{4}$ in. from the edge. The diameter of the disc is not specified, but is, in fact $3\frac{1}{2}$ in. A convenient thickness is 1/10 in. to 1/8 in.

The specimen of glass to be examined is placed in an ordinary polariscope or strain viewer fitted with a sensitive tint plate, and the intensity of the colours is compared against the intensity of the colours given by 1, 2, 3, etc., standard discs lying one above another. If the colours are matched by, say, three discs, the state of annealing is described as " Temper 3 " ; usually, however, the colours will be a little more intense than, say, two discs, but less intense than the colour given by three discs. In such a case the annealing would be described as "Temper 3."

However, a manufacturer of glass containers should aim at so controlling his annealing process that all the containers are well annealed. This is not difficult and may be expected to effect considerable saving of fuel.

Twyman's method of determining annealing temperatures

293 At the time when my experiments were undertaken it was supposed that the sole reason for annealing optical glass was to remove the double refraction due to internal strain. "Annealing" in this sense would be more correctly described as "stress-relief annealing," the phrase used by metallurgists. Although (§286) it is now known that this view is incomplete, the experiments and conclusions to be described remain of undiminished importance, since the heterogeneity of refractive index due to chilling can also be removed at the annealing temperature as determined by my method, while the cooling schedules devised by me for avoiding the introduction of impermissible strain are also appropriate for avoiding that heterogeneity, although for that purpose the soaking at the annealing temperature requires to be somewhat prolonged. There is no word in general use to describe the removal of this type of heterogeneity in optical glass; I suggest that "normalizing," the word used by metallurgists to describe a somewhat similar process, is suitable.*

It is hoped that the reader will find this somewhat lengthy preamble a sufficient excuse for this, the first published account, of the experiments and conclusions on which my system of annealing is based.

Stress-relief annealing only to be removed by flow of the material

294 We have seen in §284 how the strains originate which result in double refraction. They can only be removed by bringing the glass to such a temperature that flow can take place.

When glass is in a definitely molten condition there can be, of course, no permanent strain. Moreover, it can be shown, by keeping a suitable glass object under observation in a tube furnace, that even when the glass is cool enough to be practically solid under such stresses as are

---

* I am indebted to Mr S. M. Cox of the James A. Jobling Company's Glass Work for the following note—

Modern ideas on annealing include as you say the process of normalizing as well as of stress release, but I believe that in the case of self-sustained stresses in glass the two are mutually dependent in that the "normal" state of the glass at any temperature depends on the stress upon it as well as upon the temperature; thus the release of stress as evidenced by birefringence is not entirely a viscous process (§294) but is due in part to concurrent normalizing under the varying stress system.

occasioned by its own weight, it may yet be mobile enough for all strain internal in origin to disappear in a few minutes. On the other hand, at ordinary air temperature glass is almost (though not quite) perfectly elastic; even if kept for days or weeks under stress produced by external forces it will not become progressively deformed, and on the removal of the external forces producing the stress it will spring back to its original shape, the strain produced by the external forces disappearing at the same time.

Thus it is neither at the lower nor at the higher temperatures that permanent internal stress can originate. Furthermore, if it were possible to cool any glass object from the molten condition with the temperature uniform throughout, then, no matter how fast the rate of cooling, no strain would be introduced.

Between this high temperature, where the glass is so mobile that internal strain is evanescent in a few seconds, and the low temperatures, at which the glass behaves as an elastic solid, there is a range where the stresses take something of the order of some minutes, or an hour, or a few hours, to die out. It is this range of temperature which is all important both in annealing and normalizing, and an accurate knowledge of the properties of the glass throughout this region is necessary if we are to attain any desired perfection in a minimum time.

Definition of " annealing " and " annealing temperature "

295 Before we can start to find a suitable temperature at which to anneal we must decide on a definition of " annealed." Early in my experiments I fixed as the degree of annealing to be attained that in which the residual stress should not exceed $1/_{20}$th of the original stress in the glass. The same criterion was adopted later by Adams and Williamson (1920, p. 625). I further defined the annealing temperature as that at which this degree of annealing is achieved in 3 minutes; experience shows that at such a temperature the glass is not palpably soft—for example a milk bottle or jam jar will not become obviously mis-shapen even when maintained at such a temperature for an hour or more.*

---

* It is interesting to compare this definition of the annealing temperature with that of the American Ceramic Society (1948)—

" Annealing point; the temperature at which the internal stress is substantially removed in 15 minutes. It corresponds to the equilibrium temperature at which the glass has a viscosity of $10^{13.4}$ poises as measured by the loaded fibre method." This definition is accompanied by the formula from which the viscosity is calculated.

The definition of annealing temperature adopted by me implies a viscosity of 2.7×10^{13} poises (see §298), while the $10^{13.4}$ poises adopted by the American Ceramic Society is 2.5×10^{13} poises; a remarkable agreement considering the entire independence of argument by which the conclusions were presumably arrived at.

The disappearance of stress in a viscous body

296 The case of the disappearance of stress in a viscous body was considered by Maxwell (1868). He said—

A distortion or strain of some kind, which we may call S, is produced in the body by displacement. A state of stress or elastic force, which we may call F, is thus excited. The relation between the stress and the strain may be written $F = ES$, where E is the coefficient of elasticity for that particular kind of strain.

In a solid body free from viscosity, F will remain $= ES$, and

$$\frac{dF}{dt} = E\frac{dS}{dt}$$

If, however, the body is viscous, F will not remain constant, but will tend to disappear at a rate depending on the value of F, and on the nature of the body. If we suppose this rate proportional to F, the equation may be written

$$\frac{dF}{dt} = E\frac{dS}{dt} - \frac{F}{T}$$

which will indicate the actual phenomena in an empirical manner. For if S be constant,

$$F = ESe^{-t/T}, \qquad \text{(i)}$$

showing that F gradually disappears, so that if the body is left to itself it gradually loses any internal stress, and the pressures are finally distributed as in a fluid at rest.

If dS/dt is constant, that is, if there is a steady motion of the body which continually increases the displacement,

$$F = ET\frac{dS}{dt} + Ce^{-t/T} \qquad \text{(ii)}$$

showing that F tends to a constant value depending on the rate of displacement. The quantity ET, by which the rate of displacement must be multiplied to get the force, may be called the coefficient of viscosity. It is the product of a coefficient of elasticity, E, and a time T, which may be called the "time of relaxation" of the elastic force. In mobile fluids T is a very small fraction of a second, and E is not easily determined experimentally. In viscous solids T may be several hours or days, and then E is easily measured.

The evanescent term in (ii), $Ce^{-t/T}$, may be neglected in our case, since at the epoch when our observations commence t is very large compared with T.

Briefly then my argument was this. My definition of "annealing" and "annealing temperature" defines a viscosity and by my method the temperature is found at which that viscosity is attained by the glass under examination.

Twyman's apparatus for determining stress-relief annealing temperature

297 The apparatus used to determine dS/dt was as follows—

A strip of glass AB (Fig. 234) was fixed at A into a metal jig and at B into a weighted steel arm.

The length of glass exposed was 6 millimetres.

Two arrangements and methods were used.

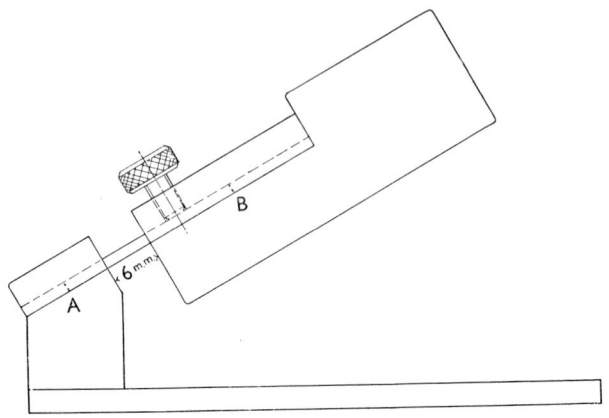

Fig. 234—Jig for measuring the viscosity of glass

Arrangement A

The jig was placed within a muffle furnace (Fig. 235) whose ends were closed by silica windows. The temperature was controlled by pyrometer. Illumination was provided by a 4-volt lamp A, the image of which was brought to a focus on the objective of the observing telescope G which was focused on the jig. The telescope could be rotated about its axis and had a fine line running across the field of view, the orientation being measured on a divided circle.

The experiment consisted of raising the temperature so that the weight began to sag, setting the line parallel to the edge of the weight from time to time, and in this way measuring the angular rate of sag, $d\varphi/dt$.

Arrangement B

In an alternative arrangement the light from the lamp passed through a polarizing prism before passing through the furnace, and the telescope was provided with a Babinet compensator* so that the double

---
* Actually Jamin's modification of Babinet's compensator (Wood, 1911).

refraction of any specimen in the furnace could be measured. For example, in Fig. 236 we see the double refraction in the weighted glass strip, while Fig. 237 shows that in two pieces of imperfectly annealed glass.

In using arrangement B the glass specimen of which two opposite surfaces were roughly polished was placed within the furnace and observed between crossed nicols. The temperature was raised slowly

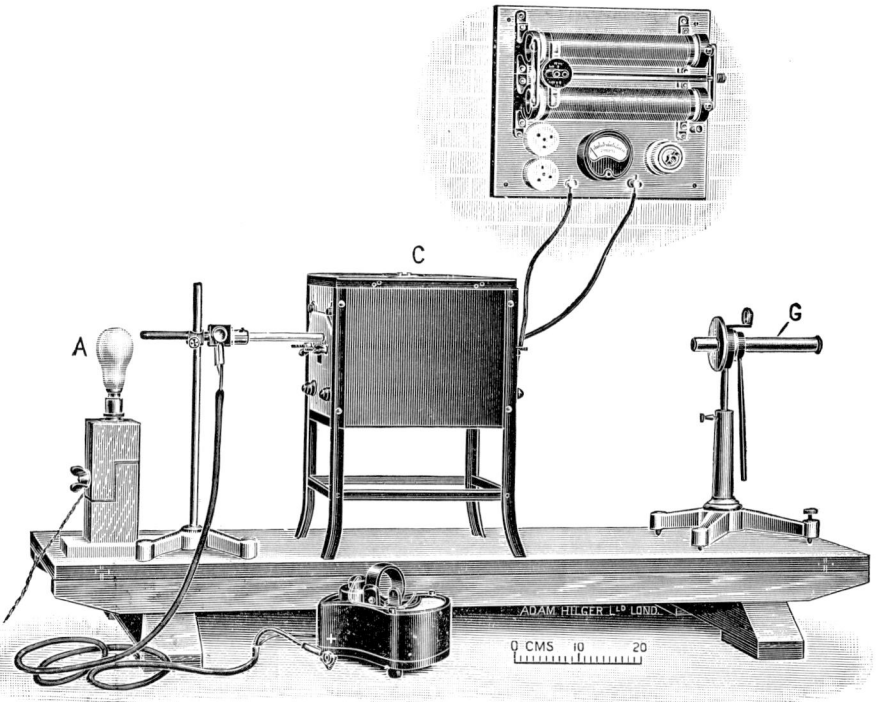

Fig. 235—Twyman's apparatus for determining annealing temperatures

until a point was reached where the double refraction disappeared quickly. It was found that the temperature then reached was very suitable for annealing optical glass. This purely empirical method (which was the first to be used) did not, however, satisfy me; it lacked quantitative precision, for an act of judgment was required to decide when the double refraction was disappearing " rapidly." I therefore determined to carry out experiments to find the temperature at which the viscosity of the glass had the value complying with my definition of " annealing " and " annealing temperature." This was the method finally adopted; it has never failed to yield good results whether in application to optical glass or to bottles, jars and other glass containers.

Arrangement A was, then, the one adopted for works control. The method consisted in raising the temperature of the furnace until

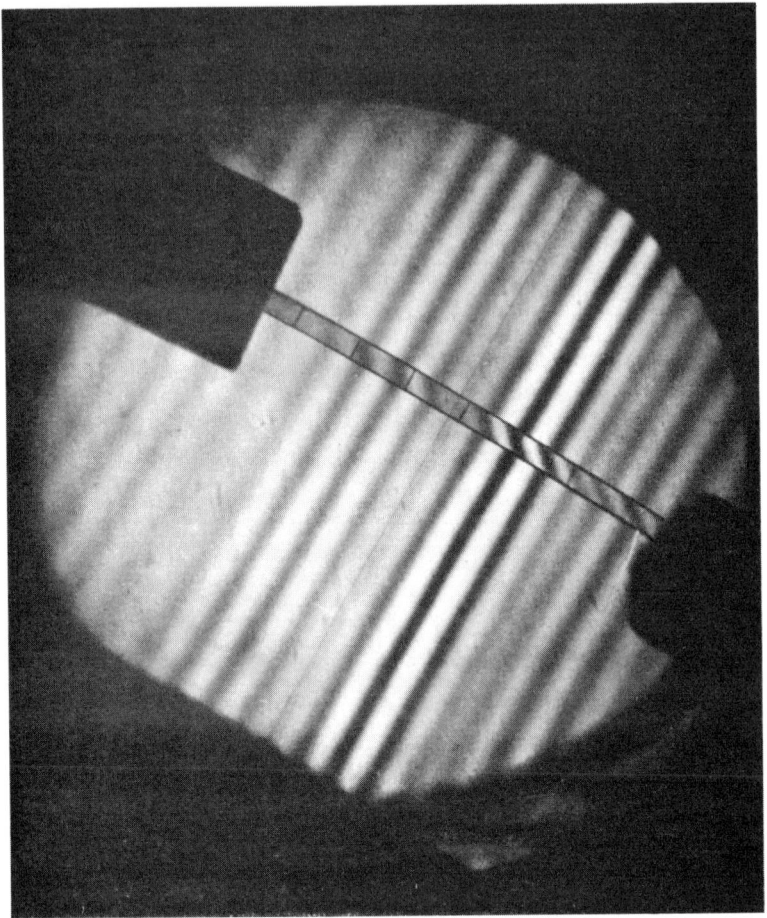

Fig. 236—Double refraction in a weighted glass strip

sagging began, and keeping it constant while measuring the rate of change of φ at appropriate intervals. The derivative $d\varphi/dt$ is a measure of the rate of deformation of the glass under the existing conditions of temperature and stress.

In the subsequent determinations, the bending moment is always taken at the middle point of the strip.

The strip is not quite horizontal, and the necessary corrections were applied in arriving at the above bending moments. The free length of strip was 6 millimetres.

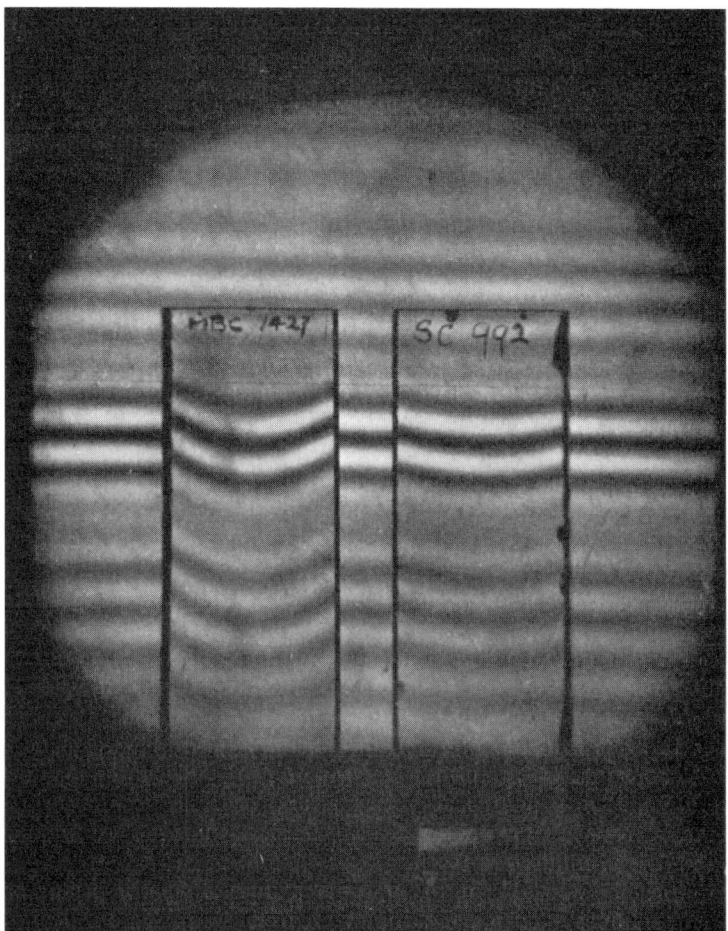

Fig 237—Double refraction in pieces of imperfectly annealed glass

Calculation of the angular sag of the weight at the annealing temperature
298 The maximum stress, F, in the strip of glass is to be found in the top and bottom surfaces, and is equal to
$$F = 6M/a^2b,$$

where M = bending moment at point considered (the middle of the free part of the glass strip)

a = thickness of strip

b = width of strip.

This stress is parallel to the length of the strip, and is unaccompanied by any stress at right-angles to the strip and to the line of sight.* Hence it is the only stress which can have any influence on the observed values of the double refraction.

In our experiments

$$a = 0\cdot173 \text{ cm}$$
$$b = 0\cdot853 \text{ cm}$$

whilst the bending moments employed were,

I. 440 gram–cm (arm and heavy weight), whence
$$F = 1040 \times 10^2 \text{ g/cm}^2$$

II. 240 gram–cm (arm and light weight), whence
$$F = 566 \times 10^2 \text{ g/cm}^2$$

III. 35 gram–cm (arm alone), whence
$$F = 83 \times 10^2 \text{ g/cm}^2$$

The first glass to be tested was Chance's B.L.F. 1336, and to determine the annealing temperature the light weight was used.

The quantities in the formulae (i) and (ii) of § 296 are to be expressed in the following units—

F in g/cm^2

E in g/cm^2

S in cm elongation per 1 cm length

t in minutes measured from the commencement of annealing.

If F_o be the initial stress at a given point and F_t the stress remaining at time t, the time when annealing has been achieved, then from (i) $F_t/F_o = e^{-t/T}$.

By the definitions of "annealing" and "annealing time" given in §295, $F_t/F_o = 0\cdot05 = e^{-3/T}$

whence $T = 1$ minute.

We must now find the rate of angular sag of the weight corresponding to this relaxation time T.

E may be taken as 600×10^6 g/cm^2

$F = 56{,}600$

and from (ii) dS/dt for the top or bottom of the strip is—

$F/ET = 56{,}500/600 \times 10^6$ cm per minute (elongation per 1 cm length)

$= 94\cdot3 \times 10^{-6}$ cm per minute.

---

* Neglecting the shear stress across the bar which is relatively very small.

The angular sag of the weight will therefore be
$(94 \cdot 3 \times 10^{-6} \times 2 \times 0 \cdot 6)/0 \cdot 173$ radians per minute
or $(94 \cdot 3 \times 10^{-6} \times 2 \times 0 \cdot 6 \times 57/0 \cdot 173$ degrees per minute
$= 0 \cdot 0373$ degrees per minute
or 2·24 degrees per hour which corresponds to $2 \cdot 7 \times 10^{13}$ poises (see footnote p. 522).

299 The curves connecting $d\varphi/dt$ and temperature for this glass are shown in Fig. 238 for all three bending moments. It was found that $d\varphi/dt . \sin \varphi$ was a constant for any one temperature and bending

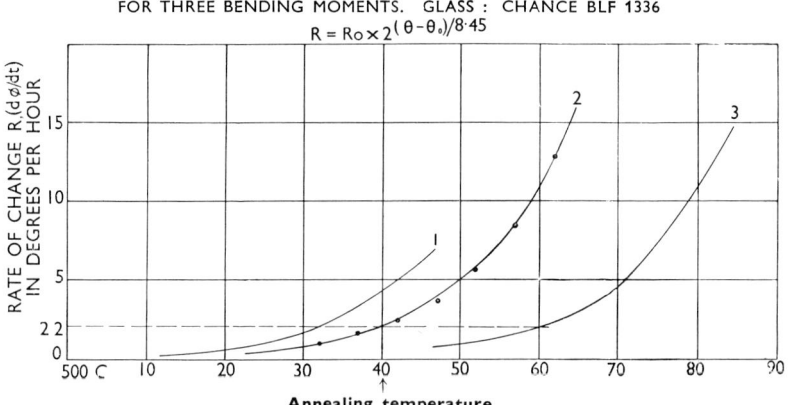

Fig. 238—Curves connecting the rate of bending of a stressed strip of glass with its temperature

moment; and there was no indication that the deformation would not continue to proceed indefinitely at the same rate.

The formula $R = R_o \times 2^{(\theta - \theta_0)/8 \cdot 45}$ obtained with the light weight was ascertained by putting the values for R, R_o, θ and θ_o for two points on curve No. 2 in the general formula $R = R_o \times 2^{(\theta - \theta_0)/k}$ and thus finding k. In order to see how closely this formula agrees with observation R was calculated from the formula for seven different temperatures and the values plotted as heavy points. It will be seen how very closely they lie to the "Observed" curve.

Curves 1 and 3 were taken with the greatest and least of the bending moments (§298) and point to the approximate truth that the rate of deformation is proportional to the stress. The evidence of curve 3, which shows a rather greater rate of strain, should not be given the same weight, having regard to the conditions of experiment.

The truth of this relation is further borne out by the fact that when the strips are examined, after cooling, at any cross section the traces of the polished faces are straight lines.* This proves that the lateral strain has been proportional to the distance from the neutral axis, and since it is also proportional to the longtitudinal strain, the latter also must bear a constant relation to the stress, which varies linearly with the distance from the neutral axis.

Application to alloys

The same argument and method of determining a stress-relief annealing temperature can be used for a number of metals including brasses, manganese-bronzes, phosphor-bronzes and a number of non-metallic substances. For example it was very rapidly found that a 70/30 brass could be stress relieved in one hour at 275°C while a 60/40 brass could be stress relieved in one half hour at 200°C.

By adopting the lowest temperature found by this method one avoids softening the brass, so that the method is useful when it is desired to flatten sheets of brass (as for shutters of camera dark slides) without the necessity of machining the metal, the risk of subsequent distortion, or the softening which would render the shutter, for example, very liable indeed to be accidentally bent.

Dependence of annealing temperature on composition

300 It may be interesting to record the composition of a few glasses whose annealing temperatures have been determined.

Maker.	Type	Melt.	Annealing Temperature.	n_D	v
Chance	H.C.	1340	595°C.	1·5158	61·0
,,	B.L.F.	1336	548°C.	1·6097	37·3
,,	M.B.C.	1494	601°C.	1·5771	56·9
,,	D.F.	2008	445°C.	?	

Compositions:

	H.C. 1340 per cent	B.L.F. 1336 per cent	M.B.C. 1494 per cent	D.F. 2008 per cent
SiO_2	71·9	59·3	44·7	41·1
K_2O	16·4	9·0	3·8	7·2
N_aO_2	—	—	—	—
C_aO	11·5	—	—	—
PbO	—	9·6	—	51·6
B_2O_3	—	—	3·7	—
BaO	—	17·2	34·4	—

* The straightness of these lines is tested by the delicate method of using a very short length of the strip as a prism, and observing the definition of the resulting image.

	H.C. 1340 per cent	B.L.F. 1336 per cent	M.B.C. 1494 per cent	D.F. 2008 per cent
Al_2O_3	—	—	4·1	—
ZnO	—	4·2	7·8	—
Sb_2O_3	—	0·7	1·3	—
As_2O_3	0·2	—	0·2	0·3

Determination of the best cooling schedule for annealing

301 The calculation of the best cooling schedule is based on the following facts—

1. As a piece of glass cools in the furnace a difference of temperature results; this causes elastic strain and, while the glass remains within the annealing range, there is a flow tending to accommodate the glass to the differences of temperature within it at the moment. When at the end of the annealing process the glass is taken out of the furnace and becomes of a uniform temperature, strain appears to the extent determined by the degree to which the glass has so accommodated itself to the temperature differences existing when it is taken out of the furnace. The process has been very clearly demonstrated by Adams and Williamson (loc. cit., pp. 601 to 604) and they sum up their conclusion in the following words—

> The strain remaining in a block of glass is equal and opposite in sign to the reverse strain lost by viscous yielding in the early stages of the cooling process.

2. Observation by double refraction of a piece of glass within the furnace, in the manner illustrated in Fig. 237, showed that the strain due to cooling was proportional to the rate of cooling, it being assumed that double refraction is proportional to the strain.

It may be admitted that there is the same doubt in this assumption as I point out at the end of §302 in the case of the Adams and Williamson observations, and that the soundest justification for the formula arrived at below is that it works.

There are, of course, other assumptions: for example, the temperature of the glass is supposed to be that indicated by the pyrometer so that the viscosity of the glass corresponds with the temperature as so shown. We need have no misgivings on this score since an error of 10°C would cause no serious mistake in our deductions.

Another assumption of prime importance in the derivation of my cooling schedule will perhaps find less ready acceptance. It is that the degree of stress-relief annealing attained in a given time by cooling from the annealing temperature to air temperature is greatest when the rate of cooling at each temperature is such that the flow (adjustment to stress) remains constant throughout the cooling process. In 1916 when I was 40 years old this was self-evident to me; now, when I am over 73, it is no longer so. Correct or not, however, it leads to a type of cooling schedule which, whether for optical glass or glassware, has never been bettered for speed and effect.

From the equations, facts, and assumptions set forth in the preceding pages I arrived at the following expression connecting temperature and time in order that the best stress relief should be attained in a given time—

$$t = \frac{1}{mb}\left(1 - \frac{1}{e^{m(\theta_0 - \theta)}}\right) *$$

where $m = 0{\cdot}0865$

b is a factor by the variation of which quicker or slower schedules of cooling are arrived at.

θ_o is the correct annealing temperature as found by the apparatus.

e is base of the Napierian logarithms.

and t is the time taken to fall from the annealing temperature to temperature θ.

It is sometimes convenient to know the rate of cooling. By differentiation of the above expression we get—

$$\frac{d\theta}{dt} = -be^{m(\theta_0 - \theta)}$$

When $\theta = \theta_o$ this reduces to $d\theta/dt = -b$.

One must repeat that this exponential law (" Twyman's law ") can be relied on only for the range which one may term " the annealing range." For example, if a glass, the annealing temperature of which was 500°C, followed this law on cooling to 20°C its viscosity would increase $1{\cdot}15 \times 10^{18}$—that is, instead of taking 3 months to anneal it would take 65×10^{10} centuries. Actually, changes in a glass plate, 2 cm thick by 5 cm square, have been detected in six months. The change amounted to $1/10$ of a wavelength change on single transmission through the plate.

Temperatures are to be measured in degrees Centigrade. If b is the rate of cooling per hour, t is in hours; if b is the rate per minute, t is in minutes.

Alternatively the formula (1) may be written

(2) $$\theta_0 - \theta = \frac{\log_{10}\left(\frac{1}{1 - mbt}\right)}{m \log_{10} e}$$

Representative cooling schedules derived from this formula are given in the table; these schedules apply to glasses whose viscosity doubles for each 8°C fall of temperature.

* For derivation of this formula see Appendix I.

Cooling Schedules

(a)		(b)		(c)		(d)		(e)	
Time in Mins.	Fall of Temperature.	Time in Mins.	Fall of Temperature.	Time in Mins.	Fall of Temperature.	Time in Mins.	Fall of Temperature.	Time in Mins.	Fall of Temperature.
0	0	0	0	0	0	0	0	0	0
1	·6	2	·6	3	·6	4	·6	5	·6
2	1·3	4	1·3	6	1·3	8	1·3	10	1·3
3	2·0	6	2·0	9	2·0	12	2·0	15	2·0
4	2·7	8	2·7	12	2·7	16	2·7	20	2·7
5	3·5	10	3·5	15	3·5	20	3·5	25	3·5
6	4·3	12	4·3	18	4·3	24	4·3	30	4·3
7	5·2	14	5·2	21	5·2	28	5·2	35	5·2
8	6·2	16	6·2	24	6·2	32	6·2	40	6·2
9	7·3	18	7·3	27	7·3	36	7·3	45	7·3
10	8·5	20	8·5	30	8·5	40	8·5	50	8·5
11	9·8	22	9·8	33	9·8	44	9·8	55	9·8
12	11·3	24	11·3	36	11·3	48	11·3	60	11·3
13	13·0	26	13·0	39	13·0	52	13·0	65	13·0
14	15·0	28	15·0	42	15·0	56	15·0	70	15·0
15	17·5	30	17·5	45	17·5	60	17·5	75	17·5
16	20·5	32	20·5	48	20·5	64	20·5	80	20·5
17	24·7	34	24·7	51	24·7	68	24·7	85	24·7
18	31·4	36	31·4	54	31·4	72	31·4	90	31·4
19	44·1	38	44·1	57	44·1	76	44·1	95	44·1
19¼	80·8	38½	80·8	57¾	80·8	77	80·8	96¼	80·8

Other cooling schedules, quicker or slower, can be deduced from (a) by multiplying all the times by a constant factor; the schedules in the above table are appropriate for bottles and other glass containers, one or other table being taken according to the sizes of the articles; the prisms and other comparatively massive pieces require much longer, at least 20 times as long as the slowest cooling schedule for right-angled prisms of 4 in. cathetus.

It must be remembered that some glasses, particularly some of the modern ones have very different viscosity-temperature relationships. Some examples are given in appendix I, together with the method of calculating appropriate cooling schedules for annealing them.

Results of Adams and Williamson (1920)

302 The apparatus of these authors was in effect the same as one of mine (§297 arrangement *B*). The first stage of their experiment consisted in getting strain into glass. This was accomplished by raising the temperature about 50° above the temperature at which strain disappears almost instantaneously and then dropping the temperature rapidly by cutting off the current and opening the door of the furnace. When the temperature had fallen about 300° the current was turned on and the furnace brought up to the temperature at which observations were to be made. At least 20 minutes were allowed for the glass to attain a constant temperature throughout. The rate of release of stress at constant temperature was then determined by successive observations of double refraction. The interval between readings depended on the rapidity with which the stress was released, so that at the higher temperatures readings were taken a few minutes apart, while at lower temperatures several hours might elapse between two successive observations.

From these experiments they derived, from the arbitrary assumption that the rate of release of stress is proportional to the square of the stress, a formula which represented with satisfactory precision the release of stress in glass—at any rate, for the temperatures and with the amounts of stress with which it is convenient to measure the speed of annealing. Their results confirmed my conclusion that the mobility of glass is a simple exponential function of the temperature.

Tool and Valasek (1920) determined the rate of relaxation of stress at various temperatures both by the optical method (diminution of birefringence) and by mechanical methods (stretching or bending of rods), and calculated the "relaxation time," T, from Maxwell's relation (see equation (ii), §523). They found that the values of T obtained by the two methods do not agree and that in general T is not independent of the initial stress ; *i.e.* larger stresses relax more quickly and smaller stresses more slowly than would be expected from the extrapolation.

It will be noted that neither Adams and Williamson nor Tool and Valasek agreed with my conclusion that the rate of deformation is simply proportional to the stress. It must be pointed out that my statement could have no meaning except as to the connection between the rate of deformation and stress at the *given point* of the glass. The conclusions of Adams and Williamson were, however, derived from observation of the total double refraction of light passing through a slab of glass. Such double refraction is the integrated result arising from passage through a piece of glass in a highly complex condition of stress, and it cannot be taken, I think, as indicating the relation between stress and rate of deformation at a point.

Preston (1924) confirmed the conclusions of Adams and Williamson on a theoretical basis, although expressing the relation between stress and the rate of its release in a different but practically equivalent way— briefly that the rate of release of stress is proportional to the square of the latter. Later (1952) he re-examined the Adams–Williamson law of annealing and concluded that " when restricted to what was actually observed, the law becomes intelligible and its seeming inconsistencies are straightened out ".

Temperatures for moulding

303 Closely linked to the annealing temperatures are the temperatures adequate for adhesing, slow moulding and quick moulding and these are related to the annealing temperature in the following way. If T be the annealing temperature as described in the preceding pages, then, for ordinary glasses and with a fair approximation—
(1) the adhesing temperature is $T + 65°C$ for two hours.
(2) Self-sagging temperature (*i.e.* for $\frac{1}{8}$ inch strip to bend as shown under its own weight in two hours) $T + 157°C$.
(3) the slow-moulding temperature is $T + 145°C$.
(4) the quick-moulding temperature is $T + 225°C$.

By the adhesing temperature is meant the temperature at which polished glass pieces, if put into optical contact, will adhere to each other—becoming as it were one piece.

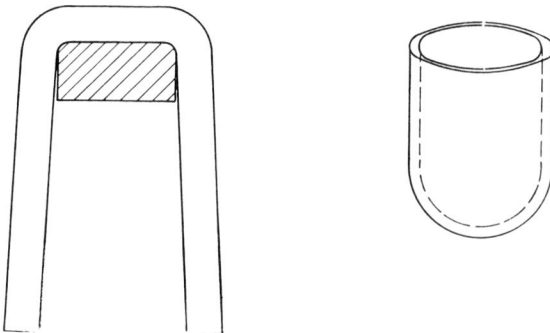

Fig. 239—(*a*) **Self-sagging** (*b*) **Slow moulding**

The slow-moulding temperature is the temperature at which in two hours a piece of glass can be moulded into a shape such as a shallow cup (Fig. 239) while by quick moulding is meant the same process taking place in 3 seconds.

Precautions to be taken in making glass cells

303.1 Glass cells, such as those used for measuring the light transmission of liquids, are made by cementing or fusing together three or more pieces of optically worked plate glass in a temperature-controlled furnace. To avoid strain in the finished cell, the pieces of glass of which it is composed should have identical expansion. Preferably they should be cut from the same sheet, but this is, of course, impossible in mass production, where hundreds of pieces of glass are passing through the shop. Therefore, before any sheet of glass is cut for such use, a quick means must be found of comparing the expansion of every sheet of glass before it is cut, with that of a " standard sheet," so that only those sheets with identical or almost identical expansions shall be used for production.

A member of the Research Staff of Adam Hilger, Mr G. E. Fensom, devised the following simple but efficient technique. Two strips of glass, each ¼ in. wide, are taken, one from the " standard sheet " and one from the sheet to be compared. They are held, one in each hand (see Fig. 240) and the ends are introduced into a blowlamp flame. When sufficiently soft just to " sag " under their own weight, the two ends are placed in contact, face to face, removed from the flame and given a squeeze between two small carbon blocks to ensure

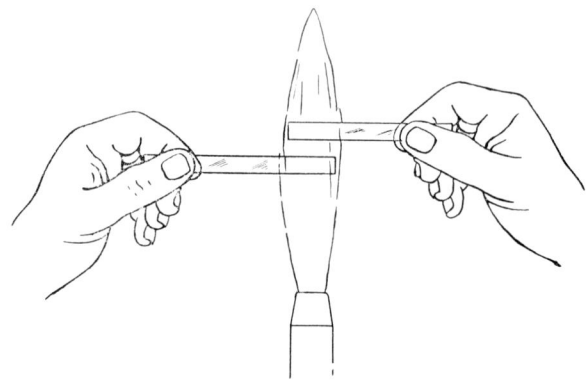

Fig. 240—Preparing for Fensom's test of coefficient of expansion

complete welding of the surfaces. The welded portion is now returned to a small hot flame and reheated until soft. Using great care not to twist the joined strips, they are again taken from the flame and the hands moved about four feet apart. A long, slightly flattened, coarse fibre of the two glasses " face to face "—in other words, a glass version of the well known bi-metal strip—is thus obtained.

If this fibre is now cut into two lengths by bringing the centre to the tip of the flame and pulling slightly, the two fibres will either remain perfectly straight or " spring " into a smooth curve, the degree of curvature depending on the difference of expansion of the glasses. If the fibres remain quite or almost straight, the two glasses can with confidence be welded or contacted without the least fear of strain or subsequent fracture.

As an indication of the sensitivity of the above method, the two borosilicate glasses Pyrex and Phoenix were compared. One authority gives the coefficient of expansion of Pyrex as $3 \cdot 2 \times 10^{-6}$ and of Phoenix as $3\cdot 3 \times 10^{-6}$. These two very similar glasses produced a curve of approximately 24 inches radius.

The technique also gives very instructive curves when applied to the " steps " of graded glass seals.

CHAPTER 15

LARGE OBJECT GLASSES AND MIRRORS

304 This Chapter deals with the working of object glasses and mirrors from $4\frac{1}{2}$ in. diameter upwards. When a lens or mirror is more than $3\frac{1}{2}$ in. in diameter, it is most successful in any case to work it as a single surface. When the diameter exceeds about $4\frac{1}{2}$ in. problems arise, chiefly due to flexure, which are not dealt with in Chapter 2. These problems naturally become more difficult as the diameter increases.

The development of astronomical telescopes

305 Sir Harold Spencer-Jones, the Astronomer Royal, in his Presidential address to Section A of the British Association in 1949, gave a sketch of the development of astronomical telescopes (Spencer-Jones, 1949).

The following passage is quoted, with permission, from this address.

Because of the invention of the achromatic object glass and because also of the difficulties of casting and figuring specula, the refractor for a time almost completely superseded the reflector. In 1828 Lord Rosse wrote, ' Many practical men whom I have spoken to seem to think that since Fraunhofer's discoveries the refractor has entirely superseded the reflector and that all attempts to improve the latter instrument are useless.' Larger and larger objectives were made by Merz and Mahler, the successors of Fraunhofer in Germany ; by Cauchoix in France ; by Grubb and Cooke in England ; and by Alvan Clark in America. The largest objective in 1824 was the $9\frac{1}{2}$ inch made by Fraunhofer for Dorpat ; in 1839, the 15 inch, made by Merz and Mahler for Poulkova ; in 1871, the $24\frac{1}{2}$ inch made by Cooke for Newall, now at Cambridge ; in 1878 the 26 inch made by Grubb for Vienna ; culminating in 1888 with the 36 inch by Alvan Clark for the Lick Observatory, and in 1897 with the 40 inch, also by Alvan Clark, for Yerkes Observatory. One or two attempts to make larger objectives have not been successful.

The development of the silver-on-glass mirror, by avoiding the difficulties inherent in specula, began to bring the reflecting telescope back into favour in the seventies and eighties of the last century. Lassell's 47-inch reflector, made in 1860, was the last large reflector with a speculum mirror to have been used, as far as I have been able to ascertain ; the 48-inch speculum reflector made in 1867 by Grubb for Melbourne was never used. The idea of making mirrors of glass was not new; Short in 1734 had made some glass mirrors, silvered on the *back*, which Maclaurin declared to be excellent. In 1827 Airy had proposed that a layer of silver should be deposited on the *surface* of a figured glass disk. But the method did not become practicable until Liebig discovered a simple chemical process for dispositing a thin film of silver on a glass surface. The methods of figuring and testing glass disks were due primarily to Foucault who, in 1857, presented his first telescope with a silvered mirror to the French Academy. His knife-edge test is still a valuable tool to the optician. Further developments in technique were made by Draper in America, by Common in England and by Ritchey in America.

In recent years aluminium has almost entirely replaced silver as the reflecting coating on a figured glass surface.

No material has yet been found which is more suitable than glass as a medium for supporting the reflecting film. Glass is capable of taking a high polish and is free from any tendency to warping or distortion with age.* But its relatively large coefficient of expansion and poor thermal conductivity are disadvantages. Assuming the temperature to be uniform throughout the disk, the position of the focus of the mirror will depend upon the temperature; for a given change of temperature, the displacement of the focus will be greater the larger the coefficient of expansion. If the temperature is not uniform throughout the disk, distortion of the figure of its surface will occur, which will impair the quality of the images; the effect will be greater the thicker the disk and the lower its thermal diffusivity. Pyrex glass is now used in preference to plate glass because of its smaller coefficient of expansion and its larger thermal

---

* Ten years ago I should have agreed with this, but we now know that it is not strictly true. Interferometers of the Twyman and Green type, described in Chapter 12, have been in use in our optical workshops for 35 years and I have at length had to agree with the opinion, held by our optical foremen for a number of years, that the glass of the mirrors of these instruments is sometimes liable either to warp or to suffer a change in heterogeneity in the course of time. In one recent instance an interferometer with a 5 in. square aperture, originally made practically perfect, developed a cylindrical curvature of 1/5 of a band in 6 months. In attempts to make careful observations of the interference pattern—

(1) the air temperature must be very constant; even in a temperature-controlled room a change of position of the interferometer in the room may cause the pattern to change somewhat.
(2) the temperature at which the test is made must be the same as that at which the correction was effected. A change of 7°C has been known to cause about the same change of curvature as that referred to above.

Despite the greatest care in the above two respects, a change can occur in the course of time and it is customary with our opticians to recorrect the smaller size interferometers (3-in. aperture) when they develop these small defects, which they sometimes do to an observable extent in a year or two (Author).

diffusivity. The thinner the disk the more rapidly it will reach a state of temperature equilibrium but then also the more flexible it is, and the more elaborate the support system must be. The requirements are therefore to some extent self contradictory and the choice of thickness is a matter of compromise.

For the 200-inch Hale telescope at Mount Palomar it was decided to reduce the weight of the disk by constructing it with a thin face supported on a honeycomb structure, consisting of 36 cylindrical pockets, lying on five concentric circles, each pocket being connected to each of the adjacent pockets by a glass rib, thereby providing stiffness. These pockets were designed to form part of the support system. Each was ground accurately cylindrical and fitted with a steel sleeve, gaskets being inserted between the bearing faces and the glass. Each sleeve is connected to the cell-frame by an integral two-component system of balance weights and levers. Double gimbal bearings on the steel sleeve and on the support fixed to the cell frame prevent constraints. Three of the lever systems are tied down by circumferential springs to constrain the disk against rotation about the axis, while three others are tied down as axial defining points. A central tube passes through the 40-inch diameter central hole in the mirror; four central radial jacks, made of invar and steel to compensate for expansion of the central steel tube, define the radial location of the mirror, no edge band being required. These centre jacks and 12 edge radial squeeze arcs, to remove gravitational astigmatism, are provided with ball-bearing faces to allow freedom in the plane normal to the applied reaction. A number of fans have been mounted in the back of the mirror cell to draw air from the front to the back of the mirror in order to secure better temperature equilibrium. (*A description and illustrations of the mode of handling and working the* 200-*inch mirror appear in* §335.)

Sir William Herschel's telescopes

306 Sir Frederick William Herschel (1738–1822), the father of modern astronomy, may with equal justice be called the father of the modern large telescope. With mirrors made by himself, improved throughout a long term of years by the polishing of hundreds of specula of increasing size and perfection, he built telescopes with which he attacked the gigantic project of surveying systematically the entire heavens with the purpose of ascertaining their general structure. His largest telescope had a mirror of no less than 48 ins.

Between the completion in 1774 of a Newtonian telescope of 6 ft. focal length and 1822, when he was 84 years of age, he communicated to the Royal Society a long series of papers recording pioneer discoveries, the development of which constituted the chief astronomical triumphs of the 19th century. (*Enc. Britt.*, 11th ed., article " Herschel, Sir F. W.")

Much has been written about making large object glasses and mirrors, but there are three papers which are particularly informative. One by Draper (1864) on the construction of a silvered glass telescope $15\frac{1}{2}$ in. in aperture ; the second by Howard Grubb (1886) on " Telescope Objectives and Mirrors, their preparation and testing " and the third by Ritchey (1904). This last named paper describes the methods employed in making and mounting the 5 ft reflector of the Mount Wilson Observatory. Ritchey later made a 100 inch reflector for Mount Wilson Observatory but I have been unable to find any description of the working of the mirror.

It is unfortunate that Dr J. A. Anderson has not published a description of the methods which he used in figuring the large optics of the 200-in. telescope on Mount Palomar. Through the kindness of Mr J. V. Thomson, of Messrs Cox, Hargreaves & Thomson Ltd., I am however, able to give some notes about the work on the mirror illustrated by photographs supplied by him.

Of this telescope it is stated (*Popular Astronomy*, 1949) that the mirror was ready for aluminizing by the end of September, 1949, and was then good enough to take full advantage even of exceptionally good seeing nights and that it was believed that the surface was as nearly perfect as possible—the caution being added however that " We cannot be absolutely certain until the telescope has been tested under actual operating conditions."

It has been stated that the final difficulty was due to thermal distortion due to the continuous ring rib in which the honeycomb structure terminates.

Methods of making and correcting large mirrors for astronomical telescopes

This Chapter consists largely of verbatim extracts made by permission from the three publications mentioned above.

The three authors cited dealt only with the simpler types of figuring e.g. parabolizing of mirrors. In modern systems, of large aperture, still further problems arise ; for example, the making of non-spherical surfaces as described in Chapter 10. These will doubtless require much study for their solution when the Isaac Newton Telescope to be installed at Herstmonceux Castle, Sussex, is made. This will be the largest Schmidt telescope yet attempted. Work on the 98-in. diameter glass disc, cast in the United States in 1936 and presented by the Board of Trustees of the McGregor Fund, The University of

Michigan, has now been started by Messrs. Howard Grubb, Parsons Ltd.

Another problem is that of avoiding the scattered light resulting from surface irregularities so small that, until recently, they were considered immaterial; their importance was brought to notice by Lyot and they were studied by himself and others (§§227.1 and 336).

The work of Draper

307 Draper's work was directed towards making mirrors of up to $15\frac{1}{2}$ in. diameter for astronomical telescopes. He says (Draper, 1864)—

In the summer of 1857, I visited Lord Rosse's great reflector, at Parsonstown, and, in addition to an inspection of the machinery for grinding and polishing, had an opportunity of seeing several celestial objects through it. On returning home, in 1858, I determined to construct a similar, though smaller instrument; which, however, should be larger than any in America, and be especially adapted for photography. Accordingly, in September of that year, a 15-inch speculum was cast, and a machine to work it made. In 1860, the observatory was built, by the village carpenter, from my own design, at my father's country seat, and the telescope with its metal speculum mounted. This latter was, however, soon after abandoned, and silvered glass adopted. During 1861, the difficulties of grinding and polishing that are detailed in this account were met with, and the remedies for many of them ascertained. The experiments were conducted by the aid of three $15\frac{1}{2}$-inch disks of glass, together with a variety of smaller pieces. Three mirrors of the same focal length and aperture are almost essential, for it not infrequently happens that two in succession will be so similar, that a third is required for attempting an advance beyond them. One of these was made to acquire a parabolic figure.

. . . .

During the winter of 1862, the art of local corrections was acquired, and two $15\frac{1}{2}$-inch mirrors, as well as two of 9 inches for the photographic enlarging apparatus, were completed. The greater part of 1863 has been occupied by lunar and planetary photography, and the enlargement of the small negatives obtained at the focus of the great reflector.

GRINDING AND POLISHING MIRRORS

(1) *Experiments on a metal speculum*

308 My first 15 inch speculum was an alloy of copper and tin, in the proportions given by Lord Rosse. His general directions were closely followed, and the casting was very fine, free from pores, and of silvery whiteness. It was 2 inches thick, weighed 110 pounds, and was intended to be of 12 feet focal length. The grinding and polishing were conducted with the Rosse machine. Although a great amount of time was spent in various trials, extending over more than a year, a fine figure was never obtained— the principal obstacle to success being a tendency to polish in rings of different focal length. It must, however, be borne in mind that Lord Rosse had so thoroughly mastered the peculiarities of his machine as to produce with it the largest specula ever made and of very fine figure.

During these experiments there was occasion to grind out some imperfections, $8/100$ of an inch deep, from the face of the metal. This operation was greatly assisted by stopping up the defects with a thick alcoholic solution of Canada balsam, and, having made a rim of wax around the edge of the mirror, pouring on nitro-hydrochloric acid, which quickly corroded away the uncovered spaces. Subsequently an increase in focal length of 15 inches was accomplished, by attacking the edge zones of the surface with the acid in graduated depths.

An attempt also was made to assist the tedious grinding operation by including the grinder and mirror in a Voltaic circuit, making the speculum the positive pole. By decomposing acidulated water between it and the grinder, and thereby oxidizing the tin and copper of the speculum, the operation was much facilitated, but the battery surface required was too great for common use. If a sufficient intensity was given to the current, speculum metal was transferred without oxidation to the grinder, and deposited in thin layers upon it. It was proposed at one time to make use of this fact, and coat a mirror of brass with a layer of speculum metal by electrotyping. The gain in lightness would be considerable.

During the winter of 1860 the speculum was split into two pieces, by the expansion in freezing of a few drops of water that had found their way into the supporting case.

(2) *Silvering glass*

309 At Sir John Herschel's suggestion (given on the occasion of a visit that my father paid him in 1860), experiments were next commenced with silvered glass specula. These were described as possessing great capabilities for astronomical purposes. They reflect more than 90 per cent. of the light that falls upon them, and only weigh one-eighth as much as specula of metal of equal aperture.

.   .   .   .

In order to guard against tarnishing, experiments were at first made in gilding silver films, but were abandoned when found to be unnecessary. A partial conversion of the silver film into a golden one, when it will resist sulphuretted hydrogen, can be accomplished as follows : Take three grains of hyposulphite of soda, and dissolve it in an ounce of water. Add to it slowly a solution in water of one grain of chloride of gold. A lemon yellow liquid results, which eventually becomes clear. Immerse the silvered glass in it for twenty-four hours. An exchange will take place, and the film become yellowish. I have a piece of glass prepared in this way which remains unhurt in a box, where other pieces of plain silvered glass have changed some to yellow, some to blue, from exposure to coal gas.

. . . .

(3) *Grinding and polishing glass*

310 Some of the facts stated in the following paragraphs, the result of numerous experiments, may not be new to practical opticians. I have had, however, to polish with my own hands more than a hundred mirrors of various sizes, from 19 inches to $\frac{1}{4}$ of an inch in diameter, and to experience very frequent failures for three years, before succeeding in producing large surfaces with certainty and quickly. It is well nigh impossible to obtain from opticians the practical minutiæ which are essential, and which they conceal even from each other. The long continued researches of Lord Rosse, Mr Lassell, and M. Foucault are full of the most valuable facts, and have been of continual use.

The subject is divided into : (*a*) The peculiarities of Glass ; (*b*) Emery and Rouge ; (*c*) Tools of Iron, Lead and Pitch ; (*d*) Methods of Examining Surfaces ; (*e*) Machines.

(*a*) *Peculiarities of glass.*

311 *Effects of pressure.*—It is generally supposed that glass is possessed of the power of resistance to compression and rigidity in a very marked manner. In the course of these experiments it has appeared that a sheet of it, even when very thick, can with difficulty be set on edge without bending so much as to be optically worthless. Fortunately in every disk of glass that I have tried, there is one diameter on either end of which it may stand without harm.

In examining lately various works on astronomy and optics, it appears that the same difficulty has been found not only in glass but also in speculum metal. Short used always to mark on the edge of the large mirrors of his Gregorian telescopes the point which should be placed uppermost, in case they were removed from their cells. In achromatics the image is very sensibly changed in sharpness if the flint and crown are not in the best positions; and Mr Airy, in mounting the Northumberland telescope, had to arrange the means for turning the lenses on their common axis, until the finest image was attained. In no account, however, have I found a critical statement of the exact nature of the deformation, the observers merely remarking that in some positions of the object glass there was a sharper image than in others.

. . . .

I am led to believe that this peculiarity results from the structural arrangement of the glass. The specimens that have served for these experiments have probably been subjected to a rolling operation when in a plastic state, in order to be reduced to a uniform thickness. Optical glass, which may be made by softening down irregular fragments into moulds at a temperature below that of fusion, may have the same difficulty, but whether it has a diameter of minimum compression can only be determined by experiment. Why speculum metal should have the same property might be ascertained by a critical examination of the process of casting, and the effect of the position of the openings in the mould for the entrance of molten metal.

. . . .

The quantity of the finer emeries consumed in smoothing a $15\frac{1}{2}$-inch surface is very trifling—a mass of each as large as two peas sufficing

. . . .

The pair of iron tools for my large mirrors are $15\frac{1}{2}$ inches in diameter, and were cast $\frac{3}{8}$ of an inch thick, radiating from a solid centre two inches in diameter. They weighed 26 pounds apiece. Four ears, with a tapped hole in each, project at equal distances round the edge, and serve either as a means of attachment for a counterpoise lever, or as handles.

At first no grooves were cut upon the face, for in the lead previously employed for fining they were found to be a fruitful source of scratches, on account of grains of emery embedded in them, and gradually breaking loose as the lead wore away. Subsequently it appeared that unless there was some means of spreading water and the grinding powders evenly, rings were likely to be produced on the mirror, and the iron was consequently treated as follows :—

A number of pieces of wax, such as is used in making artificial flowers, were procured. The convex iron was laid out in squares of $\frac{3}{4}$ of an inch on the side, and each alternate one being touched with a thick alcoholic solution of Canada balsam, a piece of wax of that size was put over it. This was found after many trials to be the best method of protecting some squares, and yet leaving others in the most suitable condition to be attacked. A rim of wax, melted with Canada balsam, was raised around the edge of the iron, and a pint of aqua regia poured in. In a short time this corroded out the uncovered parts to a sufficient depth, leaving an appearance like a chess-board, except that the projecting squares did not touch at the adjoining angles.

. . . .

312 Draper corrected his mirrors by local retouching. He used two distinct modes of examination : first, observing with an eye-piece of power 20 the image of an illuminated pin-hole at the focus, and the cone of rays inside and outside that plane : second, Foucault's test.

He also seems to have used Foucault's method of controlling the retouching, namely, as Draper says, first to bring the mirror to a spherical surface, and then " by moving the luminous pin-hole toward the mirror, and correspondingly retracting the eye-piece or opaque screen, to carry the surface—avoiding aberration continually by polishing—through a series of ellipsoidal curvatures, advancing step by step towards the paraboloid of revolution."

. . . .

The machine, for polishing was a simplification of Lord Rosse's but had no lateral motion, and *the mirror was always uppermost while polishing*, and, not being counterpoised, escaped to as great an extent as possible from the effects of irregular pressure.

The operation of retouching

The mode of practising the retouches by Draper was as follows : Several disks of wood, varying from 8 inches to ½ an inch in diameter, are to be provided, and covered with pitch or rosin of the usual hardness, in squares on one side. On the other side a low cylindrical handle is to be fixed. The mirror having been fined with the usual succession of emeries before described, is laid face upward on several folds of blanket, arranged upon a circular table, screwed to an isolated post which permits the operator to move completely round it.

The large polisher is first moved over the surface in straight strokes upon every chord, and a moderate pressure was exerted. As soon as the mirror was at all brightened, perhaps in five minutes, the operation was suspended, and an examination at the centre of curvature made, the curve of the mirror being ascertained. If it is nearly spherical, as will be the case if the grinding has been conducted with care and irregular heating avoided, it is to be replaced on the blanket support, and the previous action kept up until a full polish is attained. This stage occupies three or four hours. Another examination should reveal the same appearances as the preceding. It is next necessary to lengthen the radius of curvature of the edge zones, or what is much better shorten that of the centre, so as to convert the section curve into a parabola. This is accomplished by straight strokes across every diameter of the face, at first with a 4-inch, then with a 6-inch, and finally with the 8-inch polisher. Examinations must, however, be made every five or ten minutes, to determine how much lateral departure from a direct diametrical stroke is necessary, to render the curve uniform out to the edge. Care was always taken to warm the polisher, either in front of a fire or over a spirit lamp, before using it.

313 The retouching described above was first carried out by hand. It was then found that two effects presented themselves which prevented the best results being obtained. In the first place the edge parts of such mirrors, for more than half an inch all around, bend backwards and become of too great focal length, and the rays from these parts cannot be united with the rest forming the image. In the second place, the surface, when critically examined by the second test, is found to have delicate wavy or fleecy appearance, not seen in machine polishing. Although the variations from the true curve implied by these latter greatly exaggerated imperfections are exceedingly small, and do not prevent a thermometer bulb in the sunshine appearing like a disk surrounded by rings of interference, yet they must divert some undulations from their proper direction, or else they would not be visible. All kinds of strokes were tried, straight, sweeping circular, hypocycloidal, etc., without effecting their removal. M. Foucault, who used a paper polisher, also encountered them. Eventually they were imputed to the unequal pressure of the hand, and in consequence, to overcome the two above mentioned faults of manual correction, a machine was constructed which performed similar movements.

314 Draper's paper contains full descriptions and illustrations of his machines, of the mounting of the mirrors for testing, and of his telescope and observatory. It is full of careful observations and ingenious ideas, and should be read carefully by anyone who intends to embark on the labour of making a large astronomical mirror.

The work of Sir Howard Grubb

315 Sir Howard Grubb's article (Grubb, 1886) is chiefly concerned with the working of achromatic object glasses, which he carried up to a diameter of $27\frac{1}{2}$ inches. No one before seems to have given so much attention to the effect of flexure both in the production of optical surfaces and in its effect on the performance of the finished article.

316 Speaking of the spherometer in its then customary form with three legs, he says—

I do not find the points satisfactory for regular work. They are apt to get injured or worn, and for ground surfaces are a little uncertain, as one or other of the feet may find its way into a deep pit. This particular spherometer has three feet, of about half an inch long, which are hardened steel knife-edges forming three portions of an entire circle. In using this it is laid on the surface to be measured, and the screw with micrometer head is turned till the point is felt to touch the surface of glass. This scale and head can then be read off. The screw in this instrument has fifty threads to the inch, and the head is divided into 100 parts, so that each division is equal to 1/5000th of an inch. With a little practice it is easy to get determinate measures to 1/10th of this, or 1/50,000th of an inch, and by adopting special precautions even more delicate measures can be taken, as far probably as 1/100,000th or 1/150,000th of an inch, which I have found to be practically the limit of accuracy of mechanical contact.

To give an idea of the delicacy of the instrument, I bring the screw firstly into contact with the glass. Now the screw is in good contact; but there is so much weight still on the three feet, that, if I attempt to turn it round, the friction on the feet oppose me, and it will not stir except I apply such force as will cause the whole instrument to slide bodily on the glass. Now, however, I raise the whole instrument, taking care that my hands touch none of the metal-work, and that the screw is not disturbed. I lay my hands for a moment on part of the glass where centre screw stood, and thus raise its temperature slightly, and on laying the spherometer back in the same place, you now see that it spins on the centre screw, showing how easily it detects what to it is a large lump, caused by expansion of the glass from the momentary contact of my hand.

Flexure

317 One of the greatest difficulties to be contended with in the polishing of large lenses is that of flexure during the process.

It may appear strange that in disks of glass of such considerable thickness as are used for objectives, any such difficulty should occur; but a simple experiment will demonstrate the ease with which such pieces of glass can be bent, even under such slight strain as their own weight.

We again take our spherometer and set it upon a polished surface of a disk of glass of about $7\frac{1}{2}$ inches diameter and $\frac{3}{4}$ inch thick. I set the micrometer head as in the former experiment to bear on the glass, but not sufficiently tight to allow the instrument to spin round. This has now been done while the glass, as you see, is supported on three blocks near its periphery. I now place one block under the centre disk and remove the others thus, and you see the instrument now spins round on centre screw.

It is thus evident that not only is this strong plate of glass bending under its own weight, but it is bending a quantity easily measurable by this instrument, which, as I shall presently show, is quite too coarse to measure such quantities as we have to deal with in figuring objectives.

After this experiment no surprise will be felt when I say that it is necessary to take very special precautions in the supporting of disks during the process of polishing to prevent danger of flexure ; of course if the disks are polished while in a state of flexure, the resulting surface will not be true when the cause of flexure is removed.

For small-sized lenses no very special precautions are necessary, but for all sizes over 4 inches in diameter I use the equilibrated levers devised by my father, and utilised for the first time on a large scale in supporting the 6-foot mirror of Lord Rosse's telescope. These have been elsewhere frequently described, but I have a small set here as an example.

I have also sometimes polished lenses while floating on mercury. This gives a very beautiful support, but it is not so convenient, as it is difficult to keep the disk sufficiently steady while the polishing operation is in progress without introducing other chances of strain.

So far I have spoken of strain or flexure during the process of working the surface ; but even if the surface be finished absolutely perfectly, it is evident from the experiment I showed you that very large lenses when placed in their cells must suffer considerable flexure from their own weight alone, as they cannot then be supported anywhere except round the edge.

To meet this I proposed many years ago to have the means of hermetically sealing the tube, and introducing air at slight pressure to form an elastic support for the objective, the pressure to be regulated by an automatic arrangement according to the altitude. My attention was directed to this matter very pointedly a few years ago from being obliged to use for the Vienna 27-inch objective a crown lens which was, according to ordinary rules, much too thin.

I had waited some years for this disk, and none thicker could be obtained at the time. This disk was very pure and homogeneous, but so thin that, if offered to me in the first instance, I would certainly have rejected it. Great care was taken to avoid flexure in the working, but to my great surprise, I found no difficulty whatever with it in this respect. This led me to investigate the matter, with the following curious results.

If we call f the flexure for any given thickness t, and f_1 the flexure for any other thickness t_1, then $f/f_1 = t^2/t_1^2$ for any given load or weight approximately. But as the weight increases directly as the thickness, the flexure of the disks due to their own weight, which is what we want to know, may be expressed as $f/f_1 = t/t_1$.

Let us now consider the effect of this flexure on the image. In any lens bent by its own weight, whatever part of its surface is made more or less convex or concave by the bending has a corresponding part bent in the opposite direction on the other surface, which tends to correct the error produced by the first surface. This is one reason why reflectors which have not this second correcting surface are so much more liable to show strain than refractors. If the lens were infinitely thin, moderate flexure would have no effect on the image. The effect increases directly as the thickness. If then the flexure, as I have shown, decreases directly as the thickness, and the effect of that flexure increases directly as the thickness, it is clear that the effect of flexure of any lens due to its own weight will be the same for all thicknesses; in other words no advantage is gained by additional thickness.

This has reference, of course, only to flexure of the lens in its cell after it has been duly perfected, and has nothing to do with the extra difficulty of supporting a thin lens during the grinding and polishing processes.

318 The precise form of Grubb's polishing machine is uncertain; the article in *Nature* is a report of a lecture delivered by him at the Royal Institution, and although he showed a lantern slide of his machine, no reproduction of it is included in the printed description, and no note of its exact nature is extant at the Royal Institute.

His reference to the machines of the Earl of Rosse and of Mr. Lassel and of the modification designed by his father which is described in Nichol's *Physical Science* make it clear that his own machine was built on much the same lines, except that he says—

> Like all machines, however, which give a series of strokes constantly recurring of the same amplitude, it is apt to polish in rings. It is impossible to obtain absolute homogeneity in the pitch patches, and if any one square be a shade harder than the general number, and that square ends its journey at each stroke at the same distance from the centre of speculum or glass, there will almost surely be a change of curvature in that zone. To avoid this I have made a

slight modification in the machine, which has increased its efficiency to a great extent.

The reader must be left to guess the nature of the modifications referred to, but may like to consider, by reference to §14, whether the machine there described—(the disc to be polished being underneath, and the local polisher on top)—might be expected to avoid this objectionable characteristic of polishing in rings.

I think that Grubb can scarcely have fully imbibed the substance of the paper of Draper, who—so far as written evidence goes—seems to have brought the art of polishing large surfaces to a higher level, except in appreciation of the effects of flexure. Grubb continues—

Figuring and testing

319 By the figuring process I mean the process of correcting local errors in the surfaces, and the bringing of the surfaces to that form, whatever it may be, which will cause the rays falling on any part to be refracted in the right direction. When an objective has undergone all the processes I have described, and many more which are not so important, and with which I have not had time to deal, and when the objective is centred and placed in its cell, it is, to look at, as perfect as it will ever be, but to look through and use as an objective it may be useless. The fact is that when an objective has gone through all the processes described, and is in appearance a finished instrument, I look upon it as about one-fourth finished. Three-fourths of the work has probably to be done yet. True, sometimes this is by no means the case, and I have had instances of objectives which were perfect on the first trial; but this is, I am sorry to say, the exception and not the rule.

This part of the process naturally divides itself into two distinct heads :—

(1) The detection and localization of faults—what may, in fact, be termed the diagnosis of the objective.

(2) The altering of the figures of the different surfaces to cure these faults. This may be called the remedial part.

It may be well here to try to convey some idea of the quantities we have to deal with, otherwise I may be misunderstood in talking of great and small errors.

I have before mentioned that it is possible to measure with the spherometer quantities not exceeding 1/50,000th of an inch or with special precautions much less even that that; but useful as this instrument is for giving us information as to the general curves of the surface, it is utterly useless in the figuring process; that is, an error which would be beyond the power of the spherometer to detect, would make all the difference between a good and a bad objective.

Take actual numbers and this will be evident. Take the case of a 27-inch objective of 34 feet focus; say there is an error in the centre of one surface of about 6 inches diameter, which causes the focus of that part to be 1/10th of an inch shorter than the rest.

For simplicity's sake, say that its surface is generally flat; the centre 6 inches of the surface therefore, instead of being flat, must be convex and over 1,000,000 inches radius. The versed sine of this curve, as measured by spherometer, would be only about 1/250,000th, 4 millionths of an inch, a quantity mechanically unmeasurable, in my opinion.

If that error was spread over 3 inches only instead of 6 inches, the versed sine would only be about 1/1,000,000th. Probably the effect on the image of this 3-inch portion of 1/10th inch shorter focus would not be appreciable on account of the slight vergency of the rays, but a similar error near the edge of objective certainly would be appreciable. Until, therefore, some means be devised of measuring with certainty quantities of 1 millionth of an inch and less, it is useless to hope for any help from mechanical measurement in this part of the process.

For concave surfaces, Foucault's test is useful. I shall not trespass on your time to explain this in detail, as it is described very fully in many works, in none better than in Dr. Draper's account of the working of his own reflecting telescope.

320 If an objective have but one single fault, its detection is easy; but it generally happens that there are many faults superposed, so to speak. There may be faults of achromatism, and faults of figure in one or all of the surfaces; faults of adjustment, and perhaps want of symmetry from some strain or flexure; and the skill of the artist is often severely taxed to distinguish one fault from the rest and localize it properly, particularly if, as is generally the case, there be also disturbances in the atmosphere itself, which mask the faults in the objective, and permit of their detection only by long and weary watching for favourable moments of observation.

It would be impossible in one or a dozen of such lectures as this to enumerate all the various devices that are practised for the localization of errors, but a few may be mentioned, some of which have never before been made public.

For detection of faults of symmetry, it is usual to revolve one lens on another and watch the image. In this way it can generally be ascertained whether it is in the flint or crown lens.

With some kinds of glass the curves necessary for satisfying the conditions of achromatism and spherical aberration are such that the crown becomes an equi-convex and the flint a nearly plano-concave of same radius on inside curve as either side of the crown. This form is a most convenient one for the localization of surface errors in this manner.

The lenses are first placed in juxtaposition and tested. Certain faults of figure are detected. Now calling the surfaces $A\ B\ C\ D$ in the order in which the rays pass through them, place them again together with Canada balsam or castor-oil between the surfaces B and C, forming what is called a cemented objective. If the fault be in either A or D surface, no improvement is seen; if in B or C, the fault will be much reduced or modified. Now reverse the crown lens, cementing surfaces A and C together. If same fault still shows, it must be in either B or D. If it does not show, it will be in either A or C. From these two experiments the fault can be localized.

It often happens that a slight error is suspected, but its amount is so slight that it appears problematical whether an alteration would really improve matters or not. Or the observer may not be able to make up his mind as to the exact position of the zone he suspects to be too high or too low, and he fears to go to work and perhaps do harm to an objective on which he has spent months of labour and which is almost perfect. In many such cases I have wished for some means by which I could temporarily alter the surface and see it so altered before actually proceeding to abrade and perhaps spoil it.

During my trials with the great objective of Vienna, I thought of a very simple expedient, which effects this without any chance at all of injuring the surface. If I suspect a certain zone of an objective is too low, and that the surface might be improved by lowering the rest of it, I simply pass my hand, which is always warmer than the glass, some six or eight times round that particular zone. The effect of this in raising the surface is immediately apparent, and is generally too much at first, but the observer at the eye end can then quietly watch the image as the effect goes off, and very often most useful information is thus obtained. The reverse operation, that of lowering any required part of the surface, is equally simple. I take a bottle of sulphuric ether and a camel's-hair brush, and pass the brush two or three times round the part to be lowered, blowing on it slightly at the same time; the effect is immediately perceived, and can always be overdone if required.

321 So far then for the diagnosis. Now for the remedy. When the fault has been localized, the lens is again put upon the machine and the polisher applied as before, the stroke of the machine and the size of the pitch patches being so arranged as to produce, or tend to produce, a slightly greater action on those parts that have been found to be too high (as before described while treating of the polishing processes).

The regulation of the stroke, eccentricity, speed, and general action of the machine, as well as the size and proportion of the pitch squares, and the duration of the period during which the action is to be continued, are all matters the correct determination of which depends upon the skill and experience of the operator, and concerning which it would be impossible to formulate any very definite rules. All thanks are due to the late Lord Rosse and Mr. Lassell, and also to Dr. de la Rue for having published all particulars of the process which they found capable of communication; but it is a notable fact that, as far as it is possible to ascertain, every one who has succeeded in this line has done so, not by following written or communicated instructions, but by striking out a new line for himself; and I think I am correct in saying that there is hardly to be found any case of a person attaining notable success in the art of figuring optical surfaces by rigidly following directions or instructions given or bequeathed by others.

There is one process of figuring which is said to be used with success among Continental workers. I refer to the method called the process of local touch. In this process those parts, and those parts only, which are found to be high, are acted upon by a small polisher.

This action is of course much more severe; and if only it were possible to know exactly what was required, it ought to be much quicker; but I have found it a very dangerous process. I have sometimes succeeded in removing a large lump or ring in this way (by large I mean 3 or 4 millionths of an inch), but I have also and much oftener succeeded in spoiling a surface by its use. I look upon the method of local touch as useful in removing gross quantities, but for the final perfecting of the surface I would not think of employing it.

In small-sized objectives the remedial process is the most troublesome, but in large-sized objectives the diagnosis becomes the more difficult, partly on account of the rare occurrence of a sufficiently steady atmosphere. In working at the Vienna objective it often happened when the figure was nearly perfect that it was dangerous to carry on the polishing process for more than ten minutes between each trial, and we had then sometimes a week to wait before the atmosphere was steady enough to allow of an observation sufficiently critical to determine whether that ten minutes' working had done harm or good. It must not be supposed either that the process is one in which improvement follows improvement step by step till all is finished. On the contrary, sometimes everything goes well for two or three weeks, and then from some unknown cause, a hard patch of pitch perhaps or sudden change of temperature, everything goes wrong. At each step, instead of improvement there is disimprovement, and in a few days the work of weeks or months perhaps is all undone. Truly anyone who attempts to figure an objective requires to have the gift of patience highly developed.

322 In view of the extraordinary difficulty in the diagnostic part of the process with large objectives, it is my intention to make provision which I hope may reduce the trouble in the working of the new 28-inch objective for the Royal Observatory Greenwich.

Two of the greatest difficulties we have to contend with are: (1) the want of homogeneity in the atmosphere, through which we have to look in trials of the objective, due to varying hydrometric and thermometric states of various portions; and (2) sudden changes of temperature in the polishing-room. The polisher must always be made of a hardness corresponding to the existing temperature. It takes about a day to form a polisher of large size, and if before the next day the temperature changes $10°$ or $15°$, as it often does, that polisher is useless, and a new one has to be made, and perhaps before it is completed another change of temperature occurs. To grapple with these two difficulties I propose to have the polishing-chamber under ground, and leading from it, a long tunnel formed of highly glazed sewer-pipes about 350 feet long, at the end of which is placed an artificial star illuminated by electric light; on the other side of the polishing-chamber is a shorter tunnel, forming the tube of the telescope, terminating in a small chamber for eye-pieces and observer. About half-way in the long tunnel there will be a branch pipe connected to the air-shaft of the fan, which is used regularly for blowing the blacksmith's fire, and through this, when desired, a current of air can be sent to " wash it out " and mix up all currents of varying temperature and density. It may be found necessary even to keep this going during observations.

By this arrangement I hope to be able to have trials whenever required, instead of having to wait hours and days for a favourable moment.

Figuring of plane mirrors

323 There is a general idea that the working of a plane mirror or one of very long radius is a more difficult operation than those of more ordinary radii. This is not exactly the case. There is no greater difficulty in figuring a low curve than a deep one, but the difficulty in the case of absolutely plane mirrors consists simply in the fact that in their figuring there is one additional condition to be fulfilled, viz. that the general radius of curvature must be made accurate within very narrow limits. In figuring a plane mirror to use, for instance, in front of even a small objective, say 4-inch aperture, an error in radius which would cause a difference of focus of 1/100th of an inch would seriously injure the performance. This would be about equivalent to saying that the radius of curvature of the mirror was about 8 miles, the versed sine of which, with the 6-inch spherometer, would be about 1/50,000th of an inch. Now what I mean to convey is this : that it would be just as difficult to figure a convex or concave lens of moderate curvature as a flat lens of the same size if it were necessary to keep the radius accurate to that same limit, i.e. one-tenth of a division of this spherometer. Of the speculum metal for mirrors Grubb says—

The composition of metallic mirrors of the present day differs very little from that used by Sir Isaac Newton. Many and different alloys have been suggested, some including silver or nickel or arsenic ; but there is little doubt that the best alloy, taking all things into account, is made with 4 atoms of copper, and 1 of tin, which gives the following proportions by weight : copper, 252, tin, 117·8.

The work of Ritchey

324 Ritchey's paper (Ritchey, 1904) describes the methods employed by him at the Yerkes Observatory in making and testing spherical, plane, paraboloidal, and (convex) hyperboloidal mirrors up to 5 feet in diameter. Ritchey's own wording is used throughout except for the substitution of oblique for direct narration.

He used disks made at the glass-works of St. Gobain, near Paris, of sizes from 8 inches in diameter and 1½ inches thick, to the great one which is 5 feet in diameter and 8 inches thick, and which weighs a ton, in every case it was specified that great care should be given to thorough stirring and annealing.

For mirrors of 24 and 30 inches diameter, a thickness of one-sixth of the diameter was preferred. In the cases of the large paraboloidal mirror of a reflecting telescope, and the large plane mirror of a coelostat or heliostat, which should always be supported at the back to prevent flexure, the thickness should be one-seventh of the diameter.

All mirrors should be polished (not figured) and silvered on the back as well as on the face, in order that both sides shall be similarly affected by temperature changes when in use in the telescope*; for the same reason the method of supporting the large mirror at the back, in its cell, should be such that the back is as fully exposed to the air as possible.

. . . .

325 The grinding and polishing machines were similar in principle to Dr Draper's, but were more elaborate. The machine used in making the 5-foot mirror is shown in Figs. 241 and 242.

Fig. 241—Ritchey's machine for making a five-foot mirror

The turntable upon which the glass rests consists of a vertical shaft five inches in diameter, carrying at its upper end a heavy triangular casting, upon which, in turn, is supported the circular plate upon which the glass lies. This plate is of cast-iron, weighs 1,800 pounds, is 61 inches in diameter, is heavily ribbed on its lower surface, and is connected to its supporting triangle by means of three large levelling screws. The surface of the large plate was turned and then ground approximately flat; two thicknesses of Brussels carpet are

---

* For protecting an optically worked surface of a mirror, during the polishing of the other surface, a methacrylate adhesive can be used. This can easily be dissolved off by trichlorethylene (F.T.).

laid upon this, and the glass, with its lower surface previously ground flat, rests upon the carpet.

Three adjustable iron arcs at the edge of the glass serve for centering the latter upon the turntable, and prevent it from slipping laterally.

The entire turntable, with the heavy frame of wood and metal which supports it, can be turned through 90° about a horizontal axis, thus enabling the optician to turn the glass quickly from the horizontal

Fig. 242—Ritchey's machine for making a five-foot mirror
Method of turning the mirror into the vertical position for testing

position which it occupies during grinding and polishing, to a vertical position for testing. This is shown in Fig. 242.

The large pulley below the turntable slowly rotates it on its vertical axis by means of belting from the main vertical crank-shaft. At the upper end of this shaft is the large crank, with adjustable throw, which moves the large main arm to which the grinding and polishing tools are connected, and by means of which they are moved about upon the glass. This arm is a square tube of oak wood, and is strong enough to carry a counterpoising lever and the weight of any of the grinding tools, when counterpoised. This main arm also carries the system of pulleys and belts by which the slow rotation of the grinding and polishing tools is controlled.

. . . .

A long transverse slide on the secondary arm allows the grinding and polishing tools to be placed so as to act on any desired zone of the glass, from the centre to the edge ; and this setting can be changed as desired while the machine is running. The secondary crank, which turns at the same speed as the large one which drives the main arm, enables the optician to change as desired the width of the (approximately) elliptical stroke or path of the tool with reference to the length of this stroke ; this change is especially desirable when figuring the glass ; it is, of course, impossible when only one driving-stroke is used.

. . . .

To recapitulate briefly : this method of connecting the grinding and polishing tools allows them to be controlled in all of the following ways simultaneously : (1) the stroke of the tool is given by the motion of the main arm ; (2) the slow rotation of the tool is rigorously controlled by the belting above ; (3) the tool is allowed to rock or tip freely by means of the universal coupling, in order that it may follow the curvature of the glass ; (4) the tool rises and falls freely by means of the sliding of the $1\frac{3}{8}$-inch vertical shaft in its bearings, in order that it may follow the curvature of the glass ; (5) the tool is counterpoised by means of the lever on the main arm, through the medium of the same vertical shaft and universal coupling.

In Fig. 241 is shown the large lever by which the 5-foot glass, which weighs a ton, is lifted on and off the machine, and by means of which, also, the large grinding tools are handled. One of the full-size grinding tools, weighing 1,000 pounds, is shown suspended by the lever. The arrangements are such that the optician alone can do all parts of the work.

. . . .

Grinding tools

326 While grinding tools of glass were used in much of his earlier work, and are still used for small work, he later used cast-iron grinding tools for all large work. These are cast very heavy, with ribs on the back. For large work iron tools are cheaper than glass ones ; they are more easily prepared ; they are more easily and safely counterpoised, which is always necessary in the fine-grinding of large work ; and they produce on the glass a fine-ground surface fully as smooth and perfect as can be obtained with glass tools.

An important question concerns the size of grinding tools—should they be of the same diameter as the mirror? For mirrors up to 24 or 30 inches in diameter full-size tools are generally used. For concave mirrors larger than 30 inches in diameter Ritchey used grinding tools whose diameter is slightly more than half that of the glass, i.e., a 16-inch tool for a 30-inch glass ; a 32-inch tool for a 60-inch glass. Full-size tools are, of course, much more expensive and difficult to make ; they are many times heavier than half-size tools of equal stiffness ; and they require a much stronger grinding machine to counterpoise them properly ; grinding can be done with them, however, more quickly than with the smaller tools. Half-size tools are economical and are quickly prepared ; they are easily counterpoised ; and a

Fig. 243—Ritchey's spherometer

much greater variety of stroke can be used with them, so that with a well-designed grinding machine he found it easier to produce fine-ground surfaces, entirely free from zones, with half-size than with full-size tools. If temperature conditions and uniform rotation of the glass are carefully attended to, the surface of revolution produced by the smaller tools is fully as perfect as that given by the larger ones.

· · · · · · · · ·

327 The curvature of the tools and of the glass is measured by means of a large spherometer. The spherometer (Fig. 243) is of the usual three-leg form ; the legs terminate in knife-edges, the lines of which are parts of the circumference of a 10 inch circle. The screw is of ½-millimetre pitch, and the head, which is 4 inches in diameter, is graduated to 400 divisions. On fine-ground surfaces settings can be made to one-half or one-third of a division, corresponding to a depth of 1/40,000th or 1/60,000th of an inch, approximately.

. . . .

Polishing tools

328 The tools adopted after trying various kinds consisted of a wooden disk or basis covered on one side with squares of rosin faced

Fig. 244—Ritchey's arrangement for grinding the edge of the five-foot mirror

with a thin layer of beeswax. The wooden disk may be replaced, in the case of small polishing tools up to 12 or 15 inches diameter, by a ribbed cast-iron plate so designed as to be extremely light and rigid ; the bases of larger tools may be made of cast aluminium alloy ; such a basis for a 30-inch polishing tool weighs about sixty pounds. It is possible that a metal basis possesses an advantage over a wooden one in that its surface is less yielding. Tools properly constructed of wood, however, are light and extremely rigid, are easily made, and are economical in cost. Their proper construction is a matter of importance, and is fully described in Ritchey's paper.

. . . .

Grinding edge of glass—rounding of corners

329 In order that an efficient edge-support may be given to the glass it is desirable that the edge of the latter be ground truly circular and square with the face, and Ritchey's paper gives a clear and complete description of the way this is accomplished, means being provided in the machine shown in Fig. 244.

. . . .

330 It is interesting to compare the pressure per square inch used in grinding by Ritchey with those given in §80.

His full-size iron tools for a 24-inch mirror weigh about 150 pounds, or 1/3rd pound for each square inch of area. This weight, or even 1/2 pound to the square inch, is not objectionable with emeries down to 5-minute or 10-minute washed; but when this weight is allowed with finer emeries, scratches are liable to occur. The pressure on the glass is therefore decreased, by counterpoising the tool, to approximately 1/5th pound to the square inch for 12 to 20 minute emeries, 1/8th pound per square inch for 30 to 60 minute emeries, and about 1/12th pound per square inch for 120 to 240 minute emeries. *This rule is followed, approximately, in all fine-grinding, whether of back or face.* This obviates, to a great extent, the danger of scratches in grinding, provided that thorough cleanliness is practiced in the preparation and use of fine emeries.

In fine-grinding a 24-inch glass, the 2 minute and 5 minute emeries are used for three-quarters of an hour each; the 12 and 30 minute emeries for one hour each, and the 60, 120, and 240 minute emeries for one and one-half hours each. The fine-ground surface resulting takes a full polish very readily.

. . . .

Polishing

331 The preparation of polishing tools has already been described. The polishing rouge which Ritchey used was of the quality which is used in large quantities commercially in polishing plate-glass. This grade of rouge is not expensive but contains hard, sharp particles which may cause scratches. It must therefore be thoroughly washed; particulars of the process are given in the paper.

. . . .

The pressure per square inch used by Ritchey in polishing is of interest. His 24-inch polishing tool, with its wooden basis, weighs about 25 pounds; this is not heavy enough for the best action in polishing; so about 50 per cent additional weight is put on in the form of 12 lead blocks which are distributed uniformly and screwed to the back of the tool. This gives a weight of about 1/12th pound for each square inch of area, which is found to work well for all large tools. For tools 18 inches or less in diameter somewhat greater pressure per square inch of area may be used. A 36-inch tool, with wooden basis $3\frac{3}{4}$ or 4 inches thick, weighs 75 or 80 pounds, and needs no additional weighting.

. . . .

Testing and figuring spherical and plane mirrors
332 Ritchey, like Draper and Grubb, used Foucault's Test which was originally described in Vol. V of the *Annals of the Paris Observatory*.

. . . .

The making of large plane mirrors of fine figure is usually regarded as much more difficult than that of large concave mirrors. The difficulty has been, in the past, largely one of testing. With a satisfactory method of testing the large plane surface *as a whole*, in a rigorous and direct manner, the problem is greatly simplified. So far as Ritchey was aware, no such test had hitherto been fully developed. In *Monthly Notices*, Vol. 48, p. 105, Mr Common suggests, very briefly, the testing of plane mirrors in combination with a finished spherical mirror, and gives a diagram in illustration. This method was developed and used by Ritchey in testing plane mirrors up to 30 inches in diameter. When this test is used, the difficulty of making a 24-inch plane mirror which shall not deviate from perfect flatness by an amount greater than 1/500,000 inch is neither greater nor less than that of making a good spherical mirror of 2 feet aperture and 50 feet radius of curvature, when it is required that the radius of curvature shall not differ from 50 feet by a quantity greater than 1/100 inch.

. . . .

Testing and figuring paraboloidal mirrors
333 The work of changing a spherical mirror to a paraboloidal was accomplished entirely by the use of polishing tools, by shortening the radii of curvature of the inner zones, instead of by increasing or lengthening those of the outer zones. The methods of effecting this change of curvature will be described after the methods of testing a paraboloid have been discussed.

Such testing can be done at the centre of curvature, by determining there the foci or the radii of curvature of successive zones of the mirror it may be done at the focus of the paraboloid, by the aid of a finished mirror which should be at least as large as the paraboloidal one ; and it may be done directly on a star. The first two methods named have the very great advantage that they may be conducted without interruption, under the practically perfect atmospheric and temperature conditions of the optical laboratory.

Testing a paraboloid at the centre of curvature

A knowledge of the properties of the parabola enables the optician to compute the positions of the centres of curvature of successive, definite, narrow zones of the mirror, and the surface must be so figured that the radius of curvature of each zone agrees with the computed value. In testing, each zone in succession is exposed by means of a suitable diaphragm, all of the rest of the surface being covered. In practice, two different formulae are used, depending upon the position of the illuminated pinhole.

. . . .

The diaphragms which Ritchey used do not expose entire zones, but only pairs of arcs on the right and left sides of the mirror. This ends the account of Ritchey's work contained in the reference cited.

. . . .

The method of testing a paraboloid at its focus was briefly described by Ritchey in the *Astrophysical Journal*, November, 1901. It is more simple, direct, and rigorous than the test at the centre of curvature. A well-figured plane mirror, which should not be smaller than the paraboloidal one, is necessary in order that the testing may be done in the optical laboratory.

. . . .

Changing a spherical surface to a paraboloid

This is accomplished by shortening the radii of curvature of all of the inner zones of the surface, leaving the outermost zone unchanged. There are two distinct methods of accomplishing this : (1) by the use of full-size polishing tools, the surfaces of which are cut away in such a manner as to give a large excess of polishing surface near the central parts of the tool ; (2) by the use of small polishing or figuring tools worked chiefly upon the central parts of the mirror, and less and less upon the zones towards the edge.

(1) *Parabolizing with full-size tools.* The rosin surface can be trimmed in a variety of ways to give a great excess of action on the central parts of the mirror. A form of tool used for this purpose was very similar to that shown in Fig. 146 (Chapter 10). The form of the

polisher areas can be altered as desired, and thus the amount of action on any zone can be in some measure controlled. Length of stroke and amount of side-throw are also very important factors in controlling the figure of the mirror. Tools of this kind serve admirably in parabolizing mirrors up to 36 or 40 inches in diameter, when the angular aperture is not very great.

(2) *Parabolizing with one-third-size and smaller tools.* In the case of very large mirrors, when full-size tools are almost unmanageably heavy, and in the case of mirrors of great angular aperture, in which the departure from a spherical surface is great and is effected with difficulty with full-size tools, one-third-size and smaller figuring tools may be used. The machine should invariably be employed in this work, the transverse slide being used to place the tool in succession upon the various zones. In order to preserve the surface of revolution the setting of the transverse slide should be changed only at the end of one or more complete revolutions of the glass.

. . . .

The mirror should be tested very often, and the utmost care taken to keep the apparent curve of the surface, as seen with the knife-edge test, a smooth one, i.e. free from small zonal irregularities, at all stages of the parabolizing; this is not extremely difficult when the optician has become experienced in the use of the transverse slide. This ends the account of Ritchey's work.

Hartmann test

334 The Hartmann test has already been described in §217 but since it has been chiefly used for the purpose for which Prof. Hartmann designed it, namely the testing of large astronomical telescopes, supplementary details are given here.

An account was published by Hartmann (1908) of the use of his method of test for the correction of the 80-cm objective of the Potsdam objective. The following is an abstract from an account of Hartmann's paper given by Lehmann (1909).

Most of the article is concerned with the use of Professor Hartmann's extra-focal method of measurement in testing the 80-cm Potsdam objective, then the largest outside America. The method has been fully described by Hartmann and others and its utility widely accepted. (A number of references are cited in the abstract.)

It should be remembered that the Hartmann test is the only one which determines the exact course of the rays and so yields a quantitative statement of the true residual aberrations of optical systems.

It was concluded from the tests that the zonal aberrations could be reduced by local retouching; in such local polishing only those zones are worked which have too great a curvature, the work being usefully carried out on the last surface.

The following interesting table compares the results obtained with the Hartmann method when applied to various large refractors—

Observatory	Diameter cm	Optician	T
Yerkes	102	Clark	0·16
Pulkowa	76	Clark	0·18
Ottawa	38	Brashear	0·36
Potsdam	80	Steinheil (III)	0·34
Potsdam	30	Steinheil (III)	0·44
Vienna	70	Grubb	0·46
Vienna	32	Clark	0·84
Potsdam	23	Grubb	0·93
Leipzig	30	Reinfelder & Hertel	0·95
Potsdam	50	Steinheil	1·08
Potsdam	32·5	Steinheil	1·30
Potsdam	80	Steinheil (1)	3·04

The paper in which Lehmann defines his *technische-Konstante*, T, (Lehmann 1902) should be read by anyone intending to use the Hartmann method.

To obtain a single number which would express the quality of the correction arrived at he divides the size of the mean diameter of the focus due to the mechanico-spherical aberration for monochromatic light by the focal length of the objective. This number he calls the technische-Konstante of the objective. The implication of the word " mean " in the phrase " mean diameter of the focus " is not too precise. It is arrived at by taking the full diameter and reducing it to make allowance for the fact that the intensity of illumination is greatest in the middle. Obviously to give this quantitative meaning would entail taking into account the gamma of the photographic plate and defining what weight to give to the respective intensities of the various zones of the image.

Without this elaboration, however, it does seem from the examples given to afford a simple and sufficiently discriminating means of comparing the degree of correction of the objective.

As Kingslake (1928) points out the assumption that is made in the ordinary method of performing the Hartmann test that all the rays cross the optical axis is only true if both surfaces and homogeneity of the glass are perfect. " Technical " aberrations which arise from the manufacture of the glass or in the working or figuring of the surfaces may result in rays which do not do so. This is especially the case in the large astronomical objectives and mirrors to which the Hartmann test is most generally applied. In the paper cited a description is given in which the deviation of each ray from the ideal path is determined. The paper and the discussion stress the limitation of the Hartmann test to the actual parts of the lens or mirror which happen to fall under the holes in the diagram.

This limitation was appreciated by Prof. Hartmann. The measurements on the 40-in. (102 cm) objective at the Yerkes Observatory (Philip Fox, 1908) were made at Prof. Hartmann's suggestion and the results of the measurements communicated to him. He said in reply " So far as one can see from the few points observed the objective is excellent . . . " Hartmann, however, considered the investigation incomplete because so few zones and points were included and therefore advocated the desirability of applying to the objective, in addition to his own test, the modification of the Foucault test which he had recently applied to the 80-cm objective at Potsdam. The improvement consisted merely in replacing the eye by a camera the objective of which throws upon a photographic plate a sharp image of the objective to be investigated with the knife edge in a suitable position.

This application of the Foucault test revealed an astonishing amount of detail in the surface which, of course, was not revealed by the Hartmann test. It was even possible to see traces of the epicycloidal motion of the polishing tool. The paper cited above is immediately followed by one by Prof. Hartmann in which, speaking of the Foucault test, he says, " This method of Foucault has been introduced into many optical works and constitutes the principal means of the more accurate testing of objectives and mirrors. The method has, however, hitherto never been employed for testing on the sky an already mounted astronomical refractor."

Other tests

It seems to the present writer that the methods described in the foregoing pages and in §335 leave nothing to be desired as regards control of flexure, general handling and polishing of large objectives and mirrors, but he is strongly of the opinion that, in future work of the kind help should be sought from additional methods of testing and correcting the surfaces of the finished mirrors or object glasses. There is, for example, the method of deposition in vacuo described by Strong and Gaviola (§192). Interferometers also should be brought into play, as for instance the Shearing Interferometer of Bates (Bates, 1947), described in §252.1.

The Hartmann test will doubtless remain of service, notwithstanding the incompleteness pointed out by Kingslake.

Notes on the 200-inch mirror at Mount Palomar
by Mr J. V. THOMSON of Messrs Cox, Hargreaves & Thomson Ltd

Dr J. A. Anderson who was in charge of the grinding and figuring of the mirror would have been the logical person, in fact the only person, who could tell with full authority the whole story of the work on this instrument. Dr Anderson felt, however, that he could not undertake to provide me with a description. I am very grateful, therefore, for the following account written by Mr J. V. Thomson, and for the photographs, also supplied by him, from which the illustrations were made.

Fig. 245—Honeycomb structure of the 200-inch mirror

335 The illustrations, Figs. 245 to 252, make the construction and handling clear. Fig. 245 shows the honeycomb structure of the 200-inch mirror referred to in §305. Fig. 246 shows the mirror in the testing position tilted 11° from the vertical. A closer view, Fig. 247, gives a better idea of its dimensions. In Fig. 246 in the foreground somewhat to the left is the Foucault testing apparatus set up at the focus 55 ft from the mirror. On the left of the mirror right at the back of the shop stands the 200-inch tool used during the latter stages of the grinding and the early stages of the polishing. The square facets of the reticulation, of which there are 1938, can be seen. Fig. 248 shows a 68-inch diameter tool after a polishing run. A few of the 36 supports are shown in Fig. 249. Fig. 250

Fig. 246—The 200-in. mirror in the testing position

shows the reciprocating machine attached to the spindle with a 12-inch tool in operation; the normal speed of rotation of the mirror was one revolution every 159 seconds. The usual method of pressing the 50-inch polishing tool is shown in Fig. 251. Finally, Fig. 252 shows the mirror ready to be taken off the turntable.

Work on the 200-inch mirror began in 1936 in the specially built optical shop at the California Institute of Technology, Pasadena.

After the front surface of the 20 ton disc had been ground flat the disc was turned over and the back surface dealt with in the same manner, the pockets and cavities being filled with wooden plugs and plaster of paris to prevent chipping of the ribs. The 36 pockets were then accurately ground to receive the steel sleeves of the support mechanism.

Fig. 247—Close-up view of the 200-in mirror

The work of grinding out the curvature of the mirror to a depth of $3\frac{3}{4}$ inches at the centre occupied three months, during which operation 10 tons of coarse abrasive grit were used, removing a little over 5 tons of glass. A grinding tool 100 inches in diameter was used for this part of the work. When the approximate depth had been reached a full-sized 200 inch tool was constructed for grinding the surface spherical and establishing a true surface of revolution. This tool, of welded sheet steel construction, was surfaced with about 1950 Pyrex glass blocks and the abrasive grits were fed to the surface of the mirror through funnels in the upper surface of the hollow tool.

After the last grade of fine abrasive had left the surface ready for polishing, $3\frac{1}{2}$ months were spent in cleaning up the optical shop and removing all trace of grit from the grinding machine, in order to reduce

Fig. 248—The 68-in diameter polisher after a run

the danger of scratching the optical surface during the polishing and figuring.

Altogether 31 tons of abrasives, including rouge, were used in the work on the mirror, which, first to last, took 180,000 man-hours of labour.

The full-sized tool was used during most of the initial polishing and for this purpose squares of a special pitch mixture were cemented to the Pyrex glass blocks, each square being cut and channelled into smaller squares, so that the 200 inch polisher comprised a mosaic of some 8,000 pitch facets which distributed the rouge equally over the whole surface of the mirror. The pitch used was a mixture of rosin, paraffin wax and cylinder oil, later replaced by pine tar.* The temperature in the shop ranged from 65 to 85 degrees over the year,

Fig. 249—A few of the thirty-six supports of the 200-in. mirror

making it necessary to vary the proportion of the ingredients according to the existing temperature. Air conditioning and temperature control plant had been installed but its action disturbed the air more than enough to make optical testing impossible, and its use was

* " Pine tar," or " pine tar oil," is a treacly black liquid similar to gas tar, but with a pungent odour resembling that of hot Swedish wood pitch. It can be supplied by Dussek Bros, Crayford, Kent.

therefore abandoned. After the pitch had been melted it was poured into specially constructed wooden trays. When cool the trays were taken apart and the pitch cut into uniform squares ready for cementing to the under surface of the polishing tools. After a polisher had been made up in this way it was first suspended over gas stoves to soften the pitch facets, and then lowered on to a thick cream of rouge and

Fig. 250—Reciprocating machine with 12-in. tool in operation

water on the mirror's surface, additional weight being then applied to press all the facets into perfect contact with the optical surface. The tool was then lifted off and the facets trimmed and bevelled. The smallest polisher used on the 200 inch mirror measured 8 inches in diameter.

It had originally been planned to test the 200 inch with parallel light, employing a 120 inch plane mirror, offset laterally, for this purpose. Although this mirror had been fine ground ready for polishing it was decided to abandon the original plan for certain reasons. It was estimated that far less time would be occupied in parabolizing the 200 inch and testing it by other methods, than would be the case if it had first to be polished and figured truly spherical in order to test the 120 inch flat mirror, during the figuring of which no

Fig. 251—Pressing the 50-in. polisher into contact with the mirror

further work could be done on the 200 inch itself. The work of parabolizing the mirror was therefore begun as soon as it had received its initial polish, and as the surface at the centre had to be deepened by the relatively large amount of 1/200 inch it was accomplished by alternately grinding and polishing until a stage was reached when the final figuring could be done by polishing alone.

The mirror, tipped up on its turntable into a vertical position, was tested at its radius of curvature 110 feet away. A Foucault screen, dividing each radius of a diameter into 13 six-inch apertures, was wheeled in front of the mirror, and the measured zonal aberrations compared with the calculated parabolic aberrations. From these differences a graph would be drawn indicating in wavelengths the

Fig. 252—The 200-in. mirror ready to be taken off the turntable

differences in height between the actual surface and the required parabolic surface, showing clearly where the next spell of figuring should be done.

For the very final figuring the Strong and Gaviola system was used (§192).

The 160 ton grinding and polishing machine consisted of three main parts; the horizontal beam, the spindle section and the turntable. The heavy horizontal beam, called the bridge, moved back and forth on wheels above the surface of the mirror. This bridge supported a carriage, also on wheels, which held the spindle, at the lower end of which were attached the grinding and polishing tools. The directions of motion of the bridge and of the spindle being at right-angles to each other, a variety of different strokes could be obtained. The turntable was designed to play the dual role of turntable and mirror cell, the 36 support mechanisms being secured in their pockets and supporting the mirror during the optical work. When the mirror was finished the turntable, carrying the mirror, was detached from the machine and transported as a complete unit to the observatory on Palomar Mountain where it was bolted to the base of the tube after the mirror had been aluminized.

A further period of figuring, confined to the outermost 18 inches of the mirror, was carried out at the observatory during 1949, since when the performance of the telescope has exceeded the hopes of all who were concerned with its design and construction.

At the time of writing (July 1950) the most recent report concerning the performance of the telescope (Bowen 1950) describes extensive tests carried out to disentangle the complex effects of temperature changes and flexure due to gravity and to correct them. As a result a system of fans has been installed to accelerate adjustment to temperature changes of the central portion combined with extra insulation to retard the effect of temperature changes around the rim.

The criterion of performance was the proportion of light from a point source which was concentrated at the focus within a circle of specified diameter. On the three best nights the average figure was 84 per cent within a circle of 0·05 mm diameter.

The importance of minor irregularities of optical surfaces

336 We have seen in §227.1 how, in constructing his coronograph, M. Lyot found it necessary to test for surface irregularities which until then were considered negligibly small even for optical work of high quality. The lenses for these were polished by M. Couder, as described in §337. Pursuing this subject, Texereau (1949) studied and analysed such faults arising from various polishing techniques. The remainder of this section is abstracted from Texereau's paper.

1. The search for a criterion of "definition" of objectives

To characterize by a single figure the quality of an objective is a convenient and serviceable simplification, but in many cases it is insufficient. One knows the discussions, interpretations and criticisms centred round the famous criterion of Hartmann which is supposed to sum up the importance of transverse aberrations.

The celebrated quarter-wavelength rule of Lord Rayleigh has been much more useful because it is founded on a physical property and not on a simple and more or less artificial mathematical treatment. It may be formulated thus—

> The wavefront emerging from a good objective is contained completely between two spheres whose radii differ by a maximum of a quarter wavelength.

Without wishing to detract from the practical comprehensiveness of this rule, which is often sufficient, one must issue a warning against the widely relied upon supposition that it is a criterion of quality so unsurpassable that the efforts of the optician to do better are childish ; one says " qu'il a dépassé le but " (*he has overshot the mark*). Occasionally, some opticians have maintained that approximately half a wavelength is good enough.

Actually, an optical surface (and consequently the emergent wavefront) of an objective is a very complex physical entity, and there can be no question of predicting the full extent of the influences of its imperfections on the image in terms of a single parameter.

Before attempting a more detailed analysis some facts may be called to mind which may reduce our enthusiasm for the quarter wavelength limit.

First of all, there is a factor which does not come within the optician's province, but may well spoil the result of his work—atmospheric disturbances. Following upon a programme of investigations into the quality of imagery in different places, A. Danjon (*Réunions Inst. d'Optique*, 1933, 2nd, 20) has formulated an important law : To the faults of an objective must be added those atmospheric disturbances which change the incident wavefront (and also the emergent wavefront if there are thermal heterogeneities in the air within the tube of the instrument). Naturally, if the objective has faults of its own approaching the quarter wavelength limit, the total effect frequently exceeds this limit. Such an objective is therefore much more sensitive to atmospheric agitation.

The smallest perceptible alteration of the focal diffraction pattern of a star, which gave Rayleigh the basis for formulating his rule, is not the most sensitive test for determining the faults of an objective and their effect upon the images. The effect on the diffraction patterns near to the focal plane becomes apparent for a phase difference of $\lambda/10$.

Astronomical observers often speak of small contrasts as being better or less well seen with different objectives of comparable powers. This effect goes so far that, in special circumstances, observers can recognise the objectives used merely by inspecting the image of Mars (see for example *Annales des Observationes Jarry—Desloges*, 1907, p. 115 ; 1913, p. 296).

M. Françon has shown convincingly that for the faintest perceptible contrasts ($\lambda = 0.03$), the efficiency of an instrument corrected to $1/4\lambda$ falls to 0·62, while it remains at 0·92 if the residual imperfections do not exceed $1/16\lambda$.

The most satisfactory criterion is, without doubt, that of A. Danjon and A. Couder (*Lunettes et telescopes*, p. 522), which combines the rule of Lord Rayleigh with the requirement that the transverse aberrations shall be smaller than the radius of the diffraction pattern.

A. Couder has for a long time, been observing and describing (*l'A.* **50**, Feb. 1936, p. 65) particular defects of the wavefront, often very

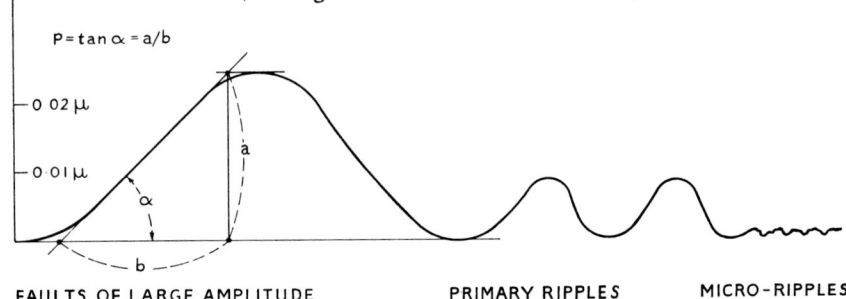

Fig. 253—Wavefront imperfections

small, but nevertheless sometimes serious because of their slope or periodicity. In 1944 he summarised many fruitful ideas in a note (*Cahiers de Physique*, **26**, 27) which has directly inspired the present work.

F. Zernicke (1935) in describing the method of phase-contrast, drew attention to the " phase-mosaic," which is a wavefront disturbed by small periodic undulations (*M.N.*, 1934, p. 377).

The study of the diffracted light at a large distance from the central image has more recently led B. Lyot (*Compter Rendues*, **222**, 765) to evolve a phase-contrast method of another form, that reveals a whole class of defects which were comparatively unknown before, but which appear to deserve attention.

2. Classification of the imperfections of a wavefront

It is not intended to consider here the faults caused by refraction but only those of the optical surfaces themselves.

Let us call a the height of a fault that represents the phase-difference, and b the width of the element of the inclined wave; hence $p = a/b$ is the corresponding slope.

With these three parameters we can lay down a satisfactory classification of the imperfections. This is more important to the optician than to the user, because every form of imperfection generally originates in the method of working employed. The conditions which make elimination of the faults possible are not always compatible and one has to decide on the compromise to be aimed at for each particular case.

We shall at once distinguish between two completely different and clearly defined classes of faults. The " goose-flesh " (or " orange peel ") effect on cloth polished surfaces would form an intermediate class, but it would be of no interest for the optics of astronomy.

A. *Surface faults*: a and b are usually small and of the same order, the slope p is therefore considerable and of the order of unity.

B. *Faults of shape*: In the case of finished surfaces a is restricted to a maximum of $\lambda/4$, but may be perceptible when as small as 1/10,000 of a wavelength; b also may vary within wide limits, from the radius of the surface down to 1/10 mm. The slope of these imperfections is always very small, in a good surface it is of the order of 10^{-6} to 10^{-5}.

The amount of light scattered can be estimated quantitatively by the use of simplifying assumptions which are more or less unsatisfactory. It will be enough to mention the general effect for the main categories, with the reminder that a direct investigation into the energy distribution away from the main image is always advisable.

A. *Surfacing faults*
   (i) In the case of spots of " grey " a and b are of the order of one micron. The stray light is diffracted within a solid angle of considerable extent, of the order of several tens of degrees. There are also micro-spots which are only visible with a powerful ultra-microscope, and which diffract very little light away from the main image.
   (ii) In scratches, a and b are of the order of 1/10 mm. In order to realise their seriousness, other parameters are necessary. If a scratch is straight and sufficiently long, a diffraction flare will be visible with a sufficiently bright star, but if the scratch is curved, the effect will be scattered and mostly not noticeable.
   (iii) With sleeks, a and b are of the order of one micron, and the length of the order of one centimetre. These very fine

marks may accur sporadically (only a few over the whole surface), and then the effect will mostly be negligible, or they may be present in great numbers, in which case the effect will be similar to that of greyness. One could also mention micro-sleeks, which are invisible by ordinary means, and which may occur in very great numbers on a surface polished with little pressure but whose optical effect is in most cases negligible.

B. *Faults of shape*

(i) Faults of large amplitude in which a may be up to $\lambda/4$ and b may be of the order of several centimetres, the slope being 10^{-5} to 10^{-6} at most, are the classical faults, the most serious ones, because of their direct bearing on the nucleus of the diffraction pattern in its immediate neighbourhood. It is also necessary to distinguish between zonal and astigmatic defects. A test report including the analysis of these defects is given with every well-made astronomical objective.

(ii) Faults of medium amplitude comprise those in which a is several tenths of a wavelength and b of the order of a centimetre. Local faults, such as superficial veins in the glass (the slope can exceed 10^{-5} even in a good mirror) or traces left by local polishing, are generally of negligible effect.

(iii) The result of more or less periodic faults of fairly great number (primary ripples) can be serious, in extreme cases (*e.g* not properly tended machine work) the difference in the diffraction pattern is clearly visible. The diffraction pattern rings from the second and third outward are disturbed and appear as a luminous haze. A similar specially disastrous effect is produced by turbulence, which can break the wavefront into minute elements of greater inclination. Contrast in the telescopic images of planets are reduced and the image is blurred.

(iv) Faults of small amplitude (micro-ripples) in which a is of the order of 1/1,000 of a wavelength and b is of the order of a millimetre, the slope being about 10^{-6} to 10^{-5}, cannot be considered as local imperfections, but rather as a general condition of the surface, there being millions of them on a surface. It is not astonishing that, even if the distribution of these faults is not very regular, a considerable part of the mirror can be in phase well away from the main image and produce a diffraction spectrum. The diffracted light, especially up to 1° off axis, can in bad cases ($a \neq 30\text{A}, b \neq 1$ mm) be as much as several thousandths of the total energy at 2° off axis. It shows up specially clearly if one moves a bright object out of the field, and it affects the detection of the solar corona with a Lyot coronograph much more seriously than do many defects of surfacing.

Finally there are the elementary faults which could be investigated at grazing incidence by the electron microscope, but information is not available to permit the classification or even detection of these imperfections.

3. The surfaces investigated

During the vacations of 1948 and 1949 we have worked and investigated four similar spherical concave mirrors, of 125 mm diameter and 2,000 mm radius of curvature.

The choice of these dimensions was the result of a compromise between our wish to extrapolate, with a fair degree of certainty, the standard polishing procedure for the working of large pieces of astronomical optics, and the multitude of variations which the working and investigating of large mirrors would have allowed. Most of the characteristics of the working operations, carried out by hand on a fixed post, were, however, chosen according to previous experience based on the working of some 50 pieces of optic of 20–60 cm diameter.

It is clearly preferable to study the surface defects on mirrors rather than on lenses; not only is the wave-disturbance due to the faults four times as large (an advantage partially cancelled out by the necessity of using a glass reflection only) but one is also sure of avoiding confusion with the effects of other surfaces and with the heterogeneity of the material.

The mirrors were numbered 1, 2, 3, 4. Numbers 1, 2 and 4 were of St. Gobain glass of 15 mm thickness, and the birefringence does not exceed 30-40 mμ (special annealing) : the network of internal veins certainly has little or no connection with strikingly noticeable " veins " which were visible on the photographs. Number 3 was of first-quality optical glass, ordinary boro-silicate crown (B.1864 of Parra-Mantois) of 15 mm thickness, specially chosen for this experiment. Its strain did not exceed 10 mμ. Number 2 was reworked six times, by different methods, for sufficiently long periods to be sure of the correctness of the result in each case.

No attempt has been made to formulate rules sufficiently general so far as defects of large amplitude are concerned—such defects depending on inexplicable factors; only those methods have been retained which are suitable for obtaining, with a large tool, surfaces of some 20 cm not too badly deformed, to an accuracy of about 1/10 λ.

With these experimental mirrors, especially intended for the investigation of medium and small imperfections, large amplitude errors of $1/4$ λ were often allowed to remain, in order to avoid any disturbance of the working process—which would have reduced the clarity of the results. Effectively, the small defects depend only on the technique and the polishing method, therefore the cases we are going to examine can be accurately reproduced.

The author then proceeds to describe the various materials and methods of polishing tried on these four mirrors, the methods of test (including M. Lyot's improvement of the Zernicke phase-contrast method) and the results obtained.

The following are briefly the tests applied.

Test for faults of surface

An achromatic condenser focuses on to the test surface the image of the coiled filament of an overrun car headlamp, the image being about 4 mm in extent. The illuminated area is observed with a binocular microscope (magnification about $\times 36$) so arranged that it receives only the light diffracted at about $20°$ from the main image.

Tests for large and medium faults of shape

Foucault's knife edge test and Zernicke's phase-contrast test were both used.

Tests for small faults of shape

For these the modified Zernicke test of Lyot was used; the method is fully dealt with in the paper.

Effect of various methods of polishing

Guided by these tests the authors tried the effect of polishing with—

(a) Beeswax ("honeycomb foundation" as described by A. W. Everest, *Advanced Telescope Making*) and rouge.
(b) Wax coated pitch and rouge.
(c) Pitch and cerium oxide.
(d) Local retouching with rubber covered with silk—taffeta and rouge.
(e) Local retouching with the thumb and rouge.
(f) Pitch and rouge.

The last-named method, when carried out with the technique described by Texereau (1949, *l'A*, **63**, 42 and 85) gave the best results. The author says (translation)—

Surfacing faults. When the surface is cleaned with boiled ammoniated calcium carbonate, well dried and dusted, one sees absolutely nothing in the illuminated area—not even under the microscope.

Overall shape. Spherical to a very high degree, but the method was not pushed so far as to eliminate the slight turned-down edge—about $1/5$ λ at the rim. But this is a question of dimensions, viz. diameter and thickness of the lens. On a lens of 20-cm diameter and 35-mm thickness, the same method would have left no perceptible turned-down edge.

Primary ripples. Hills and valleys are less regularly distributed than with a hard-working process, the slopes are 2×10^{-6} and the

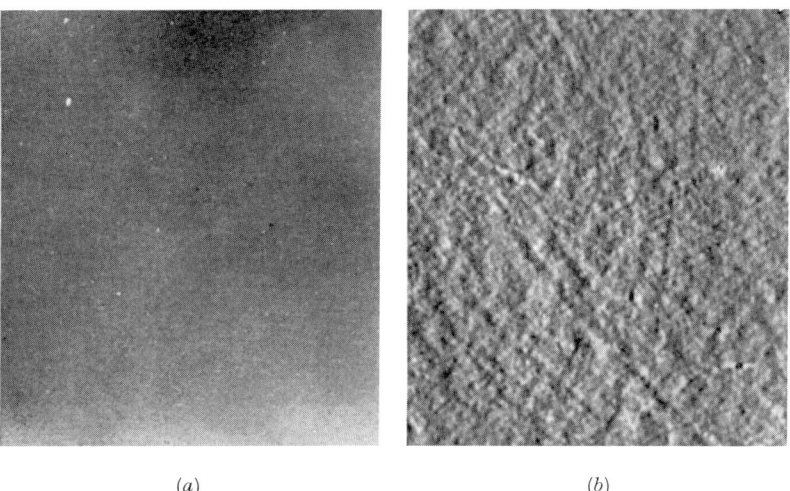

(a) (b)

Fig. 254—Two tests compared

The same mirror tested (a) by the Foucault method with a slit 22μ wide, (b) by Lyot's modification of the phase contrast method. Defects of 1 or 2 angstroms, which are shown by Lyot's test, are quite invisible by the Foucault test.

deviations less than $1/40$ λ. One can reduce these defects to about $1/100$ λ by more careful gentle-working, or by varied strokes (several persons working in turn).

Micro-ripples. This is the smoothest surface we could obtain. The plate (Fig. 254), in spite of its low contrast, has been taken with the densest phase plate we had ($D = 2.81$) and with an exposure of 15 minutes of high contrast collodion process plates by Guilleminot, as also were the other photographs. The polishing defects measure 0·2 to 0·5 mm across, the extreme amplitude—repeated with a fairly regular pseudo-period of 1·6 mm—reaches just on 2A on the wavefront, but over three-quarters of the surface the deviations from the mean background do not exceed 1A (0·5A on the surface) and many of the little hills on the microphotometer record correspond to only 0·2A. But there are more extended and irregularly distributed defects which are due to heterogeneities of the material. There is no hope of levelling these no matter how long polishing is continued, and whatever technique is used. In the case of this mirror, worked from first quality glass, selected specially for this experiment, the defects tend to assume the form of fairly large grooves, the most serious ones having a width of 1·6 mm, an elevation of 4A, and having a straight or slightly curved shape of length up to 10 mm. No doubt these arise from internal heterogeneities of the glass which remain primarily a problem for the glass manufacturer. This study of mirrors, made from glass of the best quality, shows a new class of defect which will have to be borne in mind if noticeable progress is to be made in the production of coronograph objectives.

Figures 254a and b show two of the photographs obtained.

Nothing but a careful study of the entire text will do adequate justice to this valuable investigation. The paper is an invaluable guide to the advantages and limitation of these various methods of working, all of which were carried out by M. Couder.

Superfine polishing

337 The removal of such minor faults as those described in the previous section had already been undertaken by Couder (1944). He says in the paper cited—

I had the opportunity of polishing two lenses for Coronographs, one 20 cm diameter for M. Lyot, the other 10 cm diameter for M. Waldmeyer. The difficulty arises entirely in that one is forced to abandon the procedure suitable for large pieces and to act like a manufacturer of small lenses. One must finish at one and the same time getting the surface to the correct figure and wearing out the abrasive. Actually the finest polish appears at the moment when the rouge being completely crushed and driven out, the tool threatens to adhere to the glass. If one interrupts the operations sooner there is a marked increase of the diffused light. The customary method of working astronomical objectives, which leads to a practically perfect figure by alternate examination interspersed with renewals of polishing of shorter and shorter duration is no longer permissible. It is necessary to watch from the outset for an acceptable figure at the end of each wet (" séchée ") which leaves a perfect polish. It is obvious that repeated attempts

are necessary before this happy coincidence takes place. Here are numbers which make clear what results can be obtained. Let us take as the unit of illumination that of the solar image formed by the objective of the Lyot coronograph, diaphragmed to 13 cm. The illumination in the focal plane at 5 minutes of arc outside the solar image due to the internal defects of the glass (principally to veins) amounts to ... ... 3×10^{-7}
Illumination due to surface defects ... ... $0{\cdot}8 \times 10^{-7}$
Illumination due to dust which even the most minute precautions cannot avoid is of the order ... ... 10×10^{-7}
Light diffused by the atmosphere at the top of the Pic-du-Midi in fine weather ... ... 4×10^{-7}

Total parasitic illumination approximately ... ... 18×10^{-7}

This is almost twice the illumination produced at the same point by the light of the solar corona itself. The diffusion by the surfaces of an ordinary astronomical objective of modern construction would be several tens of times as great, while that of objectives made, say, 60 years ago would be tens of times greater still. No example could illustrate better than does the coronograph the investigation of the minor defects of optical surfaces. When one undertook to make daily observations of the solar corona—an important date in the history of solar physics—it became necessary to begin by a notable change in the sense which one gives to the words " to polish the glass."

I am able to supplement the above remark by information which I received in a private letter from M. Couder dated 27th December, 1950. He says—

I did not describe the methods of test because I consider that they concern particularly M. Lyot who had already made a small coronograph before asking for my collaboration. The two methods of examination which I adopted are (a) for diffusion over a wide angle, examination of the surface with a weak magnifier at the region on which one projects the image of an arc with a good condenser, (b) for small angle diffusion strioscopic examination of the image of the objective with an eyepiece of small power, the set up being such that the lens can be carried in the hand.

After several trials I adopted pitch polishing, the polisher carrying 25-mm squares of pitch 4 mm thick. The rouge was produced by calcination of ferric oxalate washed and levigated. Test (a) is made after each complete wet and one stops when by good luck no scratch is seen. Test (a) is the more difficult to satisfy as far as the optician is concerned, test (*b*) is more difficult for the glass manufacturer. When I carried out this work neither M. Lyot nor myself used the phase contrast method which is now found so valuable.

338 It is interesting to compare the optical means of producing polished surfaces and measuring their irregularities with those developed of recent years in connection with automobile manufacture. The latter surfaces are produced by *superfinishing machines*, as they are called. Such machines are described in " *The Story of Superfinish* "[1] and " *Surface Finish* "[2] Their action is based on the application to the surface to be finished, as it rotates or traverses, of bonded abrasive stones, oscillating rapidly in random directions. (See reference[3], pages 164 *et seq.*) For other useful references on the subject of these machines see footnote 3.

The instruments for *measuring* the irregularities, as for example, the Profilometer[4], the Talysurf[5] and others, will detect irregularities even in the best superfinished surfaces, which may be no greater than 0·5 micro-inch, or as the engineers write it 0·5 mu.

The ordinary interferometric methods of the optician can easily detect irregularities of this amount on polished surfaces and surfaces of a like accuracy are customarily produced on glass by the optician. Improved interferometric methods have been developed and are described in §227 which can detect differences of 1 molecular layer of mica—namely about 0·02 mu. Surfaces approaching this accuracy can be produced on glass by the superfine polishing techniques described in §337.

Briefly then, as regards accuracy the available methods of optical polishing and testing of glass surfaces can, though with considerable labour and expenditure of time, achieve a freedom from irregularities 25 times as great as those of the superfinishing engineer. The latter methods, however, can produce on metal in a few seconds surfaces comparable in appearance with those which the optician takes hours to produce on glass, and of a perfection adequate for the engineer's purpose.

1. Arthur M. Swigert, Jr. (Detroit, U.S.A., Lynn Publishing Company.)
2. " *Surface Finish* " (Report of the Research Department, The Institution of Production Engineers; Published by the Institution January, 1942.)
3. Conference on Superfinishing (Proc. Inst. Mechanical Engineers, **151**, 1944.) *American Machinist*, 13 January, 1949. *Aircraft Production*, 11 November-December.
   How to Select superfinishing abrasives. (*The Machinist*, 5 November, 1949.)
4. Abbott, Bousky and Williamson, *Mechanical Engineering*, March, 1938.
5. Made by Taylor, Taylor & Hobson.

Whereas the optician's surfaces are brought about by pressure and heat, whereby the glass or metal surface assumes the vitreous or semi-vitreous condition described by the metallurgist as "the Beilby layer" (see §38) or "smear metal," the superfinished surfaces have the metallurgical condition of the interior of the metal, the crystal structure being in no way disturbed or destroyed from its initial condition, at least when it leaves the superfinishing machine. Further, however successful orthodox methods may be in polishing glass, they cannot be applied so easily to metals because the latter are relatively soft. In an instance within my experience a workman highly skilled in polishing metal could not produce in several days by the optical methods a surface so free from scratches and other defects as were produced on a superfinishing machine in a few minutes.

The results obtained are very striking—for example on page 169 of "*Surface Finish*" it is stated that a typical piston and cylinder with ordinary machine finish shows a total wear of 0·0018 in. in the initial running-in period, whereas superfinished surfaces reduce the wear in 10,000 miles operation to not more than 0·0002 in. Doubtless this running-in period produces Beilby layers, covered, as such surfaces invariably are, with an amorphous oxide layer. Beilby layers are eventually formed on bearing surfaces whatever the machining method employed, but with superfinished surfaces they must cover the complete surface in a very short time, after which, of course, the wear is very small. The load carrying capacity of such a bearing thus rises to its maximum very quickly.

These considerations, however, are leading us rather too far from the subject of this book. The reader interested in the friction of bearing surfaces will find much up-to-date and interesting information in the report of the discussion on the subject of *Friction*, which took place at the Royal Society on the 19th April, 1951.

CHAPTER 16

REFERENCE BOOKS ON OPTICS AND OPTICAL GLASSWORKING

Theoretical

The reader must not expect to find here guidance to works on advanced optics, but some grounding in theory all opticians should aim to acquire.

To get a foundation of knowledge on the theoretical side the practical optician may study with advantage *Light for Students*, by Edwin Edser (London : Macmillan & Co, 1915). I would also recommend the reader to peruse repeatedly, until he has thoroughly absorbed it, the few pages on optical instruments in the article *Optics* by Lord Rayleigh in the 9th edition of Encyclopaedia Britannica, 1884.

R. W. Wood's *Physical Optics* is a mine of useful information on the experimental side and will give the reader a grounding in the principles employed in the optical testing room.

Practical

The section on laboratory optical work in *Modern Laboratory Practice*, by John Strong can be recommended (London : Blackie & Son Ltd, 1940). It deals with a variety of optical operations including some of those described here, all treated with reference to manufacture in a laboratory workshop. The descriptions are excellent and illustrated with numerous clear and well-selected illustrations, and many of the hints will be found useful to learners and experts alike. The article by J. W. French on *The Working of Optical Parts* in Volume IV of Glazebrook's Dictionary of Applied Physics (1923) is the best condensed summary I know from the point of view of manufacture.

Another book full of useful information is *Le Travail des Verres d'Optique de Précision*, by Col. Charles Dévé (Paris, 1936 : Revue d'Optique theorique et instrumentale). The character of this book is well described by the author in his " avertissement ".

It is not a manual for beginners. It is a résumé of the technological training which is given to the apprentices of the Trade School for Optical Glass Work during their three years' course, and to the foremen and workmen who come to the Institut d'Optique. It is rather a book for the master than for the pupil, for certain questions which are treated in it require more elementary lessons adapted to the degree of general education of the pupil. The second part of the work is particularly intended for optical engineers in charge of precision optical workshops. (Free translation).

An excellent account of the normal precedure in 1949, including details of the manufacture of optical glass, is to be found in *The Manufacture of Optical Glass and of Optical Systems* (publication No. 2037 U.S.A. Ordnance Department), now (Dec 1949) out of print.

Die Optische Werkstatt by W. Ewald (Berlin : Gebrüder Borntraeger, 1930) is singularly little known in Great Britain ; prepared as it was with the assistance of more than a dozen well-known firms it reveals that at the time of publication Germany was in many respects years ahead of this country in the development of glass working machinery. The value of the book lies chiefly in the descriptions of grinding, polishing and edging machines and testing and measuring appliances then in use in Germany.

Bernard Halle's book (1913) *Handbuch der Praktische Optik* (*der Mechaniker*, Berlin-Nikolasse) gives sound practical instruction for high quality optical glassworking.

Amateur Telescope Making—Advanced (Munn & Co. Inc., U.S.A., 1946) is a collection of contributions to amateur precision optics by numerous authorities. Each contributor deals adequately with his own special subject. Scarcely any of the operations involved in telescope making are omitted and, although directed to amateurs, the book is of value to those engaged in the optical instrument industry.

La Construction du Téléscope d'Amateur by J. Texereau (1951 Société de France) is outstandingly clear and complete.

Engineering Optics (London ; Pitman, 1948) gives good descriptions of the optical systems used for engineering measurements.

The theoretical aspects of ophthalmic lenses are dealt with in Emsley and Swaine's book on *Ophthalmic Lenses* (London, Hatton Press, 1946)

Tables of lens curves

The computation of the curves of achromatic objectives is a highly specialised art, but many opticians working in research or development departments will value the tables of curves for such lenses by Brown & Smith ; they may often save him the trouble of referring to the lens designers for help on experimental jobs (*Systematic Constructional Tables for Thin Cemented Aplanatic Lenses*, by E. D. Brown and T. Smith, F.R.S. (1946, *Phil. Trans.*, **240A**, 59)

The refractive index range covered by the tables is 1·333 to 2,000. Two types of lens combinations are described. Triple cemented objectives working at unit magnification—such lenses being used as telescope erector assemblies and the like, and double cemented objectives working with objects at infinity—such lenses being used as telescope objectives, etc.

All the lenses given in the table are aplantic in the complete sense of the word, that is, correction is made for both spherical aberration and the sine condition. This ensures that the images will be good for a small and limited field about the axis.

The procedure for using the tables consists essentially of searching for actual glasses which conform to the table entries, thereby giving chromatic correction.

In using Brown and Smith's tables it might be well to refer to Kingslake (1927) who has determined experimentally the best minimum wavelength for visual achromatism which may differ according to the nature of the type of object to be observed and the nature of its illumination. Earlier tables by Smith and Cheshire (*N.P.L. Collected Researches*, 1915 and 1920) have some advantages for amateurs whose choice of glasses may be restricted and who can sometimes tolerate more coma.

APPENDIX I

DERIVATION OF THE FORMULA FOR THE TWYMAN ANNEALING SCHEDULES

As I had lost the calculations by which I arrived at this formula, its derivation was kindly checked for me by Mr W. Weinstein this year, as follows—

As was stated in the text three assumptions are made.
(1) From observations of double refraction the stress due to cooling is found to be proportional to the rate of cooling (§301)
(2) The degree of annealing attained in a given time is greatest when the rate of flow (adjustment to stress) is constant (§301)
(3) The rate of flow (A) is given by
$$A = kF2^{\theta/8}$$
where θ is the temperature, F is the stress and k is a constant.

Now from (3) the rate of adjustment to stress is
$$A = kF2^{\theta/8} = kFe^{m\theta}$$
where $m = (1/8)\log_e 2 = 0 \cdot 0865$.

This can be written
$$A = kFe^{m(\theta - \theta_0)}$$
where θ is the annealing temperature on choosing a suitable value for k. Also from (1) we can put
$$F = C_1 (d\theta/dt)$$
where C_1 is a constant and t is the time. Hence eliminating F
$$A = kC_1(d\theta/dt)e^{m(\theta - \theta_0)}, \text{ which}$$
according to (2) must be constant with respect to time, so that
$$e^{m(\theta - \theta_0)}d\theta = C_2 dt$$
where C_2 is another constant.

Integrating both sides
$$(1/m)e^{m(\theta - \theta_0)} = C_2 t + C_3.$$
To determine the constant of integration C_3 we note that $\theta = \theta_0$ when $t = 0$, so that $C_3 = 1/m$, hence
$$(1/m)e^{m(\theta - \theta_0)} = C_2 t + 1/m,$$
or $$t = -(1/C_2 m)\{1 - 1/e^{m(\theta_0 - \theta)}\}$$
The constant C_2 governs the scale of the schedule since during annealing $\theta_0 - \theta$ is always positive, it will be seen that C_2 must be negative; hence putting $-C_2 = b$
$$t = (1/bm)\{1 - 1/e^{m(\theta_0 - \theta)}\}$$
This is the required formula. It can easily be seen that the total cooling time is given by $1/bm$, i.e. as the temperature drop $(\theta_0 - \theta)$ tends to infinity, t tends to $1/bm$ thus for a total annealing time of T hours $b = 1/mT$ and t is given from the above formula in hours.

It must be remembered that although the expression for the relative mobility as a function of temperature given in (3) above holds approximately for most of the ordinary optical glasses, yet for some glasses, particularly of the newer varieties, the behaviour may be very different. Fig. 255 gives curves for some of these glasses from which appropriate cooling schedules can be derived in the manner indicated above.

To take a particular example, the $\log_{10}$ viscosity-temperature curve was supplied by Messrs Chance Bros for their glass No. 13993, from which it can be seen that a factor of 10 is introduced into $1/M$ ($=V$, the viscosity) by a fall in temperature of 16°C approximately.

Hence $1/M = K.10^{\theta/16} = K.10^{0.063\theta} = Ke^{0.144\theta}$.

The value of 'm' for Melt 13993 in the cooling schedule formula is, therefore, 0·144.

V will thus be doubled in a fall of temperature θ^1 given by the equation $2 = e^{0.144\theta}$,

or $\theta^1 = 4.8$°C approximately.

The following is a cooling schedule for Melt 13993 worked out for $m = 0.144$ and for 20 time intervals. The value of 'b' (rate of cooling at the start of the run) is, therefore, $1/20m = 0.35$°C per hour. This follows from the cooling schedule equation §301 by substituting the conditions at the end of the run when $t = 20$ hours and $e^{m(\theta_0 - \theta)}$ is infinite, so that

$$20 = 1/mb, \text{ or } b = 1/20m.$$

Time interval (t) after start of cooling in hours.	Fall in temperature from annealing temperature (°C).
0	0
1	0·35
2	0·73
3	1·14
4	1·55
5	2·00
6	2·50
7	3·00
8	3·57
9	4·15
10	4·82
11	5·55
12	6·35
13	7·30
14	8·30
15	9·60

Time interval (t) after start of cooling in hours.						Fall in temperature from annealing temperature (°C).
16	...	...	...	...	...	11·20
17	...	...	...	...	...	13·10
18	...	...	...	...	...	16·00
19	...	...	...	...	...	20·60
$19\tfrac{3}{4}$	...	...	...	...	...	30·45
19·99	...	...	...	...	...	52·82
20	...	...	...	...	...	∞

Mobility-temperature tests on special glasses

Figs. 255a and 255b show the mobility-temperature relationship for the following glasses—

Chances SBF 717479
,, SBC 691548
,, SBF 744447
,, SBC 651586 (special)
,, Borate Flint 614440 ;

Kodak Melt No. 32/1009 ($EK.320$), $N_D = 1·7452$ $\nu = 45·8$ (made in Rochester, N.Y., U.S.A.).

Specimens of cross section 6×2 mms were cut and ground from each of the above glasses. Each specimen was placed in the annealing jig, 6 mm length being free, and heated to a temperature which produced slight sagging ; the temperature was held constant and the rate of fall of the jig arm was measured for this temperature. After this the furnace temperature was raised by a constant amount and the new rate of fall determined. More than one specimen had to be used to get the fast rates of fall at the higher temperatures.

The formula for each glass shown in the figures was derived from the general formula (see §301)

$$R = R_0 2^{(\theta - \theta_0)/k}$$

by substituting the values of R, R_0, θ and θ_0 for two points on each curve, thus obtaining the value of the constant k.

In order to ascertain the agreement with the observed values from which the curves were plotted a number of values were calculated from the formula of each curve and these are indicated by heavy points. It will be seen that they mostly lie very close to the "observed" curve.

The following are the approximate compositions of the four glasses for which mobility curves have been drawn.

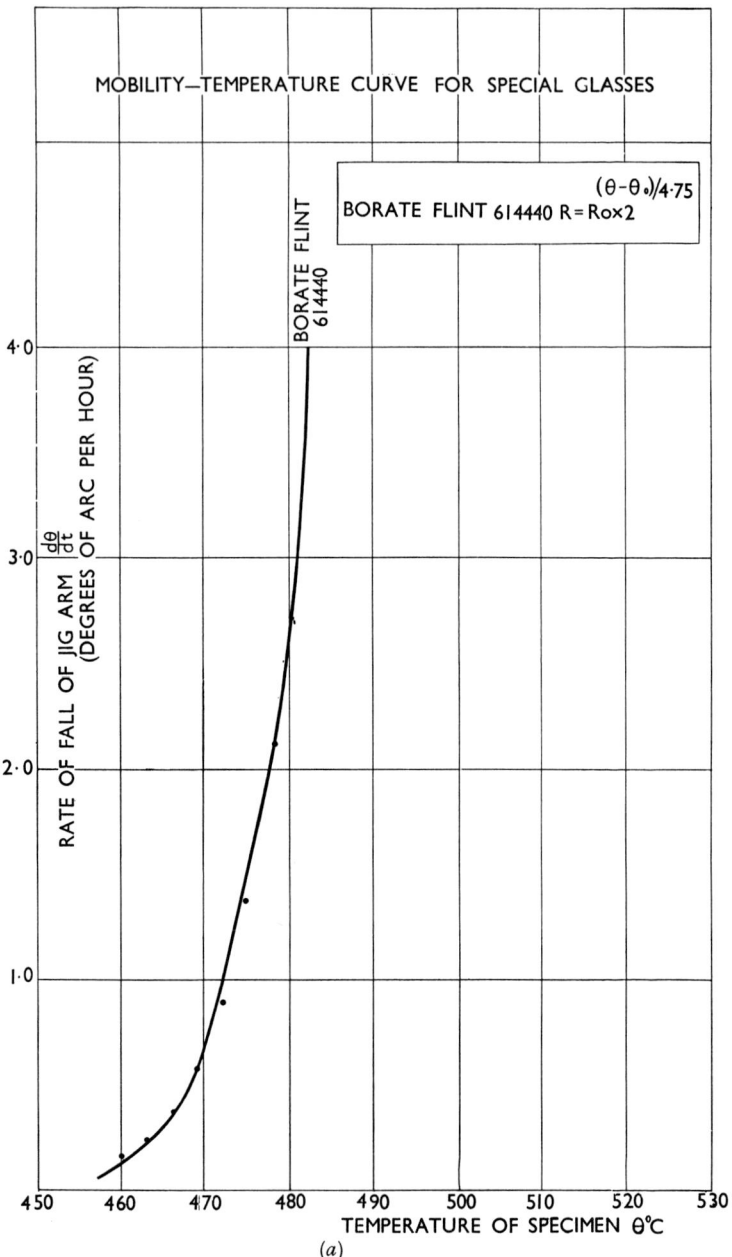

Fig. 255—Mobility–temperature relationship

APPENDIX I

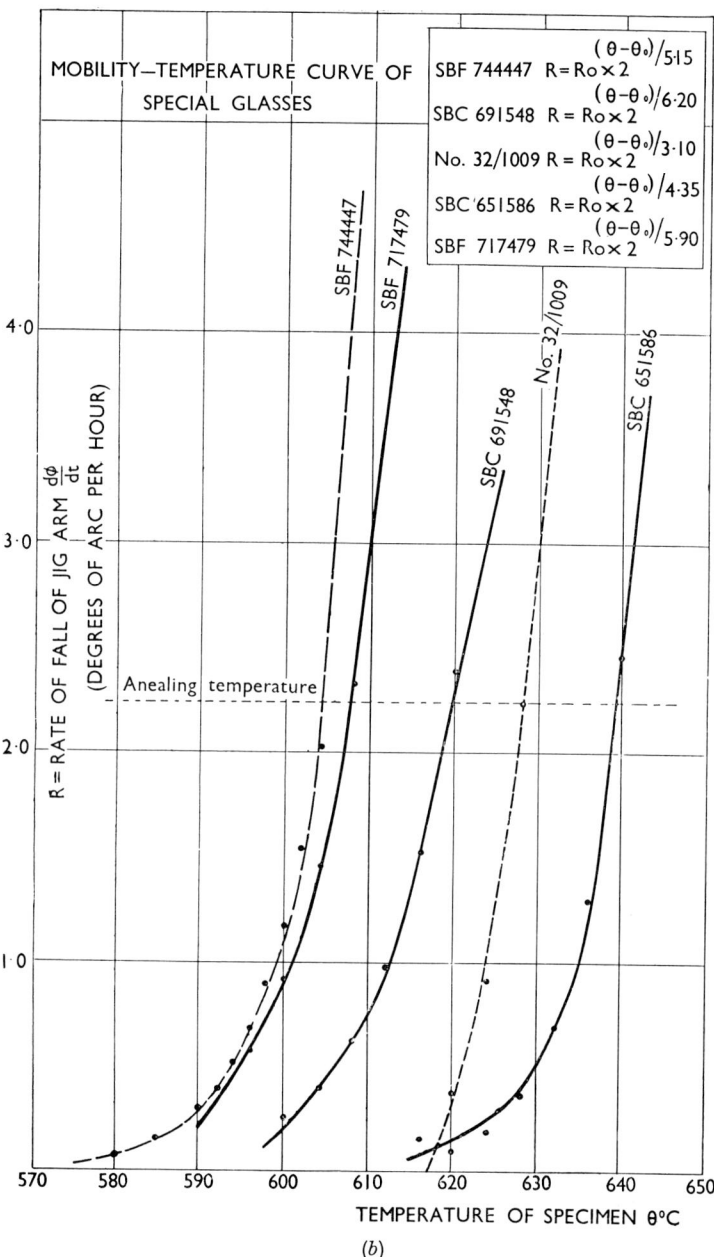

(b)

for some modern glasses

Oxide	Borate	Flint	SBF 744447	SBC 691548	SBC 651586	SBF 717479
SiO_2	9		11	11	21	16
B_2O_3	36		20	25	19	15
Al_2O_3	10		—	—	—	—
K_2O	2		—	—	—	—
Na_2O	1		—	—	—	—
PbO	36		13	—	—	9
BaO	—		11	26	45	31
ZnO	6		5	4	—	3
ThO_2	—		17	14	15	11
La_2O_3	—		16	17	—	13
ZrO_2	—		7	2	—	2

Some of these glasses have when molten a special tendency to form molecular aggregates of refractive index different from the mother mass, with the result that the glass becomes veined; even at or near the annealing temperature this process can occur and devitrification set in. Thus the annealing must be carried out rapidly and with the accelerating rate of cooling thus entailed the differences of temperature within the block must result in heterogeneity of refractive index, as described in §114.

To combine rapid cooling with small differences of temperature within the block will require a little ingenuity, but I do not doubt that it can be achieved:—for example, let us suppose that from a given temperature within the annealing range, one drops the furnace temperature rapidly a few degrees, raising it almost immediately to an intermediate temperature and continuing this process while broadly following the appropriate cooling schedule, the differences of temperature within a block could, I imagine, be reduced to something like one quarter of those which would result from faithfully following the cooling schedule.

The determination of the steps in this process and the intervals of time between them could be easily ascertained by observation of a specimen within a furnace.

APPENDIX II

MAKING POLARIZING PRISMS

by S. J. UNDERHILL, Technical Adviser on Optics to the Hilger Division of Hilger & Watts Ltd.

Types of polarizing prisms

For the production of polarized light of the highest purity, it is customary to insert into the light train a birefringent crystal whereby the incident beam is split into two beams which are polarized at right angles. All else being equal, a crystal having a higher birefringence than the others is to be preferred as wider fields of polarized light would become available and the optical path length of existing instruments could be reduced. In view of the comparatively recent progress made in the production of artificial crystals, it is to be expected that within a few years the monopoly held by Iceland spar (calcite) for well over a hundred years may be concluded and another crystal come into favour.

On looking through a cleft crystal of Iceland spar one cannot fail to see the double beams, one of which must be deleted and the other transmitted to give plane polarized light. Attempts to isolate one of the beams were not very successful until about 1830 when Nicol (1766–1851) made his famous discovery that, by taking a cleft crystal having a ratio of the long edge of the crystal to the short end face of about 3 to 1, and cutting it symmetrically from blunt corner to blunt corner, polishing the cut faces and cementing them together with Canada balsam, the ordinary ray was reflected from the balsam film and the extraordinary transmitted.

The field of light thus polarized does not extend over the entire aperture as can be seen by tilting the prism when a blue band will be seen on the side, marking the limit of transmission of the extraordinary, and on the other side by a sudden increase in intensity, marking the limit of reflection of the extraordinary and beyond which double images are seen. The internal field is about 14 degrees and the external about 24 degrees, but the middle line of the field is inclined to the longitudinal axis about 10° unsymmetrically and, for this reason, the end faces of the prisms have been trimmed to various angles.

It was early realized that where a smaller field would suffice the length of prism could be reduced. Nicol himself in 1839 found that he might use shorter pieces, and in 1857 Foucault reduced the overall length to little more than the width by substituting a thin air film for the balsam film.

The chief disadvantage of the oblique ended form of prism is that on rotation of the prism the transmitted beam moves around the axis of rotation. In conical light this is particularly objectionable and also introduces astigmatism, so a series of square ended prisms appeared such as Thompson's, the Glan-Foucault and the Soret, which latter have the advantage that they can be used over the region which would be cut out by a balsam film viz. from 2950A to the limit of transmission of the spar 2150A.

Several forms of glass-spar prisms have been developed or proposed notably, Abbe's, Ahrens' Leiss's and Feussner's. This last has a thin parallel plate of the double refracting material cemented between two glass prisms and it is interesting to note that Feussner pointed out that, if sodium nitrate were used, the field would be increased to 56 degrees and that this could be further extended to 90 degrees by substituting potassium chromate. Other successful essays to widen the field were made by Glazebrook, Hartnack and Thompson; some of which consume a large amount of material. All these prisms are described in *Proc. Optical Convention*, 1905, 216-235. In the same paper a useful table of the percentages of spar utilized if cut from equal pieces to begin with is given.

From the above it will be seen that there is a wide variety of prisms to choose from but it should be noted that some types consume much more spar than others and it must be borne in mind that the once prolific supply of large and often beautifully clear blocks of spar from Iceland no longer exists.

Manufacturers of a wide variety of instruments can sometimes effect economy by reviewing the variety of types and sizes of prisms in use to see whether a less costly type of prism cannot be substituted in some of the instruments. It may be that a new technique or development has reduced the cost of one type of prism and with little or no alteration this same prism will serve for another instrument. Not only is the cost of this instrument reduced but the standardization enables larger batches of prisms to be put in hand and may reduce costs still further. As an example of evolution leading to standardization, and as no record of this has been published hitherto, it may be of interest to readers to state a method of manufacture introduced in 1916 by Mr A. Green, then Chief Foreman of the Hilger Optics Department.

A plate of spar is cleft to a thickness a little more than the length of side of aperture required. Assuming that the plate will cut more than one prism it is marked to the best advantage with parallel lines, perpendicular to the bisector of the blunt corner, separated from each other either by the same distance as the thickness of the plate plus 1 cut width or this same distance plus a distance which allows of the diagonal cut (balsamed plane) being made. The determination as to which of the two distances shall be marked off for this first cutting depends on the relative length to width of the cleft plate; the space for the diagonal cut must either detract from length or width. The plate is then cut along these lines producing sticks of spar with two cut sides, two cleft sides and cleft ends, the optical axis being perpendicular to the length and diagonally across the square section.

All sticks of equal width are assembled together and the narrow group cut to the final length of the prism plus sufficient allowance for the diagonal cut (balsam plane cut) to be made; the wider group is cut to the final length. In both cases these cuts are perpendicular to the length of the sticks (as the prisms when completed will be square ended).

All rectangular pieces are assembled so that a cleft side is face upwards and arranged in groups of about six at a time in echelon formation so as to bring all balsam cut planes into line. The balsam cut is then made through each group in turn resulting in the production of as many pairs of single prisms as there were blocks. These pairs are then numbered so that pairs can be identified at all further stages. This is important as however carefully the marking out is made and the cuts taken the axis direction is sure to vary slightly between pair and pair and, if the pairs are mixed, they will suffer from a defect which we call " squirm," *i.e.* a line object viewed through the completed prisms will oscillate as the prism is turned about the line of sight. After polishing, the pairs are cemented together with Canada balsam and a polarizing prism results.

It was found that the field of such prisms was a little too small for some instruments and Green therefore introduced a corrective for this in the form of a thin parallel polished plate of fluor crown, which has a refractive index nearly the same as the extraordinary ray of spar ; this when balsamed between the two halves increased the field to the required amount.

Prisms thus constructed were found to meet the majority of our needs at the time and have been incorporated in many instruments up to the last war period when n-Butyl Methacrylate cement became available ; and as this, when baked, has a refractive index of about 1·46 the fluor crown plate was dispensed with and the prisms were affixed together directly with this cement.

The particulars and performance of this last named prism are as follows—

> Ratio of length to square aperture face : 3 to 1.
> Angle between cemented face and square aperture face : 72°.
> External field (D lines) : cut-off begins at 15° to axis, total transmission at 14° to axis on the other side.

Polishing Iceland spar

Notes on polishing materials other than glass will be found in §§33, 73, 88 and 119.

Examination of the raw material

Iceland spar cleaves so easily that either nature or man has cleft it before it reaches the test room.

When cleft by nature the faces are sometimes covered with an opaque deposit which must be removed by further cleaving. This is done by scoring the four surfaces along the same cleavage plane or as nearly on the one plane as can be estimated with a sharp knife such as a safety razor blade set in a handle. This produces a strain on the cleave plane around the crystal such that, with sufficiently small pieces, a continuation of the scoring may cause the obscure face to flake off leaving a transparent cleft face on the main block. If it does not part readily, then the knife is placed on and in a line with the scored line and given a smart tap with a hammer. If on looking through this face sufficient imperfections are seen the crystal is rejected; if not, the opposite face is cleft in the same way and the crystal again inspected for imperfections. This process of cleaving off a face from a side is repeated till the crystal is rejected or passed.

The method of inspection is to project a concentrated light beam through the spar and look for reflections of this light from particles at all possible angles, if great clearness is desired, as in the polarizers of colorimeters, using a magnifying lens. The apparatus consists of a box from which one side and one end has been removed. In the remaining end a hole is made and a large angle condensing lens inserted. On the axis of this lens and outside the box is mounted a powerful light source, at such distance from the lens that the conjugate focus is within the box, and the divergent beam therefrom emerges through the space at the other end without infringing on the side of the box. The light which passes through the end of the box should be absorbed by black material, allowed to pass into another room, etc, so as to leave the room in darkness save for the unavoidable scatter of light from dust in the atmosphere and on the lens, which latter should be reduced by a mask between lens and observer's eye.

The spar is held in the hand in the narrowest part of the beam and tilted through all possible orientations while the eye scans the interior watching for any signs of a reflection.

It is worth while spending some time over this examination, especially if the spar is to be used for prisms in which the slightest blemish will be a cause for rejection (the half shadow prism of a polarimeter for example). There is an optimum distance for the eye from the crystal since, on the one hand, the multiple reflections within the crystal flash intermittence beams into the eye which cause the pupil to close and, on the other hand, the 8 mm diameter of the pupil has to scan a very large part if not the whole of the envelope of a sphere of radius equal to the distance between crystal and eye, and the eye must be sighted more or less on that part of the crystal in which the defect exists.

Vision into crystals through a translucent surface can sometimes be clarified by dipping the crystal into liquid, but this should not be done impulsively as the liquid will run into any crevice and may stay

APPENDIX II   601

there for weeks, rendering its detection the more difficult. This caution does not apply so much to Iceland spar as to crystals having lower or no birefringence.

Sodium nitrate

As was pointed out earlier, Feussner (1884) described a polarizing prism using a plate of the double refracting material cemented between two identical glass prisms, and particular mention was made of sodium nitrate.

It is possible now to obtain synthetic crystals of sodium nitrate, and Feussner's suggestion has been revived.

In 1947 an experimental prism was designed and made by Adam Hilger Ltd with the purpose of studying the effect of atmospheric and water vapour attack. Sodium nitrate is more seriously affected by these conditions than is Iceland spar, and is furthermore somewhat more difficult to work to a fine polish. The finished prism requires complete sealing for protection against water vapour, and this may set a limit to extensive use.

The prism requires a cement of minimum refractive index 1·587. The refractive index of the glass is more critical, and should lie between 1·587 and 1·581.

It is interesting to note that the sodium nitrate prism described transmits the ordinary beam, whereas the calcite or Nicol prism transmits the extraordinary beam.

A reference to this form of prism is made by Huot de Longchamp (1947) in a paper which gives appropriate angles for the prisms and also makes some comments on the growing of sodium nitrate crystals.

A sodium nitrate polarizer was also designed later by Baubet and Lafont (1949), these authors adopting a solid sodium nitrate prism with a diagonal cut cemented as in the traditional Nicol prism. A correct choice of glass and cement makes the use of this form unnecessary.

A further development of the Feussner type was investigated by Adam Hilger Ltd in 1947. The maximum field of either form of prism described above is about 53°. This was extended in the Hilger prism to 60° by replacing the glass prisms by calcite prisms of appropriate angle and cut, the conditions for the cement being that the refractive index must be equal to or greater than 1·587.

Polaroid

(I am indebted to the Polaroid Corporation of Cambridge, Mass., U.S.A., for the major part of this description)

A wider range of uses of polarimetry was made possible by the introduction of Polaroid. The invention of this material is based on

u

the property of some doubly refracting crystals, known as "pleochroism," of absorbing one of the polarized beams more strongly than the other, with the result that the transmitted beam can, under favourable conditions, become completely polarized. The best known of these crystals is tourmaline, in which the property was discovered by Biot in 1815. In this substance the ordinary ray, which vibrates perpendicular to the principal axis, is much more strongly absorbed for the whole of the spectrum than is the extraordinary ray which, although subject to absorption, is comparatively transmissive for the blue-green. Thus a piece of tourmaline 1 or 2 mm thick, sliced in the direction of the principal or optical axis, transmits almost perfectly polarized bluish-green light.

It was discovered by an English physician, Dr W. B. Herapath, that sulphate of iodo-quinine, popularly known as herapathite, in honour of its discoverer, transmits polarized light of all colours with astonishingly high relative intensity and hence yields almost white polarized light. The importance of the discovery was fully realised by Dr Herapath and his contemporaries, but the crystals are so unstable mechanically as to shatter into useless powder at the slightest impact, and the problem of making useful polarizers of Herapathite was abandoned for the time being.

About 1927 Edwin H. Land attacked the problem from a completely new point of view. Instead of aiming to cover a given area with a single crystal he used innumerable small ones packed close together, and after getting these into the necessary uniform alignment, he embedded them in a transparent covering material to prevent them from breaking up into a useless powder.

The numerous problems involved in this work were solved one by one, so that today the commercial grades of Polaroid sheeting are produced by almost completely automatic machines in pieces 30 in. wide and of any desired length. The polarization of the best quality Polaroid (Type H) today is so complete, and the transmission so good, that it is used by Messrs Hilger & Watts Ltd in their newly introduced "Small Polarimeter, M.417," which is designed for use in medicine, industry and by students, and reads to an accuracy of $0 \cdot 05°$ to $0 \cdot 10°$. Hilger & Watts Ltd also now use Polaroid, type H, for strain viewers and photo-electric apparatus. In these instruments the Polaroid is only cemented in the case of small sizes up to about $1\frac{1}{2}$ in. diameter; larger sizes of 3 in. and 6 in. diameter are not cemented but merely held flat between glass plates or between a glass plate and plain face of a lens.

The search for further polarizing materials continued steadily and has yielded other synthetic films; one of these now widely used contains no preformed crystals at all, its polarizing action resulting on the linear control of the molecular structure of the material to form a homogeneous and completely haze-free sheet. Doubly refracting film produced in this way has, of course, been known for a very long time and is exhibited strongly in the ordinary commercial cellophane wrapping paper, but in this both beams are equally transmitted. The perfected material has been described by Mr Land in the following words—

In making actual polarizing sheets we create a brush-like structure inside a plastic sheet—a clear tough plastic called polyvinylalcohol. First it is stretched in one direction so that the long tangled molecules straighten out and all become parallel to the direction of stretch; then the sheet is dipped into a solution like the ordinary tincture of iodine in your medicine chest. The rusty brown iodine is instantly and rather miraculously transformed. It now has two different colours. To polarized light vibrating in one direction it is perfectly white and clear; to polarized light vibrating in the other direction it is black. The sheet has become a light polarizer.

The transmission of crossed Polaroid filters, Type HN 24, is stated to be between one hundred-thousandth and one millionth of that of the incident light.

APPENDIX III

THE PRESENT STATE AND FUTURE TRENDS OF THE MANUFACTURE OF OPTICAL GLASS

As pointed out on page 519, the 1914–18 war resulted in great development of optical glass-making in Great Britain. During the 1939–45 war this development was intensified not only in Great Britain but in other countries*, and noteworthy improvements, originating chiefly in the U.S.A., have been made in methods of manufacture. The following survey of methods is abstracted by permission from a report by Dr W. M. Hampton of Messrs Chance Bros Ltd.

Introduction

Optical glass is known to be made at present by four different methods—(a) the so-called classical method, (b) the poured pot process, (c) the platinum pot and electric furnace, and (d) the continuous method.

* For an account of the Japanese industry see *Japanese Optics*, 1945 (Report B.I.O.S./J.A.P./P.R./1308 H.M. Stationery Office). This gives particulars of (a) methods of manufacture, (b) types of glass developed, (c) information received by the Japanese from Germany, (d) research laboratories or institutions, (e) types of raw materials, and (f) special ingredients or substitutions.

Each process has advantages and disadvantages. The following notes point out the pros and cons, and indicate the substantial change in emphasis on the different processes which is now apparent.

The classical process

This consists of melting together the batch materials with broken glass of a similar composition from previous manufacture in a pot made of refractory materials. After melting is complete and the glass

Fig. 256—Pot and contents "broken down"

considered free from bubble, it is stirred with a refractory rod. The pot is removed from the furnace, transferred to a cooling arch, and subsequently its contents are broken down (Fig. 256) and the lumps of glass remoulded into sizes and shapes appropriate for the intended use. This method is suitable for any type of optical glass for which sufficiently insoluble clay containers can be provided.

The early optical glasses were simple crowns and flints, which are relatively inactive in contact with clay, and the problem of pot solution was not of serious importance. Newer types of glass, notably the barium crowns, involved more corrosive batches, and the variety of these that could be made was limited. Even with the glasses suitable for this method, sizes were fortuitous and the yield of a particular size could not be forecast.

The poured pot process

In this process the contents of the pot, on removal from the furnace, are poured into a large mould (Fig. 257) so that a high proportion of the pot contents is obtained in one large slab. At the present time slabs of glass up to 1,000 lb. in weight and say 40 in. square by up

Fig. 257—Casting from large scale pot

to 10 in. thick are produced. After pouring, these slabs are annealed as a single block sufficiently well to enable the glass to be processed. Thus the glass is obtained with fair certainty in large pieces which can be polished and examined so that the defective portions of the pot can be marked and cut away. The slabs provide a stock from which plates of pre-determined size can be sawn or broken. The process also gives the possibility of producing large slabs or discs of astronomical quality, and recently astronomical objectives of 25 in. diameter have been produced by this means.

The use of the platinum pot

Some ten years ago interest began to develop in new types of optical glass containing the rarer elements and among the problems encountered was that of the attack on the container. The only material which withstands the corrosive action of these glasses is platinum, and therefore platinum containers had to be used. The process has been developed until at the present time pots holding upwards of 100 lb of glass and measuring 11 in. in diameter are in regular use. After stirring, the contents of the platinum pot are poured into a mould of appropriate size (Fig. 258) essentially on the same lines as by the method described above, but, of course, on a smaller scale.

Fig. 258—Casting from platinum pot

The yield of good glass in these slabs is very high. This use of platinum results in glass very free from bubbles and colour, but it is only applicable to glasses which do not involve a melting temperature much in excess of 1,400°C, since beyond that temperature the mechanical strength of the platinum fails.

The continuous method

This has been developed to a high degree of perfection in the United States. Based on the tank method used in making commerical bottles, it provides a continuous supply of glass of optical quality. This necessitates the use of a tank furnace, into which the raw materials are fed continuously at one end of the unit, and the good glass is bled out continuously at the other. Further details have not

APPENDIX III   607

been published. It is known that this method can (Fig. 259) produce long slabs of glass of a quality at least as high as that given by the platinum pot method described above, but owing to the fact that the cost of installation is very high and, too, the considerable time it must take to clean out the system to change from one glass to another, the method appears to be suitable only for glasses required in very large quantities.

Future trends

It appears inevitable that the classical process will ultimately disappear. The uncertainty in the size and shape of the pieces

Fig. 259—Glass produced by continuous process—American

obtained makes it unattractive for commercial manufacture. Since the bulk of the ordinary type of glass at present cannot be successfully made in platinum, refractory pots will continue to be used; but, since the economic advantages consequent on pouring the pot contents into one slab are so great, it seems that the bulk of ordinary types will be made by this poured-pot method. Such slabs will provide the raw materials for the manufacture of large discs or slabs where necessary, and will provide a quarry from which plates of any size can be cut. For the corrosive types of rare-earth glasses, and semi-corrosive types of glass which have now become ordinary, *e.g.* DBC's of high index and high v-value (Fig. 260), the platinum pot method will continue to be developed, especially as it does seem to be possible to provide virtually 100 per cent. yield of colourless bubble-free glass. Such pots will provide slabs of considerable size from which lens blanks can either be cut or moulded. For the future, this development is probably as important as the continuous process, as the quality is as high and the flexibility of manufacture much greater.

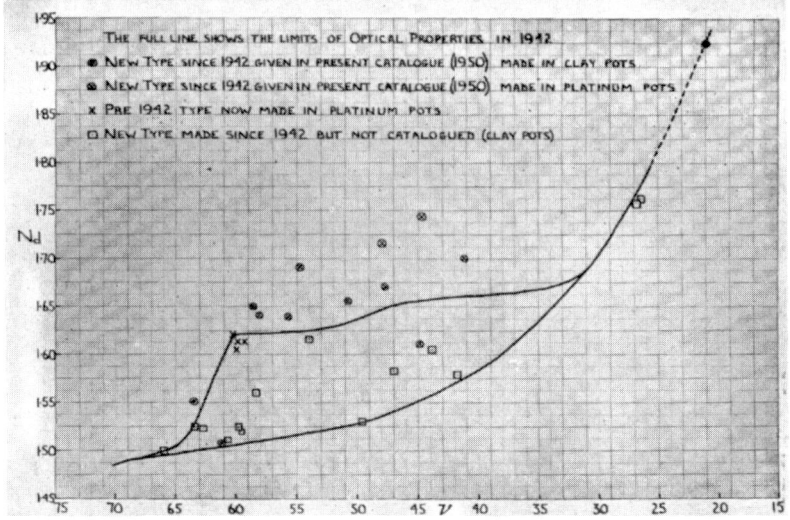

Fig. 260—Development of optical glass types since 1942

The continuous process, owing to the large quantities of any one type which it has to deliver if it is to be worked economically, will be limited to the production of glasses which are wanted in a very substantial quantity, *e.g.* binocular prisms and the like, and possibly for glasses which are wanted in large quantity but of a little less than optical homogeneity. It does not appear likely that the development of the continuous process will solve all the problems of manufacture.

APPENDIX IV

GLOSSARY OF TERMS USED IN THE OPTICAL INDUSTRY

English	French	German
abrasive	abrasif	Schleifmaterial
acetone	acétone	Azeton
adhese. To join two pieces of glass by heat treatment without distortion; a method developed in the Research Department of Adam Hilger Ltd.	——	Schweissen ohne Erweichung
alcohol	alcool	Alkohol
Aloxite	Aloxite	Aloxite
Alundum	Alundum	Alundum
angling	——	Winkel, an einen Block genau anschleifen
Angle Dekkor	cales d'angle	Autokollimationsfernrohr
angle gauges. Prisms of polished glass or metal of accurately known angle	cales étalons	Winkel-Normalien
anneal (to stress, relieve) ...	recuire	Umformungspunktes schnell erreichen. Vorwärmen
anneal (to normalize)	tremper	tempern
aspherize. To depart deliberately from sphericity in working a surface	faire des surfaces aspherique	deformieren
astig. Abbreviation of *astigmatic*; denotes the shape of a surface which has different curvatures, or the performance of a lens system which has different foci, in two axial planes at right angles to each other	cylindre ou tore	astigmatisch
beeswax	cire d'abeilles	Bienenwachs
benzene	benzine	Benzol
bevel gauge	fausse équerre	Winkellehre
block; blocking	bloc, blocage	Block (Gipsklotz); in einen Block setzen
blocking tool, tool on which optical pieces are placed for making a plaster block	outil à bloquer	——
block holder (§131)	——	Tragkörper
bort (a form of diamond)	bort	Diamantbort
bottle, drop. Stoppered bottle for allowing a drop of liquid to fall at a time	——	Tropfflasche

609

English	French	German
boxwood sticks. Name erroneously applied in the Hilger workshops to the thin orangewood sticks recommended for cleaning the slits of spectrographs and as vehicles for swabs of cotton wool	fusain	Orangenbaumstab
breath on, to	(contrôler) à la buée	anhauchen
bruiser (for breaking down and spreading abrasive)	brisoir	Schleifmittel, Werkzeug zum Zerkleinern und Verteilan
brush	——	Pinsel, Wischer
bubbles in glass brought to the surface in grinding or polishing	points crevés	Blasen
bubbles, air balls in glass ...	bulles	Blasen
bubble, natural, within a crystal	vacuole	Lufteinschluss
buff, to ; to buff up. To polish free from grey but without regard to surface flatness	——	blankpolieren, ohne Rücksicht auf Flächengüte
caliper with scale but no vernier used in measuring thickness. Also double-ended caliper used in conjunction with a gauge for the same purpose (§28)	nonius	——
cave. Abbreviation of *concave*, synonymous with *shallow*. Used to indicate that a surface, if concave, is more, or, if convex, less so than it should be	concave	flach
carborundum	carborundum	Karborundum
cement	ciment	Kitt
Cenco Sealstix (formerly called Khotinski cement)	Silastic	Cenco Sealstix
centre, to. To set a lens for edging so that the optical axis shall be central with the edge	centrer	zentrieren
cerium oxide	rose à polir oxyde de cérium	Ceroxyd
chamfer, to. To grind away sharp edges	chanfreiner	Kanten brechen ; facettieren; fasen
chamois leather	peau de chamois	Putzleder
chromic acid	acide chromique	Chromsäure
chromium oxide	oxyde de chrome	Chromoxyd
chuck	mandrin	Bohrfutter
clean	nettoyer	sauber
cleave, to	cliver	spalten
colophony ; rosin (the solid residue after distillation of oil of turpentine from crude turpentine)	colophane	Colophonium (or Kolophonium)

APPENDIX IV

English	French	German
colour, to put down in. To clean and lay two surfaces together so that they show Newton's fringes.	mettre aux couleurs	auf Passe prüfen
contact. To put surfaces in optical contact, viz. so cleanly that they adhere together without reflection at the interface ; a process first used by Adam Hilger Ltd.	adhérer	ansprengen
contrivance, apparatus ...	—	Vorrichtung
convex, convex surfaces ...	bosse	Konvex, Konvexflächen. konvexe Flächen
correct, to	corriger	korrigieren
corundum	corindon	Korund
cotton wool	ouate	Baumwolle Watte
creosote (in U.S.A. *creosote oil* or *liquid pitch oil*)	—	Kreosotöl
crown (of a lens)	couronne	Linsenscheitel
crystal, cloudy defects within	neige	Wolken
crystal, twin or other defect in ; such as a small crystal embedded in the mass	macle	Einschluss
crystolon (an artificial corundum)	—	—
curvature, radius of	rayon de courbure	Krümmungsradius
curve, to	—	krümmen
curve ; shaping cloth, felt, pitch saturated taffeta etc. to curvature for a polisher	gaufrage	—
cut, to	scier	schneiden
cut, severe scratch on a polished surface	—	Kratzer
cutter, sharp blade for trimming and grooving a polisher (§23)	tranchet	Messer
dead-metal. Small pieces of opaque material in optical glass	—	Gallen, Steine
deep (used of a surface whose curvature is too great)	courte (surface)	(zu stark gekrümmt)
depth gauge	jauge de profondeur	Tiefenmass
Diamantin	—	Diamantine
diamond	diamant	Diamant, Diamantbort
diamond, cutting. Glazier's diamond	—	Glaserdiamant
diamond dust (generally in olive oil or vaseline)	egrisé	Diamantstaub
diamond, machine for cutting circles by	tournette	Brillenglasschneidemaschine
diamond, marking. Diamond Chip set in metal for engraving or writing on glass	pointe à graver	Schreibdiamant
digs on an optical surface ...	échignures	Grübchen
disk, slice	disque, feuille de glace	Scheibe

611

English	French	German
double image	———	Doppelbild
drill, to. See also *trepanning*	saccade, par	bohren
to drill small holes in glass	perçage	bohren klein Löcher
dry, to	sécher	trocknen
dust	poussière	Staub
edge, to	déborder	rundieren
edging with a peripheral groove to take a special mounting, etc.	rainer	Nut einschleifen
elutriation (of emery, etc.) ...	minutage	Schlämmen
emery	———	Schmirgel
emery stone, natural ; Turkey stone	pierre de Levant	Schleifstein
emery, working down ; increase of fineness of abrasive during working	raffnage	———
errors, imperfections ...	erreurs	Fehler
eyeglass, eyeglasses ...	———	Augenglas, Augengläser
faults	défauts	Fehler
fluor spar	———	Fluss-spat
feathers ; feather shaped defects within transparent materials. The term is applied also to collections of bubbles or flecks	feuillets	Federn
figuring. Altering the shape of a surface to improve definition	rétouche	retouchieren, korrigieren
flake. A shallow chip broken out of an optical part	———	Muschel
flonk. A percussion tool whereby the surface of a piece of glass is flaked roughly to curvature to save grinding	fioner (pince à)	———
focus	foyer	Brennpunkt
forceps, spring	brucelles	Federzange
form, to (sometimes called *to press out*)	mouler	formen
former. Tool for forming polisher (sometimes, misleadingly, called *polishing tool*)	———	Drückkörper
Fuller's Earth (used with rouge in polishing on cloth polisher)	terre à foulon	Fullererde
gasolene (in England *petrol*)	essence	Benzin
gas pitch (in U.S.A. *coal tar pitch*)	poix	Pech
gauge (for optical tools) ...	calibre	Schablone, Lehre
glass	verre	Glas
glass, unworked. Glasses ...	verre brut, verres	Rohglas, Gläser
glass-cutting	———	Glasschneiden
goniometer	goniomètre	Goniometer
grain	grain	Korn
graticule (§267)	réticule	Glasteilung
graze, see scratch	———	Schramme, Kratzer
grey (of a ground surface).	gris	grau, narbig

APPENDIX IV

English	French	German
grains	grains	Körner
greyness of an insufficiently polished surface	chair	nicht auspoliert
grind	meuler ou ébauchage	schleifen
grinding	meulage	Schliff
grinding bench	machine à meuler	Schleifbank
grinding material, grinding powder	abrasif	Schleifmüttel, Schleitpulver
grinding tools	meules	Schleifschalen
groove, to	encocher	Rillen ; erzeugen
grooving tool	outil à rainurer	Werkzeug zum Erzeugen von Rillen
gypsum	gypse	Gips
handle, wooden, on which a small lens is worked individually	molette	Holzheft
hardness	dureté	Härte
holder for working a diamond	dopp	Diamanthalter
homogeneity	homogénéité	Gleichmässigkeit, Homogenität
hone. Oilstone or other fine sharpening stone	pièrre d'aiguiser	Abziehstein
hot plate	plaque chauffante	Heizplatte
hydrochloric acid	acide chlorhydrique	Salzsäure
ice box	glacière	Eisschrank
Iceland spar	Spath d'Islande	Kalkspat
imperfections, see errors ...	défauts	—
impregnated diamond wheel (§127)	meule diamant	Diamantscheibe
jigs	tamis	—
Khotinski cement, now sold as Cenco Sealstix	Silastic	Khotinski Kitt
knocker off. Hardwood stick or mallet for detaching "mallets" from lenses by giving a sharp blow near the junction of pitch and work piece	maillet	Schlegel
knocking-off	—	—
knocking-off mallet	deglanter, maillet à	—
lens	lentille	Linse
lens holder (small pencil shaped handle on which small lenses are cemented for working)	cotret	Kittheft (Linsenhalter)
mallet	glaud	Kittklumpen
⸝malletting (putting mallets on) lenses or discs as for blocking (§131)	glauter	aufkitten
magnetic chuck	mandrin magnétique	Magnetspannfutter
magnifier	loupe	Lupe, Vergrösserungsglas
marker, piece of glass cemented to a tool with pieces which are to be ground as an indication of the approach to the desired thickness	temoin	Richtpunkt

English	French	German
methylated spirit	alcool à brûler	Methylalkohol
micrometer gauge, "mike"	comparateur micromètre	Schraubenmikrometer
milling (by diamond wheel) ...	———	fräsen (mit Diamantwerkzeug)
mirror, speculum metal ...	miroir	Spiegel, Spiegelmetall
mortar, steel mortar ...	mortier, mortier d'acier	Mörser, Stahlmörser
Neven wheel (§127)	———	———
Newton's fringes, working to	travail au couleurs	auf Passe arbeiten
Nicol prism	Nicol	Nicol
Nicol prism, jig in which Iceland spar is cut for	hirondelle	———
nitric acid	acide nitrique	Salpetersäure
ochre, red	ocre rouge	Ocker, rot
ochre, yellow	ocre jaune	Ocker, gelb
o.g., object glass	objectif	Objektiv
oil	huile	Oel
oilstone or other fine sharpening stone	pierre d'aiguiser	Oelstein
opacity	opacité	Undurchsichtigkeit
optical bench	banc optique	optische Bank
optical contact	contact optique	optischer Kontakt
oven	four	Ofen
packing (strips of cardboard, sheet metal, etc., cemented to a curved tool to modify its curvature for use as a polisher or block holder)	cuirasse	Packung
paper for polishing (as used in France)	Berzelius	Polierpapier
paraffin wax	cire de paraffine	Paraffin wachs
paraffin (in U.S.A. kerosene)	paraffine	Paraffin
parallel up, to	parallèle à	parallel machen
petrol (in U.S.A. gasoline) ...	essence	Benzin
pimply or granular appearance of a polished surface under magnification. Greyness of an imperfectly polished surface	gris	Orangenschalen Anschein
pitch, gas (in U.S.A. coal tar pitch)	poix de Suède	Pech
pitch, pitch polishing ...	poix, polissage à poix	Pech, Pechpolitur
pitch pad (or block). Tool covered with a layer of pitch to act as backing for working pieces plane-parallel	———	Plankörper
pitch, Swedish (in U.S.A. Stockholm pitch)	———	Schwedisches Schiffspech
pits (in incompletely polished surfaces)	piqûres de "gris"	siehe digs
plaster block (§181)	blocs de plâtre	Gipsvorrichtung
press, to (e.g. to form a polisher by pressure of a warm tool)	mise en forme	abdrücken
plaster of Paris	Plâtre de Paris	Gips
plunger, steel plunger ...	piston d'acier	Stempel, Stahlstempel

APPENDIX IV

English	French	German
polish, to produce the early stage of polishing	éclaircir	anfangen zu polieren
polisher	polissoir	Polierschale, Polierwerkzeug
polisher holder	—	—
polishing machine	machine à polir	Polierbank
polishing or grinding machine	machine à polir	—
polishing powder	poudre à polir	Poliermittel
polishing tool, see also *former*	outil à polir	Polierschale Polierwerkzeug
Portland cement	—	Portland Zement
pot (as, for instance, emery pot)	pot	Topf
press out, to (see *form*)	former	—
proof plate for spherical surfaces	calibre	Probeglas
proof plate for plane surfaces	calibre	Probeplatte
protector or any piece used as a help in making anything	—	Hilfsstück
protractor	rapporteur	Transporteur
pumice powder	poudre de pierre ponce ou ponce	Bimsstein
putty powder	potée d'étain	Zinnoxyd, Zinnasche
quartz, incorrectly matched prisms of; as a Cornu prism with the two components of similar rotation	macle	—
red ochre, see ochre	ocre rouge	—
retouching. Correction of defects in a lens or mirror by local polishing	retouche locale	retouchieren
roof prism	prisme en toit	Dachkantprisma
rosin (in U.S.A. *colophony*)	résine	Kolophonium
rouge, jeweller's	rouge a polir, rouge Anglais colcotar	Polierrot
roughing	dégrossisage	schruppen
ruby powder	ébauchage	Rubinpulver
ring, brass strip or tube round the outside of a plaster block	—	Ring
roughing tool	platine	Schruppschale
rub up, to correct a polisher surface by rubbing a master (or trueing) tool on it	retoucher le polissage	durch Schaben korrigieren
ruler (with scale in inches or millimetres)	—	Lineal, Massstab
sand, or grit used for grinding, or occasionally the natural emery sands which occur in Asia Minor and Macedonia	grès	Sand
saw (circular)	scie circulaire	Kreissäge
saw, to (in general)	—	sägen
sawdust, willow	scier	Sägespäne von Weide
scratch (slight, between a sleek and a scratch)	frayure	Kratzer
sealing wax	cire à scellar	Siegellack
seeds. Opaque specks of unmelted material in glass (not infrequent in glass-ware)	points	Steine

English	French	German
set, to. To solidify	———	erstarren
shallow (of a surface ; not up to curve)	jeune	flach
shanks	équarrir, pince à ; mivoitier, pince à	Bröckelzange
shank, to	———	abbröckeln
Sira abrasive	———	Siraschmirgel
sleeks ; sleeks, heavy ...	filandres ; ragures	Wischer, Haarrisse
slit, to	———	Schneiden
slitting wheel (or blade) ...	disque à refendre ou scie	Trennscheibe
smoothing	doucissage	Feinschleifen
smoothing and polishing ...	surfaçage	———
smoothing tool	outil à doucir	Schleifwerkzeug
soft	douci	weich
spectacles (pair of)	lunettes	Brille
speculum metal, see *mirror* ...	métal à polir spéculaire	Spiegelmetall
spherical	sphérique	sphärisch
sponge	éponge	Schwamm
springing ; alteration in shape due to cementing, uncementing, etc., due to introduction or relief of strain	déformation dûe aux pressions	springen
spring off, to. To cause a protector to become detached from the work by suddenly heating or chilling it	———	absprengen
squeegee, *e.g.* for pressing out cement or adhesive under paper polisher in fixing to a tool	colloir	Presse, Drückvorrichtung
stack (of discs cemented for edging *masse*)	carrotte	Stapel, Rolle, Paket
stack, pile of discs or plates ...	colonne	———
stamping gauze (§24) ...	———	Prägegaze
stones ; dead metal	pierres	Gallen
strain	tension	Spannung
straining gauze ; closely woven muslin through which pitch is poured to remove foreign particles	tarlatane	Filtriertuch
stray light	———	Streulicht
sulphuric acid	acide sulfurique	Schwefelsäure
surfaces	surfaces	Flächen, Oberflächen
Swedish pitch, see pitch ...	poix de Suède	———
testing	contrôle	Prüfung
test plate	calibre	Probeplatte
thickness, cavity ground in spotting for,	mouche	Richtpunkt
tilting chuck	mandrin oscillant	———
tin-oxide, putty powder ...	potée d'étain	Zinnasche
tissue	chiffon	Gewebe, Stoff
tool handle (Fig. 7)	manche ou parpin	Handgriff
tool, optical, double-sided with transferable handle for grinding both convex and concave tools	outils à ébaucher	———

APPENDIX IV

English	French	German
tools, optical ; grinding tools	paire d'outils	Schalen, Schleifschalen
tools, optical, concave (for generating convex surfaces)	bassins	Schalen, Hohlschalen
tools, optical, flat	plateaux	Scheiben Planscheiben
tools, optical, convex (for generating concave surfaces)	balles	erhabene Schalen
tools, optical, pair of ...	——	Schalenpaar
tools, regrinding to bring them to correct radius and contact	réunir	Schalen korrigieren
tool, optical, with circular grooves for grinding small convex lenses	caillebotter, plateau à	Schleifwerkzeug mit Ringnuten
transparency	transparence	Durchsichtigkeit, Durchlässigkeit
trepanning tool (Fig. 58) ...	trépan	Kernbohrer
tripoli powder	tripoli	Tripel
trueing	ébauchage	schleifen
Turkey stone, natural emery stone	pièrre de Levant	Naturschleifstein
turpentine	térébenthine	Terpentin
tweezers, spring	brucelles	Kluppzange
twinning, association in one piece of crystals of the same mineral but differently orientated or of opposite optical rotation	macles	Zwilling
veils, see feathers	——	Schleier, Wolken
veins, sharply defined or diffused	fils, secou gras	Schlieren
Vienna lime	——	Wiener Kalk
warm, to	réchauffer	erwärmen
washing (of emery, etc.) ...	lavage	Schlämmung
washing, grading (emery, etc.)	minutage, tamisage	Schlämmung ; Ausschlämmung
Water-of-Ayr stone is one type	——	Graustein
wax	cire	Wachs
wet, to give a ; to polish a piece of work from the wet to the dry condition of the polisher	faire une séchée	trockenpolieren
Windolite	——	——
yellow ochre, see ochre ...	ocre jaune	——
Xylol	Xylol	Xylol

BIBLIOGRAPHY

A

Ackermann, von Ilse, 1948, *Optik*, **3**, 47
Adams, L. H. and Williamson, E. D., 1920, *J. Franklin Inst.*, **190**, 597
Airy, G., 1834, *Trans. Camb. Phil. Soc.*, **5**, 283
American Ceramic Soc. *Bulletin* (1948), **27**, [9], 353
Angus-Butterworth, L. M., 1948, *The Manufacture of Glass* (London, Sir Isaac Pitman and Sons)
Arnulph, A., 1946, *Le Vide*, **1**, 129

B

Bacon, Roger, 1542, De mirabili, *Postestate Artis et Naturae*, p. 33 (Paris)
Bates, W. J., 1947, *Proc. Phys. Soc.*, **59**, 940
Barck, C., 1905, *The History of Spectacles* (Open Court, Chicago)
Barnes, R. B., 1938, *J.O.S.A.*, **28**, 140
Barrell, H. and Marriner, R., 1948, *British Science News*, **2**, 130
Barrow, I., 1669, *Opticorum Phaenomenon*
Barth, T., and Lunde, G., 1926, *Zeit. F. Physikal Chemie*, **122**, 293
Bateson, 1947, *J. Soc. Glass. Technology* **31**, 170
Baubet, C., and Lafont, R., 1949, *Rev. d'Optique*, **28**, 490
Beck, H. C., 1928, *Antiquaries Journal*, **8**, 327
Beilby, G. T., 1903, *Proc. Roy. Soc.*, **72**, 218
Beilby, 1921, *Aggregation & Flow of Solids*. (Macmillan, London)
Blodgett, K. B., 1939, *Phys. Rev.*, **55**, 391
Bock, E., 1903, *Die Brille und ihre Geschichte*
Bois-Raymond, 1905, *Zur Geschichter der Glaslinsen*
Bourne, William, 1585, *A Treatise on the Properties and Qualities of Glasses for Optical Purposes*
Bowden, F. P. and Hughes, T.P., 1937, *Proc. Roy. Soc.*, A.**160**, 575
Bowden, F. P. and Ridler, K. E. W., 1936, *Proc. Roy. Soc.*, A. **154**, 640
Bowen, I. S., 1950, *Pub. Astronom. Soc. Pac.*, **62**, 91, and Humason, M. L., *ibid.*, 116

Bratke, E., 1924, *Z. f. Physik.*, **21**, 9
Brockman, R., 1947, *J. Brit. Astronom. Assn.*, **57**, 97
Brockwell, Maurice W., 1948, *The Times*, Feb. 4
Browning, F. P., 1944. *Proc. Roy. Soc. of N.S.W.*, **28**, 187
Bruhat, G., 1935, *Cours d'Optique a l'usage de l'enseignment superieure*
Burch, C. R., 1934, *Monthly Notices Roy. Astr. Soc.*, **94**, 384

C

Caldwell, W. C., 1941, *J. Applied Physics*, **12**, 779
Campani, 1665, *Phil. Trans. Roy. Soc.*, **1**, 2
Campbell, I. E., Powell, C. F., Nowicki, D. H., and Gonser, B. W., 1949, *Jour. Electrochem. Soc.*, **96**, 318
Carnall, R. G. R., *The Dust Problem in the Photographic Industry*, pp. 77–81 of *Dust in Industry* (Society of Chemical Industry, 1949)
Cartwright, C. H., 1940, *J.O.S.A.*, **30**, 110
Chalmers, S. D., 1912, *Proc. Opt. Convention*, 156
Chance, W. H. S. and Hampton, W. M., 1926, *Proc. Opt. Convention*, Part I, 24
Cojan, Jean, 1949, *Les relations entre les phenomènes solaires et geophysiques* (Paris, Revue d'Optique)
Crommelin, C. A., 1929, *Het Lenzen Slijpen in de 17e Eeuw.*
Couder, André, 1944, *Extraits des Cahiers de Physique*, **26**, 27

D

Davidson, C. R., 1920, *Discussion on the Making of Reflecting Surfaces*, 18 (Physical and Optical Societies, London)
Descartes, R., 1692, *Opera* 2, pt. 2 (Amsterdam)
Descartes, R., 1692, *Œuvres*
Dévé, Ch., 1936, *Le Travail des Verres d'Optique de Precision*
Digges, Thomas, 1571, *Pantometria* (A geometrical Practice, London)
Disney, A. N., Hill, C. F. and Watson Baker, W. E., 1928, *The Microscope* (Royal Microscopical Society, London)

BIBLIOGRAPHY

Doering, J., 1949, *Optik*, **5**, 167
Dollond, Peter, 1789, *Some Account of the Discovery by the late Mr John Dollond, F.R.S., which led to improvement of refracting telescope* (London)
Dourneau, F., 1949, Colloque sur la theorie des Images Optique, p. 172 (Paris)
Draper, Henry, 1864, *Smithsonian Contributions to Knowledge*, XIV (reprinted 1904 in Vol. **34**)
Dufour, C., 1950, *Jour. des Recherches du Centre Nationale de la Recherche Scientifique*, **10**, 1

E

Eberhard, G., 1903, *Z. f. Instrument.*, **23**, 82
Edser Edwin, 1915, *Light for Students*
Einsporn, E., 1949, *Optik*, **4**, 11

F

Feussner, K., 1884, *Z. f. Instrument.*, **4**, 41
Finch, G. I., Quarrell, A. G., and Roebuck, J. S., 1934, *Proc. Roy. Soc.*, **145A**, 676
Fizeau, 1862, *Ann. d. Chim. et de Phys.*, **66**, 433
Fleury, P., 1948, *Rev. d'Optique*, **25**, 195, and 1949, **28**, 410
Foucault, L., 1858, *C. R. Acad. Sci. Paris*, **47**, 958
Foucault, L., 1859, *Ann. Observatoire de Paris (Memoirs)*
Foucault, L., 1859, *M.N. Roy. Ast. Soc.*, **19**, 284
Fox, P., 1908, *Astrophys. J.*, **27**, 237
French, J. W., 1916, *Trans. Opt. Soc.*, **18**, 8
French, J. W., 1918, *Trans. Opt. Soc.*, **19**, 143
French, J. W., 1923, Article on *Optical Parts, the Working of* in Glazebrook's *Dictionary of Applied Physics*, Vol. IV
Fresnel, Augustin, 1868, *Œuvres*

G

Geffcken, W., 1942, *Deutsches Reichspatent*, 716, 153
Geffcken, W., 1944, *Lecture delivered at the Colorimetry colloquium in Jena, 8th Dec.*, 1944
Geffcken, W., 1948, *Angew. Chemie*, A.60, 1
Geffcken, W., 1949, *Ceram. Abstr.*, **32**, 87
Gen. Elec. Review, 1927, 30, 354

Glazebrook, Sir R., 1923, *A Dictionary of Applied Physics*, **4**
Goddard, A. E. and Boulind, H. F., 1934, *Properties of Matter, Part II*, 78, (London, Methuen)
Godley, P., 1948, *Iron Age*, April, 90
Grebenshchikov, I. V., 1931, *Keramika i. Steklow*, **7**, Nos. 11-12, 36
Grebenshchikov, I. V., 1935, *Sotsialisticheskaya Reconstructsiya i Nauka*, **2**, 22
Greeff, R., 1912, *Arch. Opthalmol*, 72, 44
Greeff, R., 1921, *Die Erfindung der Augengläser* (Berlin)
Greenland, K. M., and Billington, C., 1950, *Proc. Phys. Soc.*, 63B, 359
Grubb, Howard, 1886, *Nature*, **34**, 85
Guild, J., 1923, *Glazebrook's Dictionary of Applied Physics*, **4**, 786 (London, Macmillan)
Gunther, R. T., 1920, *Early Science in Oxford*, Vol. I, 266
Gurney, C. and Borysowski, Z., 1948, *Proc. Phys. Soc.*, **61**, 446
Gwynne-Jones, E. and Foster, E. W., 1936, *J. Sci. Inst.*, **13**, 216

H

Hadley, L. N. and Dennison, D. M., 1947, *J.O.S.A.*, **37**, 451 ; 1948, *ibid.*, **38**, 483
Haidinger, W., 1849, *Pogg. Ann.*, **77**, 219
Halle, B., 1921, *Handbuch der Praktischen Optik* (Berlin—Nicolasse, *Der Mechaniker*)
Hallauer, 1915, *Die Brille und ihre Geschichte*. (Cong. Int. d. Ophthalmolog. Lausanne)
Halliwell-Phillips, J. O., 1839, *Rora Mathematica*
Hansen, G., 1942, *Zeiss News*, Ser. 4, No. 5
Hampton, W. M., 1925-6, *Trans. Opt. Soc.*, **27**, 163
Hampton, W. M., 1942, *Proc. Phys. Soc.* **54**, 391
Harris, J., 1775, *A Treatise of Optics*.
Hartmann, J., 1900, *Z. f. Instrument.*, **20**, 51
Hartmann, J. 1904, *Z. f. Instrument.*, **24**, 1
Hartmann, J., 1908, *Astrophys. J.* **27**, 254
Hartmann, J., 1909, *Z. f. Instrument.*, **29**, 217
Hass and Scott, 1949, *J.O.S.A.*, **39**, 179
Hayward, R. N., 1949, *The Strength of Plastics and Glass*. (New York, Interscience Publ. Inc.)

Herschel, 1811, *The Telescope*
Herschel, 1849, *Article on "Light" in Encyclopaedia Metropolitana* (iv), 447
Herschel, 1912, *Collected Papers*
Hettner, G., and Leisegang, G., 1948, *Optik*, **3**, 305
Hiesinger, L., 1948, *Optik*, **3**, 485
Hirsch, P. B., and Kellar, J. N., 1948, *Nature*, **162**, 610
Hirschberg, 1906, *Geschichte der Augenheilkunde*, vol. 2, Part 2
Hooke, R., 1667, *Micrographia* (London)
Hooke, R., 1674, *Natural Philosophy*, Book 2, p. 5
Hopkins, Robert E. and O'Brien, Brian, 1949, *J.O.S.A. Report of 34th Annual Meeting of O.S.A.*
Hughes, J. V., 1941, *J.S.I.*, **18**, 234

J

Jensen, H., 1950, *Optik*, **6**, 261
Jones, F. L., 1941, *J. Amer. Cer. Soc.*, **24**, 119
Jones, F. L. and Homer, H.-J., 1941, *J.O.S.A.*, **31**, 34

K

Kaye, G. W. C., 1927, *High Vacua* (Longmans, London)
Kinder, W., 1946, *Optik*, **1**, 413
King, J. H., 1934, *J.O.S.A.*
Kingslake, R., 1925, *Trans. Opt. Soc.*, **27**, 94
Kingslake, R., 1926, *Trans. Opt. Soc.*, **27**, 221
Kingslake, R., 1927, *Trans. Opt. Soc.*, **28**, 173
Kingslake, R., 1928, *Trans. Opt. Soc.*, **29**, 133
Kingslake, R., 1932, *J.O.S.A.*, **22**, 207
Kingslake, R., 1936, *Proc. Phys. Soc.*, **49**, 376
Klarmann, H., 1948, *Optik*, **4**, 165
Kohn, H., 1895-6, *Optical Journal* (N.V.), **1**, 124
König, H. and Rautenfeld, F. Berens von, 1942, *Z. f. Tech. Phys.* No. 11, 273
König, A., 1923, *Die Fernrohre* (*Naturwissenschaftliche Monographien*, 5)
Koops, von R., 1948, *Optik*, **3**, 298
Kreidl, 1950, *The Glass Industry*, 409
Kremers, H. C., 1940, *Ind. and Eng. Chem.*, **32**, 1478
Kremers, H. C., 1947, *J.O.S.A.*, **37**, 337

Kyropoulos, S., 1926, *Zeit. f. Anorg. Chemie*, **154**, 360

L

Laurent, L., 1883, *Compt. Rend*, **96**, 1035
Lebedeff, A. A., 1926, *Rev. d'Optique*, No. 1, 1
Lee, H. W., 1917/18, *Trans. Opt. Soc.*, **19**, 56
Lehmann, Hans, 1902, *Z. f. Instrument.*, **22**, 103
Lehmann, Hans, 1903, *Z. f. Instrument.*, **23**, 289
Leeuwenhoek, A. van, 1719, *Epistolae Physiologicae super Compluribus Naturae Arcanis*, 167-8
Lenouvel, M. L., 1924, *Methode de determination et de mesure des aberrations des systemes optiques.* (Paris, Revue d'Optique)
Linfoot, E. H., 1946, *Proc. Roy. Soc.*, **186A**, 72
Linfoot, E. H., 1946, *Proc. Phys. Soc.*, **58**, 65
Linfoot, E. H., 1948, *Photographic J.*, **88B**, 58
Linfoot, E. H., and Wayman, P. A., 1949, *Mon. Not. R.A.S.*, **109**, 535
Linfoot, E. H. and Wolf, E., 1949, *J.O.S.A.*, **39**, 752
Longchamp, H. de, 1947, *Rev. d'Optique*, **26**, 94
Lummer, O., 1885, *Z. f. Instrument.*, **5**, 23
Lyot, B., 1937, *Bull. Soc. Astronomique*, **51**, 203
Lyot, B., 1946, *C.R. Acad. Sci. Paris*, **222**, 765
Lyot, B., and Françon, M., 1950, *Rev. d'Optique*, **29**, 499
Lyot, B., Françon, M., and Cagnet, M., 1948, *Revue d'Optique*, **27**, 657

M

Mach, E., 1926, *The Principles of Physical Optics.* (London, Methuen)
Manzini, Carlo Antonio, 1660, *L'Occiale all'Occhio, Dioptrica Pratica.* (Bologna)
Martin, L. C., 1948, *Technical Optics*, **1**, 185
Martin, Th. Henri, 1871, *Bull. Bibliog. e. Storia Scienze mat. phys.* (Roma), **4**, 165-238
Maxwell, J. C., 1868, *Phil. Mag.* (4), **35**, 129
Michelson, A. A., 1881, *American J.*, **22**, 120

Michelson, A. A., 1898, *Astrophys. J.*, **8**, 37

Michelson, A. A., 1907, *Light Waves and their Uses.* (Chicago, Univ. Press)

Minkowski, R., 1944, *J.O.S.A.*, **34**, 89

Moll, J., 1831, *Jour. Roy. Inst. G.B.*, **1**, 319 and 483

Molyneux, William, 1692, *Dioptrica Nova*

Morey, George W., 1938, *The Properties of Glass.* (New York, The Reinhold Publishing Corporation)

N

Newton, Sir Isaac, 1721, *Opticks*

Nicoll, F. H., 1942, *R. C. A. Review*, **6**, 287

O

Oberg, 1902, *Brit. Pat. Spec.*, **131**, 536/02

Oberg, 1909, *Brit. Pat. Spec.*, 3444/09

Oberg, 1918, *Machinery* (New York), **25**, 330-332

Oberg, 1926-27, *J. Brit. Astron. Assn.*, **37**

O'Neill, Hugh, 1934, *The Hardness of Metals and its Measurement.* (London, Chapman & Hall)

Oppenheimer, E. H., 1908, *Die Erfindung der Brillen.* (Zentralz. f. Optik)

P

Pansier, P., 1901, *Histoire des Lunettes.* (Paris)

Parsons, W. (3rd Earl of Rosse), 1926, *Collected Papers*, p. 92

Penny, G. W., 1937, *A new electrostatic precipitator* (Electrical Engineering)

Perry, J., 1923, *Trans. Opt. Soc.*, **25**, 97

Perry, J., 1943, *Proc. Opt. Soc.*, **15**, 257

Pfund, A. H., 1946, *J.O.S.A.*, **36**, 95

Porta, Baptista, 1591, *Magia Naturalis* (Frankfort Book 20)

Porta, Baptista, 1658 (Eng. Trans.), *Natural Magic by John Baptista Porta, a Neapolitaine, in Twenty Books.* (London)

Preston, F. W., 1922, *Trans. Opt. Soc.*, **23**, 141

Preston, F. W., *Trans. Opt. Soc.*, **23**, 150

Preston, F. W., 1926, *Trans. Opt. Soc.*, **27**, 181

Preston, F. W., Baker, T. C., and Glatshart, J. L., 1946, *J. Appl. Phys.*, **17**, 162. This reference covers four consecutive papers dealing with various aspects of the strength and fatigue of glass

R

Rantch, Kurt, 1947, *Optik*, **2**, 250

Rasmussen, Ebbe, 1945, *KGL. dansk-Videnskabernes Selskab Meddeelel ser*, **23**, 3

Ray, K., 1949, *J.O.S.A.*, **39**, 92

Rayleigh, Lord, 1901, *Proc. Roy. Inst.*, **16**, 116

Rayleigh, Lord, 1903, *Scientific Papers*, **4**, 542 (C.U.P. Cambridge)

Rayleigh, Lord, 1908, *Phil. Mag.*, (6) **16**, 444

R.C.A., 1942, *R.C.A. Review*, **6**, 287

Redding, 1916/17, *Trans. Opt. Soc.*, **18**, 56

Redmond, L., 1949, *Anales de fisica y quimica*, **45a**, 497 (Aplication de una superficie esferica a un ocular)

Ritchey, Geo. W., 1904, *Smithsonian Contributions to Knowledge*, **34**

von Rohr, Moritz, 1912, *Das Auge und die Brille*

von Rohr, 1917, *Z. Ophthalmol. Optik*, **5**, parts 1-3

von Rohr, Moritz, 1929, *Joseph Fraunhofers Leben*, 68

Ronchi, Vasco, 1926, *Z. f. Instrumenten*, **46**, 553

S

Sabine, G. B., 1939, *Phys. Rev.*, **55**, 1064

Sauerwald, 1929, *Lehrbuch der Metallkunde*, 230

Scheel, Karl, 1912, *Z. f. Inst.*, **32**, 14

Schorr, R., 1936, *Jahresber. der Hamburger Sternwarte in Bergesdorf f. das Jahr*, 1935

Schott, O., 1891, *Z. f. Instrument.*, **11**, 330

Schultz, Max, 1912, *Z. f. Instrumenten*, **32**, 258

Schulz, Hans, 1914, *Z. f. Instrumenten*, **34**, 252

Schuster, A., *Theory of Optics*, 2nd Ed.

Smakula, A., 1941, *Glastech Ber.*, **19**, 377—(*abstract in Ceramic Abs.*, 1947, **30**, 158)

Smakula, A. and Klein, M. W., 1949, *J.O.S.A.*, **39**, 445

Spencer-Jones (Sir), Harold, 1949, *The advancement of Science*, **6**, 188. (London, British Association)

Spencer-Jones (Sir), Harold, 1949, *Popular Astronomy*, **57**, 468

Stanforth, J. E., 1950, *Physical Properties of Glass* (London, O.U.P.)
Stöber, F., 1925, *Z. f. Kristall*, **61**, 299
Stockbarger, D. C., 1936, *Rev. Sci. Inst.*, **7**, 133
Strong, J., 1936, *Astrophys. J.*, **83**, 401
Strong, J., 1939, *Procedure in Experimental Physics* (New York, Prentice-Hall) published in Great Britain under the title *Modern Physical Laboratory Practice* (Edinburgh, Blackie & Son)
Strong, John and Gaviola, E., 1936, *J.O.S.A.*, **26**, 153

T

Taylor, H. Dennis, 1896, *The Adjustment and Testing of Telescope Objectives*. Reprinted 1921 (York, T. Cooke & Sons)
Taylor, H. Dennis, 1904, Brit. Patent 29561
Taylor, E. Wilfred, 1949, *Nature*, **163**, 323
Taylor, William, 1918, *British Pat. Spec.* 1,432,093
Taylor, Dennis, 1921, *The Adjustment and Testing of Telescope Objectives*, 2nd Ed.
Taylor, William, 1925, *U.S. Pat. Spec.* 1,563,068
Taylor, William, 1932, *Proc. Inst. Mech. Eng.*, **123**, 169-173
Texereau, J., 1949, *l'A.*, **63**, 42, 85
Thompson, G. P., 1930, *Proc. Roy. Soc.*, **128**, 649
Tilton, L. W., Plyler, E. K., and Stephens, R. E., 1950, *J.O.S.A.*, **40**, 540
Tolansky, S., 1945, *J.S.I.*, **22**, 161
Tolansky, S. and Wilcock, W. L., 1947, *Proc. Roy. Soc.*, A. **191**, 182
Tolansky, S., 1947, *High Resolution Spectroscopy*. (London, Metheun)
Toepler, A., 1866, *Pogg. Ann.*, **128**, 126
Toepler, A., 1867, *Pogg. Ann.*, **131**, 33
Tool, A. Q. and Valasek, J., 1920, *Bull. Bureau of Standards*, **15**, 537
Tweedale, C. L., 1943, *Reflecting Telescope Making*. (London, T. Werner Laurie Ltd)
Twyman, F., 1905, *Proc. Opt. Convention*, 78
Twyman, F., 1917, *Trans. Soc. Glass Techn.*, **1**, 61
Twyman, F., 1918, *British Pat. Spec.*, 130, 224/18
Twyman, F., 1921, *Trans. Opt. Soc.*, **22**, 174
Twyman, F., 1923, *Trans. Opt. Soc.*, **24**, 189
Twyman, F. and Dalladay, A. J., 1921, *Trans. Opt. Soc.*, **23**, 131
Twyman, F. and Green, A., 1916, *British Pat. Spec.*, 103,832/16
Twyman, F. and Green, A., 1923, *British Pat. Spec.*, 213,674/23
Twyman, F. and Perry, J. W., 1922, *Proc. Phys. Soc.*, **34**, 151
Twyman, F. and Simeon, F., 1923, *Trans. Soc. Glass Techn.*, **7**, 199

U

Upton, P. B. G. and Herrington, E. F. G., (1950), *Research*, **3**, 289
U.S.N., 1945, *Technical Report No. 455-45 of the United States Naval Technical Mission in Europe*

V

Voit, Ernst, 1887, *Bayr. Ind. u. Gew. Blatt.*, **6**, No. III

W

Watson, Baker (see Disney, etc.)
Wilde, E., 1838 and 43, *Geschichte der Optik*. (Berlin)
Williams, W. E., 1928, *Nature*, **122**, 347
Winkelmann, A., 1908, *Handbuch der Physik*, I, **1**, 859
Wood, R. W., 1911, *Physical Optics*, 333
Wormser, E. M., 1950, *J.O.S.A.*, **7** (or the design of wide-angle Schmidt optical surfaces)
Wren, Dr. (afterwards Sir Christopher), 1669, *Phil. Trans*, **4**, 1059
Wright, F. E., 1921, *The Manufacture of Optical Glass and of Optical Systems*, U.S.A. Ordnance Dept. Document, No. 2037, Washington, Gov. Printing Office
Wright, F. E., 1924, *The Manufacture of Optical Glass and of Optical Systems*. (U.S.A. Ordnance Dept., Washington, U.S.A.)

Z

Zachariasen, W. H., 1932, *J. Amer. Chem. Soc.*, **54**, 3841
Zeiss, Carl, Ltd, 1907, *Brit. Pat. Spec.*, 14,126/07
Zernike, F., 1934 (b), *Monthly Notices Roy. Astr. Soc.*, **94**, 377
Zernike, F., 1934 (a), *Physica*, **1**, 689
Zschimmer, E. and Schulz, H., 1913, *Ann. Phys.*, **42**, 345

SUBJECT INDEX

Aberration, chromatic, 120–124
  spherical, 123–124
Abrasive powders, hardness of, 159–161
Abrasives, 93–109
  aloxite, 94, 108
  approved grain sizes, 101
  boron carbide, 94
  carborundum, 93
  classification of, 98–101
  corundum, improved, 103
    preparation and grading of, 102–103
  elutriating apparatus, 102, 106, 108
  emery, 94
  for crystals, 171–174
  for ophthalmic lenses, 263
  grading, 102–105
  quality of, 102
Absorption, 124–125
Adcock & Shipley lens roughing machine, 201–205
Adhesing temperatures, 535
Alhazen, mention of lenses by, 8
Ancient lenses, 5–11
Angle, grinding to, 25–27
  by goniometer, 403–405
  grinding machines for, 301, 302
  measurements by Angle Dekkor, 299, 405–411
    by comparison with standards, 406–407, 411–413
    using sub-multiple angles, 408–11
  standards, 413
Angles, goniometer for measuring, 27
  protractors for measuring, 25–26
  tolerances on, 402–403
Aluminium deposition, 464
Annealing, Adams and Williamson's work, 533–534
  alloys, 530
  classification of, by A.S.T.M., 520–521
  cooling schedules for, 531–533
  definition of, 522
  faulty, 503
  and normalizing, 504–506
  scientific control of, 503–537
  standards, 520–521
  stress-relief distinct from normalizing, 521–522
  temperature and composition, 530–531
  Tool and Valasek's work, 534
  Twyman's methods, 521–530
  Twyman's cooling schedules, calculation of, 591–596
  without fuel, 520

Aspherical surfaces, see *Non-spherical surfaces*
Astronomical object glasses and mirrors, see under *Mirrors* and *Object glasses*

Balsam, hardness of, 238–241
  layer, thickness of, 247
  substitutes, 241–247
  of high n, 247
Balsamed lenses, annealing, 248
Balsaming and cementing lenses, 238–249
  gelatine filters, 249
  lenses in quantity, 248
  prisms, 248–9
Bates's Shearing and Rotating interferometers, 448–451
Bifocal lenses, 254, 264–277
  for cataract, 274
  various, 274–275
Bifocals, fused, 264–268
  solid, 268–275
  Univis fused, 267–268
Blocking with mallets, 207–212
  without mallets, 205, 212–213
Blooming, see *Metallizing and blooming*
Books on Optics, 588–590
Boron carbide, 94
Bowl machine for ophthalmic lenses, 256–259
Burch's aspherizing machine, 361–363
Burning glasses, early use of, 5–7
Buttons (or mallets), 207–209

Caesium bromide, 158
Camera lens interferometers, 431–437
Carborundum, 93
  and corundum, manufacture of, 108–9
Cells, glass, precautions in making, 535–537
Cement layer, thickness of, 247
Cements, plasters and varnishes, 113–116
  shrinkage of, 117
Centring, accuracy of, 236–238
  and edging lenses, 42–44, 227–236
Cerium oxide, 110–111
Chamfering lenses, 236
Chromium deposition, see *Metallizing and blooming*
Chucking by vacuum, 205
Cleaning glass surfaces for silvering, 457
Cleaning liquids, 112–113
  dermatitis due to, 116–117
Corner cube prism making, 300
  interferometric testing of, 446–448

624 INDEX

Coronograph, Lyot's, 393–394, 576
Corundum and carborundum, manufacture of, 108–109
Crown glass, 118–121
Crystals, artificial, 154–181
  binary mixed, 164–171
  caesium bromide, 158
  calcium fluoride, 155–158
  cutting, 171–181
  decolorizing, 164
  grinding, 171–181
  lithium fluoride, 155–158
  methods of preparing, 155–158
  natural calcite (Iceland Spar), 155
    fluorite, 155
    quartz, 155
    rocksalt, 155
    sylvine, 155
  polishing, 171–181
  potassium bromide, 155–158
  potassium chloride, 155–158
  purity essential, 176
  sapphire (artificial), 158–159
  silver chloride, 155–158
  sodium bromide, 175
  sodium chloride, 155–158
  sodium fluoride, 155–158
  sundry notes on, 181
  thallium bromo-iodide, 155–158

Deposition of films on optical surfaces, see *Metallizing and blooming*
Dermatitis due to cleaning liquids, 116
Devitrification, 138
Diameter, gauging, see *Gauging*
Diamond-charged laps, roughing by, 191, 198–200
Diamond cut, stress near, 52
  disc cutter, 193
  metal bonded laps, 200–202
  saws, 22–24
Diffraction pattern of a slit image, 399–401
Diopter, definition of, 254
Dispersion, 120
Dispersive power, 122, 124
Double refraction crystals, 157–158
Drikold for detaching lenses from mallets, 226
Dust, avoidance of, 84–89
  electrical filtration for removing, 87
  importance of, 84
  rate of fall, 85

Edging and centring lenses, 42–44, 227–238
  by hand, 42–44
  machines, 229–238
Elliptical surfaces, Taylor, Taylor and Hobson machine for producing, 341–342
Emery, 94

Expansion of glasses, Fensom's method of comparing, 535–537
Figuring, Grubb's method, 551–556
  lenses by depositing transparent substances, 337–338
  mirrors by depositing aluminium, 334–337
Films, thin, measuring thickness of, 130
Filters, multiple layer, 462–463
Fizeau apparatus, Michelson's modification, 387–388
  portable, 386–387
  interference test for flatness, systematic errors in , 393
  interferoscopes, Michelson's form of, 387–388
  fringes, 381–387
Flatness, absolute standard of, 413–416
  of grey surface, measuring, 29
  testing by Interferoscope, 36
  of tools, maintaining, 36
  measuring, 27–28
Flint glass, 118, 121
Foucault, mirror silvered on front, 539
Foucault's test, 372–374, 567

Gauging by air-operated gauges, 416–417
Glass, ageing of, 130
  crown, 118–120, 121
  devitrification of, 138
  durability of, 125–130
  durability of, simple test for, 129–130
  expansion of, 136–137
  flint, 118, 121
  fracture, delayed, 130
  hardness of, 131–136
  in ancient Egypt, 6
  modern manufacturing methods, 603–608
  moulding, 188–190
  nature of, 138–141
  optical, annealing, 148–154, 516–535, 591–596
    chilling lowers refractive index of, 151
    problems arising in manufacture of, 147–154
    production in Germany during War, 154
  sand for, 148
  properties of, 118
  staining of, 125–129
  strength of, 130
  structure of, 139–141
  temperature variation of refractive index, 136–137
  transmittance of, 511–512
  weathering of, 125–129
Glasses, optical, developments since 1880, 141–147

INDEX

Glass-formers, 138–141
Glassite, 45
Graticules, 483–498
  bichromated-glue method, 495–496
  composite processes, 496
  diamond engraved, 490–491
  etched, 492–494
  masks, 496–497
  photo-etching, 494–495
  photographic reproduction of, 494
  transfer methods, 497–498
  various types of, 483–488
Grinding and polishing, best pressure for, 104
  machines, early, 12, 13, 16–19
  pressure used by Ritchey, 562
  Ritchey's machine, 557–559
Grinding crystals, 171–181
Grinding to angle, 25–27
Ground surfaces, depth of, 104–105
  coarseness of, 105
  stress near, 52–53

Haidinger's fringes, 380–381
Hale telescope, honeycomb structure of, 540
Hanemann's micro-hardness tester, 134
Hardness, Bierbaum's microcharacter for measuring, 135
  Brinel, Rockwell, Vickers, Hanemann and Auerbach tests for, 132–135
  importance for the optician, 135
  micro-hardness tester, 134
  Mohs test for, 131–132
  ruled line test for, 135
Hartmann's test, 374
  limitations of, 566–567
Hilger Interferometers, 422–448
  compensated, 430
  Prism Interferometer, testing interior angles by, 446–448
H.T. cement (n-butyl methacrylate) as substitute for balsam, 242–247
  refractive index of, 244
  use of for polarizing prisms, 244

Interference colour filters, 466–467
interference tests, improved, 390
  by multiple reflection, Rasmussen's apparatus for, 390–393
  by multiple reflection, use of by Tolansky and Wilcock, 393
  by multiple reflection, use of by W. E. Williams, 390
Interferometer, rotating, 448–451
  shearing, 448–451
Interferometers, increasing contrast and light in, 430
  for testing prisms and lenses, 422–446
  thermal uniformity needed in use of, 430–431

Interferoscope, the Hilger, 385–386

Laurent's modification of Fizeau's apparatus, 382–383
Lehr, for moulding spectacle lenses, 252
Lenses, ancient, 5–11
  compound, assembly without cement, 241
  damaged, reworking without tools, 44
  detaching from mallets, 226
  machining to radius, 206–207
  moulding, 188–190, 192–193, 252–254
  number in a block, 208, 213–215
  polishing deep curves, 218–220
  process layouts for, 205
  Sixteenth to Eighteenth Century, 10–14
  tools for, machining to radius, 206–207
  wastage of glass in making, 293
  see also under *Ophthalmic lenses*
Lens interferometers, 431–446
  polishing by hand, 38–48
  roughing machines, 198–207
Light sources for optical manufacture and testing, 417–421
Liquid surface interferometer, 413–416
Load used in polishing, 218, 562–563

Magnesium fluoride, 465–466
Mallets (or buttons), 207–209
  pitch for, 92–93
Meehanite, 66
Meniscus lens, 254
Metallizing and blooming by evaporation in vacuo, 462–483
  high quality, to obtain, 473–480
  large scale industrial, 480
  origins of, 462, 482–483
  pinholes, causes of, 479–480
  plants for, 468–473
  protective coat (*of silica*), 480–481
  reduction of reflection, 482–483
  refractory materials, 480
  " sputtering ", 461–462
Metals, polishing, 173–174
Methyl-methacrylate, see *Plastics, optical*
Methyl-salicylate, for testing Quartz, 171
Michelson's test for plane parallel plates, 387–390
Micro-interferometer, Linnick (Zeiss), 451–455
Microscope cover glasses, thickness of, 280
  lenses, 278–292
    abrasives for, 282

626 INDEX

cementing, mounting and centring, 285–286
hemispherical or hyperhemispherical, 280–285
Hooke's way of making, 13
measuring thickness of, 279
mounts for, 286–287
numerical aperture, 279
proof plates for, 282–283
testing the complete lens, 288–290
tolerances in edging and centring, 278, 285–286
tolerances in thickness, 278
wavefront correction by altering draw tube, 279
wavefront tolerances, 279
lens interferometers, 290–292, 437–446
ultra-violet, 439–440
Milling glass by diamond charged laps, 198
Mirrors, aluminium coated, 539
figuring, Grubb's method, 551–556
Mirrors, large, 538–587
Draper's techniques, 542–547
flexure of, 544, 548–550
methods of making, 541, 587
Mt. Palomar, 568–576; see also *Palomar*
large speculum, 542–543
parabolizing, 546–547
paraboloidal, testing and figuring, 563–567
plane, testing, 563
to be polished on both sides, 556–557
Pyrex preferred for large, 539–540
silvered on front, prevention of tarnishing of, 544
speculum, 17, 542–543
supports for during polishing, 557–558
thickness preferred by Ritchey, 556
Mobility-temperature relationship dependent on composition, 593–596
tests on special glasses, 593–596
Moulding lenses, 188–190, 252–254
temperature 192–193, 535
Mouldings, prisms and lenses, 192–193
Multiple films for Fabry-Perot etalon plates, 467

Neven metal-bonded diamond tools, 200–202
Newton's methods of polishing, 15–16
reflecting telescope, 14
rings and proof plates, 375–380
Non-spherical lens grinding and polishing machines, 187, 355–363
applications of, 327–330
lenses, developments from 1918–1948,

elliptical, 323–327
hyperbolic, 326–327
making, various ways of, 333–342
parabolic, 326–327
position in 1918, 327–329, 331–332
Schmidt plates, departure from sphericity of, 331
Wren's suggested engine for, 326–327
surfaces, 323–363
Descartes' aplanatic, 323–326
machines for producing, 339–342
by sagging, 328, 337–339

Object glasses, definition, criterion of, 576–578
imperfections, classification of, 578–582
methods of polishing, effect of various, 582–587
tests for, 582
large, 538–587
flexure of, 548–550
flexure of, effect of thickness, 549–550
flexure of, supports to avoid, 549
Lehmann's criterion of quality, 566
scattered light, amount of, 585
technique for avoiding, 585–583
testing and correcting, Grubb's method, 551–556
testing by Foucault's method, 567
testing by Hartmann's method, 565–567
variety of faults, 552–3
Ophthalmic lenses, 251–277
blocking, 255–256
bowl machine for, 256–259
glasses used for, 251–252
grinding and polishing machines, 256–259
machines for prescription work, 276–277
production planning, 276
testing, 275–276
tools, abrasives, polishers and polishing materials for, 263–264
types of, 253–254
Optical glass, manufacture of, 603–608
testing, 499–537
tools, Ritchey's, 559–563
workshop, design of, 88–89
work, testing, 364–421

Palomar telescope, Foucault testing apparatus for, 568, 575
grinding and polishing, 569–576
200-inch mirror, structure and handling of, 568–576

INDEX

Parallelism of glass plates, notes on, 420–421
Parallel plates, Haidinger's fringes in, 380–381
Pellicles, 481
Perry and Weinstein machine (Universal) for non-spherical surfaces, 339–341
Phase contrast method, 455
   plates, preparation of, 455
   test, 393–396
Pitch for chucking lenses, 227–228
   polishing, heterogeneity of, 83
   modifications of, 83–84
   range of temperature for, 81–83
   substitute for Swedish, 83
   viscosity of, 79–83
Planeness, absolute standard of, 413–416
Plane parallel glass, interference test for when considerably out of parallel, 401–402
   making, 314–318
   polishing machines for, 318–322
   testing for curvature, 396–398
   tests for, 383–390
Plastics, optical, 182–188
Polarizing crystals, 157, 601
   prisms, making, 596–603
   Polaroid, 601–603
   sodium nitrate, 601
Polished surfaces, accuracy of, 21
   properties of, 49, 62–63
   protective coatings for, 47
   thickness of, 61
Polisher, curved, preparation of, 42
   flat, preparation of, 30–35
Polishers, 78–79
   for ophthalmic lenses, 264
   pitch, 78–81
   wax, 90–91
   wax-faced pitch, 91–92
   wax and putty powder, 91
   wood base, used by Ritchey, 561–562
Polishing, 216–220
   appearance of block during, 216–217
   correct hardness of mallets (buttons) in, 217–218
   crystals, 171–181
   due to melting, 58–60
   first use of pitch for, 15–16
   flame, 174
   and grinding machines, early, 16–19
   Ritchey's, 557–559
   hard crystals, 173
   on leather, 13
   lenses by hand, 38–48
   machines, 221–226
      controlled relative rotations of work and tools, 225–226
      Draper's, 543–546
      Grubb's, 550–551

   for lenses, W. Taylor's, 222–225
   Twyman and Underhill's, 319
   materials, 63–64, 109–111
   17th century, 12–13
   hardness of, 63–64
   melting points of, 58–60, 64
   for ophthalmic lenses, 263
   various, 111
   nature of, 49–65
   Newton's methods, 15–16
   paper, 173
   powders, 46, 111
      Glassite, 45
      hardness of, 159–161
      Indian ink, 45
      manganese dioxide, 45–46
      for metals, 46–47
      for surfaces in focus, 111
      thorium oxide, 111
   pressure used by Ritchey, 562–563
   prism by hand, 33–35
   procedure after, 226–238
   on silk, 172
   soft crystals, 172, 174–176
   superfine, 366, 584–586
   time, 218
Potassium bromide, 156–158
Prescription ophthalmic lenses, 276–277
Pressure used in polishing, 218, 562–563
Prism interferometer, 427–431
Prisms, blocks of (plaster, waxed felt), 308–313
   cutting from slab, 24–25
   machining to angle, 294–298
   milling, fixtures for, 297–307
   polishing by hand, 33–35
   quantity production of, 293–322
   roof, making, 297–299
   tetragonal, making, 300–303
   trueing to angle, 307–308
   wastage of glass in making, 293
Proof plates, Fraunhofer's, 17; see also *Test plates*
Putty powder, 111

Quarter wave criterion of optical quality, 398–401
Quartz, expansion of, 137
   selection of crystals, 171–172

Rayleigh's criterion (quarter wave rule) 398–401
Reflecting films (pellicles), 481
Reflecting films, producing without vacuum plant, 467–468
Reflecting surfaces, preparation of, 456–459
Reflection, reduction of, see under *Metallizing and blooming*

Refractive index, 122–123
  measuring, 512–516
  measuring small differences of, 508–511
  Precision Refractometer for, 513–516
Resolution of double star, 398–401
Retouching, Draper's procedure, 546–547
Rhodium deposition, 464–465, 473–480
  high quality, 473–480
Ritchey's work, 556–565
Rocksalt, 155–158
  durability of, 175
Rouge, 109–110
  grain size, 62
  preparation of, 18–19
  " white ", 110
Roughing by abrasive wheels, 191, 198
  diamond-charged laps, 191, 198
  hand, 25–30, 40–41, 196–197
Roughing machines, 198–205
  see also the headings under *Shaping prisms and lenses*
  trueing and smoothing, emery series for, 101
Royal Microscopical Society's screw, 287

Sagging temperature, 535
Sapphire, artificial, 158–159
Sawing glass, 22–24
  substitute for, 22
Schmidt camera, 342–355
  correcting plates for, 345–353
  spectrographic, 353–354
  systems, curvature of field in, 353
  for television, 353
  various, 355
Scratches, avoidance of, 36, 38, 41
Shanking, 38–40
Shaping prisms and lenses, see under :
  Diamond charged laps
  Diamond disc cutter
  Diamond metal bonded laps
  Milling
  Moulding
  Roughing
  Roughing by hand
  Trepanning
Silica, melting point of, 139
Silver chloride, 155–158
  die casting, 176
Silvering, 456–461, see also *Metallizing and blooming*
  chemical, 456–461
  half-, 459–460
  Liebig's invention of, 539
  physical, 461–481
  rapid high-efficiency, 460–461
  silver fulminate, danger of, 457

Smoothing, 215–216
  change of curvature during, 215–216
  drying in, avoidance of, 105–106
  recessed tools for, 545–546
  series of emeries for, 101
  and trueing, 27–30
Snell's law, 122
Sodium nitrate, 157–158
Society screw, see *Royal Microscopical Society's screw*
Spectacles, invention of, 9–10
Speculum metal, alloy preferred by Grubb, 556
Spherometer, 70–76
  Guild's, 73–74
  ring form, 70–72
  Ritchey's, 560–561
  various, 76
  Watts' Precision, 74–75
Spheroscope, 75–76
Sputtering, see under *Metallizing and Blooming*
Sticking lenses on blocks (malleting), 207–209
Strain viewers, 501–503
Superfine polishing and superfinish, comparison of, 586–587
Superfinished surfaces and the Beilby layer, 587
Surface reflection, reduction of, see *Metallizing and blooming*

Taylor, Taylor and Hobson machine for elliptical surfaces, 341–342
Telescope, reflecting, first made by Newton, 14–16
  Gregory's proposed, 14
Telescopes, large, Sir Wm. Herschel's, 540–541
Terms used in the Optical Industry, glossary of, 609–617
Testing bench for optical work, 371–372
  lenses, 367–375
  optical glass, 499–537
    bubbles, 365–366
    for definition, 366–367
    for double refraction, 501–503
    Foucault's test, 372–374
    Hartmann's test, 374
    interference methods, 374–380
    for surface defects, 364–366
    veins, 499–501
  surfaces, 364–366, 375–380
    by Fizeau's fringes, 381–390
Test plates, 36–37, 375–380
  flat, Whitworth's method of making, 37
Tetragonal prisms, see *Corner cube prisms*

INDEX

Thermodyne, the, 127–129
Thickness, calipers for measuring, 40
  gauging, see *Gauging*
Tolerance specifications, 398–401
Tools, maintaining flatness of, 36–37
  measuring flatness of, 27–28
  optical, 66–84
    aluminium, 67
    brass and gunmetal, 67
    chamfering, 77
    flat, 77
    iron, 66–67
    preparation of, 67–84
Toric lenses, 254, 259–263
  machines for roughing, smoothing and polishing, 260–263
Transmittance of glasses, 511–512
Trepanning, 193–195

Trueing machines, 197–198
  and smoothing, 27
  series of emeries for, 101
  tools, flat, 27–29
Tufnol tubing, 236
Two-hundred inch mirror at Mt. Palomar, 568–576
Twyman and Green Interferometers, 422–488
  see also *Hilger Interferometers*
Twyman's Law, 517
Twyman effect, 318

Zernike phase contrast test, 393–396
  compared with Foucault's test, 394
  Lyot's modification of, 394–396
Zinc oxide as polishing powder, 46
  sulphide coating for interferometer plates 430